FINITE MATHEMATICS

FINITE MATHEMATICS

Models and Applications

CARLA C. MORRIS
University of Delaware

ROBERT M. STARK
University of Delaware

WILEY

Published by John Wiley & Sons, Inc., Hoboken, New Jersey
Published simultaneously in Canada

Library of Congress Cataloging-in-Publication Data:

Morris, Carla C.
Finite Mathematics : Models and Applications / Carla C. Morris and Robert M. Stark.
pages cm
Includes bibliographical references and index.
ISBN 978-1-119-01550-5 (cloth)
1. Mathematics–Textbooks. I. Stark, Robert M., 1930- II. Title.
QA39.3.M68 2015
511′.1–dc23

2014042187

Printed in the United States of America

10 9 8 7 6 5 4 3 2 1

CONTENTS

PREFACE

Mathematical development has recently grown at an exponential rate. The mathematics of computers, statistics, matrices, linear programming, and game theory, to name a few, has changed and expanded the landscape of mathematical influence. Many view films such as *A Beautiful Mind*, and *Pi*, plays such as *Proof*, and TV shows such as *Numb3rs* to witness a wider recognition of mathematics and enhance its awareness in daily lives.

Prior to World War II, basic college and university mathematics consisted mainly of college algebra and calculus. The enormous success of those Newtonian era mathematical discoveries in the development of physical sciences and engineering continues today.

War often brings scientific and technological advances, and World War II was no exception. Developments in the immediate post-World War II years and remarkable advances in automated computation led to new mathematics and to new uses for older mathematics.

Among the many post-World War II developments: three stand out that have become a basis for finite mathematics courses:

- ***Linear Programming***
  Linear programming, the mathematics for the optimal allocation of resources, has found wide usage in the cutting and planting of forests, automobile production lines, oil refining, hospital menus, and much more. A Google search actually returns millions of references.
- ***Matrix Algebra***
  The algebra of this mathematical subject is now vital to the operation of computers that store, move, manipulate, and analyze huge arrays of data and variables. It has wide usage in engineering, social, business, and health contexts. Leontief's Nobel Prize winning input – output economic models of firms, governments, and nations are well known.
- ***Chance and Probability***
  With origins predating the 17th century, mathematical probability matured in the 20th century and with related topics has found applications in virtually every branch of science, technology, and commerce. Examples include the proliferation of medical

tests with false negatives and positives, risk assessments, artificial intelligence usages of Bayes' Rule, tracking demographics and behavior with Markov chains, and the development of mathematical statistics.

Recognition of the fuller import of these three primary topics led to their aggregation into a new course, "Finite Mathematics", in the 1960's. The course and its topics have these distinctions:

- Topics are mostly independent of one another.
- Topics often differ from pre World War II curricula.
- Topics deal with discrete or finite events from which the course and text derive their title. This contrasts with Calculus, which is based upon continuous variation.
- Finite mathematics is more a course title than a branch of mathematics marked by an intrinsic coherence. In a sense, it is a hybrid branch of mathematics.

While calculus remains fundamental to physical science and technology, its impact upon the post-World War II social, behavioral, life, health, business, management, and economic sciences has been markedly less. However, it is finite mathematics topics that have become more important to post-World War II business and social sciences applications. One expects finite mathematics topics to find a niche in college curricula.

While a finite mathematics course is often required at many colleges and universities, students in our classes range from freshmen to seniors. Although business, economics, finance, and management students are often the primary enrollment, other students often enroll and can easily benefit as the text and exercises span many disciplines in the life, natural, and social sciences.

The course typically features matrix algebra, linear programming, probability, and related topics. For a one-semester course, this text has ample material for teachers to enrich or offer a two-semester course, perhaps for honors students.

Many institutions have shown an interest in developing General Education guidelines to encourage student development in oral, written communication, and quantitative reasoning skills. Several of the exercises in this text were included with the idea that they would be well suited for instructors looking to aid students in developing skills that suit General Education guidelines. The exercises in this text may be completed with or without the use of technological aids. That decision is for the teacher.

ABOUT THE AUTHORS

Carla C. Morris has taught courses ranging from college algebra to calculus and statistics since 1987 at the Dover Campus of the University of Delaware. Her B.S. (Mathematics) and M.S. (Operations Research & Statistics) are from Rensselaer Polytechnic Institute, and her Ph.D. (Operations Research) is from the University of Delaware.

Robert M. Stark is Professor Emeritus of Mathematical Sciences at the University of Delaware. His undergraduate and doctoral degrees were awarded by, respectively, Johns Hopkins University and the University of Delaware in physics, mathematics, and operations research. Among his publications is the 2004 Dover Edition of Mathematical Foundations for Design with R.L. Nicholls.

CHAPTER CONTENT

ALGEBRA SKILLS

Chapter 1, ***Linear Equations and Mathematical Concepts*** is provided for students not fully prepared for a finite mathematics course. The chapter reviews algebra skills and emphasizes solving linear equations and basic graphing skills. In addition, exponential and logarithmic functions, π and e, variation and dimensional analysis are briefly considered.

BASIC FINANCE

Chapter 2, ***Mathematics of Finance*** addresses the importance of complex finance considerations of modern life. Here the basics underlying consumer financial products and more advanced topics are included. This chapter, requiring only elementary mathematics, appears early in the text for the benefit of some curricula and can be omitted.

MATRICES

Chapter 3, ***Matrix Algebra*** is one of the pillars of finite mathematics courses. Matrices have gained a vital role in the operations of large computer databases, manipulating large systems of equations, as in linear programs, and in the increasing quantification of economics, finance, and social sciences. The text covers the encoding of data into matrices, their characteristics, and appropriate algebraic manipulations as addition, multiplication, and inversion. The application to solving systems of equations using Gauss-Jordan elimination schemes is an entire section.

Still, teachers find higher quality examples of actual applications of matrices to be elusive. Examples from cryptology (a major government and industrial security usage) and Input-Output analysis (a fundamental application for economics) are included. A novel section in Chapter 11 has examples of sociological usage in studies of kinship structures through generations and of occupational mobility among populations.

LINEAR PROGRAMMING

Linear optimization, or linear programming as it is better known, is another pillar of finite mathematics courses. It is the subject of the second portion of the text in Chapters 4, 5, and 6 and in game theory and finance in Chapter 11. This text mirrors others and has added features and enrichment. In Chapter 4, for ***Geometric Solutions*** to linear optimization problems, graphing and solving simple systems of linear equations, some of the introductory skills of Chapter 1 are utilized. However, graphical solutions are limited to problems of few variables.

The ***Simplex Method*** appears in Chapter 5. Linear programming is likely the most widely used mathematics to allocate scarce resources in industry and government. Here the matrix algebra skills of Chapter 3 aid in solutions. However, the use of LINDO™ the widely available popular software, enables students to solve larger problems if an instructor wishes to take a more technological approach to optimization. The basic Standard Maximization Problem (SMP) with "less than" constraints is in Sections 5.1 – 5.3. A flow chart aids students to master the simplex iterations.

Section 5.4 on duality and Section 5.5 on non – SMP programs have an extended treatment of linear programming. Two solution methods are given for minimization objectives and excess variable situations. One uses artificial variables and the other algebraic manipulation. A unique flow chart guides students.

Chapter 6, **Application Models** enhances the valuable skill of formulating linear programs. The use of linear programming for diverse industrial and governmental purposes is widespread. Google has well over a million references! The many prototype models presented in Chapter 6 such as the knapsack, trim waste, and caterer problems are classics of linear programming literature. Virtually all applications require automated computation and computer software is widely available.

Two additional linear programming features are its mathematical history and application to game theory in Chapter 11. Game theory is important to contemporary economics and industrial competition. In addition to the elementary methods usual in finite texts, this text places emphasis on the connection between game theory and linear programming. This aids applications that require automated computation as well as understanding of mathematical theory.

PROBABILITY AND STATISTICS

Mathematical probability and statistical topics usually form another pillar of a finite mathematics course. It comprises the third major part of the text. Chapter 7, ***Set and Probability Relationships***; Chapter 8, ***Random Variables and Probability Distributions***; Chapter 9, ***Markov Chains***; Chapter 10, ***Mathematical Statistics***; and a section on *Monte Carlo Method*, unique to this text, in Chapter 11.

Chapter 7, ***Set and Probability Relationships*** has ample examples to illustrate the basics of sets and the relations to probabilities. One section emphasizes tree diagrams both as an aid to teaching probability and as an introduction to their widespread use in managing large projects from building construction to varied components of space ship design. A unique feature of combinatorics in this text is a table aid to help with these challenging exercises. Also included is a fourth combinatoric possibility, usually omitted, that samples with replacement but without regard to order. Newer applications of combinatorics in computer program speed estimates are included.

Bayes' Rule has earned a niche in much of everyday life in applications from spam filters to medical tests and oil exploration. The text provides a robust treatment of the Total Probability Theorem and Bayes' Rule. Both algebraic and tree diagrams are used to aid students' skill and appreciation.

While Chapter 8, ***Random Variables and Probability Distributions*** has more material than can usually be included in a semester course, it can be used for enrichment or a second semester. Aside from the customary basic topics of the distributions, the hypergeometric and Poisson distributions are included. The hypergeometric distribution, easily taught, has a role in industrial quality control and estimation of sizes of animal populations that our students find interesting. The Poisson's Law's ubiquitousness in everyday situations is briefly explored.

Chapter 9, ***Markov Chains*** is a staple of many finite mathematical courses. This text uniquely pairs matrix and diagram solutions to each example. Students are aided to grasp matrix methods with a diagrammatic solution alongside. Students learn to form and execute transition matrices in regular and absorbing Markov chains.

Chapter 10, ***Mathematical Statistics*** an occasional topic, has become fundamental in many social science and business contexts. Some feel that statistics should be a part of a collegiate education. Included here are topics in statistical graphs, and surveys, basic definitions and interpretations of mean and variance. The normal distribution receives robust attention in view of its basic theoretical and practical importance.

ENRICHMENT

Chapter 11 ***Enrichment in Finite Mathematics*** may be unique among finite mathematics texts. Each section of this chapter deals with a nontraditional and useful topic related to finite mathematics. One section has the elements of game theory and its connection to linear programming. Other sections, somewhat novel to finite mathematics texts, are both interesting and useful in seeking fresh topics.

There is a section on financial and economic applications and another that uses matrix applications to the social and life sciences. The last two sections involve the Monte Carlo Method and Dynamic Programming. Monte Carlo method, often called simulation, is a staple of planning, research, and operations in a huge variety of contexts. A popular topic with students it is easily mastered. Dynamic Programming, a relatively new optimization format, relies on mathematical recursion. A useful topic in its own right, additionally, it alerts students to the limits of automated computation even for seemingly "small" problems.

SUPPLEMENTS

A modestly priced companion "Student Solutions Manual" has detailed solutions to the odd numbered exercises and is recommended.

For teachers, a complete solutions manual, PowerPoint® slides by chapter, and a test bank are available from John Wiley & Sons.

SUGGESTIONS

Suggestions for improvements are welcome.

ACKNOWLEDGEMENTS

We have benefitted from advice and discussions with Professors Louise Amick, Washington College; Nancy Hall, University of Delaware Associate in Arts Program-Georgetown; Richard Schnackenberg, Florida Gulf Coast University; Robert H. Mayer, Jr, US Naval Academy, Dr Wiseley Wong, University of Maryland; and Carolyn Krause, Delaware Technical and Community College-Terry Campus.

We acknowledge the University of Delaware's Morris Library for use of its resources during the preparation of this text.

ABOUT THE COMPANION WEBSITE

This book is accompanied by a companion website:

http://www.wiley.com/go/morris/finitemathematics

The website includes:

- Instructors' Solutions Manual
- PowerPoint® slides by chapter
- Test banks by chapter
- Teacher Commentary

1 Linear Equations and Mathematical Concepts

Finite Mathematics: Models and Applications, First Edition. Carla C. Morris and Robert M. Stark.

Companion Website: http://www.wiley.com/go/morris/finitemathematics

1.1 SOLVING LINEAR EQUATIONS

Mathematical descriptions, often as **algebraic expressions**, usually consist of alphanumeric characters and special symbols.

Physicists describe the distance, s, that an object falls under gravity in time, t, by $s = (1/2)gt^2$. Here, the letters s and t are **variables** since their values may change, while, g, the acceleration of gravity is considered constant. While any letters can represent variables, typically the later letters of the alphabet are customary. The use of x and y is generic. Sometimes, it is convenient to use a letter that is descriptive of the variable, as t for time.

Earlier letters of the alphabet are customary for fixed values or **constants**. However, exceptions are widespread. The equal sign, a special symbol, is used to form an **equation**. An equation equates algebraic expressions. Numerical values for variables that preserve equality are called **solutions** to the equations.

For example, $5x + 1 = 11$ is an equation in a single variable, x. It is a **conditional** equation since it is only true when $x = 2$. Equations that hold for all values of the variable are called **identities**. For example, $(x + 1)^2 = x^2 + 2x + 1$ is an identity. By solving an equation, values of the variables that satisfy the equation are determined.

An equation in which only the first powers of variables appear is a **linear equation**. Every linear equation in a single variable can be solved using some or all of these properties:

Substitution – Substituting one expression for an equivalent one does not alter the original equation. For example, $2(x - 3) + 3(x - 1) = 21$ is equivalent to $2x - 6 + 3x - 3 = 21$ or $5x - 9 = 21$.

Addition – Adding (or subtracting) a quantity to each side of an equation leaves it unchanged. For example, $5x - 9 = 21$ is equivalent to $5x - 9 + 9 = 21 + 9$ or $5x = 30$.

Multiplication – Multiplying (or dividing) each side of an equation by a nonzero quantity leaves it unchanged. For example, $5x = 30$ is equivalent to $(5x)(1/5) = (30)(1/5)$ or $x = 6$.

To Solve Single Variable Linear Equations

1. **Resolve fractions.**
2. **Remove grouping symbols.**
3. **Use addition and/or subtraction to move variable terms to one side of the equation.**
4. **Divide the equation by the variable coefficient.**
5. **Verify the solution in the original equation as a check.**

Example 1.1.1 Solving a Linear Equation

Solve $(3x/2) - 8 = (2/3)(x - 2)$.

Solution:
To remove fractions, multiply both sides of the equation by 6, the least common denominator of 2 and 3. The revised equation becomes

$$9x - 48 = 4(x - 2).$$

Next, remove grouping symbols to yield

$$9x - 48 = 4x - 8.$$

Now, subtract 4x and add 48 to both sides to yield

$$9x - 4x - 48 + 48 = 4x - 4x - 8 + 48 \quad or \quad 5x = 40.$$

Finally, divide both sides by 5 (the coefficient of x) to attain $x = 8$. *The result,* $x = 8$ *is checked by substitution in the original equation:*

$$(3(8)/2) - 8 = (2/3)(8 - 2)$$
$$12 - 8 = (2/3)(6)$$
$$4 = 4 \quad checks!$$

The solution $x = 8$ *is correct!*

Equations often contain more than one variable. To solve linear equations in several variables simply bring the variable of interest to one side. Proceed as for a single variable, considering the other variables as constants for the moment.

Example 1.1.2 Solving for y

Solve for y: $5x + 4y = 20$.

Solution:
Move terms with y to one side of the equation and any remaining terms to the opposite side. Here, $4y = 20 - 5x$. *Next, divide both sides by 4 to yield* $y = 5 - (5/4)x$.

Example 1.1.3 Simple Interest

"Interest equals principal times rate times time" expresses the well-known simple interest formula, $I = prt$. *Solve for time t.*

Solution:
Clearly, $I = (pr)t$ *and pr becomes the coefficient of t. Dividing by pr gives* $t = I/pr$.

Mathematics is often called "the language of science" or "the universal language." To study phenomena or situations of interest, mathematical expressions and equations are used to create a **mathematical model**. Extracting information from the mathematical model provides solutions and insights. These suggestions may aid in modeling skills.

To Solve Word Problems

1. **Read the problem carefully.**
2. **Identify the quantity of interest and possibly useful formulas.**
3. **A diagram may help.**
4. **Assign symbols to variables and other unknown quantities.**
5. **Translate words into an equation(s) using symbols for variables and unknowns.**
6. **Solve for the quantity of interest.**
7. **Check the solution and whether the proper question has been answered.**

Example 1.1.4 Investment

Ms. Brown invests \$5000 to yield 1% annual interest. What will she earn in 1 year?

Solution:
Here, the principal (original investment) is \$5000. The interest rate is 0.01 (expressed as a decimal) and the time is 1 year.

⟶

Using the simple interest formula, $I = prt$, Ms. Brown's interest is

$$I = (\$5000)(0.01)(1) = \$50$$

After 1 year, her capital becomes $p + prt = \$5000 + \$50 = \$5050$.

Example 1.1.5 Gasoline Prices

The June, 2014, East Coast regular grade gasoline average price (including tax) was about \$3.64 per gallon. The comparable West Coast average was about \$4.00 per gallon.

a) What was the average regular grade gasoline price on the East Coast for 12 gallons of fuel?
b) What was the average regular grade gasoline price on the West Coast for 25 gallons of fuel?

Solution:

a) On average, on the East Coast 12 gallons cost $(12)(3.64) = \$43.68$.
b) On average, on the West Coast 25 gallons cost $(25)(\$4.00) = \$100.00$.

♦ The famous yesteryear comedy team of Bud Abbott and Lou Costello used arithmetic shenanigans as the basis for many of their routines. The duo are probably best known for their "Who's on first" baseball routine. Google Ivars Peterson's "Math Trek" for some fun!

Example 1.1.6 Breaking a Habit

One theory for breaking an adverse habit (smoking, snacking, childish behavior, etc.) is to delay successive gratifications. Suppose a wait time of w hours before gratifying a desire. Next, an increment of v hours to $w + v$ hours to gratification. On the next occasion, the wait time is $w + 2v$, and so on. Determine the wait time before gratification for the nth time.

Solution:
The first wait occurs at time w, the next v hours later so that the nth time is

$$w + (n - 1)v, \quad n = 1, 2, \ldots$$

EXERCISES 1.1

In Exercises 1–6 determine whether the equation is an identity, a conditional equation, or a contradiction.

1. $3x + 1 = 4x - 5$
2. $2(x + 1) = x + x + 2$
3. $5(x + 1) + 2(x - 1) = 7x + 6$
4. $4x + 3(x + 2) = x + 6$
5. $4(x + 3) = 2(2x + 5)$
6. $3x + 7 = 2x + 4$

In Exercises 7–27 solve for the variable.

7. $5x - 3 = 17$
8. $3x + 2 = 2x + 7$
9. $2x = 4x - 10$
10. $x/3 = 10$
11. $4x - 5 = 6x - 7$
12. $5x + (1/3) = 7$
13. $0.6x = 30$
14. $3x/5 - 1 = 2 - (1/5)(x - 5)$
15. $2/3 = (4/5)x - (1/3)$
16. $4(x - 3) = 2(x - 1)$
17. $5(x - 4) = 2x + 3(x - 7)$
18. $3x + 5(x - 2) = 2(x + 7)$
19. $3s - 4 = 2s + 6$
20. $5(z - 3) + 3(z + 1) = 12$
21. $7t + 2 = 4t + 11$
22. $(1/3)x + (1/2)x = 5$
23. $4(x + 1) + 2(x - 3) = 7(x - 1)$
24. $1/3 = (3/5)x - (1/2)$
25. $\dfrac{x + 8}{2x - 5} = 2$
26. $\dfrac{3x - 1}{7} = x - 3$
27. $8 - \{4[x - (3x - 4) - x] + 4\} = 3(x + 2)$

In Exercises 28–35 solve for the variable indicated.

28. Solve: $5x - 2y + 18 = 0$ for y.
29. Solve: $6x - 3y = 9$ for x.
30. Solve: $y = mx + b$ for x.
31. Solve: $3x + 5y = 15$ for y.
32. Solve: $A = p + prt$ for p.
33. Solve: $V = LWH$ for W.
34. Solve: $C = 2\pi r$ for r.
35. Solve: $Z = \dfrac{x - \mu}{\sigma}$ for x.
36. The sum of three consecutive positive integers is 81. Determine the largest integer.
37. Sally purchased a used car for \$1300 and paid \$300 down. If the car is to be paid off in five equal monthly installments, what are her monthly payments?

38. A man's suit, marked down 20%, sold for \$120. What was the original price?

39. If the marginal propensity to consume $m = 0.75$ and consumption, C, is 11 million units when disposable income is \$2 million, find the "consumption function." (Hint: use $C = mx + b$, where b is a constant).

40. An extension to a fire station costs \$100,000. The annual maintenance cost increases with the number of fire engines housed by \$2500 each. If \$115,000 has been allocated for the first year, how many additional fire engines can be housed?

41. The speed of light is much greater than the speed of sound, so lightning is seen before the sound of thunder is heard. An observer's distance from the flash can be calculated from the time between the sight of lightning and the sound of thunder. The distance, d (in miles), from the storm can be modeled as $d = 4.5t$, where time, t, is in seconds.
    a) How far is a storm if thunder is heard 2 seconds after the lightning is seen?
    b) If a storm is 18 miles away, how long before the sound of thunder is heard?

42. A worker has 40 hours to produce two types of items, A and B. Each unit of A requires 3 hours to complete and each item of B, 2 hours. After completing eight units of B, the remaining time was spent on units of A. How many units of A were produced?

43. An employee's share of Social Security Payroll Tax was 6.2% in 2003 on the first \$87,000 of earnings. This amount was matched by the employer. Develop a linear model for an employee's Social Security Payroll Tax.

44. An employee works 37.5 hours at a \$10 hourly wage. Federal tax deductions are 6.2% to Social Security, 1.45% to Medicare Part A, and 15% for income tax. What is the after tax take-home pay?

45. The body surface area (BSA) and weight (Wt) in infants and children weighing between 3 and 30 kilograms has been reported to follow the linear relationship

    $$\text{BSA} = 1321 + 0.3433\text{Wt (BSA is in square centimeters and Wt is in grams)}$$

    a) Determine the BSA for a child weighing 20 kilograms.
    b) If a child's BSA is 10,325 square centimeters, estimate its weight in kilograms.
    Current, J.D., "A Linear Equation for Estimating the Body Surface Area in Infants and Children.", The Internet Journal of Anesthesiology 1998; 2(2).

1.2 EQUATIONS OF LINES AND THEIR GRAPHS

Mathematical models express features of interest. In the managerial, social, and natural sciences and engineering, linear equations are often used to relate quantities of interest. Therefore, a thorough understanding of linear equations is important.

The standard form of a linear equation is $ax + by = c$, where a, b, and c are real-valued constants. It is characterized by the first power of the exponents.

Standard Form of a Linear Equation

$$ax + by = c$$

a, b, and c are real numbered constants, a and b not both zero.

Example 1.2.1 *Ordered Pair Solutions*

Do the ordered pairs (3, 5) and (1, 7) satisfy the linear equation $2x + y = 9$?

Solution:
An ordered pair satisfies an equation if equality is preserved. Substituting the point (3, 5) yields $2(3) + 5 \neq 9$. The ordered pair (3, 5) is not a solution to the equation. For (1, 7), the substitution yields $2(1) + 7 = 9$. This is true.

A **graph** is a pictorial representation of a function. It consists of points that satisfy the function. **Cartesian coordinates** are used to represent the relative positions of points in a plane or in space. In a plane, a point P is specified by the coordinates or ordered pair (x, y) representing its distance from two perpendicular intersecting straight lines, called the x-axis and the y-axis, respectively (see figure).

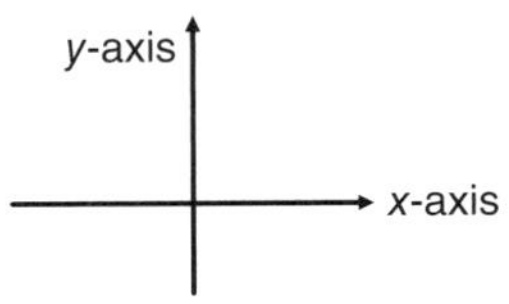

To determine the ***x*-intercept** of a line (its intersection with the x-axis), set $y = 0$ and solve for x. Likewise, for the ***y*-intercept** set $x = 0$ and solve for y. Two distinct points uniquely determine a line in a graph of a linear equation. An additional point can be a check as the three points must be **collinear**, that is, lie on the same line. The coordinate axes may be differently scaled.

Cartesian coordinates are so named to honor the mathematician **René Descartes** (Historical Notes).

◆ In Navajo religion, medicine men heal by "balancing of forces." Sand paintings in healing ceremonies use reflection symmetry to show paired forces.

Sometimes "fourfold" symmetry is found. Fourfold symmetry is reflection both horizontally and vertically as in the four quadrants of a Cartesian system. Such symmetry arises in many Native American cultures as an organizing principle. For example, prayers are to "the four winds" and teepees are made with four base poles, each placed in one of the four compass directions.

Example 1.2.2 Intercepts and Graph of a Line

Find the x- and y-intercepts of the line $2x + 3y = 6$ and graph its equation.

Solution:
When $x = 0$, $3y = 6$; the y-intercept is $y = 2$. When $y = 0$, $2x = 6$; the x-intercept is $x = 3$.

The two intercepts, (3, 0) and (0, 2), being two points of $2x + 3y = 6$, uniquely determine the line.

As a check, arbitrarily choose a value for x, say $x = -3$. Then, $2(-3) + 2y = 6$ or $3y = 12$, so $y = 4$. Therefore, $(-3, 4)$ is another point on the line. Note the three points on the graph.

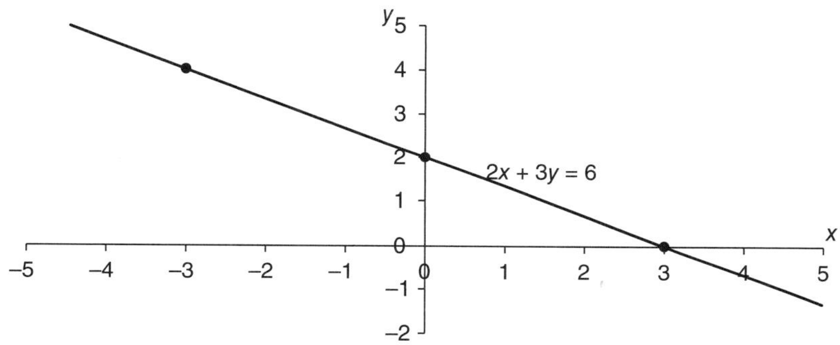

When either a or b in the standard equation $ax + by = c$ is zero, it reduces to a single value for the remaining variable. So, if $y = 0$, $ax = c$ and $x = c/a$, a vertical line. If $x = 0$, $by = c$ and $y = c/b$, a horizontal line.

Vertical and Horizontal Lines

For $ax + by = c$
The graph of $x = c/a$, ($b = 0$), is a vertical line.
The graph of $y = c/b$, ($a = 0$), is a horizontal line.

It is often useful to express the equations of lines in other (and equivalent) formats. The **slope**, m, of a line can be described in several ways, for example, as "the rise divided by the run" or as "the change in y, denoted by Δy, divided by the change in x, Δx." A line with a positive slope increases (rises) from left to right (/), while a line with negative slope decreases (falls) (\ ).

Slope

$$\text{Slope} = m = \frac{\text{Rise}}{\text{Run}} = \frac{\text{Change in } y}{\text{Change in } x} = \frac{\Delta y}{\Delta x} = \frac{y_2 - y_1}{x_2 - x_1}$$

The **slope-intercept form** of a line is $y = mx + b$, where m is the slope and b its y-intercept. A horizontal line has a slope of zero. A vertical line has an infinite (undefined) slope, as there is no change in x for any change in y.

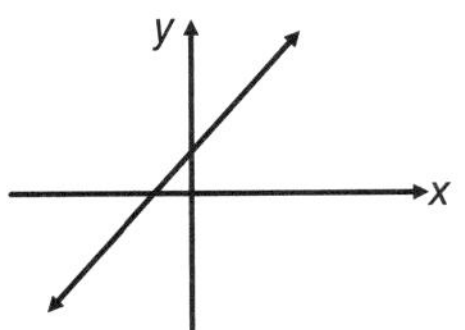

Positive slope

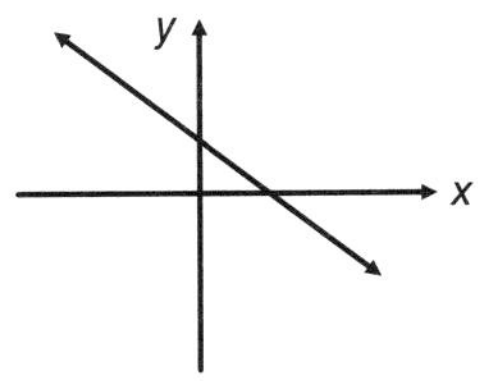

Negative slope

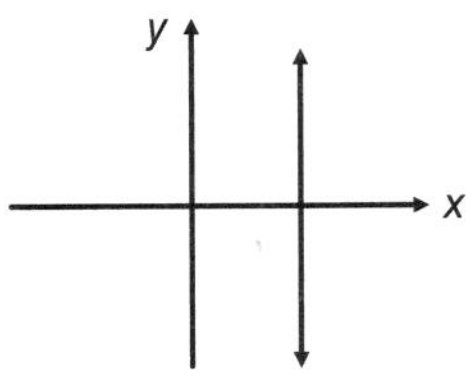

Undefined slope

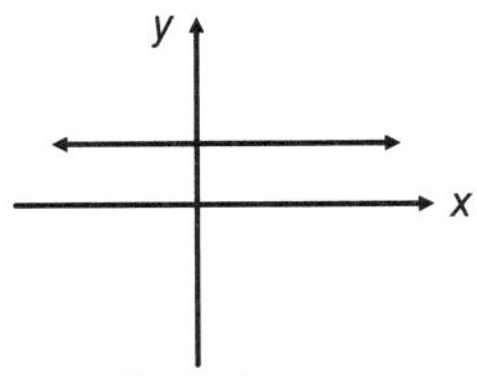

Zero slope

Slope-Intercept Form of a Linear Equation

$$y = mx + b$$

where m is the slope and b the y-intercept.

A linear equation in standard form ($ax + by = c$) is written in slope-intercept form by solving for y.

Example 1.2.3 Slope-Intercept Form

Write $2x + 3y = 6$ in slope-intercept form and identify the slope and y-intercept.

Solution:

Solving for y, $y = (-2/3)x + 2$. By inspection, the slope is $-2/3$ ("line falls") and the y-intercept is (0, 2), in agreement with the previous example.

A linear equation can also be written in a **point-slope form:** $y - y_1 = m(x - x_1)$. Here, (x_1, y_1) is a point on the line and m is the slope.

Point-Slope Form of a Linear Equation

$$y - y_1 = m(x - x_1)$$

where m is the slope and (x_1, y_1) a point on the line.

Example 1.2.4 Point-Slope Form

Find the equation of a line passing through (2, 4) and (5, 13) in point-slope form.

Solution:
First, the slope $m = \frac{13 - 4}{5 - 2} = \frac{9}{3} = 3$. *Now, using (2, 4) in the point-slope form, we have* $y - 4 = 3(x - 2)$. *(Using the point (5, 13) yields the equivalent* $y - 13 = 3(x - 5)$.*) For the slope-intercept form, solving for y yields* $y = 3x - 2$. *The standard form is* $3x - y = 2$.

Incidentally, as noted earlier, while the generic symbols x and y are most common, other letters are used to denote variables. Earlier $I = prt$ was used to express accrued interest. Economists use q and p for quantity and price, respectively, and scientists often use F and C for Fahrenheit and Celsius temperatures or p and v for gas pressure and volume, and so on.

Example 1.2.5 Temperature Conversion

Water boils at 212 degrees Fahrenheit or 100 degrees Celsius and freezes at 32 degrees Fahrenheit or 0 degrees Celsius. Find a linear relation between Celsius and Fahrenheit temperatures.

Solution:
Let the Celsius temperature be the input variable, denoted by C. The Fahrenheit temperature is the output variable, F, in a linear relation. The ordered pairs are (100, 212) and (0, 32). For the slope-intercept form, $m = \frac{32 - 212}{0 - 100} = \frac{-180}{-100} = \frac{9}{5}$. *The second ordered pair provides the vertical y-intercept by inspection. Therefore,* $F = (9/5)C + 32$ *is the widely used relation to easily convert Celsius to Fahrenheit temperatures. Similarly, there is a relation to connect Fahrenheit to Celsius temperatures.*

A common mathematical model for depreciation (equipment, buildings, etc.) relates current value "y" (dollars) to age "x" (years). Straight Line Depreciation (SLD) is a common choice. In an SLD model, annual depreciation is the same each year of useful life. Any remaining value is the "salvage value."

Example 1.2.6 Salvage Value

Equipment value (dollars) at time t (years) is $V(t) = -10,000t + 80,000$ and its useful life expectancy is 6 years. Determine the original value, salvage value, and annual depreciation.

Solution:
The original value, at $t = 0$, was the salvage value, at $t = 6$ years, the end of useful life, is \$20,000 = (−10,000(6) + 80,000). Finally, the slope, the annual depreciation is \$10,000.

Note that parallel lines have the same slope. Two lines are perpendicular if their slopes are negative reciprocals.

Parallel and Perpendicular Lines

Two lines are parallel if their slopes are equal

$$(m_1 = m_2)$$

Two lines are perpendicular if their slopes are negative reciprocals

$$(m_1 = -1/m_2)$$

Example 1.2.7 Parallel or Perpendicular Lines

Are these pairs of lines parallel, perpendicular, or neither?

a) $3x - y = 1$ *and* $y = (1/3)x - 4$
b) $y = 2x + 3$ *and* $y = (-1/2)x + 5$
c) $y = 7x + 1$ *and* $y = 7x + 3$

Solution:

a) *Two slopes are required. By inspection, the slope of the second line is 1/3. The line $3x - y = 1$ in slope-intercept form is $y = 3x - 1$ so the slope is 3. Since the slopes are neither equal nor negative reciprocals, the lines must intersect.*
b) *The slopes are 2 and $(-1/2)$. Since these are negative reciprocals, the two lines are perpendicular.*
c) *The lines have the same slope (with different intercepts), so they are parallel.*

♦ "What Makes an Equation Beautiful," the title of an article in *The New York Times* of October 24, 2004, is not likely to excite everyone's interest, especially for the linear equations you studied in this chapter. Fair enough!

However, if linear equations are fairly new to you, be assured that they are a building block for more advanced – and more interesting equations.

Some physicists were recently asked: "Which equations are the greatest?" According to the article, some were nominated for the breadth of knowledge they capture, for their historical importance, and for reshaping our perception of the universe.

EXERCISES 1.2

1. Find the x- and y-intercepts for the following:

   a) $5x - 3y = 15$
   b) $y = 4x - 5$
   c) $2x + 3y = 24$
   d) $9x - y = 18$
   e) $x = 4$
   f) $y = -2$

2. Find slopes and y-intercepts for the following:

   a) $y = (2/3)x + 8$
   b) $3x + 4y = 12$
   c) $2x - 3y - 6 = 0$
   d) $6y = 4x + 3$
   e) $5x = 2y + 10$
   f) $y = 7$

3. Find the slopes of lines defined by these points:

   a) (3, 6) and (−1, 4)
   b) (1, 6) and (2, 11)
   c) (6, 3) and (12, 7)
   d) (2, 3) and (2, 7)
   e) (2, 6) and (5, 6)
   f) (5/3, 2/3) and (10/3, 1)

4. Find equations for the lines
   a) with a slope of 4 passing through (1, 7);
   b) passing through (2, 7) and (5, 13);
   c) with undefined slope passing through (2, 5/2);
   d) with x-intercept 6 and y-intercept −2;
   e) with slope 5 and passing through (0, −7);
   f) passing through (4, 9) and (7, 18).

5. Plot graphs of

   a) $y = 2x - 5$
   b) $x = 4$
   c) $3x + 5y = 15$
   d) $2x + 7y = 14$

6. Plot graphs of

a) $2x - 3y = 6$
b) $y = -3$
c) $y = (-2/3)x + 2$
d) $y = 4x - 7$

7. Are the pairs of lines parallel, perpendicular or neither?
   a) $y = (5/3)x + 2$ and $5x - 3y = 10$
   b) $6x + 2y = 4$ and $y = (1/3)x + 1$
   c) $2x - 3y = 6$ and $4x - 6y = 15$
   d) $y = 5x - 4$ and $3x - y = 4$
   e) $y = 5$ and $x = 3$

8. Find equations for the lines
   a) through (2, 3) and parallel to $y = 5x - 1$.
   b) through (1, 4) and perpendicular to $2x + 3y = 6$.
   c) through (5, 7) and perpendicular to $x = 6$.
   d) through (4, 1) and parallel to $x = 1$.
   e) through (2, 3) and parallel to $2y = 5x + 4$.

9. Can a linear equation not have an x-intercept? Have more than one x-intercept? Have no y-intercept? Have more than one y-intercept?

10. Find a formula to convert Fahrenheit to Celsius temperatures. Hint: use the freezing and boiling temperatures of water.

11. A new machine costs \$75,000 and has a salvage value of \$21,000 after 9 years. Find a linear equation to model its SLD.

12. A new car costs \$28,000 and after 5 years it has a trade-in value of \$3000. Find a linear equation to model its SLD.

13. A car requires 7 gallons of gasoline to travel 245 miles and 12 gallons to travel 420 miles. Determine a linear relationship to express miles traveled as a function of the gasoline usage.

14. A piece of office equipment was purchased for \$50,000 and after 10 years had a salvage value of \$5000. Express its depreciation by a linear equation.

15. A firm pays \$1100 monthly rent on a building (a fixed cost). Each unit of product costs \$5 (a variable cost of operation). Form a linear model for the total monthly cost to produce x items.

16. A skateboard sells for \$24. Determine the revenue function for sales of skateboards.

17. A car rental company charges \$50 per day for a medium-sized car and 30 cents per mile of travel.
    a) Find the cost equation for renting a medium-sized car for a single day.
    b) How many miles can be traveled in a day on a budget of \$110?

18. At the ocean's surface, water pressure equals an air pressure of about 15 pounds per square inch. Below the surface, water pressure increases by 4.43 pounds per square inch for each 10 foot descent.
    a) Express water pressure as a function of ocean depth.
    b) At what depth is water pressure 80 pounds per square inch?

19. A department store priced a man's suit at \$84 that wholesaled at \$70. Also, a woman's dress priced at \$48 wholesaled at \$40. If the markup policy is linear and is reflected in the prices of these items, relate the retail price, R, to the store's cost, C.

20. The demand for a certain product is linearly related to its price. The product was priced at \$1.50 and 40 items were sold. When it was priced at \$6, only 22 items were sold. Determine a linear relationship between the price, x, and the sales, y.

1.3 SOLVING SYSTEMS OF LINEAR EQUATIONS

To solve systems of two or more linear equations is to find common (intersection) points among them. The following methods are available:

(1) Graphing (2) Substitution (3) Elimination (addition)

To solve a system of linear equations in two variables graphically, they are graphed on common axes. If there is a common solution, it is at their intersection.

The graph of a system of two linear equations in two variables has one of these properties: they intersect at a point, are parallel, or are collinear. These three possibilities are illustrated in the following figures.

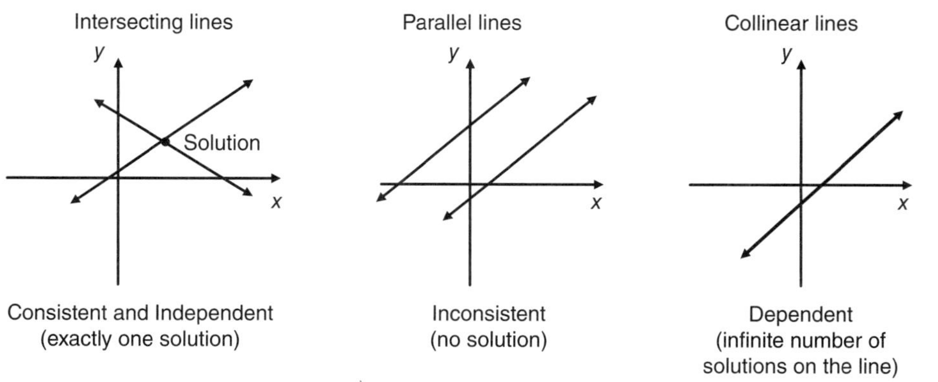

Whether a system of two linear equations is **consistent, inconsistent,** or **dependent** is apparent from the slope-intercept form $y = mx + b$.

- When the slopes differ, the system has a unique solution.
- When the slopes are equal and the y-intercepts differ, the lines are parallel. Such systems are inconsistent (have no solution).

- When the slopes and y-intercepts are equal, the lines are collinear (identical). There are an infinity of solutions that satisfy both equations.

Example 1.3.1 Graphical Solutions to a System of Linear Equations

Solve this system graphically:

$$x + y = 3$$
$$2x - y = 9$$

Solution:
As seen from the graph, the lines intersect and the solution to this system is at (4, −1).

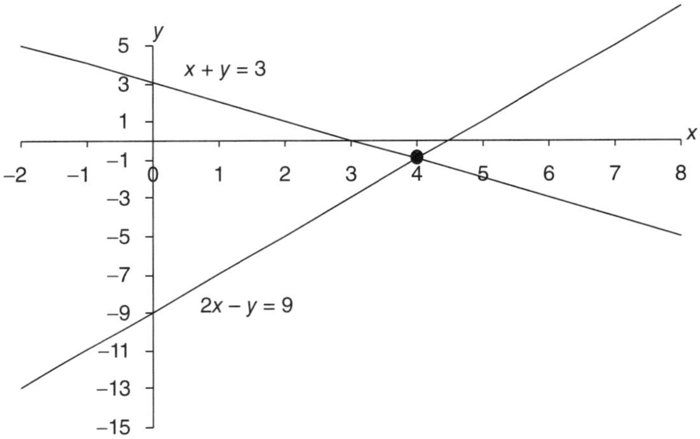

Graphing is not usually practical because of the need for precision in plotting.

An algebraic method of **substitution** can be used to solve systems of equations. In this method, one equation is solved for a variable and substituted in the other equation. It can be helpful to minimize calculations by thoughtful choice of substitution.

Solving by Substitution

1. **Solve for a variable in either equation as convenient. (For example, choose a variable with an integer coefficient, preferably ± 1, if possible.)**
2. **Substitute the chosen variable in the remaining equation. The result is an equation in one variable.**
3. **Solve the equation in one variable. (This variable cannot be the variable from Step 1.)**
4. **Substitute the value determined in Step 3 into the equation from Step 1 to determine the value of the remaining variable.**
5. **Check your solution in each of the original equations.**

Example 1.3.2 Substitution – Consistent System

Solve the following system by substitution:

$$\begin{aligned} 2x + y &= 9 \\ 3x + 5y &= 17 \end{aligned}$$

Solution:

First, observe coefficients. Since the coefficient of y in the first equation is unity, it is chosen to solve for y and to substitute into the second equation. Solving for y in the first equation yields $y = 9 - 2x$. *Substituting this into the second equation gives* $3x + 5(9 - 2x) = 3x + 45 - 10x = 17$. *Simplifying,* $7x = 28$ *and* $x = 4$. *Since* $y = 9 - 2x$, *we conclude that* $y = 1$. *The ordered pair solution is, therefore, (4, 1). This solution satisfies both equations.*

In the preceding example, we solved for y in the first equation. One could have solved for y in the second equation or for x in either equation to start the process. In solving the system, we found a single solution. The system was consistent and independent. A dependent system leads to an identity. An inconsistent system always yields a false condition.

Example 1.3.3 Substitution–Inconsistent System

Solve by substitution:

$$\begin{aligned} 4x - 2y &= 6 \\ 2x - y &= 1 \end{aligned}$$

Solution:

It is preferable to solve for y in the second equation: $y = 2x - 1$. *Substituting into the first equation yields* $4x - 2(2x - 1) = 4x - 4x + 2 = 6$. *The result "*$2 = 6$*" is false, and the system is inconsistent and lacks a solution.*

A third method to algebraically solve a system of equations is the **addition** or **elimination method**. Here, one seeks to multiply the equations by factors so that when they are added one of the variables is eliminated.

Solving by Elimination (Addition)

1. **Write each equation in standard form $Ax + By = C$.**
2. **If necessary, multiply one or both equations by a constant, so their sum contains only one variable (the other variable having been eliminated).**
3. **Solve for the remaining variable in the resulting equation.**
4. **Substitute the value from Step 3 into either of the original equations to find the value of the remaining variable.**
5. **Check your solution in the original system.**

Example 1.3.4 Elimination Method – Consistent System

Solve by elimination:

$$3x - y = 8$$
$$x + 2y = 5$$

Solution:

Sometimes it is easier to eliminate a variable whose coefficients have opposite signs. Here, y is the preferred variable for elimination. This is accomplished by multiplying the first equation by 2, as

$$6x - 2y = 16$$
$$x + 2y = 5$$

Now, adding the two equations yields $7x = 21$ *or* $x = 3$. *Using* $x = 3$, *we determine that* $y = 1$ *to give (3, 1) as the ordered pair solution to the system.*

Economists know that consumer demand for a commodity is related to its price. According to the *Law of Demand*, the quantity demanded increases as price decreases. Just as a consumer's willingness to buy is related to price, a manufacturer's willingness to supply goods is related to the realized price. In the *Law of Supply*, the quantity supplied increases as price increases. *Market equilibrium* occurs when the demand equals supply. The following figure illustrates the intersection when supply and demand are equal.

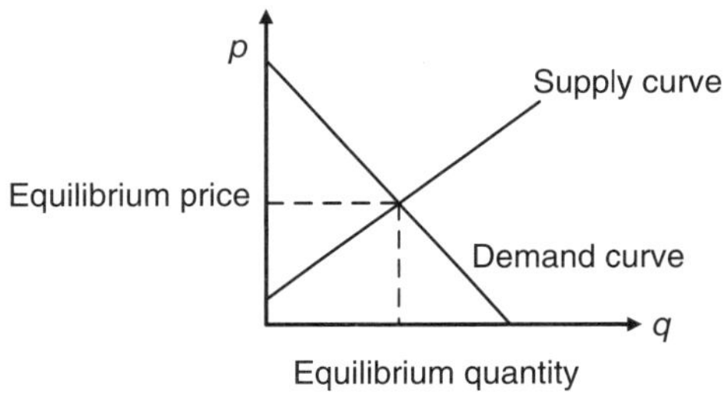

Example 1.3.5 Market Equilibrium

Determine the equilibrium quantity, q, and price, p, when the demand function for a commodity is $p = -5q + 20$ *and the supply function by* $p = 3q + 4$.

Solution:

Using substitution to solve the system of equations yields $-5q + 20 = 3q + 4$.
So, $8q = 16$ *or* $q = 2$ *and* $p = 10$.

Total cost consists of two components, a fixed cost and a variable cost. A *fixed cost* as, say, monthly rent, is independent of the "level of production." *Variable cost* changes with production level. *Revenue* equals the price of an item multiplied by its demand. *Profit* is the difference between revenue and total cost. Solving the equations for total cost and for revenue simultaneously, their intersection yields the *break-even* point, where profit is zero.

♦ In Keynesian economic theory, consumption, C (in dollars), is the linear relation $C = mx + b$, where x is disposable income in dollars; m, marginal propensity to consume; and b, a scaling constant.

Marginal cost (*MC*) is the cost associated with one additional unit of production. When cost is a linear function, MC is the slope. When revenue is linear, its slope is the *marginal revenue* (*MR*). That is, MR is the revenue from an additional unit of sales.

EXERCISES 1.3

1. Which ordered pairs are solutions to the system $2x + y = 7$ and $x + y = 5$?

   a) (3, 1) b) (2, 3) c) (4, −1)

2. Which ordered pairs are solutions to the system $2x + 2y = 4$ and $x + y = 2$?

   a) (2, 0) b) (0, 2) c) (1, 1)

3. Write each equation in slope-intercept form. Determine by inspection whether the system is consistent, inconsistent, or dependent.

   a) $3y = -x + 8$, $x + y = 6$

   b) $x + 2y = 7$, $2x = -4y + 14$

   c) $3x + 2y = 7$, $y = (-3/2)x + 5$

4. Determine the solution graphically:

   a) $x + y = 4$, $2x + 2y = 6$

   b) $x + 2y = 5$, $2x + y = 4$

   c) $y = (1/3)x - 2$, $x - 3y = 6$

5. Determine the solution graphically:

   a) $x + 2y = 5$, $3x + y = 5$

   b) $2x + 3y = 12$, $2x - y = 4$

   c) $y = (1/4)x + 1$, $2x - 3y = 2$

6. Use the substitution method to solve:

   a) $y = 5$, $2x + y = 7$

   b) $x - 1 = 0$, $x + 3y = 4$

   c) $2x + 3y = 8$, $x + 2y = 5$

7. Use the substitution method to solve:

a) $x = 3$
$x + 3y = 9$

b) $y - 2 = 0$
$x + 3y = 9$

c) $x + y = 5$
$y = -x + 3$

8. Use the elimination (addition) method to solve:

a) $x + y = 2$
$x - y = 4$

b) $2x + 3y = 6$
$4x - y = 5$

c) $2x + y = 5$
$2y = -4x + 10$

d) $y = -x + 4$
$2x + 2y = 7$

9. Use the elimination (addition) method to solve:

a) $-x + 2y = 5$
$x + y = 4$

b) $4x + 3y = 35$
$2x - y = 5$

c) $x + 4y = 13$
$2y = -4x + 10$

d) $y = x + 4$
$2x + 3y = 12$

10. A real estate agent receives a weekly salary plus a sales commission. If the agent earns \$550 in a week when a \$50,000 home is the only sale and \$850 in a week when a \$150,000 home is the only sale, calculate the weekly salary and percentage commission.

11. An elementary school plans a fund-raiser. Students are to sell boxes of cookies for \$4 each and boxes of candies for \$5 each. The students sold 2400 items for \$10,500. How many boxes of each were sold?

12. A chemist requires 500 milliliters of a 10% solution. Solutions of 5% and 25% are available. How many milliliters of each solution should be mixed?

13. Octane is a measure of a gasoline's ability to resist "knock" or "pinging." Gasoline pumps offer three unleaded octane grades: regular (87), mid-grade (89), and premium (93). How many gallons of regular should be mixed with premium to have 100 gallons of 92% octane?

14. A student works two part-time jobs and earned \$185 by working a total of 25 hours at jobs that pay \$7 and \$8 per hour. How many hours did the student work at each job?

15. Given the demand and supply functions

Demand $p = -2q + 320$

Supply $p = 8q + 20$, where p is the price and q the quantity.

a) Sketch the supply and demand curves on a coordinate system.
b) Label market equilibrium on your sketch.
c) Solve algebraically for the coordinates of market equilibrium.

16. A vendor pays \$250 monthly rent for product assembly space. Assembled items cost \$10 and are sold for \$20 each. Determine the revenue function, total cost function, profit function, and the break-even point.

17. On the graph

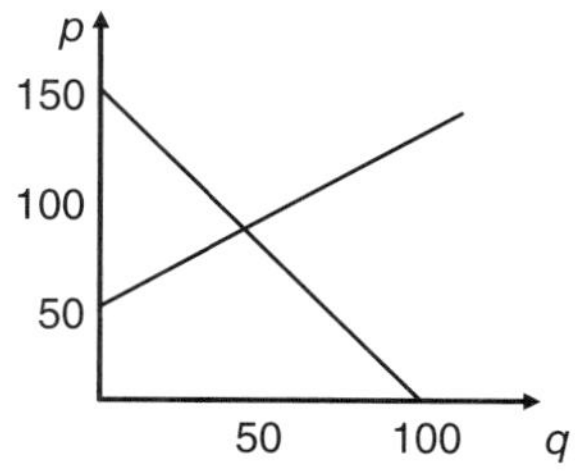

a) Label both supply and demand lines.
b) Label the market equilibrium point and estimate its coordinates.

18. Find the market equilibrium for these supply and demand functions

$$\text{Demand}: \ p = -0.5q + 54$$
$$\text{Supply}: \ p = 1.2q + 37$$

1.4 THE NUMBERS π AND e

"There are numbers and there are numbers!" to coin a cliché. Integers are familiar to you along with fractions and decimals to comprise **rational numbers**. The square roots, cube roots, and so on of numbers that are not perfect squares, cubes, and so on are called **irrational numbers** and **complex numbers**. They arise as the roots of linear, quadratic, and other equations known as **polynomials**.

However, there are other numbers in nature that are not the roots of polynomial equations – they are known as **transcendental numbers**. Every transcendental number is irrational. Prominent among these amazing transcendental numbers are π and e. Both make appearances in the text. While we briefly introduce them here, an Internet search yields much fascinating information about them.

♦ The 1998 movie "Pi" features a mathematician obsessed with finding numerical patterns in life. A list of films in which mathematics seems to play some role may be found at http://world.std.com/~reinhold/mathmovies.html.

The Number π

The Greek letter "pi," written $\pi = 3.14159\ \ldots$, is likely already familiar to you. The circumference of a circle, C, of radius r and diameter d, is $2\pi r$ or πd. Therefore, $\pi = C/2r$ or C/d.

The ancients sought to construct a square, having the same area as a given circle, using only a compass and a straight edge. Known as "squaring the circle," it was one of the most baffling challenges in geometry. Centuries of trials concluded that the area of any circle can never exactly equal the area of a square. For the area of a circle of radius r, πr^2, to equal the area of a square of side, s, requires that numbers r and s exist so that $\pi r^2 = s^2$ or $\pi = (s/r)^2$. We now know that π can never be expressed as the ratio of two numbers.

The symbol π, first used in 1706, became popular after its adoption by **Leonhard Euler** in 1737 (Historical Notes).

The Babylonians, in 2000 B.C.E., are believed to have been the first to estimate π. **Christiaan Huygens**, an early Dutch physicist, sought to carefully construct inscribed polygons of increasing numbers of sides in circles hoping to estimate π from careful measurements of the radius and polygonal area.

Newspapers periodically report extensions of the value of π. The quest to express π in increasing numbers of decimal places has challenged mathematicians for centuries and continues to this day. The 1999 Guinness Book of World Records notes that a team of Tokyo researchers had computed about 206 billion digits of π. By 2002, the record was 1.24 trillion digits for pi. A Hitachi supercomputer, capable of over 2 trillion calculations per second, took over 400 hours to calculate the digits. Such achievements are not only mathematically noteworthy, they also have implications in cryptography, random numbers, and in the security of Internet transmissions.

♦ You can estimate π on a computer or programmable hand calculator using a simple Monte Carlo Method described in Chapter 11.

Example 1.4.1 A Biblical π

A biblical reference to Solomon is written, "And he made a molten sea, ten cubits from one brim to the other. It was round all about, and his height was five cubits: and a line of thirty cubits did compass it round about."

Use this information, as the ancients, to estimate π.

Solution:
This quote implies that a 10-cubit diameter circle has a circumference of 30 cubits. Therefore, $\pi \approx 30\ \text{cubits}/10\ \text{cubits} = 3$.

Pi appears virtually everywhere in the sciences including the DNA double helix, in disks of the moon and sun, rainbows, spreading pond ripples, Einstein's gravitational field equation, the normal distribution, prime number distributions, waves, navigation, spectra and much more. In geometry, as you know, π appears in the circumference, area, and volume of circles, spheres, cones, cylinders, or circular polygons.

The Number *e*

The number *e* (a notation credited to Leonhard Euler) is another curious number that arises from a peculiar limit. Consider the quantity

$$\left(1 + \frac{1}{n}\right)^n$$

Clearly, when $n = 1$, its value is 2. When $n = 2$, the value is 2.25. Its value increases as n increases. Is there a limit? Or does it increase beyond bound as n becomes larger and larger, tending to infinity? Actually, its value is $e = 2.71828\ldots$, and while it continues to increase as n increases, it is bounded above by, say, 2.72. You can easily calculate e on a computer or handheld calculator.

One of the properties of e is as the base of a system of **natural logarithms**. As such, it has a role in the calculation of compound interest, growth or decay of natural species, or situations when a quantity increases at a rate proportional to its value. The number e also appears in probability theory, statistics, trigonometry, physical laws, and many branches of engineering.

◆ {first 10-digit prime found in consecutive digits of e}.com

That message appeared on a California highway billboard and in a Cambridge, Massachusetts subway station. People who knew the correct 10-digit prime number were invited to apply for employment at Google Inc. The solution was reported on the Web by some puzzle solvers.

EXERCISES 1.4

1. The area of a circle is π. What is its diameter?
2. The circumference of a circle is π. What is its radius?
3. Conduct an Internet search to answer these questions:
   a) Pi actually has a day in its honor. When is pi day?
   b) Who was the Tokyo professor once credited with the world record for the number of digits of pi?
   c) Who was the noted scientist born on pi day? (Hint: His initials are A.E.).
   d) How does the illness Morbus Cyclometricus relate to pi and to squaring the circle?
   e) How is pi known in Germany?
4. Cite five additional facts about pi from the Internet to share with your class.
5. Use the expression $\left(1 + \frac{1}{n}\right)^n$ to approximate the value of e when

   a) $n = 5$
   b) $n = 10$
   c) $n = 25$.
6. An alternative way to calculate e is the series

$$1 + \frac{1}{1} + \frac{1}{2} + \frac{1}{2\cdot 3} + \frac{1}{2\cdot 3\cdot 4} + \frac{1}{2\cdot 3\cdot 4\cdot 5} + \cdots$$

   Find the approximate value of these six terms.

1.5 EXPONENTIAL AND LOGARITHMIC FUNCTIONS

Exponential Functions

Functions of the type $f(x) = b^x$ are **exponential functions**. Their distinguishing feature is that the variable, x, is an exponent. The base, b, is any positive real valued number (other than 1). The rates of change of exponential functions are greater than those for polynomial functions.

Recall the *Laws of Exponents* (see the following text)!

Laws of Exponents

1. $b^0 = 1$	**5.** $(b^x)^y = b^{xy}$
2. $b^{-x} = \dfrac{1}{b^x}$	**6.** $(ab)^x = a^x b^x$
3. $b^x b^y = b^{x+y}$	**7.** $\left(\dfrac{a}{b}\right)^x = \dfrac{a^x}{b^x}$
4. $\dfrac{b^x}{b^y} = b^{x-y}$	**8.** $\left(\dfrac{a}{b}\right)^{-x} = \left(\dfrac{b}{a}\right)^x = \dfrac{b^x}{a^x}$

Example 1.5.1 Using Laws of Exponents

Simplify these expressions using Laws of Exponents

a) $(2^{3x}2^{5x})^{1/2}$ *b)* $27^{x/3}9^{2x}$ *c)* $\dfrac{5^{2x-1}5^{4x+5}}{5^{3x+1}}$ *d)* $\dfrac{14^{2x}}{7^{2x}}$

Solution:

a) Since the same base appears twice within the parenthesis, add the exponents to yield $(2^{8x})^{1/2}$*. Next, multiply the exponents to yield* 2^{4x}*.*

b) There is a common base of 3, as 9 and 27 are powers of 3. Therefore, $(3^3)^{x/3}(3^2)^{2x} = 3^x 3^{4x} = 3^{5x}$*.*

c) First add the numerator exponents to yield $\dfrac{5^{6x+4}}{5^{3x+1}}$*. Next, subtract the exponents to simplify the expression to* $5^{(6x+4)-(3x+1)} = 5^{3x+3} = 5^{3(x+1)}$*.*

d) Since the exponents are the same, the expression can be rewritten as $\left(\dfrac{14}{7}\right)^{2x} = 2^{2x}$*.*

Often, when solving exponential equations, one either tries to equate bases or exponents. If there is a common base, exponents can be equated. Likewise, if the exponents are equal, their bases can be equated. The additional laws that follow are useful for solving exponential equations.

Additional Laws of Exponents

If $b^x = b^y$ then $x = y$.
If $a^x = b^x$ then $a = b$.

Example 1.5.2 Simplifying Exponential Equations

Solve for x in the following:

a) $4^x = 128$ b) $3^{-x} = 27$ c) $7^{2x-3} = \sqrt[3]{49}$ d) $125 = x^3$

Solution:

a) *Because the variable is an exponent, first equate bases. Rewriting the equation as $(2^2)^x = 2^7$ or $2^{2x} = 2^7$ yields $2x = 7$ or $x = 7/2$.*

b) *Rewriting the equation as $3^{-x} = 3^3$ yields $-x = 3$ or $x = -3$.*

c) *First, rewrite as $7^{2x-3} = (7^2)^{1/3} = 7^{2/3}$. Therefore, $2x - 3 = 2/3$ and $x = 11/6$.*

d) *Here, we seek to equate exponents. The equation is rewritten as $(5)^3 = (x)^3$, which yields $x = 5$.*

Graphs of exponential functions have several properties of interest. Their domain is the real numbers, while their range is the positive real numbers. There is no x-intercept and the y-intercept is at $y = 1$. There is a horizontal asymptote at $y = 0$ (the x-axis). When the base is between 0 and 1 the function is decreasing, and when the base exceeds 1, the function is increasing. The graph of the exponential functions $y = 2^x$ and $y = (1/2)^x$ shows these properties.

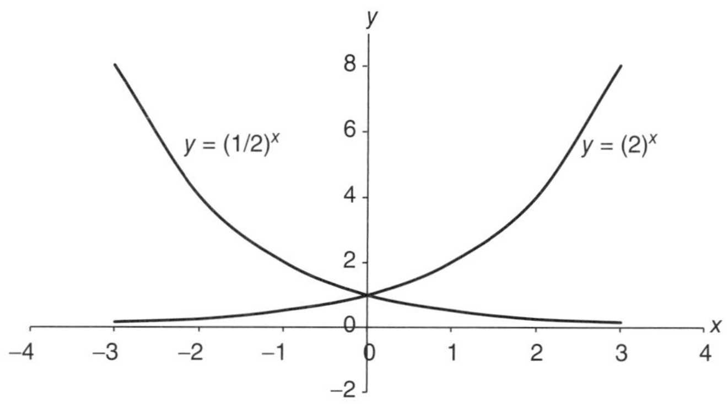

Recall that the transcendental number e is approximately 2.718. The special exponential function e^x has the characteristics discussed.

For the graph of e^x:

- the domain is all real numbers;
- the range is positive real numbers;
- the x-axis is an asymptote;
- the y-intercept is 1;
- there is no x-intercept;
- it is an increasing function ($e > 1$).

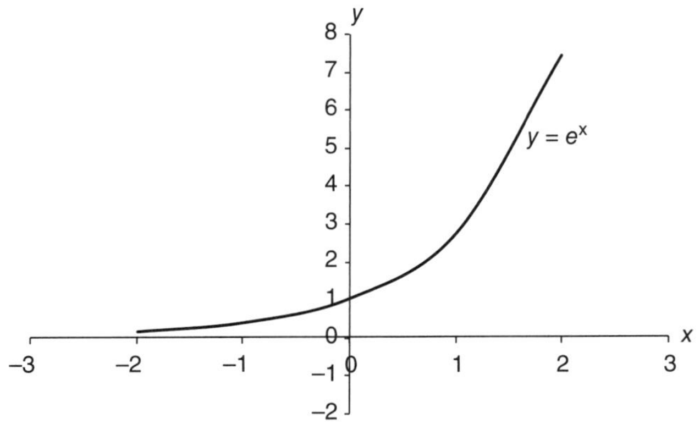

The Laws of Exponents apply to the transcendental number e, as Example 1.5.3 illustrates.

Example 1.5.3 Simplifying Exponential Exponents

Simplify

a) $(e^3)^x$ *b)* $\dfrac{e^{5x-1}}{e^{x+2}}$ *c)* $\left(\dfrac{1}{e}\right)^{5x}$

Solution:

a) Multiplying the exponents yields e^{3x}.
b) Subtracting the exponents yields $e^{(5x-1)-(x+2)} = e^{4x-3}$.
c) Rewrite as $(e^{-1})^{5x} = e^{-5x}$.

Logarithmic Functions

Logarithmic functions are inverses of exponential functions. The following formulas convert logarithmic into exponential forms and vice versa.

Logarithmic and Exponential Forms

$\log_b x = y$ means $b^y = x$ and vice versa.

♦ **John Napier** discovered logarithms in the latter part of the sixteenth century. Logarithms are useful for constructing mathematical models and for solving equations with exponential functions.

Any logarithm can be expressed as an exponential and vice versa. For instance, to graph $\log_2 x = y$ it is expressed equivalently as $2^y = x$ as shown:

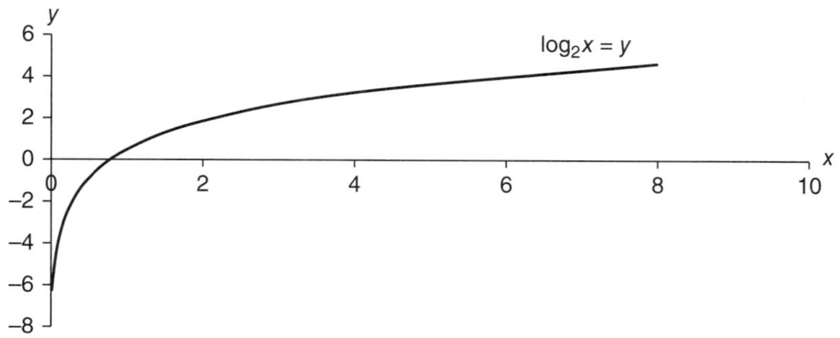

While the domain for logarithms is $(0, \infty)$, the range is all real numbers. The x-intercept of a logarithmic function is 1, while the y-intercept does not exist.

Example 1.5.4 Exponential and Logarithmic Forms

Express exponential forms in logarithmic forms and vice versa.

a) $3^4 = 81$ *b)* $2^{11} = 2048$ *c)* $\log_4 1024 = 5$ *d)* $\log_2 1/8 = -3$

Solution:

The equivalent logarithmic forms are:

a) $\log_3 81 = 4$

b) $\log_2 2048 = 11$

The equivalent exponential forms are:

c) $4^5 = 1024$

d) $2^{-3} = 1/8$

Most calculators have two logarithm keys: $\log x$ and $\ln x$. The first is for base 10 $(\log_{10} x)$ and the second is for base e $(\log_e x)$. The logarithms that are base 10 are called **common logarithms,** while the logarithms that are base e are called **natural logarithms**. For example, the common logarithm of x is written $\log x$, while the natural logarithm is written $\ln x$. Note that the bases (10 and e) are usually omitted and distinguished by the spelling.

♦ Originally known as "*log naturalis,*" the natural logarithm acquired its name from studies of natural phenomena. Common logarithms are likely named for their association with the decimal system.

Logarithms can also have bases other than 10 and e. Useful properties of these logarithms and natural logarithms are:

Properties of Logarithms

Logarithms	***Natural Logarithms***
$\log_b b^x = x$	$\ln e^x = x$
$b^{\log_b x} = x$	$e^{\ln x} = x$
$\log_b 1 = 0$	$\ln 1 = 0$
$\log_b b = 1$	$\ln e = 1$
$\log_b PQ = \log_b P + \log_b Q$	$\ln PQ = \ln P + \ln Q$
$\log_b (P/Q) = \log_b P - \log_b Q$	$\ln(P/Q) = \ln P - \ln Q$
$\log_b P^r = r \log_b P$	$\ln P^r = r \ln P$

Example 1.5.5 Simplifying Logarithms

Simplify these logarithms to eliminate products, quotients, or exponents.

a) $\log_3(x+1)(3x-5)$

b) $\log_7\left(\dfrac{(x+2)^3(x-4)}{(x-5)^2}\right)$

c) $\ln (x^2+2)^3(2y-1)^4(z+3)^2$

Solution:

a) *The logarithm of a product is written as a sum of logarithms,* $\log_3(x+1) + \log_3(3x-5)$.

b) *First, write the logarithm as* $\log_7(x+2)^3 + \log_7(x-4) - \log_7(x-5)^2$. *Next, the exponents appear as coefficients as* $3\log_7(x+2) + \log_7(x-4) - 2\log_7(x-5)$.

c) *First, expand the logarithms as* $\ln (x^2+2)^3 + \ln (2y-1)^4 + \ln (z+3)^2$. *Then, using the exponents as coefficients yields* $3\ln(x^2+2) + 4\ln(2y-1) + 2\ln(z+3)$.

Example 1.5.6 Simplifying Logarithmic Expressions

Write as a single logarithm:

a) $\log_3(x+1) + 2\log_3(x+4) - 7\log_3(2x+5)$

b) $5\ln(x+3) - 3\ln(y+2) - 2\ln(z+5)$

Solution:

a) *The coefficients, being exponents, yield* $\log_3(x+1) + \log_3(x+4)^2 - \log_3(2x+5)^7$ *Next, the sum (difference) of logarithms becomes a product (quotient) as* $\log_3\dfrac{(x+1)(x+4)^2}{(2x+5)^7}$. *(Caution: combining logarithms requires a common base.)*

b) *The coefficients become exponents when rewriting to yield* $\ln(x+3)^5 - \ln(y+2)^3 - \ln(z+5)^2$. *The two terms with negative coefficients are placed in the denominator of the single logarithm to yield* $\ln\dfrac{(x+3)^5}{(y+2)^3(z+5)^2}$.

♦ The Richter scale is one example of a logarithmic measure. It was devised by Charles Richter at the California Institute of Technology in 1935 to compare magnitudes of earthquakes. An earthquake of magnitude 5 on the Richter scale is ten times as strong as a magnitude 4 on the scale and one-tenth as strong as a magnitude 6.

A major earthquake is of magnitude 7, while a magnitude 8 or more is called a great quake. The great quake can destroy entire communities. The Indian Ocean Quake of December 2004 measured 9.0 on the Richter scale. The resulting tsunami caused massive destruction in the surrounding areas.

The strongest recorded earthquake was the Great Chilean Earthquake of 1960. It measured 9.5 on the Richter scale.

Logarithms and exponentials, being inverses of each other, enable one to solve equations by exploiting common properties as shown in Examples 1.5.7 and 1.5.8.

Example 1.5.7 Base 10 Exponents

Solve the following using logarithms.

a) $10^x = 9.95$ b) $10^{2x+1} = 1050$

Solution:

a) *Since* $10^0 = 1$ *and* $10^1 = 10$, *it follows that* $0 < x < 1$ *and that x is very close to 1. Taking the logarithm of the equation (base 10, here) yields*

$$\log_{10}10^x = \log_{10}9.95$$

$$x = \log_{10}9.95 \approx 0.9978 \text{ (as obtained by calculator)}$$

⟶

which is in agreement with our estimate.

b) *Since* $10^3 = 1000$ *and* $10^4 = 10,000$*, an estimate of x can be obtained from* $3 < 2x + 1 < 4$*, which yields* $1 < x < 1.5$*. Taking the logarithm of the equation (base 10, here) yields*

$$\log_{10} 10^{2x+1} = \log_{10} 1050 \quad so$$

$$2x + 1 = \log_{10} 1050$$

$$x = \frac{-1 + \log_{10} 1050}{2} \approx 1.0106$$

Note this is easily estimated since $\log_{10} 1000 = 3$.

Example 1.5.8 Base e Exponents

Solve using logarithmic properties:

$$5e^{4x-1} = 100$$

Solution:

First, note that $e^{4x-1} = 20$*. Since* $e^3 \approx 20$*, estimate* $4x - 1$ *as 3 (with x near 1). Taking the natural logarithm of both sides (base is e, here) yields*

$$\ln e^{4x-1} = \ln 20$$

$$4x - 1 = \ln 20$$

$$x = \frac{1 + \ln 20}{4} \approx 0.9989$$

The value for x agrees with our preliminary estimate.

EXERCISES 1.5

In Exercises 1–6, rewrite the expressions in the form of 2^{kx} ***or*** 3^{kx}***.***

1. a) $(8)^{3x}$ b) $(27)^{2x}$ c) $(16)^{5x}$

2. a) $(32)^{-2x}$ b) $(16)^{-x}$ c) $(81)^{-3x}$

3. a) $\left(\frac{1}{8}\right)^{-4x}$ b) $\left(\frac{1}{9}\right)^{6x}$ c) $\left(\frac{1}{27}\right)^{-2x}$

4. a) $\left(\frac{1}{4}\right)^{3x}$ b) $\left(\frac{1}{16}\right)^{x}$ c) $\left(\frac{1}{81}\right)^{-3x}$

5. a) $\frac{10^{5x}}{5^{5x}}$ b) $\frac{32^{2x}}{16^{2x}}$ c) $\frac{4^{3x}}{12^{3x}}$

6. a) $\frac{20^{3x}}{5^{3x}}$ b) $\frac{18^{7x}}{6^{7x}}$ c) $\frac{6^{x}}{24^{x}}$

In Exercises 7–12, use the Laws of Exponents to simplify the expressions.

7. $\frac{7x^3x^5y^{-2}}{x^2y^4}$

8. $\left(\frac{2x^3}{y^4}\right)^{-2}$

9. $\frac{x^3}{y^{-2}} \div \frac{x}{y^5}$

10. $\left(\frac{5a^3b^4c^0}{10abc^2}\right)^{-3}$

11. $\frac{2^{5x+3}4^{x+1}}{8(2^{3x-1})}$

12. $\frac{9^{4x-1}27^{x+2}}{81^{x+3}}$

In Exercises 13–20, solve for x.

13. $7^{3x} = 7^{15}$

14. $10^{-3x} = 1,000,000$

15. $2^{7-x} = 32$

16. $3^{2x}3^{x+1} = 81$

17. $(1+x)5^x + (3-2x)5^x = 0$

18. $(x^2)4^x - 9(4^x) = 0$

19. $(x^2+4x)7^x + (x+6)7^x = 0$

20. $(x^3)5^x - (7x^2)5^x + (12x)5^x = 0$

In Exercises 21–25, find the missing factor.

21. $2^{3+h} = 2^h\ (\quad)$

22. $3^{5+h} = 3^5\ (\quad)$

23. $7^{x+5} - 7^{2x} = 7^{2x}\ (\quad)$

24. $5^{2h} - 9 = (5^h - 3)\ (\quad)$

25. $7^{3h} - 8 = (7^h - 2)\ (\quad)$

26. Graph $y = \log_3 x$

27. Graph $y = \ln x$

In Exercises 28–37, evaluate the logarithms.

28. $\log_{10} 1,000,000$

29. $\log_3 243$

30. $\log_2 64$

31. $\log_5 125$

32. $\log_2 \frac{1}{32}$

33. $\log_3 \frac{1}{81}$

34. $\ln e^3$

35. $\ln e^7$

36. $\ln e^{7.65}$

37. $\ln e^{-3.4}$

In Exercises 38–45, evaluate the expressions.

38. $\ln(\ln e)$
39. $e^{\ln 1}$
40. $\log_9 27$
41. $\log_{25} 125$
42. $\log_4 32$
43. $\log_5 625$
44. $\log_2 128$
45. $\log_4(1/64)$

In Exercises 46–53, solve for x.

46. $\log_x 27 = 3$
47. $\log_x 64 = 3$
48. $\log_3(5x + 2) = 3$
49. $\log_2(x^2 + 7x) = 3$
50. $\ln 5x = \ln 35$
51. $e^{3x}\ e^{2x} = 2$
52. $\ln(\ln 4x) = 0$
53. $\ln(7 - x) = 1/3$

54. Write as a set of simpler logarithms: $\log_4 \dfrac{(x+1)^2(x-3)^6}{(3x+5)^3}$.

55. Write as a set of simpler logarithms: $\ln \dfrac{(x-1)^4}{(2x+3)^5(x-4)^2}$.

56. Write as a single logarithm: $2\ln x - 3\ln(y+1) + 4\ln(z+1)$.

57. Write as a single logarithm: $\ln 2 - \ln 3 + \ln 7$.

In Exercises 58–61, solve for x.

58. $10^{2x-1} = 105$
59. $10^{3x-1} = 100{,}100$
60. $3e^{x-1} = 4$
61. $e^{3x+1} = 22$

62. In June of 2004, an earthquake of magnitude 4.1 on the Richter scale struck northern Illinois. Its effects were felt from Wisconsin to Missouri and from western Michigan to Iowa. Another earthquake in the Midwest, the 1895 "Halloween Earthquake," is estimated to be 6.8 on the Richter scale. How much stronger was the Halloween quake than the northern Illinois quake?

1.6 VARIATION

Scientific formulas can express many types of variation. In **direct variation**, two related quantities can both increase or both decrease. For example, hourly pay, y, varies directly with the number of hours worked, x. In symbols, $y = kx$, where k, the hourly wage, is a **proportionality constant**. A graph of $y = kx$ is a line passing through the origin.

Direct Variation

$$y = kx$$

where k is a proportionality constant.

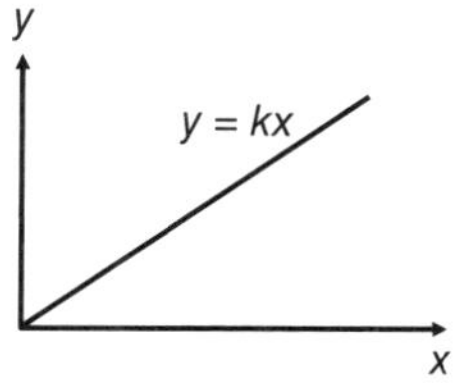

Example 1.6.1 Summer Wages

A student's summer job pays \$9.50 an hour. Write an equation for total wages earned. How much is earned in 30 hours?

Solution:
Let y represent total wages and x the hours of work. Here, the constant of proportionality is the hourly wage of \$9.50, so total wages is $y = 9.50x$. The student earns \$285 for 30 hours of work.

Physicists use *Hooke's law* for a spring to relate the force, y, directly to its elongation, x within its elastic limit. That is, $y = kx$, where k, called the *spring constant*, is the constant of proportionality. Architectural design has used proportionality since antiquity.

Example 1.6.2 Hooke's Law

If a force of 250 Newtons stretches a spring 20 centimeters (within the elastic limit), how much will it be stretched by a 100 Newton force?

Solution:
Here, the force y is 250 and the elongation x is 20, so $250 = 20k$. The spring constant k is 12.5 Newtons per centimeter. Hooke's law for this spring is $y = 12.5x$. Therefore, $100 = 12.5x$ implies that $x = 8$ centimeters.

Geometric similarity, related to proportionality, can usefully simplify mathematical modeling. In geometric similarity, there is a one-to-one correspondence among objects such that the ratio of distances between their corresponding points is constant. Recall that triangles are similar if their angles have the same measure. For the similar triangles shown: $\dfrac{a}{a'} = \dfrac{b}{b'} = \dfrac{c}{c'}$.

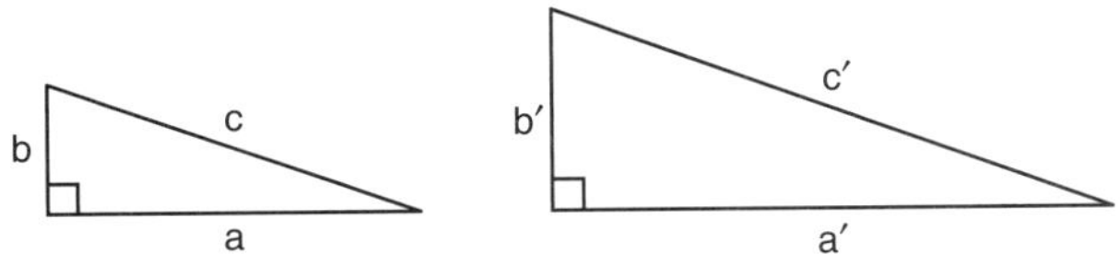

Example 1.6.3 Geometric Similarity

For the above similar triangles, $a = 2$, $b = 3$, $b' = 9$. Find the length a'.

Solution:
Here, the proportion is $\frac{2}{a'} = \frac{3}{9}$. Therefore, $(a')(3) = (2)(9)$. Solving, the length $a' = 6$.

Direct proportionality need not be linear. That is, it may **vary directly as a power** of x. A mathematical "power model" takes the form

$$y = kx^n, \quad n > 0$$

For example, the area of a circle, A, varies directly as the square of its radius, r, as $A = \pi r^2$.

Example 1.6.4 Volume of a Sphere

The volume of a sphere, V, varies directly as the cube of its radius, r. If $V = 36\pi$ when $r = 3$, determine the proportionality constant.

Solution:
Since volume varies as the cube of r, $V = kr^3$. Therefore, $36\pi = k(3)^3$ implies $k = 36\pi/27$ or $(4/3)\pi$ (the volume of a sphere is well-known as $V = (4/3)\pi r^3$).

In another form of variation, **inverse variation**, one quantity increases as the other decreases, and it is expressed as $y = k/x$ or $xy = k$.

Inverse Variation

$$y = k/x \qquad (\text{or} \quad xy = k)$$

where k is a proportionality constant.

While a graph of simple direct variation was a line through the origin, simple inverse variation is a *rectangular hyperbola* in the first quadrant when x and y are positive valued (shown).

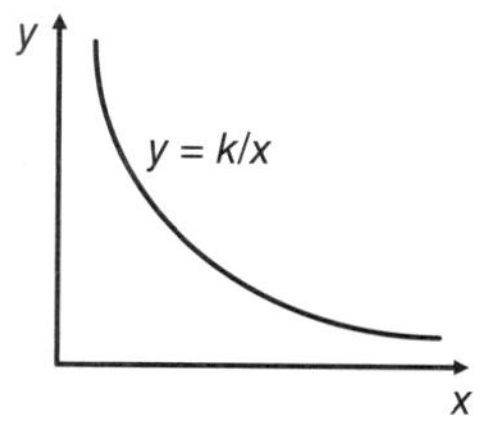

As examples, for light and sound, wave frequency, v, varies inversely with wavelength, λ, so λv is a constant; the pressure, P, of a gas at constant temperature varies inversely with its volume, V, so $PV = R$, a constant, expresses Boyle's law; and so on.

Example 1.6.5 Seawater Density

Chemists use the inverse variation of volume with mass to calculate density, defined as mass/volume. Calculate the density when 257.5 grams of seawater occupy a volume of 250 milliliters.

Solution:
Recall that a milliliter is equivalent to a cubic centimeter so the volume of seawater is 250 milliliters = 250 cubic centimeters. Its mass is 257.5 grams so its density is

$$\textit{Density} = 257.5 \textit{ grams} / 250 \textit{ cubic centimeters} = 1.03 \textit{ grams per cubic centimeter.}$$

More general inverse variation, say, as the nth power of x, can be expressed as

$$y = \frac{k}{x^n}, \quad n > 0.$$

Newton's law of gravitation has the force of attraction, F, between two masses varying inversely as the square of the distance between them, x. That is, $F = k/x^2$, where k is the constant of proportionality. Similarly, illumination from a light source varies as the inverse square of distance from the light source.

Example 1.6.6 Illumination

If illumination, I, is 75 units at a distance d = 6 meters from a light source, express the relation between illumination and distance from the source.

Solution:
The inverse square relationship takes the form $I = k/d^2$, where k is a proportionality constant. Here $75 = k/(6)^2 = k/36$. Therefore, $k = 75(36) = 2700$ illumination units per square meter. The formula relating illumination and distance here is $I = 2700/d^2$.

There are other and more complicated forms of variation. Many chemical and physical relationships exhibit some type of variation. For example, the strength of a beam is related to the amount and kind of material used and to the shape of the beam. Rectangular beams add significant strength to structures so they are often used in bridges and buildings.

Example 1.6.7 *Beam Strength*

The strength of a rectangular beam varies jointly as its width and the square of its depth. If the strength of a beam 2 inches wide by 8 inches deep is 768 pounds per square inch, what is the strength of a beam 4 inches wide and 6 inches deep?

Solution:
Let S represent the beam's strength, w its width, and d its depth. Accordingly, the model is $S = kwd^2$*, where k is a proportionality constant. From the data* $768 = k(2)(8)^2$*, from which* $k = 6$*. Therefore, for the second beam,* $S = 6(4)(6)^2 = 864$ *is its strength in pounds per square inch.*

Empirical economic relationships of supply and demand; money supply and inflation; short- and long-term interest rates; and so on often have complex forms of variation.

EXERCISES 1.6

1. Describe the indicated variation:
   a) Speed of a runner and distance traveled.
   b) Speed of a Nascar driver and time to complete a race.
   c) A person's adjusted gross income and the income tax rate.
   d) The area of a circle and its diameter.
   e) The distance between two rooms on a blueprint and their actual distance.
2. Let y vary directly as x. If x is 3 when y is 6, find x when y is 18.
3. Let y vary directly as the cube of r. If y is 32 when r is 4, find y when r is 6.
4. Let y vary inversely as the square of x. If y is 18 when x is 2, find y when x is 3.
5. Let y vary directly as x. When y is 15, x is 10. Determine x when y is 22.5.
6. Let y vary inversely as x in the previous exercise, determine x when y is 22.5.
7. Let x vary directly as y as depicted in the table. Determine the relation between x and y.

x	2	4	8
y	12.5	25	50

8. Let x vary inversely as y. Complete the following table.

x	1	2		4
y	36		12	

9. The area, A, of a circle is directly proportional to the square of its diameter, d. Write an equation relating area and diameter. Determine the proportionality constant if an area is 36π square inches when the diameter is 12 inches.

10. The circumference, C, of a circle varies directly with its diameter, d. If the diameter is 9, the circumference is 9π. What is the relation between the area of the circle and its diameter?

11. The volume, V, of a right circular cone is $(1/3)\pi r^2 h$. Find the volume when the height, h, is 10 centimeters and the radius, r, is 3 centimeters.

12. Suppose drug dosage is directly proportional to a patient's weight. A dose of 30 milligrams is prescribed for a patient weighing 50 kilograms.
    a) Find the constant of proportionality.
    b) What is the appropriate dosage for a 70-kilogram patient?

13. A force of 400 Newtons stretches a spring 10 centimeters (within the elastic limit). What is the stretch by a force of 250 Newtons?

14. The intensity of light, I, on a screen varies inversely as the square of the distance, d, from the source. If the intensity is 31.25-foot candles at a distance of 12 feet, find the intensity at 15 feet.

15. Consider the similar triangles. Find the length of the hypotenuse and the shorter side of the larger triangle.

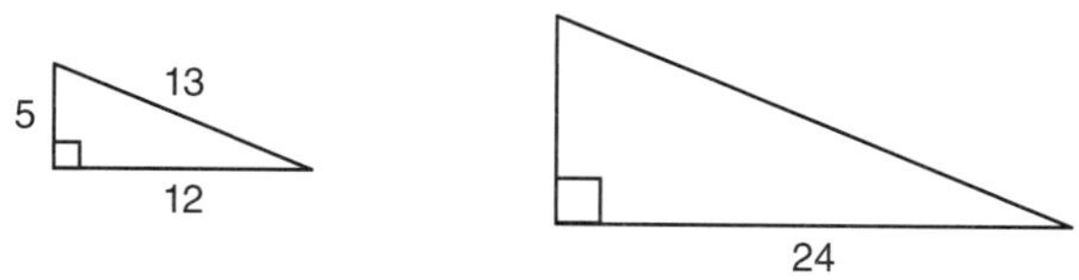

16. Consider the similar triangles. Find the length of the hypotenuse and the shorter side of the larger triangle.

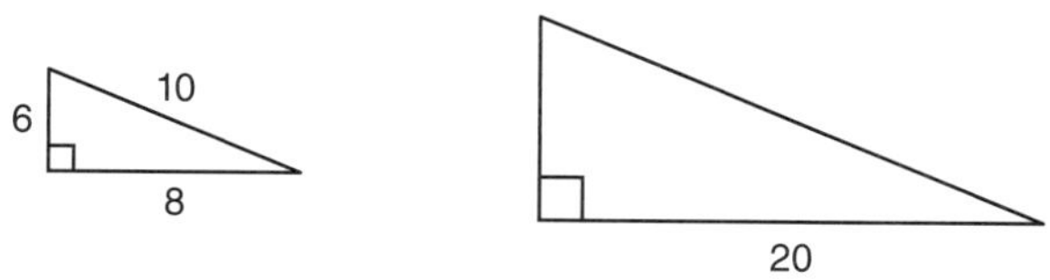

17. For the pair of similar figure, determine the length of the unknown side, x.

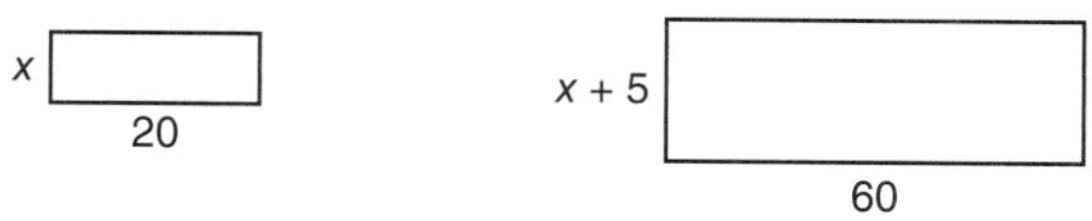

18. For the pair of similar figure, determine the length of the unknown side, x.

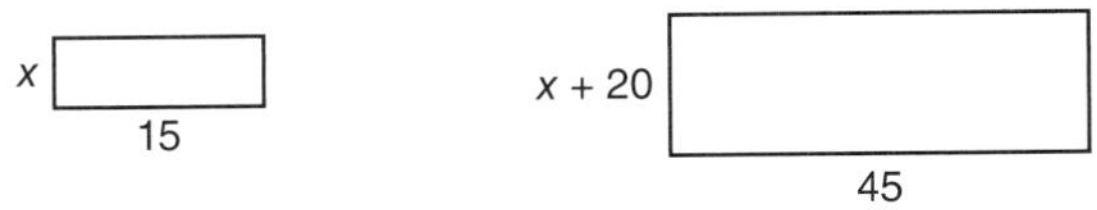

19. An area of a triangle, A, varies jointly as the length of its base, b, and its height, h. If the area of a triangle is 50 square inches when the base is 20 inches and the height is 5 inches, find the area of a triangle whose base is 16 inches and height is 8 inches.

20. Identify several examples of variation encountered in your courses this semester.

21. Conduct an Internet search for NASA CONNECT. How does this program deal with proportionality?

22. How does the "principle of humanitarian law" relate to proportionality and military actions? Hint: use an Internet search.

23. Have a friend measure the distance between the tips of the middle fingers of your outstretched arms. Compare this distance to your height. Are you more a "rectangle" or more a "square"?

    When the ratio of the fingertip to height measurement is about one, a person is classified as a "square", otherwise as a "rectangle."

    a) Repeat the measurements for five individuals noting their gender. Find each person's constant of proportionality.
    b) Share the data with class members to obtain a larger data set. Calculate the average constant of proportionality for the group. What are the largest and smallest values for the group?
    c) Reevaluate your class data separating individuals by gender. Does the average proportionality constant differ with gender? What about the largest and smallest values?

1.7 UNIT CONVERSIONS AND DIMENSIONAL ANALYSIS

While many quantities of interest can be expressed in different systems of units, each is considered to be of the same **dimension**. Mass or length, for example, can be expressed in units of grams, ounces, pounds, and so on, or centimeters, inches, feet, and so on, respectively. However, mass (M) and length (L) are fundamental dimensionalities. Many, if not most, physical and chemical quantities can be expressed in fundamental dimensions of mass, length, and time (T).

Numbers, including transcendental numbers such as π and e, and trigonometric quantities are considered to be dimensionless.

Example 1.7.1 Dimensional Correctness

Show that $I = prt$ *and* $s = (1/2)gt^2$ *are dimensionally correct.*

→

Solution:
Interest, I, has dimensions of dollars or, in symbols, \$. Similarly, principal is in \$; and interest, r, in dollars per dollar per unit time, \$/\$T

Therefore, $I = prt$, $(\$ = \$\frac{\$}{\$T}T)$ *is dimensionally correct.*

Similarly, s has the dimension of a length L, while gravitational acceleration g is dimensionally L/T^2. *The coefficient of 1/2 is dimensionless. Therefore,*

$$L = \frac{L}{T^2} \cdot T^2 = L,$$

and is dimensionally correct. Note that dimensional correctness does not necessarily mean that the relation is correct since it may lack dimensionless quantities.

Called **dimensional analysis**, the dimensionality of each side of an equation must be the same or the expression cannot possibly be correct. Scientists use this property of dimensional correctness to aid in the exploration and development of relationships. For example, if one has written the law of freefall as $s = (1/2)t^2$, it is immediately clear that this cannot be correct since the length on the left side could not possibly be equivalent to units of time squared. Note, again, that dimensional analysis, while valued and useful, cannot vouch for the complete correctness of a relation since numeric coefficients are dimensionless.

Example 1.7.2 Speed Conversion

Express 55 miles per hour in centimeters per second.

Solution:
Several conversion factors are needed:
1 mile = 5280 feet, 1 foot = 12 inches, 1 inch = 2.54 centimeters,
1 hour = 60 minutes, and 1 minute = 60 seconds. *Therefore,*

$$55\ \text{mph} = \frac{55\ \text{miles}}{\text{hour}} \cdot \frac{5{,}280\ \text{feet}}{1\ \text{mile}} \cdot \frac{12\ \text{inches}}{1\ \text{foot}} \cdot \frac{2.54\ \text{cm}}{1\ \text{inch}} \cdot \frac{1\ \text{hour}}{60\ \text{min}} \cdot \frac{1\ \text{min}}{60\ \text{sec}} = 2458.72\ \frac{\text{cm}}{\text{sec}}$$

Example 1.7.3 Time Conversion

Express 2.5 years in hours.

Solution:

$$2.5\ \text{years} \cdot \frac{365\ \text{days}}{\text{year}} \cdot \frac{24\ \text{hours}}{1\ \text{day}} = 21{,}900\ \text{hours}$$

Since mass is conserved in a chemical reaction, establishing a scale of atomic masses for the elements is a necessity. A *mole* of a substance contains $6.022 \cdot 10^{23}$ atoms of the substance. This number is *Avogadro*'s number. This, of course, is an immense quantity. For example, $6.022 \cdot 10^{23}$ spheres each about 14 centimeters in diameter would fill a volume as large as the earth.

Example 1.7.4 Atoms of Nitrogen

How many atoms of nitrogen (N) are present in 21 grams of nitrogen?

Solution:

The periodic table lists the weight of a mole of nitrogen as approximately 14 grams.
Changing units from grams to atoms:

$$21\cancel{g\,N} \cdot \frac{1\ \cancel{mole\,N}}{14\ \cancel{g\,N}} \cdot \frac{6.022 \cdot 10^{23}\ atoms\ N}{1\ \cancel{mole\,N}} = 9.033 \cdot 10^{23}\ atoms\ of\ nitrogen.$$

Example 1.7.5 Atoms of Hydrogen

How many atoms of Hydrogen (H) are present in 36 grams of water (H_2O) molecules?

Solution:

From the periodic table, a mole of water weighs 18 grams since the chemical formula indicates that each molecule of water has two atoms of hydrogen and one atom of oxygen:

$$36g\ \cancel{H_2O} \cdot \frac{1\ \cancel{mole\ H_2O}}{18g\ \cancel{H_2O}} \cdot \frac{6.022 \cdot 10^{23}\ \cancel{molecule\ H_2O}}{1\ \cancel{mole\ H_2O}} \cdot \frac{2\ atoms\ H}{1\ \cancel{molecule\ H_2O}} = 2.41 \cdot 10^{24}\ atoms\ of\ hydrogen.$$

◆ In 2002, a black hole streaked through the Milky Way galaxy. In about 200 million years, it is predicted to be as close as 1000 light years from our solar system. How many miles correspond to 1000 light years? (A light year is the distance a light beam travels in 1 year.)

Felix Mirabel of the French Atomic Energy Commission estimated:

"The probability of a catastrophic event on the Earth from a black hole moving at high speed is almost zero compared with the probability of a catastrophic event from asteroids or comets."

EXERCISES 1.7

1. Express 5 miles in meters. (Use 1 meter = 39.37 inches, 1 mile = 5280 feet))
2. Express 1 yard in centimeters (Use 2.54 centimeters = 1inch)
3. Express 100 kilometers per hour in miles per hour.
4. A car travels at 50 miles per hour . What is its speed in yards per second?
5. Express 50 miles per hour in centimeters per second.
6. What is the weight of 75-kilogram person in pounds? (Use 453.6 grams = 1 pound)
7. A watermelon weighs 4.3 pounds. What is its weight in grams? (Use 1 pound = 453.6 grams)
8. How many ounces are in a 2-liter soda bottle? (Use 1 ounce = 29.6 milliliters)
9. A block of zinc occupies a volume of 45 milliliters and weighs 320.85 grams. What is the density of zinc?
10. A gold ingot is 6 centimeters × 3.5 centimeters × 2.7 centimeters. If the ingot weighs 1095.44 grams, what is the density of gold?
11. What is the weight in pounds of a liter of mercury? The density of mercury is 13.55 grams per cubic centimeter.
12. If the density of copper is 559 pounds per cubic foot, express its density in grams per cubic centimeter.
13. How many atoms are there in 30 grams of carbon (C)?
14. How many atoms of oxygen (O) are there in 100 grams of carbon dioxide (CO_2)?
15. A gas company is to store 5 tons of a gas with a density of 5.83 pounds per gallon. What tank volume is needed to store the gas?
16. Glucose is 40% C, 6.71% H, and 53.29% O (by weight). Determine its chemical formula if its weight is 180 grams.
17. Laughing gas is 63.6% nitrogen and 36.4% oxygen by weight. Determine its chemical formula if its weight is 44 grams.
18. A keg of beer contains 15.5 gallons. How many cases of twenty-four 12-ounce cans can be filled? Use dimensional analysis.
19. Can the linear dimensions of an acre be used to determine the number of square inches in 5 acres? Why? Hint: use the definition of an acre.
20. The velocity of light (in vacuum) is 186,000 miles per second. If light from another planet requires 45 minutes to reach earth, how many kilometers distant is the planet? What is the planet's name? Hint: use the Internet to find interplanetary distances.

HISTORICAL NOTES AND COMMENTS

According to the Oxford Dictionary of the English Language (OED), the word **algebra** was widely accepted by the seventeenth century. It is believed to have origins in early Arabic language from root words as "reunion of broken parts" and "to calculate." The word "algebra" has contexts other than in mathematics as, for example, in the surgical treatment of fractures and bone setting and as a branch of mathematics that investigates the relations and properties of numbers by means of general symbols; and, in a more abstract sense, a calculus of symbols combined according to certain defined laws.

♦ Kinship relationships, architecture, sewing, weaving, agriculture, and spiritual or religious practices are examples of the human activities that can be expressed in mathematical expressions. The study of mathematics in the context of different cultures is called *Ethnomathematics.*

René Descartes (1596–1650) – Descartes, born in France in 1596, the son of an aristocrat, traveled throughout Europe studying a wide variety of subjects including mathematics, science, law, medicine, religion, and philosophy. He was greatly influenced by other thinkers of the Age of Enlightenment.

Descartes ranks as one of the most important and influential thinkers in history and is sometimes called the founder of modern philosophy. In addition to his accomplishments as a philosopher, he was an outstanding mathematician, founding analytic geometry and seeking simple universal laws that governed all physical changes.

Mathematics was probably Descartes greatest interest; building upon the works of others, he originated Cartesian coordinates and Cartesian curves. To algebra, he contributed the treatment of negative roots and the convention of exponent notation.

Leonhard Euler (1707–1783) – Euler is often considered one of the most prolific mathematicians of his time. Author of 866 books and papers, he won the Paris Academy Prize 12 times. Born in Switzerland, he was the son of a Lutheran minister. Originally a theology student, he changed to mathematics under the influence of Johann Bernoulli. He made significant contributions in differential calculus, mathematical analysis, and number theory. He introduced the symbols e, i, $f(x)$, π, and the sigma summation sign. Remarkably, most of his publications appeared in the last 20 years of his life when he was blind.

Christiaan Huygens (1629–1695) – Huygens, born in the Netherlands in 1629, was the son of an important diplomat. At age 16, he entered Leiden University where he studied mathematics and law.

His first published work in 1651 displayed his geometrical skills by showing the fallacy in methods that had been proposed that claimed to square the circle. Huygens was also interested in astronomy and turned his attention to lens grinding and telescope construction. He discovered Saturn's major satellite Titan. As work in astronomy required accurate time keeping, he took an interest and patented the first pendulum clock. Huygens' mathematical education was influenced by Descartes occasional visits to his home.

John Napier (1550–1617) – born in Edinburgh (Scotland) little is known of his early years. He entered St. Andrews University at age 13. He is later believed to have completed his studies elsewhere in Europe. Napier's study of mathematics as a hobby led to his invention of logarithms and introducing decimal notation for fractions. He is also known for his "Napier's Bones," used to mechanically multiply, divide, and to obtain square roots and cube roots.

♦ The web site, "The Sound of Mathematics" has files of mathematically related algorithmic music. Files for e, π, and other mathematical numbers are posted.
http://www.geocities.com/Vienna/9349

SUPPLEMENTARY EXERCISES

1. To earn a B grade, a student must receive an 82.6 average on three exams and a final exam score, which is weighted as two exams. If three test scores are 76, 89, and 70, what is the lowest final exam score for a B grade?
2. The price of a car including a 4% sales tax is \$19,500. What is the price excluding the tax?
3. Federal personal income tax rates depend on marital status and adjusted taxable income. Some rates for 2014 are:

Single Filers Between, \$	Married Filing Jointly, \$	Married Filing Separately, \$	Tax Rate, %
0–9,075	0–18,150	0–9,075	10
9,075–36,900	18,150–73,800	9,075–36,900	15
36,900–89,350	73,800–148,850	36,900–74,425	25
89,350–186,350	148,850–226,850	74,425–113,425	28
186,350–405,100	226,850–405,100	113,425–202,550	33
405,100–406,750	405,100–457,600	202,550–228,800	35
40,6750 or more	457,600 or more	228,800 or more	39.6

   a) What Federal tax is due for a single filer with an adjusted taxable income of \$250,000?
   b) What Federal tax is due for a married couple filing jointly with an adjusted taxable income of \$250,000?
   c) A couple's taxable income is \$250,000. If one person earns \$200,000 and the other \$50,000, is it advantageous for them to file separately?
4. The 2014 Federal Minimum Wage is \$7.25 per hour. Oregon and Washington annually index their state's minimum wage to inflation. Several other states also have minimum wages that exceed the Federal rate. For instance, the minimum wage in Connecticut is \$8.70, Alaska \$7.75, California \$8.00, Maine \$7.55, and Delaware \$7.75. How much more will a person who works for 37.5 hours per week at minimum wage in Connecticut earn in a year (52 weeks) than a person
   a) in Maine?
   b) for the Federal minimum wage?

5. Find an equation of the line below:

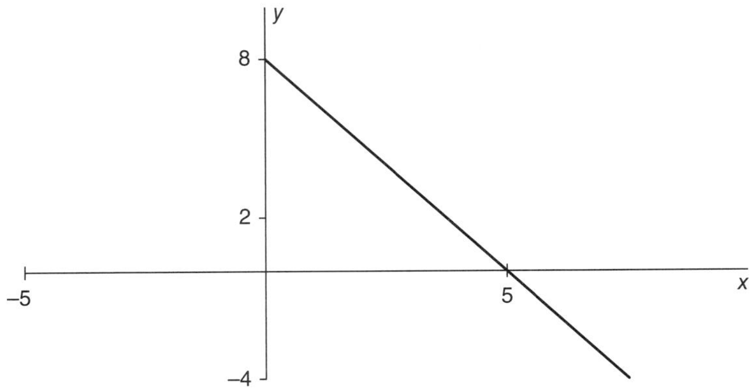

6. Determine the slope-intercept form for these lines:

a) $3x + 4y = 6$.

b) $y - 3x = 2$

c) $y - 5 = 2(x - 3)$

d) $4y = 8x + 12$

7. a) Find the equation of a line passing through (3, 5) and perpendicular to the line with x-intercept of 2 and y-intercept of 5.

   b) Find an equation of a horizontal line passing through (1, 4).

   c) Find an equation of a line passing through (2, 7) and perpendicular to the x-axis.

8. The graph shown is most likely which of the following? Explain your reasoning.

   a) $3x + 4$ b) $-3x + 4$ c) $3x - 4$ d) $-3x - 4$

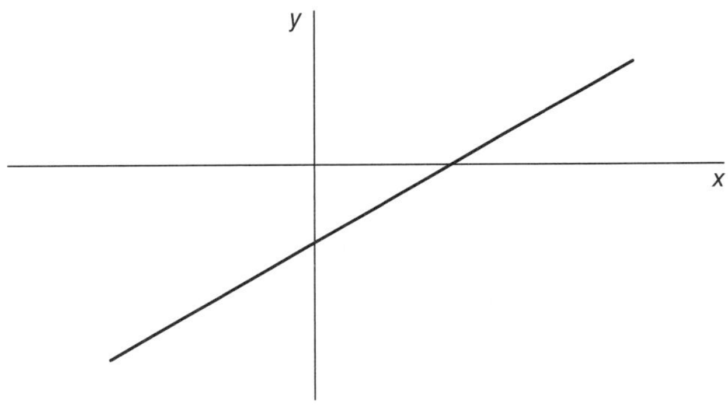

9. The graph shown is most likely which of the following? Explain your reasoning.
   a) $2x + 3$ b) $-2x + 3$ c) $2x - 3$ d) $-2x - 3$

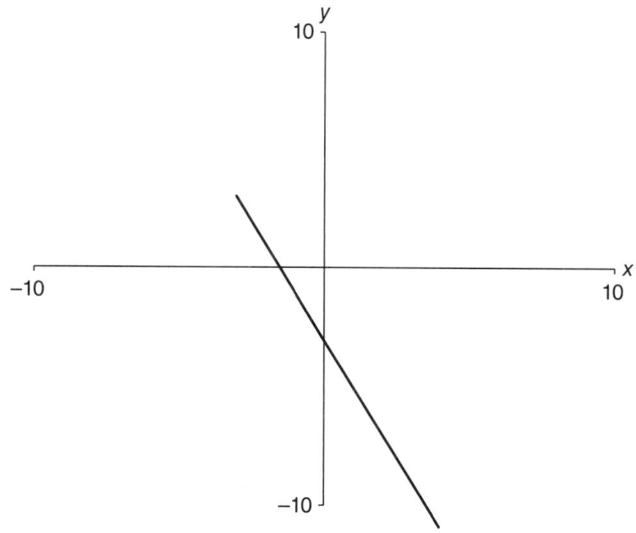

10. An electric utility's bill adds an energy charge of 6.45 cents per kilowatt hour to its base charge of \$7.95 per month. Write an equation for the monthly charge, y, in terms of, x, the number of kilowatt hours.
11. A water utility's quarterly bill has a fixed charge of \$40 and \$2.25 for each thousand gallons of water used. What is the bill for a customer who uses 12,000 gallons of water during the quarter?
12. A chemistry student is to make 20 milliliters of an 8% solution from two solutions that are of 12.5% and 5% concentrations. How many milliliters of each solution should be mixed?
13. How should 50 milliliters of 20% sulfuric acid be made by mixing 100% sulfuric acid with water?
14. How many liters of a 50% solution of acid should be added to 10 liters of a 20% solution to obtain a 30% solution?
15. At a charity event, 25 donated items were sold for either \$50 or \$100 each. A total of \$1650 was raised. How many more \$50 items than \$100 items were sold?
16. A concert promoter raises \$600,000 on the sale of 25,000 tickets. If the tickets were \$20 in advance sale and \$25 at the door, how many were sold in advance?
17. A business pays a monthly rent of \$2000. It costs \$2 for each item produced. The items are sold for \$6 each. Find the
    a) revenue function;
    b) cost function;
    c) break-even point;
    d) profit/loss if 1000 items are sold this month.

18. The demand function for a commodity is shown in the graph.

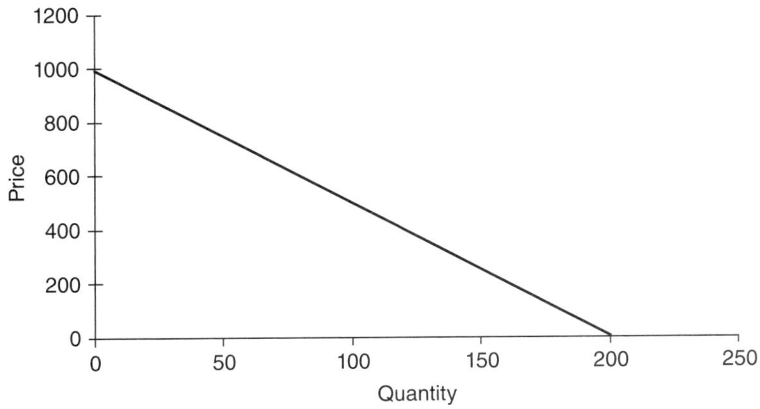

a) Determine the demand function.
b) What is the price when demand is 50 units?
c) What quantity is demanded when the price is 600?

19. There are many interesting facts about pi on the Internet. Use websites to answer these questions.
a) How many 0s, 1s, … , 9s are in the first million digits of pi?
b) Do any of the digits seem to appear more frequently than others?

20. Rewrite 125^{2x} in the form 5^{kx}.

21. Use the Laws of Exponents to simplify $\dfrac{125^{2x-3}25^{x+1}}{5^{4x-5}}$.

22. Solve $x^3(4^x) - 13x^2(4^x) + 36x(4^x) = 0$.

23. Evaluate $\log_4 128$.

24. Write as a single logarithm:

$$4\ln(x+1) - 2\ln(y+3) - (1/2)\ln(z+4).$$

25. Solve $4e^{x-1} = 20$.

26. Let y vary directly as the cube of x. If $y = 128$ when $x = 4$ find y when $x = 5$.

27. Let y vary inversely as x. Complete the table.

x	2	4		12	
y	24		8		48

28. Convert 5 yards to centimeters.

29. How many atoms of carbon (C) are present in 54 grams of carbon?

2 Mathematics of Finance

2.1 SIMPLE AND COMPOUND INTEREST

Simple Interest

You have heard the refrain: "A dollar today is not the same as a dollar tomorrow." Used in many contexts, here we mean that money earns more money!

Finite Mathematics: Models and Applications, First Edition. Carla C. Morris and Robert M. Stark.

Companion Website: http://www.wiley.com/go/morris/finitemathematics

Interest is monies paid or received for the use of money. When you borrow money, your promise to pay interest is the inducement for lenders to part with their money. In your savings or money market accounts, you are the lender, and your accounts are credited with interest.

The money borrowed is the **principal**. The **interest rate** is expressed in dollars per dollar per unit time. Interest rates are often expressed in decimal format. For example, an annual interest rate of 6%, say, means that 6 cents is paid on each dollar of principal after 1 year. To help compare interest rates, it is customary to express them on an annual basis.

In the simplest situation, accrued interest is proportional to principal, time, and interest rate.

Simple Interest

$$I = p \times r \times t$$

I = interest **r = interest rate as a decimal**
p = principal **t = time**

The total amount accumulated, A, principal plus interest is

$$A = p + prt = p(1 + rt).$$

Example 2.1.1 Simple Interest

a) What simple interest accrues on $5000 at 6% interest for 3 years?

b) What simple interest accrues on $10,000 at 5% interest for 48 months?

Solution:
Assume that the interest rate, r, is expressed as an annual rate.

a) Here, $p = \$5000$, $r = 0.06$, and $t = 3$ so the interest $I = 5000(0.06)(3) = \$900$.

b) Here, express 48 months as 4 years. Then, $I = (10,000)(0.05)(4) = \$2000$.

Example 2.1.2 A Savings Account

Jane Doe deposits $1000 in an account at 2% simple interest. What is the account balance after 18 months?

Solution:
First, simple interest is determined by $I = 1000(0.02)(18/12) = \$30$. (Note that 18 months is divided by 12 to convert into years.) Jane has a total of $1000 + 30 = \$1030$ in her account at the end of the 18 months.

Interest is paid periodically. Sometimes, loan interest is received (or paid) at various time periods other than at maturity dates. Therefore, when funds (interest plus principal) remain invested, at the next payment period interest is paid not only on the principal, as before, but also on the interest earned earlier. This is **compound interest**.

Compound Interest

Letting p_0 be an initial principal invested at rate r for a time t, in years, the principal, p_1, after 1 year is

$$p_1 = p_0 + rp_0 = (1 + r)\ p_0.$$

In words, the (new) principal after 1 year is the original principal plus interest. Similarly, successive principals are

$$p_2 = p_1 + rp_1 = (1 + r)\ p_1 = (1 + r)\ [(1 + r)\ p_0] = (1 + r)^2\ p_0 \qquad \text{after 2 years}$$

$$p_3 = p_2 + rp_2 = (1 + r)\ p_2 = (1 + r)\ [(1 + r)^2\ p_0] = (1 + r)^3\ p_0 \qquad \text{after 3 years}$$

A pattern is clear, the principal at t years is

$$p_t = (1 + r)^t\ p_0, \qquad t = 1, 2, \ldots$$

(This result is easily proved by a mathematical induction.)

Annual Compound Interest

$$\mathbf{A = p_t = (1 + r)^t\ p_0, \qquad t = 1, 2, \ldots}$$

$\mathbf{A = p_t}$ **= total accumulation**

$\mathbf{p_0}$ **= initial principal**

$\mathbf{t}$ **= time (years)**

$\mathbf{r}$ **= annual interest rate (expressed as a decimal)**

Example 2.1.3 Compound Interest

a) What compound interest accrues on \$5000 at 6% annually for 3 years?

b) What compound interest accrues on \$10,000 at 5% annually for 48 months?

Solution:

a) After 3 years, the total amount is $5000(1.06)^3 = \$5955.08$. Subtract the original principal, \$5000, to yield an accrued interest of \$955.08. Notice that this exceeds the \$900 simple interest in a previous example. Compound interest always exceeds simple interest under like conditions.

⟶

b) *After 4 years, the total amount is* $10,000(1.05)^4 = \$12,155.06$ *and the accrued interest is \$2155.06; exceeding the \$2000 simple interest of Example 2.1.1.*

Example 2.1.3 assumes that interest is compounded annually. However, sometimes interest is compounded quarterly or monthly. If p_0 is the original principal; r, the annual interest rate (as a decimal); and n the number of times a year that interest is compounded, the total amount at time t, p_t, is $p_0(1 + r/n)^{nt}$.

Compound Interest

The principal p_t (p_0 plus earned interest) after t years when interest is compounded n times a year at annual rate r is

$$A = p_t = p_0(1 + r/n)^{nt}$$

(r as a decimal)

Note that when $n = 1$ interest is compounded annually and matches the earlier result.

Example 2.1.4 Effects of Compounding Periods

Returning to part (a) of Example 2.1.3,

a) *What interest accrues on \$5000 at 6% interest for 3 years when compounded quarterly?*

b) *When compounded monthly?*

Solution:

a) *Here,* $n = 4$ *and* $p_t = 5000\ (1 + 0.06/4)^{4\times3} = 5000(1.015)^{12} = \$5978.09.$
The accrued interest is \$978.09.

b) *Now,* $n = 12$ *and* $p_t = 5000\ (1 + 0.06/12)^{12\times3} = 5000(1.005)^{36} = \$5983.40.$
The accrued interest is \$983.40.

Notice that both amounts exceed the accrued interest in the two previous examples where interest was compounded annually. In general, the more frequent the compounding, the greater the accrued interest.

Example 2.1.4 illustrates the effect of increased frequency of compounding. The accrued interest increases as the frequency of compounding increases, that is, as the period between interest payments decreases. Clearly, this is due to the additional sums earned by accrued interest, that is, by the effect of compounding.

The last observation raises the question of instantaneous or continuous compounding. It is easy to jump to the guess that continuous compounding might result in infinite sums. Alas, that is not the case! The following formula applies for continuous compounding.

Continuous Compounding

A principal p_t, when p_0 dollars has continuously compounded for t years is

$$p_t = p_0 e^{rt}$$

where r = interest rate (decimal)

$e \approx 2.71828$ (Chapter 1)

Example 2.1.5 Compounding Continuously

What interest accrues on \$5000 at 6% interest for 3 years when compounded continuously?

Solution:
Using the continuous compounding formula, we have

$$p_t = 5000\, e^{0.06 \times 3} = 5000\, e^{0.18} = \$5986.09$$

As noted earlier, continuous compounding does not yield an infinite sum. However, it produces a larger return than any other compounding period.

Effective Rate of Interest

Offhand comparisons of compound interest for different terms are usually difficult. How does one compare an account with a higher rate of annual compounding to one that may offer, say, quarterly or monthly compounding? This is achieved by an **effective rate of interest,** which is the simple interest rate necessary to achieve the same results as the compounded rate.

Example 2.1.6 Effective Rate of Interest

If \$5000 is borrowed at a 10% interest rate compounded quarterly for 7 years, what is the effective simple interest rate, r?

Solution:
First, determine the amount to be repaid.

$$5000(1 + 0.10/4)^{(4)(7)} = 5000(1.025)^{28} = \$9982.48.$$

Now, to determine the simple effective interest rate, r, set $I = prt = \$9982.48 - \$5000 = \$4982.48.$

⟶

$$5000(r)(7) = 4982.48$$

$$r = 4982.48/35,000 = 0.142357$$

The simple effective interest rate is about 14.24%. This means that the loan at 14.24% simple interest for 7 years yields the same repayment as a 10% rate compounded quarterly for the same period.

Another comparison often used by financial institutions is called the "annual percentage yield" (APY). It is often used in association with CDs (certificates of deposit) and savings accounts. It seeks to determine the annual compound interest rate that would yield the same amounts as more frequent compounding.

When interest is paid rather than received, as in the case of credit cards, it is called the "annual percentage rate" (APR).

APY (= APR) = Annual Percentage Yield (or Rate)

$$\mathbf{APY = APR = 100}\left[\left(\frac{p_t}{p_0}\right)^{1/t} - 1\right]\%$$

Credit card holders pay balances in full each month or installments according to the card issuer's terms. Payments can range from a minimum amount to the entire balance. Generally, no interest is charged if the entire balance is paid within the time allotted. Interest is charged on unpaid balances at an APR set by the issuer.

A **debit card** is linked to checking, savings, or another account. Purchases are deducted directly since the debit card functions as a written check cashed instantly.

Example 2.1.7 Annual Percentage Rate

If $5000 is borrowed at 10% interest compounded quarterly for 7 years, what is the APR?

Solution:

Recall from Example 2.1.6 that the repayment amount is $9982.48. Using the formula,

$$APR = 100\left[\left(\frac{9982.48}{5000}\right)^{1/7} - 1\right]\% = 10.38\%.$$

The annual interest rate is effectively about 10.38%. This means that if the loan had been at 10.38% annual compound interest for the 7 years it would yield the same amount as the 10% rate compounded quarterly over that time.

♦ Interest rates can be reported in many ways. Not long ago, newspaper ads specified interest rates sometimes without indication of how they were to be compounded. Congress intervened and now law requires interest rates to be quoted in any form provided the APR is also specified.

Present Worth

As noted at the outset, a dollar today is worth more than a dollar later since it can earn interest in the interim (changes in buying power are not considered).

Suppose that an amount, p_t, is to be paid t periods hence. The equivalent amount now is called the **present worth** of p_t.

Imagine an amount, W, as the present worth of p_t at a rate r (\$/\$-period). Then

$$W = \frac{p_t}{(1+r)^t}.$$

In effect, "the time clock runs backward"!

Example 2.1.8 Present Worth

In 10 years you are to receive \$10,000. At a 6% compound annual rate, what is its present worth?

Solution:

Here,

$$W = 10,000/(1.06)^{10} = \$5583.95$$

That is, the \$10,000 in 10 years hence is equivalent to less than \$5600 now!

♦ State lotteries furnish an everyday example. Huge headline winnings are not paid in entirety upon presentation of the winning ticket. Winners have a choice of receiving periodic payments over, say, 25 years (whose total is the headlined amount) or a much reduced (present worth) amount now.

Example 2.1.9 Real Estate Bargains

Most of us regard the purchases of Manhattan Island from the Native Americans, the Louisiana Territory from France, and Alaska from Russia as outstanding bargains. And, indeed, they were. However, suppose the purchase prices had been invested at a 5% annual compound interest rate. What sums would have accrued by 2014?

⟶

	Year	*Original Price, $*
Manhattan Island	*1626*	*24*
Louisiana Territory	*1803*	*15,000,000*
Alaska	*1867*	*7,200,000*

Solution:

At a 5% compound rate, the present worth of Manhattan Island's purchase price is $24(1.05)^{388} = \$3,996,311,022$ *(about $4 billion).*

The present worth of the Louisiana Territory's purchase price is $15,000,000(1.05)^{211} = \$443,642,723,600$ *(about $444 billion).*

The present worth of Alaska's purchase price is $7,200,000(1.05)^{147} = \$9,379,063,144$ *(about $9.4 billion).*

EXERCISES 2.1

1. What is the simple interest after 3 years on $5000 invested at 8%?
2. What is the simple interest after 6 years on $20,000 invested at 7%?
3. What is the simple interest after 7 years on $50,000 invested at 5%?
4. What is the simple interest after 30 months on $10,000 invested at 6%?
5. What interest accrues on $50,000 at 4% compounded annually for 6 years?
6. What interest accrues on $150,000 at 6% compounded annually for 5 years?
7. What interest accrues on $500,000 at 3% compounded annually for 6 years?
8. What interest accrues on $15,000 at 12% compounded annually for 4 years?
9. What interest accrues on $250,000 at 5% compounded annually for 4 years?
10. A $10,000 bond pays 5% compounded quarterly. What is its value after 5 years?
11. A $100,000 bond pays 6% compounded quarterly. What is its value after 10 years?
12. A bond account has $250,000 at 6% compounded monthly. What is its value after 10 years?
13. A bond account has $50,000 at 3% compounded monthly. What is its value after 7 years?
14. How much is $40,000 worth at maturity if compounded continuously at 3.5% for 4 years?
15. How much is $75,000 worth at maturity if compounded continuously at 5% for 5 years?
16. How much is $300,000 worth at maturity if compounded continuously at 6% for 10 years?
17. How much is $75,000 worth at maturity if compounded continuously at 7% for 6 years?
18. What interest accrues on $5000 at 7% compounded continuously for 2 years?

19. What interest accrues on \$50,000 at 8% compounded continuously for 3 years?

20. A single dollar at a 100% annual rate for 1 year compounded n times becomes $A = \left(1 + \frac{1}{n}\right)^n$. Calculate the accrued interest compounded monthly, weekly, daily, and hourly. What is the limiting value of the expression?

21. If \$5000 is borrowed at a 6% rate compounded monthly for 9 years, what is the effective interest rate?

22. What is the effective interest rate for a sum at 8% interest compounded quarterly for 5 years?

23. What is the effective interest rate for a sum at 10% interest compounded quarterly for 8 years?

24. Suppose \$10,000 is invested at 12% compounded monthly for 5 years. What is the APY?

25. What present sum will increase to \$250,000 in 8 years at 9% if interest is compounded annually?

26. What present sum will increase to \$100,000 in 7 years at 6% if interest is compounded annually?

27. What present sum will increase to \$750,000 in 10 years at 5% if interest is compounded annually?

28. What present sum will increase to \$250,000 at 8% in 4 years if interest is compounded annually?

2.2 ORDINARY ANNUITY

In an **ordinary annuity,** periodic payments are made to an **annuitant**; usually a retiree. The monies from which the periodic payments are made usually accumulate from payroll deductions by employers and employees over many working years. The sum of the payments made and interest accrued is the amount of the annuity, its **future value (FV)**. An annuity may also be purchased by a lump sum, such as a lottery winning or inheritance.

Suppose that a \$2500 contribution is made annually to an account that paid interest at 4% compounded annually. The growth of the contributions after, say, 4 years is shown in the diagram.

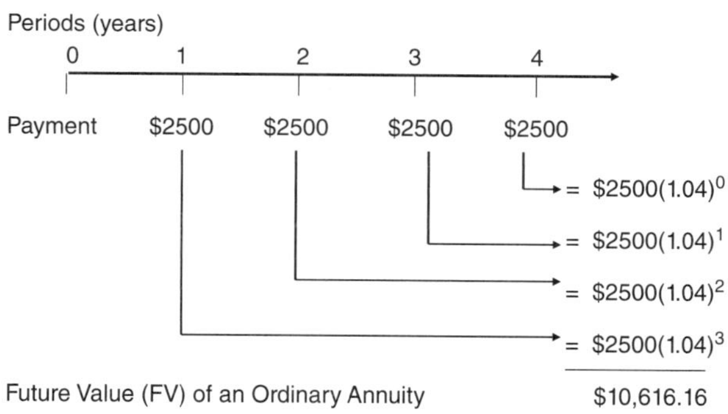

More generally, suppose that annual payments of p dollars accumulate for n years at an annual rate r. The first payment, p, becomes $p(1 + r)^{n-1}$ since interest will have been earned for $n - 1$ years. The second payment, p, a year hence, will grow to $p(1 + r)^{n-2}$ after $n - 1$ years, and so on. That is, the FV of the annuity after n years is

$$\text{FV} = p(1+r)^{n-1} + p(1+r)^{n-2} + \cdots + p(1+r) + p$$

This finite geometric sum is the FV

$$\text{FV} = \frac{p(1-(1+r)^n)}{1-(1+r)} = \frac{(1+r)^n - 1}{r}p$$

Future Value of an Ordinary Annuity

In an ordinary annuity an amount, p, deposited at n equal intervals at interest rate i per period has a Future Value

$$\textbf{FV} = \frac{p(1-(1+i)^n)}{1-(1+i)} = \frac{(1+i)^n - 1}{i}p$$

i = periodic interest rate $\quad$ n = number of periods

FV = future value

Example 2.2.1 Future Value Calculation

Apply the FV formula to an annuity payment of \$2500 for 4 years at 4% to verify the \$10,616.16 depicted in a diagram earlier in this section.

Solution:
Let FV be the value of the total accumulation. Then

$$FV = \frac{(1.04)^4 - 1}{0.04}(\$2500) = \$10{,}616.16.$$

Example 2.2.2 An Annuity Accumulation

Assuming an average interest rate of 5% over a long period, what will annual IRA contributions of \$2000 yield after 30 years?

Solution:
Let FV be the future value of the accumulation. Then,

⟶

$$FV = (2000)(1.05)^{29} + (2000)(1.05)^{28} + \cdots + (2000)(1.05) + 2000$$
$$= \frac{(1.05)^{30} - 1}{0.05}(\$2000) = \$132,877.70$$

an amount more than twice the \$60,000 in contributions.

Example 2.2.3 An Annuity Payout

What annual payment, p, for 25 years at a 5% interest rate will produce a \$2 million retirement fund?

Solution:
The first payment, p, starts the first year. A year hence, a second payment is made, which, when interest on the first year is added, becomes 1.05p. The total of the payments made plus accrued interest is to total \$2 million.

Solving for p

$$p + (1.05)p + \cdots + (1.05)^{25-1}p = \frac{(1.05)^{25} - 1}{0.05}p = 47.727p = \$2,000,000$$

and, therefore, p = \$41,905 is the annual payment. Note that only little more than one-half of the total \$2 million were "out of pocket" dollars, the remainder being accrued interest.

Often the **present value (PV)** of a series of future payments is sought, as in a lottery jackpot paid over a number of years. It is the "flip side" of the FV calculation.

An analysis similar to that for the FV of payments can be devised. Returning to the earlier diagram, now \$2500 is to be received in each of the next 4 years. What is the present worth of the \$10,000 in future payments?

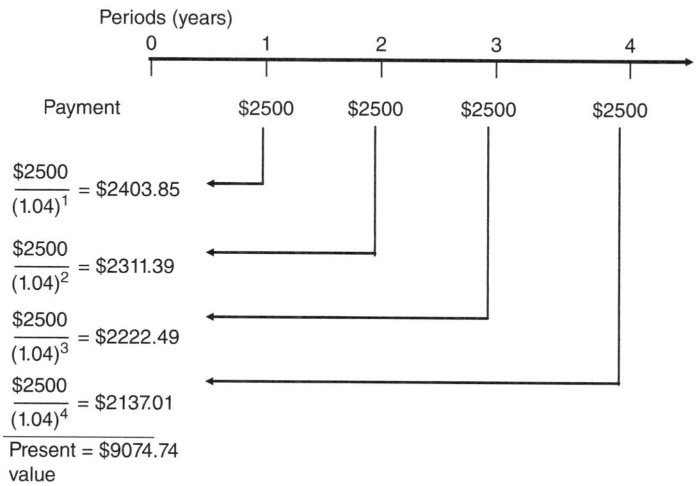

Present Value (PV) of an Ordinary Annuity

The table shows (previous page) that the present worth of four future payments of \$2500 is about \$9075.

Using the expression for a finite geometric sum leads to the following formula for the PV of an annuity.

Present Value of an Ordinary Annuity

An amount, p, deposited at the end of each period for n periods in an ordinary annuity that earns interest at a rate of i per period has a PV

$$\mathrm{PV} = \frac{p(1-(1+i)^{-n})}{1-(1+i)} = \frac{1-(1+i)^{-n}}{i}p$$

i = periodic interest rate n = number of periods

PV = present value

Example 2.2.4 Present Value of an Annuity

Show that a payment of \$2500 for 4 years at 4% has a PV of \$9074.74 as depicted in the previous diagram.

Solution:

Let PV be the present value of the total annuity. Then,

$$PV = \frac{1-(1.04)^{-4}}{0.04}(2500) = \$9074.74,$$ *in agreement with the diagram.*

◆ While the Federal Deposit Insurance Corporation (FDIC) guarantees a certificate of deposit (CD), it does not guarantee annuities. The security of an annuity usually relies on the issuing company.

EXERCISES 2.2

1. What is the value of an annuity after 20 years when \$1000 is paid yearly and earns 5% compounded annually to maturity?
2. What is the value of an annuity after 6 years when \$250 is paid monthly and earns 12% compounded monthly to maturity?
3. What is the value of an annuity after 20 years when \$3000 is paid yearly and earns 5% compounded annually to maturity?

4. What is the value of an annuity after 10 years when $100 is paid monthly and earns 6% compounded monthly to maturity?
5. What is the value of an annuity after 8 years when $2500 is paid yearly and earns 3% compounded annually to maturity?
6. What is the value of an annuity after 20 years when $250 is paid quarterly and earns 8% compounded quarterly to maturity?
7. What is the value of an annuity after 15 years when $1000 is paid yearly and earns 7% compounded annually to maturity?
8. What is the value of an annuity after 6 years when $150 is paid quarterly and earns 12% compounded quarterly to maturity?
9. What is the value of an annuity after 25 years when $1000 is paid every 6 months and earns 10% compounded semiannually to maturity?
10. Find the PV of a yearly annuity payment of $250 for 20 years if the annual compound interest rate is 10%.
11. Find the PV of an annuity of $50 paid monthly for 5 years if the interest rate is 12% compounded monthly.
12. Find the PV of an annuity of $500 paid yearly for 15 years if the interest rate is 8% compounded annually.
13. Find the PV of an annuity of $125 paid quarterly for 10 years if the interest rate is 6% compounded quarterly.
14. Find the PV of an annuity of $5000 paid yearly for 10 years if the interest rate is 5% compounded annually.
15. Find the PV of an annuity of $2500 paid every 6 months for 8 years if the interest rate is 8% compounded semiannually.
16. Find the PV of an annuity of $75 paid monthly for 15 years if the interest rate is 6% compounded monthly.
17. A lottery winner is given a choice of receiving $25,000 monthly for 20 years or a lump sum of $2.5 million. If the winner thinks a 6% rate can be consistently earned, which option is better?

2.3 AMORTIZATION

Home mortgages and other long-term loans are usually paid or "retired" by periodic payments, interest being charged on the unpaid balance at the end of each payment period. Here, the interest rate is an APR as the loan is retired by equal payments. This is called **amortization.**

The French *a mort* roughly means "at death." In finance, it means the "retirement" or "paying off" of loans. Amortization is the complement of an annuity. Here, the money or value has been advanced by a lender. It is a borrower's obligation to make periodic payments, including interest, to reimburse the lender. Installment, auto, and home mortgages immediately come to mind.

Consider a loan, L, to be retired by periodic payments, Y, at interest rate i ($/$-period) for n periods. The present worth of the first payment, one period hence, is $Y(1+i)^{-1}$. The payment to be made, two periods, hence has a present worth of $Y(1+i)^{-2}$, and so on. The last payment, n periods, hence has a present worth of $Y(1+i)^{-n}$. The sum of all these payments must equal the loan, L. That is,

$$L = Y(1+i)^{-1} + Y(1+i)^{-2} + \cdots + Y(1+i)^{-n}.$$

This finite geometric sum, after simplification, is

$$L = \frac{Y(1+i)^{-1}(1-(1+i)^{-n})}{1-(1+i)} = \frac{Y}{i}\left(1 - \frac{1}{(1+i)^n}\right).$$

Amortization Formula

Periodic payments, Y, on a loan, L, amortized over n periods at an interest rate of $i\%$ per period is

$$Y = \frac{iL}{\left(1 - \frac{1}{(1+i)^n}\right)}$$

Y = periodic payment **L = loan amount**

i = interest rate $/$ (period) **n = number of periods**

When repaying an amortized loan, such as a mortgage or other installment loans, it may be useful to calculate the annual interest paid. The remaining loan balance may be a factor in refinancing or in seeking a home equity loan. An amortization schedule can be calculated without specialized computer software, but it is time consuming. All that need be calculated is the first period interest, which is then subtracted from the periodic payment amount (Y) to determine the new principal amount. Many amortization calculators are available on the Web.

Example 2.3.1 Amortization Schedule

A 5-year $15,000 loan calls for annual payments at a 4% rate. Prepare an amortization schedule that includes the payment, annual interest, and the year end balances.

Solution:

First, Y is calculated using the amortization formula (last box). Therefore,

$$Y = \frac{\$15{,}000(0.04)}{\left(1 - \frac{1}{(1.04)^5}\right)} = 3369.41.$$

→

The first period interest is \$15,000(0.04) = \$600. The payment on principal is \$3369.41 − \$600 = \$2769.41. Subtracting this from the original principal of \$15,000 yields \$12,230.59, the remaining loan principal.

The amortization table is formed with this information.

Period	*Beginning Balance, \$*	*Interest, \$*	*Amortized Payment (Less Interest), \$*	*PMT, \$*
1	15,000.00	600.00	2,769.41	3,369.41
2	12,230.59	489.22	2,880.19	3,369.41
3	9,350.40	374.02	2,995.39	3,369.41
4	6,355.01	254.20	3,115.21	3,369.41
5	3,239.80	129.59	3,239.80*	3,369.39
Total		1,847.03	15,000.00	16,847.03

** Adjusted by 2 cents.*

Example 2.3.2 Amortization

What periodic payments, Y, will retire a \$10,000 loan at 16% compounded quarterly in 20 equal quarterly payments?

Solution:

Here $i = 0.16/4 = 0.04$ (4% interest per quarter), $n = 20$ quarters, and $L = \$10,000$. We seek Y, the quarterly payment,

$$Y = \frac{\$10,000(0.04)}{\left(1 - \frac{1}{(1.04)^{20}}\right)} = \$735.82$$

Example 2.3.3 Mortgage Payment

A home is purchased with an \$80,000 30-year mortgage at 8.65%. What are the monthly payments?

Solution:

Here, $i = 0.0072 (= 0.0865/12)$, $n = 360 \ (12 \times 30)$, and $L = 80,000$

$$Y = \frac{\$80,000(0.0072)}{\left(1 - \frac{1}{(1.0072)^{360}}\right)} = \$623.09 \text{ monthly}$$

♦ Strategies for Financing a Home Mortgage

Small differences in interest rates can make a sizable difference in home financing costs. For example, a \$100,000 mortgage at 6% for 30 years requires monthly payments of principal and interest of about \$600. The same mortgage at 5% requires monthly payments of about \$537 for a total difference of about \$22,500 over the life of the mortgage.

Reducing the term to 25 years, a \$100,000 mortgage at 6% requires a somewhat higher monthly payment of about \$644 to yield about the same total savings of \$22,500 over the 30-year mortgage.

A popular strategy is to pay 1/2 of the monthly mortgage amount every 2 weeks. There is a dual benefit – first, some monthly interest may be saved by the midmonth payment and, second, the 26 biweekly payments result in 13 monthly payments, rather than the 12 monthly each year. The results can be surprising.

A sound business practice is to prepare for a forthcoming expense using interest bearing accounts, known as *sinking funds*. A calculation is made to determine the periodic deposits, P, to reach a desired sum, S, in n years hence. That is,

$$S = P\frac{(1+i)^n - 1}{i} \text{ or } P = S\left[\frac{(1+i)^n - 1}{i}\right]^{-1}$$

EXERCISES 2.3

1. A \$500 loan calls for repayment in 6 monthly installments at a monthly rate of 1%. Calculate an amortization schedule that includes the payment amount, the interest payment, and the balance at the end of each month.
2. A \$10,000 loan calls for repayment in 10 yearly installments at an annual rate of 5%. Calculate an amortization schedule that includes the payment amount, the interest payment, and the balance at the end of each year.
3. A car is financed for \$25,000 at 6% interest. What is the monthly payment on a 5-year loan?
4. A car is financed for \$20,000 at 5% interest. What is the monthly payment on a 5-year loan?
5. A car is financed for \$35,000 at 4% interest. What is the monthly payment on a 6-year loan?
6. A car is financed for \$25,000 at 6% interest. What is the monthly payment on a 4-year loan?
7. A car is financed for \$18,000 at 3% interest. How much can be saved by opting for a 4-year loan (at higher monthly payments) rather than a 5-year loan?
8. A home mortgage is \$250,000. What are the monthly payments on a 30-year mortgage at 5% annual interest? A 20-year mortgage?
9. A home mortgage is \$350,000. What are the monthly payments on a 30-year mortgage at 4% annual interest? A 25-year mortgage?

10. What annual sums are required to repay \$10,000 in 5 years at 4% compounded annually?

11. A town council considers two plans to provide additional water storage. In one, an elevated steel tank is estimated to cost \$900,000. The second plan calls for a lower elevation tank on a nearby hill estimated to cost \$800,000. Annual operating and maintenance costs are \$3000 and \$7000, respectively. Evaluate the annual costs for each plan assuming a 50-year tank life and a 5% interest rate on capital project bonds.

♦ This exercise typifies a frequent situation for local and state governments. Long-term and costly projects such as dams, bridges, highways, water systems, and treatment plants often involve payments and revenues over many years. In comparing viable alternatives, the monies are usually evaluated by a present worth calculator.

12. Credit card issuers require a minimum monthly payment on unpaid balances. Assuming a \$1000 unpaid balance, ascertain the minimum payment your favorite credit card requires. Determine when the balance will be retired making minimum monthly payments.

13. Obtain requisite data for the purchase of a new car and loans for 3, 4, and 5 years to payoff. Compare total costs.

2.4 ARITHMETIC AND GEOMETRIC SEQUENCES

Consider the sequence 1, 4, 7, 10, …, where each term is formed by adding 3 to the previous term. This is an example of an arithmetic sequence.

An **arithmetic sequence** (or progression) is defined by

Arithmetic Sequence

A general term of the arithmetic sequence $a_1, a_2, \ldots, a_n$ is

$$a_n = a_1 + (n-1)d, \quad n > 1$$

where d is the term difference.

It can be shown that the sum, S_n, of the first n terms of an arithmetic sequence, called an *arithmetic sum*, is

$$S_n = \frac{n(a_1 + a_n)}{2} \qquad \text{arithmetic sum}$$

When a sequence of a finite number of terms are added, it is known as a **finite sum**.

If $a_1, a_2, a_3, \ldots, a_n, \ldots$ is an infinite sequence, the expression that involves adding the terms of the sequence $a_1 + a_2 + a_3 + \cdots + a_n + \cdots$ is called a **series**.

Example 2.4.1 Arithmetic Sum

For the arithmetic series $3 + 8 + 13 + \cdots$ determine the next four terms and calculate the sum of the first seven terms.

Solution:
The common difference is 5. The next four terms are 18, 23, 28, and 33.

The sum of the first seven terms is $S_7 = \dfrac{7(3+33)}{2} = 126.$

Consider the sequence 1, 1/2, 1/4, 1/8, … , where each term is formed by multiplying the previous term by 1/2. This is an example of a geometric sequence.

The general term of a geometric sequence is

$$a_n = ra_{n-1} = a_1 r^{n-1}, \quad n \geq 1$$

where a_1 is the first term of the sequence and r, the term ratio. This can be verified by induction.

A **geometric sequence** (or progression) is defined by

Geometric Sequence

The general term of the geometric sequence $a_1, a_2, \ldots, a_n$ is

$$a_n = r\,a_{n-1} = a_1 r^{n-1}, \quad n > 1$$

where r is the term ratio.

It can be shown that the sum of the first n terms of a geometric sequence with first term a_1 and common ratio r is

$$S_n = \frac{a_1(1 - r^n)}{1 - r} \qquad \text{geometric sum}$$

Example 2.4.2 Geometric Series

For the geometric series $2 + 1 + 1/2 + 1/4 + \cdots$ determine the term ratio and calculate the sum of the first seven terms of the series.

Solution:
The term ratio is 1/2. The first seven terms of the geometric series above sum to

$$\frac{2(1 - (1/2)^7)}{1 - (1/2)} = 3.969$$

to three decimal places.

Example 2.4.3 Compound Interest

One hundred dollars is placed at an annual 6% rate. What sum accrues after 20 years?

Solution:

The principal at the end of any year is that year's starting principal plus annual interest. Therefore, letting p_n denote the principal after n year at rate r,

$$p_1 = p_0 + rp_0 = (1 + r)p_0$$

$$p_2 = p_1 + rp_1 = (1 + r)\ p_1 = (1 + r)\ [(1 + r)\ p_0] = (1 + r)^2 p_0$$

$$p_3 = p_2 + rp_2 = (1 + r)\ p_2 = (1 + r)\ [(1 + r)^2\ p_0] = (1 + r)^3 p_0$$

$$\vdots$$

$$p_n = p_{n-1} + rp_{n-1} = (1 + r)\ p_{n-1} = (1 + r)\ [(1 + r)^{n-1}\ p_0] = (1 + r)^n p_0$$

An induction can verify that $p_n = (1 + r)^n p_0$, $n = 1, 2, \ldots, n$.
Substituting $n = 20$, $r = 0.06$, and $p_0 = \$100$, to find $p_{20} = p_0(1 + 0.06)^{20} = \$320.71$.

Example 2.4.4 Benjamin Franklin's Bequest

Benjamin Franklin bequeathed 1000 pounds sterling to both Boston (his birthplace) and to Philadelphia (his home) to accumulate at compound interest for 100 years. By then, he calculated that the fund would reach 131,000 pounds. He specified that 100,000 pounds be used to fund projects and the remaining 31,000 pounds be left to accumulate with interest for another 100 years.

What interest rate did Franklin assume his monies would earn?

Solution:

Here, solve for r, the interest rate, in $p_n = p_0(1 + r)^n$, the principal after n years.

$$1000(1 + r)^{100} = 131,000$$

$$(1 + r)^{100} = 131$$

$$100\ \log(1 + r) = \log\ 131$$

$$\log(1 + r) = 0.021172173$$

$$(1 + r) = 10^{0.021172173}$$

$$r = 0.0499599$$

Apparently, Franklin assumed his monies would earn about 5% annually.

A Web search for "Benjamin Franklin's Bequest" yields a trove of interesting reading and varied calculations of accrued interest.

Fibonacci Numbers

Leonardo Fibonacci, a thirteenth-century merchant-scholar queried:

"How many pairs of rabbits can be produced from a single pair in a year if every month each pair begets a new pair which from the second month on becomes productive?"

The answer to this query generates the famed Fibonacci numbers.

The Fibonacci Numbers

$$F_n = F_{n-1} + F_{n-2}, \quad n > 2$$

$$F_1 = F_2 = 1$$

F_n is the nth Fibonacci number, $n = 1, 2, \ldots$

Example 2.4.5 Fibonacci Sequence

Answer the query posed by Fibonacci by indicating the monthly sequence of paired rabbits.

Solution:

Let F_n denote the number of rabbit pairs in the nth month, $n = 1, 2, \ldots$

Clearly $F_1 = F_2 = 1$ since there is only an immature (nonproductive) pair in the first month and remains a single (now productive) pair in the second month.

The third month has two pairs of rabbits, the original pair and its offspring. Therefore, $F_3 = 2$. That is, $F_3 = F_2 + F_1$.

The fourth month has the original pair, a new offspring pair, and the last month's offspring pair (now productive). That is, $F_4 = F_2 + F_3$ pairs.

More generally, the number in the nth generation, F_n, is the number in the last generation, F_{n-1}, to which is added the number in the generation of 2 months ago (since each pair of that generation is productive). In symbols,

$$F_1 = F_2 = 1$$

$$F_n = F_{n-1} + F_{n-2}, \quad n > 2$$

Substituting various values of n yields the famed ***Fibonacci sequence***

$$1, 1, 2, 3, 5, 8, 13, 21, 34, 55, 89, \ldots$$

This remarkable sequence arises in surprising places in nature.

Fibonacci numbers are related to another interesting number called the **Golden Ratio.** The Golden Ratio was known to Euclid some 2000 years ago. Studying the proportion

$$\frac{x}{1} = \frac{x+1}{x}$$

derived from the segmented line

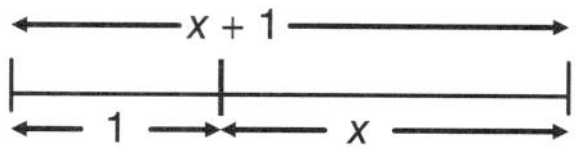

led him to the quadratic $x^2 = x + 1$.

Its roots are $x = \dfrac{1 \pm \sqrt{5}}{2}$ and its positive root is approximated by 1.6180339887 … , the Golden Ratio, Φ.

The Golden Ratio emerges as the limiting ratio of the Fibonacci numbers:

$$1/1,\ 2/1,\ 3/2,\ 5/3,\ 8/5,\ 13/8,\ 21/13,\ 34/21,\ 55/34$$

$$1,\ 2,\ 1.5,\ 1.667,\ 1.60,\ 1.625,\ 1.615,\ 1.619,\ 1.618$$

These ratios oscillate above and below their limiting value, Φ.

♦ The Fibonacci numbers are related to the Golden Ratio, Φ, which has been called "the world's most astonishing number." It has been associated with art and architecture, galaxies and poetry, mollusk shells and phyllotaxis – in much the manner of the Fibonacci numbers.

The Golden Ratio

$$\boldsymbol{\Phi = \lim_{n\to\infty} \frac{F_n}{F_{n-1}}}$$

$$\boldsymbol{\Phi = \frac{1+\sqrt{5}}{2} \approx 1.618}$$

F_n is the nth Fibonacci number, $n = 1, 2, 3, 5, 8, \ldots$

Curiously, the reciprocal of Φ is its decimal portion 0.618. A rectangle whose sides are in the "golden proportion," that is, 1:1.618 or the equivalent 0.618:1, has been known for centuries. It occurs in proportions of the Parthenon in Greece and other classic architectural structures and designs.

♦ The Golden Ratio also appears in the works of Leonardo da Vinci; of Stradivari, the famed violin maker of the 1500s; and in Mozart's sonatas that divide into two parts exactly at the point of the golden section in almost all instances.

EXERCISES 2.4

1. Write the next four terms of the arithmetic sequence: 2, 7, 12, …
2. Write the next three terms of the arithmetic sequence: 1/8, 3/8, …

3. Write the next five terms of the arithmetic sequence: 20, 35, 50, …

4. Write the next three terms of the arithmetic sequence: 12, 9, …

5. Sum the first seven terms of the arithmetic series whose first term is 3 and common difference is 6.

6. Sum the first five terms of the arithmetic series whose first term is 5 and common difference is 7.

7. Write the first five terms of the sequence whose nth term is $a_n = \dfrac{3n+1}{2n+3}$.

8. Write the first four terms of the sequence whose nth term is $a_n = \dfrac{2^n+5}{n^3+1}$.

9. Write the next three terms of the geometric sequence 2, 6, 18 …

10. Write the next two terms of the geometric sequence 5, 20, 80 …

11. Write the next three terms of the geometric sequence 729, 243, 81 …

12. Write the next four terms of the geometric sequence 4, 3, …

13. Sum the first five terms of the geometric sequence 4, 16, 64 …

14. Sum the first six terms of the geometric sequence 5, 10, 20 …

15. A ball is dropped from a height of 200 feet. If it rebounds to one-half height from which it falls each time it hits the ground, how high will it be on the third bounce?

16. In 1990, 200 years after Benjamin Franklin's death, reports on his 1790 bequest appeared. Boston received about 5 million dollars, while Philadelphia received about 2.25 million dollars. In January 1894, after about 100 years had passed, Boston's fund had grown to about 90,000 pounds instead of the predicted 131,000 pounds. Assuming, \$100,000 of the Boston fund was deposited for 100 years and was then worth 5 million dollars, verify that the fund earned about 4% interest (rather than the 5% Franklin may have anticipated). *More information on Franklin's will and its codicil can be found with a Google search or by reading*
    Benjamin Franklin and the Invention of Microfinance by Bruce Yenawine

17. How much interest will \$5000 earn at 6% compounded annually after 4 years?

18. *Rapid addition* – Verify that the sum of the first n Fibonacci numbers is one less than the $(n+2)^{\text{th}}$ Fibonacci number.

19. *Squaring rectangles* – Verify that the sum of an odd number of products of successive Fibonacci numbers is the square of the largest Fibonacci number employed. For example, for the first three Fibonacci numbers

$$1 \times 1 + 1 \times 2 + 2 \times 3 = 9 = 3^2$$

For the first five Fibonacci numbers

$$1 \times 1 + 1 \times 2 + 2 \times 3 + 3 \times 5 + 5 \times 8 = 64 = 8^2, \text{ etc.}$$

♦ A geometric consequence is that any odd number of rectangles with sides equal to two successive Fibonacci numbers fits exactly in a square of side equal to the largest Fibonacci number used.

20. Show that $\Phi^2 = \Phi + 1$
21. Show that $\Phi^n = F_n\Phi + F_{n-1}$

HISTORICAL NOTES

Leonardo Pisano (Fibonacci) (ca.1175–1250) – Regarded as the most talented mathematician of the Middle Ages, a span of 900 years, Fibonacci introduced Hindu-Arabic numerals and decimal notation to Europe. Born to a mercantile family in Pisa, Italy, a great commercial center at the time, he was afforded opportunities to travel in Asia and the Middle East where his interest in practical mathematics was stimulated.

CHAPTER 2 SUPPLEMENTARY EXERCISES

1. What is the simple interest after 30 months on $25,000 at 8%?
2. What interest accrues on $20,000 at 6% compounded annually for 3 years?
3. What interest accrues on $10,000 at 7% compounded continuously for 15 years?
4. What present sum will increase to $60,000 in 10 years at 4% if interest is compounded quarterly?
5. What present sum will increase to $175,000 in 20 years at 8% if interest is compounded continuously?
6. A bond account has $250,000 at 6% compounded monthly for 10 years. Another has $300,000 at 5% annually for 8 years. Which is worth more at the end of those times?
7. Suppose $1500 is paid yearly into an annuity at 7% compounded annually. What is the FV of the annuity after 10 years?
8. Suppose $200 is paid monthly into an annuity that earns 6%. What is the FV of the annuity after 4 years?
9. Find the PV of an annuity of $100 paid monthly for 10 years if the interest rate is 6%.
10. What is the PV of an annuity of $2000 paid yearly for 20 years if the interest rate is 8%?
11. A $25,000 loan calls for repayment in 10 yearly installments at 6%. Calculate an amortization schedule that includes the payment amount, the interest payment, and the year-end balance.
12. A car is financed for $16,000 at 3% interest. Determine monthly payments for a 5-year loan.

13. Write the next three terms of the arithmetic sequence 2, 5, 8, …

14. Write the next two terms of the geometric sequence 4, 20, 100, …

15. *A Round Robin Party Game*

    A person calls out an arbitrary number followed by another person who calls another arbitrary number. The next person adds the two prior called numbers and calls their sum. That person is followed by another person who calls the sum of the last two spoken numbers, and so on, each person continues by calling the sum of the last two spoken numbers.

    At any point in the game, as it progresses, calculate the ratio of the last number to the preceding one. As the game continues, the successive ratios approximate the Golden Ratio regardless of the choice of the initial arbitrary numbers. Explain why!

    Hint: if $\frac{x}{w} = \frac{y}{v}$ then $\frac{x+y}{w+v}$ for arbitrary positive numbers v, w, x, and y.

16. It is estimated that 90% of plants exhibit Fibonacci number patterns. Find other examples of Fibonacci numbers in nature. Hint: try an Internet search.

17. Examine the ratio of the sides of rectangles that you encounter in daily life. For example, the ratio of the sides of a picture frame, TV, or computer screen and cabinet. Do you notice that many are in a proportion approximating the Golden Mean?

3 *Matrix Algebra*

Finite Mathematics: Models and Applications, First Edition. Carla C. Morris and Robert M. Stark.

Companion Website: http://www.wiley.com/go/morris/finitemathematics

Roughly said, a **matrix** is an array of numbers called **elements** or **entries**. They permit the algebraic movement of numbers in a "bulk" format by prescribed rules. As such they are well suited to automated computation in which a particular command is executed repetitively on the elements that make up the matrix. As a result, the growing importance of matrices has paralleled the increasing capability and lower cost of computers. Matrices are vital to the ability of computers to process information on a large, even humongous, scale in sorting and solving large systems of equations in engineering, cryptography, and high-tech medical and other imaging. Besides, as cited in the Historical Notes, matrices are important to the social sciences.

3.1 MATRICES

Introduction

This section introduces the use of matrix methods in the solution of systems of equations.

Recall that in Chapter 1, systems of two equations were solved by simple algebraic manipulation. For example, the system

$$
\begin{aligned}
2x - 3y &= 2 \\
-x + 2y &= 3
\end{aligned}
$$

was solved by multiplying the second equation by +2 and adding it to the first to yield:

$$
\begin{aligned}
x &= 13 \\
y &= 8
\end{aligned}
$$

Another method, whose principles are useful later, is needed to allow automated computation. We demonstrate using a simple illustration.

Divide the first equation by 2, the coefficient of x, to obtain the new form $x - 3/2y = 1$.

Next, use this equation to eliminate x from the second equation. Here, it is only necessary to add the new form since $+x$ and $-x$ cancel. The result is

$$
\begin{array}{r}
x - (3/2)y = 1 \\
\underline{-x + \qquad 2y = 3} \\
(1/2)y = 4
\end{array}
$$

Now, multiply the last equation by 2 to yield $y = 8$. Finally, "back substituting" $y = 8$ in the first equation yields $x = 13$. The ordered pair solution to the system is (13, 8).

Example 3.1.1 *"3 × 3 System"*

Solve the system:

$$\begin{aligned} 2x + 3y - z &= 5 \\ -2y + 3z &= 4 \\ 3x + y &= 2 \end{aligned}$$

Solution:

It is convenient to keep an alphabetical order, although it is not necessary.

In the following table, the first boxed column lists the equations in algebraic format; the second, in matrix format; and the third, indicates operations step-by-step. The operations indicated are performed in order to make the coefficients along the diagonal unity and all coefficients below the diagonal zero.

0: Write the original system and its coefficient matrix.

1: The coefficient of x in the first equation of the original equations must become unity. This operation, denoted by $R_1 \rightarrow (1/2)R_1$, means "replace row 1 by 1/2 of row 1."
This convenient notation aids in tracking operations.

2: The x is then eliminated from the third equation as the coefficient for x is already zero in the second equation.

3: The coefficient of y in the second equation must become "1" as the goal is to have the coefficient on the diagonal become unity.

4: The y is eliminated from the third equation as entries below the unity on the diagonal must become zero.

5: The coefficient of z is made to become unity to complete the "top-down" approach.

0	$2x + 3y - z = 5$ $-2y + 3z = 4$ $3x + y = 2$	$\left[\begin{array}{ccc\|c} 2 & 3 & -1 & 5 \\ 0 & -2 & 3 & 4 \\ 3 & 1 & 0 & 2 \end{array}\right]$	$R_1 \rightarrow (1/2)R_1$
1	$x + 3/2y - 1/2z = 5/2$ $-2y + 3z = 4$ $3x + y = 2$	$\left[\begin{array}{ccc\|c} 1 & 3/2 & -1/2 & 5/2 \\ 0 & -2 & 3 & 4 \\ 3 & 1 & 0 & 2 \end{array}\right]$	$R_3 \rightarrow R_3 - 3R_1$
2	$x + 3/2y - 1/2z = 5/2$ $-2y + 3z = 4$ $-7/2y + 3/2z = -11/2$	$\left[\begin{array}{ccc\|c} 1 & 3/2 & -1/2 & 5/2 \\ 0 & -2 & 3 & 4 \\ 0 & -7/2 & 3/2 & -11/2 \end{array}\right]$	$R_2 \rightarrow -1/2R_2$

⟶

3	$x + 3/2y - 1/2z = 5/2$ $y - 3/2z = -2$ $-7/2y + 3/2z = -11/2$	$\left[\begin{array}{ccc\|c} 1 & 3/2 & -1/2 & 5/2 \\ 0 & 1 & -3/2 & -2 \\ 0 & -7/2 & 3/2 & -11/2 \end{array}\right]$	$R_3 \to R_3 + 7/2R_2$
4	$x + 3/2y - 1/2z = 5/2$ $y - 3/2z = -2$ $-15/4z = -25/2$	$\left[\begin{array}{ccc\|c} 1 & 3/2 & -1/2 & 5/2 \\ 0 & 1 & -3/2 & -2 \\ 0 & 0 & -15/4 & -25/2 \end{array}\right]$	$R_3 \to -4/15R_3$
5	$x + 3/2y - 1/2z = 5/2$ $y - 3/2z = -2$ $z = 10/3$	$\left[\begin{array}{ccc\|c} 1 & 3/2 & -1/2 & 5/2 \\ 0 & 1 & -3/2 & -2 \\ 0 & 0 & 1 & 10/3 \end{array}\right]$	

Now, it is clear that $z = 10/3$. This value is "back substituted" into the second equation so $y - 3/2(10/3) = -2$ implies $y = 3$. Finally, substitute the values for z and y into the first equation to yield $x + (3/2)(3) - (1/2)(10/3) = 5/2$, so $x = -1/3$. The solution is $(x, y, z) = (-1/3, 3, 10/3)$.

Example 3.1.2 **"Another 3 × 3 System"**

Solve the system:

$$\begin{aligned} 2x_1 - 4x_2 + 2x_3 &= 2 \\ 3x_1 + x_2 - x_3 &= 4 \\ 2x_1 + 5x_2 - 2x_3 &= 9 \end{aligned}$$

Solution:

Here, numerical subscripts rather than letters distinguish variables. Such subscripts are useful for computer programs and for systems with more variables than an alphabet allows.

Again, coefficient matrices appear in the second column, and a vertical line is used to substitute for equal signs. The sequence of steps to obtain x_3 appears in the table.

0	$2x_1 - 4x_2 + 2x_3 = 2$ $3x_1 + x_2 - x_3 = 4$ $2x_1 + 5x_2 - 2x_3 = 9$	$\left[\begin{array}{ccc\|c} 2 & -4 & 2 & 2 \\ 3 & 1 & -1 & 4 \\ 2 & 5 & -2 & 9 \end{array}\right]$	$R_1 \to (1/2)R_1$
1	$x_1 - 2x_2 + x_3 = 1$ $3x_1 + x_2 - x_3 = 4$ $2x_1 + 5x_2 - 2x_3 = 9$	$\left[\begin{array}{ccc\|c} 1 & -2 & 1 & 1 \\ 3 & 1 & -1 & 4 \\ 2 & 5 & -2 & 9 \end{array}\right]$	$R_2 \to R_2 - 3R_1$ $R_3 \to R_3 - 2R_1$
2	$x_1 - 2x_2 + x_3 = 1$ $7x_2 - 4x_3 = 1$ $9x_2 - 4x_3 = 7$	$\left[\begin{array}{ccc\|c} 1 & -2 & 1 & 1 \\ 0 & 7 & -4 & 1 \\ 0 & 9 & -4 & 7 \end{array}\right]$	$R_2 \to (1/7)R_2$

⟶

<table>
<tr><td>3</td><td>$x_1 - 2x_2 + x_3 = 1$
$x_2 - (4/7)x_3 = 1/7$
$9x_2 - 4x_3 = 7$</td><td>$\left[\begin{array}{ccc|c} 1 & -2 & 1 & 1 \\ 0 & 1 & -4/7 & 1/7 \\ 0 & 9 & -4 & 7 \end{array}\right]$</td><td>$R_3 \to R_3 - 9R_2$</td></tr>
<tr><td>4</td><td>$x_1 - 2x_2 + x_3 = 1$
$x_2 - (4/7)x_3 = 1/7$
$(8/7)x_3 = 40/7$</td><td>$\left[\begin{array}{ccc|c} 1 & -2 & 1 & 1 \\ 0 & 1 & -4/7 & 1/7 \\ 0 & 0 & 8/7 & 40/7 \end{array}\right]$</td><td>$R_3 \to 7/8R_3$</td></tr>
<tr><td>5</td><td>$x_1 - 2x_2 + x_3 = 1$
$x_2 - (4/7)x_3 = 1/7$
$x_3 = 5$</td><td>$\left[\begin{array}{ccc|c} 1 & -2 & 1 & 1 \\ 0 & 1 & -4/7 & 1/7 \\ 0 & 0 & 1 & 5 \end{array}\right]$</td><td></td></tr>
</table>

Having the value of $x_3 = 5$, "back substitution"(as in the previous example) yields $x_1 = 2$, $x_2 = 3$, and $x_3 = 5$ or the ordered solution (2, 3, 5).

We can modify the iterative procedure and its matrix format from the previous two examples to obtain solutions without "back substituting." The method, known as **Gauss–Jordan elimination**, consists of algebraic iterations to transform the system of equations to a format having these properties:

- *Variables appear only once in the system.*
- *Variables have unit coefficients.*
- *No more than one variable to an equation.*

Example 3.1.3 Gauss–Jordan Elimination – A Preview

Use the iterative procedure and its matrix format to solve the system of Example 3.1.2.

Solution:

Starting with Step 5 of Example 3.1.2, the procedure begins with the elimination of x_3 from the first two equations by using multiples of the third equation.

Next, x_2 is eliminated from the first equation using a multiple of the second equation.

<table>
<tr><td>5</td><td>$x_1 - 2x_2 + x_3 = 1$
$x_2 - (4/7)x_3 = 1/7$
$x_3 = 5$</td><td>$\left[\begin{array}{ccc|c} 1 & -2 & 1 & 1 \\ 0 & 1 & -4/7 & 1/7 \\ 0 & 0 & 1 & 5 \end{array}\right]$</td><td>$R_1 \to R_1 - R_3$
$R_2 \to R_2 + 4/7R_3$</td></tr>
<tr><td>6</td><td>$x_1 - 2x_2 = -4$
$x_2 = 3$
$x_3 = 5$</td><td>$\left[\begin{array}{ccc|c} 1 & -2 & 0 & -4 \\ 0 & 1 & 0 & 3 \\ 0 & 0 & 1 & 5 \end{array}\right]$</td><td>$R_1 \to R_1 + 2R_2$</td></tr>
<tr><td>7</td><td>$x_1 = 2$
$x_2 = 3$
$x_3 = 5$</td><td>$\left[\begin{array}{ccc|c} 1 & 0 & 0 & 2 \\ 0 & 1 & 0 & 3 \\ 0 & 0 & 1 & 5 \end{array}\right]$</td><td></td></tr>
</table>

⟶

The final matrix format has $x_1 = 2$, $x_2 = 3$, and $x_3 = 5$, identical to the earlier solution. (Remember that the vertical line in a matrix is a shorthand for equal signs).

Note that in the last steps of this Gauss–Jordan method, variables are eliminated in a reverse numeric order from each preceding equation. That is, the solution for x_3 in the third equation is used to eliminate x_3 from the first two equations. This determines a value for x_2. Next, the value for x_2 from the second equation is used to eliminate x_2 from the first equation. A value for x_1 is at hand. This organization ensures the systematic elimination of variables.

The method used for two equations and two unknowns, (where the value of one of the variables was substituted to determine the other), is not suited for systems with many equations and variables, as noted earlier.

Gauss–Jordan elimination is further considered in Section 3.3. Note the convenience of Gauss–Jordan elimination when working with a matrix format instead of the entire equations. It is used in many computer programs. Remember, not all systems of equations have solutions and some may have multiple solutions. We consider them later. Matrix methods are invaluable for automated computation.

EXERCISES 3.1

1. Write the system of equations in the matrix format of Examples 3.1.1 and 3.1.2 in this section.

a) $$\begin{aligned} 3x_1 - 5x_2 + 4x_3 &= 5 \\ x_1 + x_2 - x_3 &= 0 \\ 2x_1 + 3x_2 - 2x_3 &= 2 \end{aligned}$$

b) $$\begin{aligned} x_1 - 3x_2 + 5x_3 &= 19 \\ 2x_1 + 4x_2 - x_3 &= -5 \\ -x_1 + 5x_2 + 3x_3 &= 3 \end{aligned}$$

2. Write the system of equations in the matrix format of Examples 3.1.1 and 3.1.2.

a) $$\begin{aligned} 4x_1 \qquad + 7x_3 &= 11 \\ 2x_1 + 3x_2 - 3x_3 &= 2 \\ 2x_1 + 3x_2 \qquad &= 5 \end{aligned}$$

b) $$\begin{aligned} x_1 - 3x_2 + 5x_3 + x_4 &= 15 \\ 2x_1 + 4x_2 \qquad - x_4 &= 6 \\ -x_1 + x_2 + 3x_3 \qquad &= 4 \\ 2x_1 - x_2 + 4x_3 - 3x_4 &= 0 \end{aligned}$$

3. The matrix format with variables x_1, x_2 is

$$\left[\begin{array}{cc|c} 5 & 3 & 13 \\ 4 & 7 & 18 \end{array}\right].$$

Write the system of equations represented.

4. The matrix format with variables x, y, and z is

$$\left[\begin{array}{ccc|c} 5 & 0 & 1 & 13 \\ 7 & 1 & -2 & 9 \\ 2 & -1 & 4 & 15 \end{array}\right].$$

Write the system of equations represented.

5. The matrix format with variables x_1, x_2, and x_3 is

$$\left[\begin{array}{ccc|c} 1 & 2 & 3 & 4 \\ 2 & 1 & 2 & 3 \\ 4 & 1 & -2 & 9 \end{array}\right].$$

Write the system of equations.

6. The final matrix for a system in the variables x_1 and x_2 is

$$\left[\begin{array}{cc|c} 1 & 0 & 12 \\ 0 & 1 & 5 \end{array}\right].$$

Write the solution.

7. The final matrix for a system in the variables x, y, and z is

$$\left[\begin{array}{ccc|c} 1 & 0 & 0 & 4 \\ 0 & 1 & 0 & 3 \\ 0 & 0 & 1 & 1 \end{array}\right].$$

Write the solution.

8. The final matrix for the solution to a system of equations is

$$\left[\begin{array}{cccc|c} 1 & 0 & 0 & 0 & 1 \\ 0 & 1 & 0 & 0 & -2 \\ 0 & 0 & 1 & 0 & 3 \\ 0 & 0 & 0 & 1 & 7 \end{array}\right]$$

Write the values of the variables x_1, x_2, x_3, and x_4.

9. At a local theater, tickets are \$4 for children and \$5 for adults. The theater sold 2400 tickets realizing revenue of \$10,500. Use a matrix format to display this information.

10. A person has 12 coins in quarters, halves, and dollars that total \$6.50. There are two more 50-cent pieces than dollars and one-half as many quarters as 50-cent pieces and dollars combined. Use a matrix format to display this information.

11. How does this final matrix indicate that there is no solution to the original system of equations?

$$\left[\begin{array}{ccc|c} 1 & 0 & 0 & 2/3 \\ 0 & 1 & 0 & 1 \\ 0 & 0 & 0 & 4 \end{array}\right]$$

12. Perform the indicated operations.

$$\left[\begin{array}{ccc|c} 1 & 2 & 3 & 4 \\ 6 & 4 & -1 & 4 \\ 2 & 5 & -2 & 9 \end{array}\right] \qquad \begin{array}{l} R_2 \to R_2 - 6R_1 \\ R_3 \to R_3 - 2R_1 \end{array}$$

13. Perform the indicated operations.

$$\left[\begin{array}{ccc|c} 1 & 1 & 2 & 9 \\ 0 & 1 & 3 & 7 \\ 0 & 0 & 1 & 1 \end{array}\right] \qquad \begin{array}{l} R_1 \to R_1 - 2R_3 \\ R_2 \to R_2 - 3R_3 \end{array}$$

14. Use the matrix technique and back substitute x_3 for Exercise 3.1.1a.

15. Use the matrix technique and back substitute x_3 for Exercise 3.1.1b.

16. Use the matrix technique and back substitute x_3 for Exercise 3.1.2a.

3.2 MATRIX NOTATION, ARITHMETIC, AND AUGMENTED MATRICES

As noted earlier, a **matrix** is a rectangular array of **elements** (or **entries**) in rows and columns. A matrix having m rows and n columns is called an $(m \times n)$ matrix. It is said to be of $(m \times n)$ **dimension**. (*Note that in specifying dimension the number of rows is cited first.*)

For instance, a (3×2) matrix, **A**, with elements or entries a_{ij}, $i = 1, 2, 3$ and $j = 1, 2$ is written as

$$\mathbf{A} = \begin{bmatrix} a_{11} & a_{12} \\ a_{21} & a_{22} \\ a_{31} & a_{32} \end{bmatrix}$$

Two matrices, **A** and **B**, are equal if their entries are equal. That is, if $a_{ij} = b_{ij}$ for $1 \leq i \leq m$, $1 \leq j \leq n$. This means that corresponding entries in matrix **B** must be identical to those of matrix **A**. Matrices of dimension $(1 \times n)$ and $(m \times 1)$ are known as **vectors**.

Matrix Addition and Subtraction

The addition or subtraction of two matrices **A** and **B** is possible only if they have the same dimension. The resulting matrix, **C**, consists of the entries $a_{ij} \pm b_{ij}$, $1 \leq i \leq m$ and $1 \leq j \leq n$. Matrices of unlike dimension can neither be added nor subtracted.

If **k** is some scalar (constant) and **A** is an $(m \times n)$ matrix, then the scalar multiple **kA** is a matrix with entries ka_{ij}, $1 \leq i \leq m$ and $1 \leq j \leq n$. That is, each element is multiplied by a scalar **k**.

Example 3.2.1 Matrix Arithmetic

$$\mathbf{A} = \begin{bmatrix} 1 & 2 & 4 & 2 \\ 0 & 7 & -6 & 5 \\ 3 & 1 & 4 & 9 \end{bmatrix} \quad \text{and} \quad \mathbf{B} = \begin{bmatrix} 3 & 4 & 7 & 11 \\ -1 & -5 & 0 & 6 \\ 2 & 1 & 4 & 8 \end{bmatrix}$$

a) Determine $\mathbf{A} + \mathbf{B}$ *b) Determine* $4\mathbf{A}$.

→

Solution:

a) *Since both matrices are (3 × 4), they can be added (or subtracted)! The addition of corresponding entries forms the matrix* $\mathbf{A}+\mathbf{B}$.

$$\mathbf{A}+\mathbf{B}=\begin{bmatrix}1+3 & 2+4 & 4+7 & 2+11\\ 0-1 & 7-5 & -6+0 & 5+6\\ 3+2 & 1+1 & 4+4 & 9+8\end{bmatrix}=\begin{bmatrix}4 & 6 & 11 & 13\\ -1 & 2 & -6 & 11\\ 5 & 2 & 8 & 17\end{bmatrix}.$$

b) *To multiply by a scalar, in this case 4, here it multiplies each entry. Therefore,*

$$4\mathbf{A}=\begin{bmatrix}4\cdot 1 & 4\cdot 2 & 4\cdot 4 & 4\cdot 2\\ 4\cdot 0 & 4\cdot 7 & 4\cdot -6 & 4\cdot 5\\ 4\cdot 3 & 4\cdot 1 & 4\cdot 4 & 4\cdot 9\end{bmatrix}=\begin{bmatrix}4 & 8 & 16 & 8\\ 0 & 28 & -24 & 20\\ 12 & 4 & 16 & 36\end{bmatrix}.$$

Example 3.2.2 Matrix Arithmetic (continued)

If $\mathbf{A}=\begin{bmatrix}2 & 0 & 5 & 4\\ 7 & 1 & 6 & -2\end{bmatrix}$ *and* $\mathbf{B}=\begin{bmatrix}3 & 1 & 0 & -3\\ 5 & 2 & 4 & -2\end{bmatrix}$, *find* $5\mathbf{A}-2\mathbf{B}$.

Solution:

To add or subtract matrices, they must be of like dimension. Here, both A and B are (2 × 4) matrices. Note that scalar multiples do not alter matrix dimension.

Therefore,

$$5\mathbf{A}=\begin{bmatrix}5\cdot 2 & 5\cdot 0 & 5\cdot 5 & 5\cdot 4\\ 5\cdot 7 & 5\cdot 1 & 5\cdot 6 & 5\cdot(-2)\end{bmatrix}=\begin{bmatrix}10 & 0 & 25 & 20\\ 35 & 5 & 30 & -10\end{bmatrix}$$

$$2\mathbf{B}=\begin{bmatrix}2\cdot 3 & 2\cdot 1 & 2\cdot 0 & 2\cdot(-3)\\ 2\cdot 5 & 2\cdot 2 & 2\cdot 4 & 2\cdot(-2)\end{bmatrix}=\begin{bmatrix}6 & 2 & 0 & -6\\ 10 & 4 & 8 & -4\end{bmatrix}$$

and

$$5\mathbf{A}-2\mathbf{B}=\begin{bmatrix}10 & 0 & 25 & 20\\ 35 & 5 & 30 & -10\end{bmatrix}-\begin{bmatrix}6 & 2 & 0 & -6\\ 10 & 4 & 8 & -4\end{bmatrix}=\begin{bmatrix}4 & -2 & 25 & 26\\ 25 & 1 & 22 & -6\end{bmatrix}$$

You may prefer to find $5\mathbf{A}-2\mathbf{B}$ *in a single step rather than the two steps above.*

Matrix Multiplication

Besides addition and subtraction, matrix multiplication can also be defined. Since vectors are also matrices, we begin with them. When a (1 × *n*) matrix (row vector) is multiplied by an $(n \times 1)$ matrix (column vector), the result is a (1×1) matrix (vector), a single element matrix.

If $\mathbf{R} = [r_1 \ r_2 \ \cdots \ r_n]$ and $\mathbf{C} = \begin{bmatrix} c_1 \\ c_2 \\ \vdots \\ c_n \end{bmatrix}$, then the multiplication of these $(1 \times n)$ and $(n \times 1)$ matrices yield $\mathbf{RC} = [r_1c_1 + \ r_2c_2 + \ \cdots + \ r_nc_n]$, a single element matrix.

Example 3.2.3 Vector Multiplication

If $\mathbf{R} = [2 \quad 0 \quad 5 \quad 4]$ *and* $\mathbf{C} = \begin{bmatrix} 3 \\ 5 \\ 4 \\ -2 \end{bmatrix}$, *what is their product* $\mathbf{RC}$?

Solution:

Here, $\mathbf{R}$ *has dimension* (1×4) *and* $\mathbf{C}$, (4×1) *so, they are dimensionally compatible and can be multiplied for a* $(1 \times 1) = [(1 \times 4) \ (4 \times 1)]$ *matrix.*

Therefore, $\mathbf{RC} = [2 \times 3 + 0 \times 5 + 5 \times 4 + 4 \times (-2)] = [18]$.

Next, we generalize to multiplication of larger matrices. Again, multiplicative compatibility or **conformability** is required. An $(m \times n)$ matrix can only multiply an $(n \times r)$ to yield and $(m \times r)$ matrix, m and r arbitrary. There is no division of matrices.

Matrix Multiplication

If A is an $(m \times n)$ matrix and B an $(n \times r)$ matrix, then they are said to be conformable for multiplication.

Their matrix product AB will have the dimension $(m \times n)(n \times r) = (m \times r)$.

The entry for the ith row and j^{th} column of the matrix product AB is found by taking the ith row of A and the jth column of B and multiplying them as row and column vectors.

Example 3.2.4 Matrix Multiplication

If $\mathbf{A} = \begin{bmatrix} 3 & 4 & 1 & 2 \\ 2 & 6 & 9 & 8 \end{bmatrix}$ *and* $\mathbf{B} = \begin{bmatrix} 1 & 4 & 7 \\ 2 & 3 & 0 \\ 0 & 1 & 0 \\ 2 & 3 & 1 \end{bmatrix}$, *what is their product,* $\mathbf{AB}$?

⟶

Solution:
These (2 × 4) and (4 × 3) matrices are conformable. The elements of **AB** *are found, as before, by appropriate multiplication of a row vector by a column vector. The matrix* **AB** *will have the dimension (2 × 3) [= (2 × 4)(4 × 3)].*

The entry in the second row and first column of **A** *is determined by multiplying the row vector (second row) of* **A** *by the column vector (first column) of* **B**. *It is* $2(1) + 6(2) + 9(0) + 8(2) = 30$. *The other entries are determined similarly to yield*

$$\mathbf{AB} = \begin{bmatrix} 3(1) + 4(2) + 1(0) + 2(2) & 3(4) + 4(3) + 1(1) + 2(3) & 3(7) + 4(0) + 1(0) + 2(1) \\ 2(1) + 6(2) + 9(0) + 8(2) & 2(4) + 6(3) + 9(1) + 8(3) & 2(7) + 6(0) + 9(0) + 8(1) \end{bmatrix}$$

$$\mathbf{AB} = \begin{bmatrix} 15 & 31 & 23 \\ 30 & 59 & 22 \end{bmatrix}$$

Unlike ordinary arithmetic, matrix multiplication is not generally commutative. That is, for conformable matrices, **AB** is not generally equal to **BA**. Multiplication is not possible if matrices are not conformable. This is illustrated in Example 3.2.5.

Example 3.2.5 Commutative Property

Suppose $\mathbf{A} = \begin{bmatrix} 1 & 2 & 7 \\ 3 & 4 & 9 \\ 1 & 5 & 1 \end{bmatrix}$ *and* $\mathbf{B} = \begin{bmatrix} 1 & 2 & 0 \\ 0 & 1 & 1 \\ 2 & 2 & 1 \end{bmatrix}$.

Verify that their product is not commutative.

Solution:
The matrix **AB** *has the dimension (3 × 3)(3 × 3) = (3 × 3). Likewise, the dimension for* **BA** *is (3 × 3)(3 × 3) = (3 × 3).*

Therefore, both pairs of matrices are conformable for multiplication. Calculating, their products yield:

$$\mathbf{AB} = \begin{bmatrix} 1(1) + 2(0) + 7(2) & 1(2) + 2(1) + 7(2) & 1(0) + 2(1) + 7(1) \\ 3(1) + 4(0) + 9(2) & 3(2) + 4(1) + 9(2) & 3(0) + 4(1) + 9(1) \\ 1(1) + 5(0) + 1(2) & 1(2) + 5(1) + 1(2) & 1(0) + 5(1) + 1(1) \end{bmatrix}$$

$$\mathbf{BA} = \begin{bmatrix} 1(1) + 2(3) + 0(1) & 1(2) + 2(4) + 0(5) & 1(7) + 2(9) + 0(1) \\ 0(1) + 1(3) + 1(1) & 0(2) + 1(4) + 1(5) & 0(7) + 1(9) + 1(1) \\ 2(1) + 2(3) + 1(1) & 2(2) + 2(4) + 1(5) & 2(7) + 2(9) + 1(1) \end{bmatrix}$$

$$\mathbf{AB} = \begin{bmatrix} 15 & 18 & 9 \\ 21 & 28 & 13 \\ 3 & 9 & 6 \end{bmatrix} \text{ and } \mathbf{BA} = \begin{bmatrix} 7 & 10 & 25 \\ 4 & 9 & 10 \\ 9 & 17 & 33 \end{bmatrix}.$$

Clearly, these are not equal and, here $\mathbf{AB} \neq \mathbf{BA}$.

Matrix multiplication can be applied to obtain information about data stored in a matrix format. Example 3.2.6 helps to illustrate this point.

Example 3.2.6 Travel Expenses

An employee is at a conference on Monday, Tuesday, and Wednesday. The company pays for meals, lodging, and travel. An expense matrix (in dollars) is:

$$\mathbf{E} = \begin{array}{r} \\ Mon \\ Tues \\ Wed \end{array} \begin{array}{c} \begin{array}{ccc} Meals & Lodging & Travel \end{array} \\ \begin{bmatrix} 27 & 60 & 50 \\ 30 & 60 & 45 \\ 35 & 75 & 25 \end{bmatrix} \end{array}$$

a) The employee's next trip results in expenses 10% higher in each category. Show the expenses in a matrix format.

b) On this trip, three employees were reimbursed for the same amounts on Monday, two on Tuesday, and four on Wednesday. All had the same expense matrix. Use a matrix multiplication for the company reimbursement to the employees in each category.

Solution:

*a) This is simply 1.10 multiplied by **E**.*

$$1.10\mathbf{E} = \begin{array}{r} \\ Mon \\ Tues \\ Wed \end{array} \begin{array}{c} \begin{array}{ccc} Meals & Lodging & Travel \end{array} \\ \begin{bmatrix} 29.70 & 66.00 & 55.00 \\ 33.00 & 66.00 & 49.50 \\ 38.50 & 82.50 & 27.50 \end{bmatrix} \end{array}$$

*b) If [3 2 4] is multiplied by **E**, the resulting matrix is the amounts reimbursed for meals, lodging, and travel for the 3-day period.*

$$[3 \quad 2 \quad 4] \begin{bmatrix} 27 & 60 & 50 \\ 30 & 60 & 45 \\ 35 & 75 & 25 \end{bmatrix} = [281 \quad 600 \quad 340]$$

That is, \$281 for meals, \$600 for lodging, and \$340 for travel.

Square, Identity, and Transpose Matrices

A **square matrix** has the same number of rows and columns, that is, $(n \times n)$.

The entries of the **main diagonal** of an $(n \times n)$ matrix **A** are double subscripted as $a_{11}, a_{22}, a_{33}, \ldots, a_{nn}$,

An **identity matrix**, **I**, is a square matrix whose entries are ones on the main diagonal and zeros elsewhere. Examples of identity matrices are

$$[1] \qquad \begin{bmatrix} 1 & 0 \\ 0 & 1 \end{bmatrix} \qquad \begin{bmatrix} 1 & 0 & 0 \\ 0 & 1 & 0 \\ 0 & 0 & 1 \end{bmatrix} \qquad \begin{bmatrix} 1 & 0 & 0 & 0 \\ 0 & 1 & 0 & 0 \\ 0 & 0 & 1 & 0 \\ 0 & 0 & 0 & 1 \end{bmatrix}$$

An $n \times n$ identity matrix is conveniently denoted by $\mathbf{I}_n$. A matrix all of whose entries are zero is called a **zero matrix** and denoted by **0**.

The **transpose** of a matrix has its rows and columns interchanged. That is, the transpose of an $(m \times n)$ matrix, **A**, is an $(n \times m)$ matrix, denoted $\mathbf{A}^t$, in which the rows and columns of matrix **A** are interchanged. (Note: t denotes a transpose. It is not an exponent.)

Example 3.2.7 Transpose of a Matrix

Find the transpose of each matrix:

$$a)\begin{bmatrix} 2 & -1 \\ 4 & 7 \end{bmatrix} \qquad b)\begin{bmatrix} 3 & 5 & 9 \\ 1 & 7 & 6 \\ 4 & 2 & 0 \end{bmatrix} \qquad c)\begin{bmatrix} 6 & 8 & 3 & 2 \\ 0 & 5 & 4 & 9 \end{bmatrix}$$

Solution:

To form the transpose, the rows and columns are interchanged. Thus,

$$a)\begin{bmatrix} 2 & 4 \\ -1 & 7 \end{bmatrix} \qquad b)\begin{bmatrix} 3 & 1 & 4 \\ 5 & 7 & 2 \\ 9 & 6 & 0 \end{bmatrix} \qquad c)\begin{bmatrix} 6 & 0 \\ 8 & 5 \\ 3 & 4 \\ 2 & 9 \end{bmatrix}$$

We have defined addition, subtraction, and multiplication of matrices. Matrix division is not defined. Some rules for matrices are similar to ordinary arithmetic; not all.

Rules for Matrix Arithmetic

1. AI = A	**6. A + B = B + A**
2. A + 0 = A	**7. A + (B + C) = (A + B) + C**
3. $\mathbf{I \cdot I = I^2 = I}$	**8. A(BC) = (AB)C**
4. 0A = 0	**9. A(B + C) = AB + AC**
5. A0 = 0	**10. k(B + C) = kB + kC**

(assuming operations are permitted.)

Matrices for Systems of Equations

Consider the system of m linear equations in n variables:

$$\begin{aligned}
a_{11}x_1 + a_{12}x_2 + \cdots + a_{1n}x_n &= b_1 \\
a_{21}x_1 + a_{22}x_2 + \cdots + a_{2n}x_n &= b_2 \\
a_{31}x_1 + a_{32}x_2 + \cdots + a_{3n}x_n &= b_3 \\
\vdots \qquad \vdots \qquad \vdots \quad &\quad \vdots \\
a_{m1}x_1 + a_{m2}x_2 + \cdots + a_{mn}x_n &= b_m
\end{aligned}$$

These equations can be written compactly in a matrix format as **AX** = **B**, where **A** is called the **coefficient matrix**; **X**, the **variables matrix**; and **B**, the **constants matrix**.

$$\mathbf{A} = \begin{bmatrix} a_{11} & a_{12} & a_{1n} \\ a_{21} & a_{22} & a_{2n} \\ a_{31} & a_{32} & a_{3n} \\ \vdots & \vdots & \vdots \\ a_{m1} & a_{m2} & a_{mn} \end{bmatrix}, \quad \mathbf{X} = \begin{bmatrix} x_1 \\ x_2 \\ x_3 \\ \vdots \\ x_n \end{bmatrix}, \quad \text{and} \quad \mathbf{B} = \begin{bmatrix} b_1 \\ b_2 \\ b_3 \\ \vdots \\ b_m \end{bmatrix}$$

An **augmented matrix**, denoted [**A**|**B**], conveniently combines in one matrix the coefficient and constant matrices comprising the entire system of equations. This is written in the matrix format used in the previous section. A vertical bar, representing equal signs, separates them as

$$\left[\begin{array}{ccc|c} a_{11} & a_{12} & a_{1n} & b_1 \\ a_{21} & a_{22} & a_{2n} & b_2 \\ a_{31} & a_{32} & a_{3n} & b_3 \\ \vdots & \vdots & \vdots & \vdots \\ a_{m1} & a_{m2} & a_{mn} & b_m \end{array}\right]$$

Note, again, that the augmented matrix neatly compacts all the information in the algebraic representation of the system of equations.

Example 3.2.8 Matrix Formats

a) Express this system in a matrix format:

$$\begin{aligned} 5x_1 - 10x_2 + 3x_3 + 4x_4 &= 15 \\ x_1 - x_2 + x_3 - x_4 &= -1 \\ 3x_1 + x_2 + 3x_3 - 2x_4 &= 4 \\ x_1 + x_2 + 2x_3 - x_4 &= 1 \end{aligned}$$

b) Write the augmented matrix for the system.

Solution:

a) Here, $\mathbf{A} = \begin{bmatrix} 5 & -10 & 3 & 4 \\ 1 & -1 & 1 & -1 \\ 3 & 1 & 3 & -2 \\ 1 & 1 & 2 & -1 \end{bmatrix}$, $\mathbf{X} = \begin{bmatrix} x_1 \\ x_2 \\ x_3 \\ x_4 \end{bmatrix}$ *and* $\mathbf{B} = \begin{bmatrix} 15 \\ -1 \\ 4 \\ 1 \end{bmatrix}$.

The system in a matrix format is:

$$\begin{bmatrix} 5 & -10 & 3 & 4 \\ 1 & -1 & 1 & -1 \\ 3 & 1 & 3 & -2 \\ 1 & 1 & 2 & -1 \end{bmatrix} \begin{bmatrix} x_1 \\ x_2 \\ x_3 \\ x_4 \end{bmatrix} = \begin{bmatrix} 15 \\ -1 \\ 4 \\ 1 \end{bmatrix}.$$

⟶

b) The augmented matrix is:

$$\left[\begin{array}{cccc|c} 5 & -10 & 3 & 4 & 15 \\ 1 & -1 & 1 & -1 & -1 \\ 3 & 1 & 3 & -2 & 4 \\ 1 & 1 & 2 & -1 & 1 \end{array}\right].$$

◆ In his famous Renaissance engraving *Melencolia I*, Albrecht Dürer displays a **magic square** (4 × 4) matrix. In magic squares, the entries of rows, columns, and diagonals each sum to the same number. Benjamin Franklin was a master of magic squares. See Historical Notes.

EXERCISES 3.2

1. If $\mathbf{A} = \begin{bmatrix} 1 & 2 & -3 \\ 9 & 4 & -1 \end{bmatrix}$, indicate values for these entries:

a) a_{12} b) a_{23} c) a_{13} d) a_{22}

2. If $\mathbf{A} = \begin{bmatrix} 4 & 3 & 5 & -3 \\ 2 & 8 & 4 & 1 \\ 7 & 0 & 2 & 6 \end{bmatrix}$, indicate values for these entries:

a) a_{32} b) a_{11} c) a_{24} d) a_{14}

3. Given $\mathbf{A} = \begin{bmatrix} 10 & 25 \\ 74 & 30 \end{bmatrix}$, $\mathbf{B} = \begin{bmatrix} 20 & 52 \\ 41 & 36 \end{bmatrix}$, and $\mathbf{C} = \begin{bmatrix} 21 & 62 & -37 \\ 19 & 40 & -91 \end{bmatrix}$
Which of these matrix operations are possible?

a) $\mathbf{A}+\mathbf{B}$ b) $\mathbf{B}+2\mathbf{C}$ c) $5\mathbf{C}$ d) $\mathbf{A}+\mathbf{B}-\mathbf{C}$ e) $3\mathbf{B}-6\mathbf{A}$

4. Given $\mathbf{A} = \begin{bmatrix} 1 & 2 \\ 7 & 3 \\ 5 & 0 \end{bmatrix}$, $\mathbf{B} = \begin{bmatrix} 0 & 2 \\ 4 & 3 \end{bmatrix}$, and $\mathbf{C} = \begin{bmatrix} 1 & 6 \\ 9 & -1 \end{bmatrix}$
Which of these matrix operations are possible?

a) $\mathbf{A}-\mathbf{B}$ b) $4\mathbf{A}$ c) $\mathbf{B}+\mathbf{C}$ d) $\mathbf{A}-3\mathbf{C}$ e) $3\mathbf{B}-6\mathbf{A}$

5. Given $\mathbf{A} = \begin{bmatrix} 4 & 1 \\ 3 & 5 \\ 6 & 1 \end{bmatrix}$, $\mathbf{B} = \begin{bmatrix} 8 & -1 \\ 12 & 3 \\ 10 & 7 \end{bmatrix}$, and $\mathbf{C} = \begin{bmatrix} 3 & 8 & 9 \\ 2 & 0 & 0 \\ 4 & 0 & 3 \end{bmatrix}$
Which of these matrix operations are possible?

a) $\mathbf{A}+\mathbf{B}$ b) $4\mathbf{C}$ c) $2\mathbf{A}+\mathbf{B}$ d) $3\mathbf{A}-2\mathbf{C}$ e) $\mathbf{B}+\mathbf{C}$

6. Given $\mathbf{A} = \begin{bmatrix} 5 & 4 & 10 \\ 3 & 6 & 15 \\ 0 & 9 & 11 \end{bmatrix}$, $\mathbf{B} = \begin{bmatrix} 1 & 5 & 8 \\ 2 & 0 & -1 \end{bmatrix}$, and $\mathbf{C} = \begin{bmatrix} -3 & 7 & 8 \\ -1 & 5 & 4 \end{bmatrix}$,

Which of these matrix operations are possible?

a) $5\mathbf{A}$ b) $-4\mathbf{B}$ c) $2\mathbf{B} + 3\mathbf{C}$ d) $\mathbf{A} - 2\mathbf{C}$ e) $\mathbf{B} - 4\mathbf{A}$

7. Carry out the valid matrix operations in Exercise 3.2.3.

8. Carry out the valid matrix operations in Exercise 3.2.4.

9. Carry out the valid matrix operations in Exercise 3.2.5.

10. Carry out the valid matrix operations in Exercise 3.2.6.

11. Given $\mathbf{A} = \begin{bmatrix} 1 & 0 & 2 \\ 3 & 4 & 7 \end{bmatrix}$, $\mathbf{B} = \begin{bmatrix} -2 & 3 & 6 \\ -5 & 1 & 4 \end{bmatrix}$, and $\mathbf{C} = \begin{bmatrix} 4 & 6 & -3 \\ 2 & 8 & -9 \end{bmatrix}$

Determine the following:

a) $\mathbf{A} + \mathbf{B}$ b) $2\mathbf{B}$ c) $\mathbf{A} + 3\mathbf{C}$ d) $3\mathbf{A} - 2\mathbf{B} + \mathbf{C}$

12. Given $\mathbf{A} = \begin{bmatrix} 2 & 3 & 1 \\ 1 & 0 & 5 \\ 4 & 2 & 3 \end{bmatrix}$, $\mathbf{B} = \begin{bmatrix} 2 & 4 & 5 \\ -3 & 1 & 0 \\ -2 & 7 & 6 \end{bmatrix}$, and $\mathbf{C} = \begin{bmatrix} 5 & 5 & -1 \\ 3 & 4 & -2 \\ 0 & 7 & -3 \end{bmatrix}$

Determine the following:

a) $\mathbf{A} + 2\mathbf{B}$ b) $2\mathbf{A}$ c) $\mathbf{B} + 5\mathbf{C}$ d) $2\mathbf{A} - \mathbf{B} + 2\mathbf{C}$

13. Given $\mathbf{A} = [1 \;\; 0 \;\; 3 \;\; 4]$ and $\mathbf{B} = \begin{bmatrix} -2 \\ -1 \\ 8 \\ 0 \end{bmatrix}$

Determine **AB** and **BA**.

14. Given $\mathbf{A} = [2 \;\; 6 \;\; 5]$ and $\mathbf{B} = \begin{bmatrix} 1 \\ 3 \\ 0 \end{bmatrix}$

Determine **AB** and **BA**.

15. Given $\mathbf{A} = \begin{bmatrix} 0 & 2 \\ 4 & 6 \end{bmatrix}$, $\mathbf{B} = \begin{bmatrix} 2 & 5 \\ 4 & 3 \\ 1 & -2 \end{bmatrix}$, and $\mathbf{C} = \begin{bmatrix} 1 & 2 & -3 \\ 9 & 4 & -1 \end{bmatrix}$

Which of these matrix multiplications are valid?

a) **AB** b) **AC** c) **BA** d) **BC** e) **CA** f) **CB**

16. Given $\mathbf{A} = \begin{bmatrix} 1 & 0 & 3 \\ 2 & 5 & 0 \end{bmatrix}$, $\mathbf{B} = \begin{bmatrix} -1 & 3 \\ 0 & 2 \\ 7 & 0 \end{bmatrix}$, and $\mathbf{C} = \begin{bmatrix} 4 & 1 & 0 \\ 2 & 3 & 5 \\ 1 & 6 & 0 \end{bmatrix}$

Which of these matrix multiplications are valid?

a) **AB** b) **AC** c) **BA** d) **BC** e) **CA** f) **CB**

17. Carry out valid matrix multiplications from Exercise 3.2.15.

18. Carry out valid matrix multiplications from Exercise 3.2.16.

19. If $\mathbf{A} = \begin{bmatrix} 1 & 4 \\ 7 & 3 \\ 8 & -5 \end{bmatrix}$, what are the dimensions of zero matrices for these operations to be valid?

    a) $\mathbf{A} + \mathbf{0}$ b) $\mathbf{A0}$ c) $\mathbf{0A}$

20. Suppose **A** is an $(m \times n)$ matrix, **B** (3×6), and **C** $(6 \times r)$ matrix. What are the values of m, n, and r for **ABC** to be a (4×8) matrix?

21. If $\mathbf{A} = \begin{bmatrix} 3 & 0 & 1 \\ 0 & 1 & 2 \\ 1 & 2 & 4 \end{bmatrix}$

    a) Find $\mathbf{A}^2$ (i.e., **AA**)

    b) What must be true about **A** in order to repeatedly multiply itself?

22. Find the transposes of the matrices

    a) $\begin{bmatrix} 3 & 7 & 11 & -2 & 9 \\ 4 & 8 & -5 & 0 & 1 \end{bmatrix}$ b) $\begin{bmatrix} 1 & 8 & 3 \\ 4 & 0 & -1 \\ 3 & 5 & 2 \end{bmatrix}$.

23. a) If **A** is a column vector, describe $\mathbf{A}^t$.

    b) If **A** is a row vector, describe $\mathbf{A}^t$.

24. Smith and Jones are neighbors. Smith gets car gasoline, home heating oil, and propane gas for cooking from the same vendors as Jones. Smith buys 10 gallons of gasoline, 75 gallons of fuel oil, and seven gallons of propane weekly. Jones buys 15 gallons of gasoline, 56 gallons of fuel oil, and 9 gallons of propane weekly. The prices per gallon for these items are \$3.70, \$4.10, and \$3.10, respectively. Determine their weekly bills using a matrix product.

25. The USDA is required to release weekly dairy product sales information. Some prices from the National Dairy Products Sales Report are:

Item	May17, 2014	May 24, 2014	May 31, 2014	June 7, 2014	June 14, 2014
Cheddar cheese/lb	2.1603	2.1128	2.0516	2.0234	2.0172
Butter/lb	2.0554	2.1189	2.1261	2.1857	2.1836
Nonfat dry milk/lb	1.8554	1.8143	1.8901	1.8427	1.8624

   a) Use matrix multiplication to determine the price of 5 pounds of cheese, 3 pounds of butter, and 2 pounds of dry milk for the week ending May 24, 2014.

   b) Use a single matrix multiplication to determine the price of 4 pounds of cheese, 1 pound of butter, and 3 pounds of dry milk for each of the three weeks May 31, June 7, and June 14.

26. A professor computes final grades for his class by using the following weights: 15% for quizzes, 20% for each of three tests, and 25% for the final exam. Compute the grades for the following four students in the class using one set of matrices.
    Student 1: 80% homework; 85%, 75%, 90% on tests; and 95% on final
    Student 2: 60% homework; 95%, 85%, 80% on tests; and 90% on final
    Student 3: 95% homework; 85%, 85%, 95% on tests; and 75% on final
    Student 4: 75% homework; 100%, 70%, 90% on tests; and 85% on final

27. A college student took five courses in her first semester. The student received an A in Biology (4 credit course), an A− in Ice Skating (1 credit), a B− in Short Story (3 credits), a B in Psychology (3 credits), and a C+ in History (3 credits).
    Use A = 4, A− = 3.67, B+ = 3.33, B = 3.00, B− = 2.67, C+ = 2.33 to calculate grade point average (GPA). Use matrices for the student's quality points (for instance, the B− in Short Story earns the student 3(2.67) = 8 quality points). Then, determine the GPA.

28. Express these equations in matrix format:

a) $$\begin{aligned} x_1 - x_2 + 3x_3 &= 5 \\ 2x_1 + x_2 - 2x_3 &= 1 \\ x_1 + 2x_2 + 3x_3 &= 20 \end{aligned}$$

b) $$\begin{aligned} 4x + 3y + z &= 7 \\ 2x - y + 3z &= 1 \\ x - 3y + 7z &= -2 \end{aligned}$$

29. Express these equations in matrix format:

a) $$\begin{aligned} 3x_1 - 7x_2 + x_3 + 3x_4 &= 7 \\ 2x_1 + 3x_2 + x_3 - 2x_4 &= 0 \\ 3x_1 + x_2 + 3x_3 + 7x_4 &= 25 \\ x_1 + 5x_2 + 2x_3 - 8x_4 &= -19 \end{aligned}$$

b) $$\begin{aligned} w + 4x + 3y + z &= 8 \\ 2w + 3x - y + 5z &= 7 \\ w + x - 2y + 4z &= 3 \\ 3w + 2x - y + 3z &= 10 \end{aligned}$$

30. Write the systems of equations for these matrices.

a) $$\begin{bmatrix} 2 & 7 & 3 \\ 1 & 1 & -1 \\ 3 & 6 & -2 \end{bmatrix} \begin{bmatrix} x \\ y \\ z \end{bmatrix} = \begin{bmatrix} 31 \\ -2 \\ 5 \end{bmatrix}$$

b) $$\begin{bmatrix} -1 & 2 & 0 \\ 3 & 1 & 2 \\ 5 & 0 & 1 \end{bmatrix} \begin{bmatrix} x_1 \\ x_2 \\ x_3 \end{bmatrix} = \begin{bmatrix} 3 \\ 16 \\ 25 \end{bmatrix}.$$

31. Write the systems of equations for these matrices.

a) $$\begin{bmatrix} 1 & 5 & 2 & 2 \\ 1 & 1 & 3 & 0 \\ 2 & 3 & -2 & 5 \\ 3 & 1 & 0 & 2 \end{bmatrix} \begin{bmatrix} w \\ x \\ y \\ z \end{bmatrix} = \begin{bmatrix} 19 \\ 14 \\ 15 \\ 17 \end{bmatrix}$$

b) $$\begin{bmatrix} 3 & 1 & 4 & 1 \\ 1 & 1 & 3 & 2 \\ 1 & 2 & 0 & 5 \\ 2 & 3 & 1 & 1 \end{bmatrix} \begin{bmatrix} x_1 \\ x_2 \\ x_3 \\ x_4 \end{bmatrix} = \begin{bmatrix} 12 \\ 13 \\ 17 \\ 8 \end{bmatrix}$$

32. Furnish augmented matrices for

a) Exercise 3.2.30a

b) Exercise 3.2.30b

33. Furnish augmented matrices for

a) Exercise 3.2.31a

b) Exercise 3.2.31b.

34. If the coefficient matrix is $\begin{bmatrix} 6 & 3 & 7 & 1 \\ 1 & 2 & 0 & 4 \\ -2 & 3 & 4 & -1 \\ 0 & 0 & 1 & -2 \end{bmatrix}$ and the solution to the system is $\begin{bmatrix} w \\ x \\ y \\ z \end{bmatrix} = \begin{bmatrix} -3 \\ 1 \\ 0 \\ 1 \end{bmatrix}$, express as a system of equations.

3.3 GAUSS–JORDAN ELIMINATION

A system of linear equations is said to be **homogeneous** if its right-hand sides are zero. That is,

$$\begin{aligned} a_{11}x_1 + a_{12}x_2 + \cdots + a_{1n}x_n &= 0 \\ a_{21}x_1 + a_{22}x_2 + \cdots + a_{2n}x_n &= 0 \\ a_{31}x_1 + a_{32}x_2 + \cdots + a_{3n}x_n &= 0 \\ \vdots \qquad \vdots \qquad\quad \vdots \quad &\ \ \vdots \\ a_{m1}x_1 + a_{m2}x_2 + \cdots + a_{mn}x_n &= 0 \end{aligned}$$

Every homogeneous system of linear equations is consistent, since there is always the **trivial** solution $x_1 = 0, x_2 = 0, \ldots, x_n = 0$. Any other solutions are called **nontrivial solutions**. For every homogeneous system, exactly one of the following statements is true:

1. *The only solution is the trivial solution.*
2. *There are infinitely many nontrivial solutions in addition to the trivial one.*

When the number of variables exceeds the number of equations, homogeneous systems of linear equations have infinitely many nontrivial solutions in addition to the trivial one.

Example 3.3.1 Homogeneous Systems

By inspection, which of the following homogeneous systems have nontrivial solutions?

a) $\begin{aligned} x_1 + 2x_2 + 3x_3 + x_4 &= 0 \\ 5x_1 - 3x_2 + x_3 - 2x_4 &= 0 \\ 5x_1 + 7x_2 - 2x_3 - x_4 &= 0 \end{aligned}$ *b)* $\begin{aligned} x_1 + x_2 &= 0 \\ 4x_1 + 4x_2 &= 0 \end{aligned}$ *c)* $\begin{aligned} x_1 + 2x_2 + x_3 &= 0 \\ x_1 + \quad\;\; 2x_3 &= 0 \\ 5x_3 &= 0 \end{aligned}$

Solution:

a) In this system of equations, there are four variables (unknowns) and only three equations. There are infinitely many nontrivial solutions.

⟶

b) *There appear to be two equations. However, one is a multiple of the other. Hence, there is only one independent equation in two variables. There are infinitely many nontrivial solutions.*

c) *There are three equations in three variables. The third equation requires* $x_3 = 0$. *It follows that* x_1 *and* x_2 *are also zero. The only solution is the trivial one.*

Sometimes, it is easier to solve a system of linear equations by replacing it by a new and less complicated **equivalent** system having the same solution. Three **elementary row operations** used to simplify the system of equations are listed as follows:

1. *Equations can be multiplied by any nonzero quantity.*
2. *Pairs of equations may be interchanged.*
3. *A multiple of one equation can be added to another equation.*

Elementary row operations can be made on augmented matrices since their rows correspond to the system of equations.

Elementary Row Operations

1. **Any row can be multiplied by c, a nonzero scalar. ($R_i \rightarrow cR_i$)**
2. **Any two rows can be interchanged, ($R_i \leftrightarrow R_j$),**
3. **A multiple of one row can be added to another row, ($R_i \rightarrow R_i + cR_j$),**

R_i denotes "row i," R_j denotes "row j"; and arrows ($\rightarrow$) are read as "becomes."

An $(n \times n)$ matrix is called an **elementary matrix** if it can be obtained from an $(n \times n)$ identity matrix, $(\mathbf{I}_n)$, by a single elementary row operation.

Example 3.3.2 Elementary Row Operations

Perform the indicated elementary row operations on $\mathbf{I}_n$ *to form an elementary matrix.*

a) On $\mathbf{I}_3$, $R_2 \rightarrow 4R_2$ b) On $\mathbf{I}_5$, $R_3 \leftrightarrow R_4$ c) On $\mathbf{I}_4$, $R_1 \rightarrow R_1 + 5R_3$

Solution:

a) *In this case, the* (3×3) *identity matrix has its second row replaced by four times itself. Therefore,*

$$\begin{bmatrix} 1 & 0 & 0 \\ 0 & 4 & 0 \\ 0 & 0 & 1 \end{bmatrix}.$$

$\longrightarrow$

b) The third and fourth rows are interchanged in a (5 × 5) identity matrix.

$$\begin{bmatrix} 1 & 0 & 0 & 0 & 0 \\ 0 & 1 & 0 & 0 & 0 \\ 0 & 0 & 0 & 1 & 0 \\ 0 & 0 & 1 & 0 & 0 \\ 0 & 0 & 0 & 0 & 1 \end{bmatrix}$$

c) In this case, a (4 × 4) identity matrix has its first row replaced by the sum of the first row and five times the third row.

$$\begin{bmatrix} 1 & 0 & 5 & 0 \\ 0 & 1 & 0 & 0 \\ 0 & 0 & 1 & 0 \\ 0 & 0 & 0 & 1 \end{bmatrix}.$$

A **submatrix** is a matrix that is formed by deleting some rows and/or columns from a larger matrix.

Example 3.3.3 Submatrices

Let $\mathbf{A} = \begin{bmatrix} 1 & 2 & 3 \\ 4 & 5 & 6 \\ 7 & 8 & 9 \end{bmatrix}$.

Determine the resulting submatrices when these operations are applied:

a) The second row is eliminated.
b) The first and third columns are eliminated.
c) The first row and second column are eliminated.

Solution:

a) After eliminating the second row, the resulting submatrix is $\begin{bmatrix} 1 & 2 & 3 \\ 7 & 8 & 9 \end{bmatrix}$.

b) After eliminating the first and third columns, the resulting submatrix is the second column only, or $\begin{bmatrix} 2 \\ 5 \\ 8 \end{bmatrix}$.

c) After eliminating the first row and second column, the resulting submatrix is $\begin{bmatrix} 4 & 6 \\ 7 & 9 \end{bmatrix}$.

When solving systems of linear equations, it is useful to apply row operations to form a matrix in a particular format called **reduced row echelon** to determine the solution (if it exists) to the system. Each row operation must be applied to the matrix that results from the previous row operation.

Row Echelon Matrices

Properties of Reduced Row Echelon Matrices

1. If a row does not consist entirely of zeros, then the first nonzero entry in the row is 1 (called a "leading" one).

2. Rows that consist only of zeros are placed at the bottom of the matrix.

3. In any two successive rows that do not consist entirely of zeros, the leading one in the lower row occurs farther to the right than the leading one in any higher row.

4. Each column containing a leading one has zeros elsewhere.

Matrices having properties 1, 2, and 3 are in *row echelon form*.

Example 3.3.4 Reduced Row Echelon Properties

Identify properties of reduced row echelon form that these matrices lack.

$$a)\ \begin{bmatrix} 1 & 0 & 0 & 0 \\ 0 & 0 & 0 & 0 \\ 0 & 0 & 1 & 0 \\ 0 & 0 & 0 & 1 \end{bmatrix} \qquad c)\ \begin{bmatrix} 1 & 0 & 0 & 2 \\ 0 & 0 & 1 & 1 \\ 0 & 1 & 0 & 3 \end{bmatrix}$$

$$b)\ \begin{bmatrix} 1 & 0 & 2 & 0 \\ 0 & 1 & 3 & 1 \\ 0 & 0 & 1 & 1 \end{bmatrix} \qquad d)\ \begin{bmatrix} 2 & 0 & 0 & 0 \\ 0 & 1 & 0 & 0 \\ 0 & 0 & 1 & 0 \\ 0 & 0 & 0 & 1 \end{bmatrix}$$

Solution:

a) The row of zeros should be the last row of the matrix. This matrix violates the second property.

b) The third row has a leading one but zeros are missing elsewhere in that third column. This matrix violates the fourth property.

c) The second and third rows should be interchanged, so the leading 1 is farther to the right in each row. This matrix violates third property.

d) In the first row, the first nonzero entry is a 2 and it should be a 1. This matrix violates the first property.

Now, we have the tools to properly consider **Gauss–Jordan elimination**. It is a step-by-step procedure to solve a system of linear equations by rearranging its associated augmented matrix in reduced row echelon form as illustrated in Example 3.3.5.

Gauss–Jordan Elimination

1. **Obtain the augmented matrix for the system of equations so that the first entry is not zero. (If necessary, interchange the top row with another row.)**
2. **Multiply the first row by the reciprocal of the first entry to yield a leading 1.**
3. **Add (subtract) suitable multiples of the top row with the leading 1 to each succeeding row so that all other entries in that first column are zero.**
4. **Repeat the above steps for the submatrix formed by deleting the first row and column of the matrix.**
5. **Continue this procedure until the entire matrix is in *row echelon form*.**
6. **Beginning with the last nonzero row, "back substitute" adding suitable multiples of each row to the rows above it to create zeros above each leading 1. The matrix is then in *reduced row echelon form*.**

Note: with practice Steps 3–6 may be combined by causing zeros for all entries above and below the leading 1 without the matrix being in row echelon form.

Example 3.3.5 Gauss–Jordan Elimination

Use Gauss–Jordan elimination to solve the system of equations

$$\begin{aligned} 2x_1 - 8x_2 + 4x_3 &= -6 \\ 5x_1 - 2x_2 - 3x_3 &= 8 \\ 2x_1 + 3x_2 + 5x_3 &= 17 \end{aligned}$$

Solution:
First, write the augmented matrix:

$$\left[\begin{array}{ccc|c} 2 & -8 & 4 & -6 \\ 5 & -2 & -3 & 8 \\ 2 & 3 & 5 & 17 \end{array}\right]$$

No row interchanges are necessary as the far left column does not consist entirely of zeros and the lead entry (top row) is not zero. To have the lead entry become a leading 1, multiply the first row by 1/2. (The current row operation is indicated in brackets at the right.)

$$\left[\begin{array}{ccc|c} 2 & -8 & 4 & -6 \\ 5 & -2 & -3 & 8 \\ 2 & 3 & 5 & 17 \end{array}\right] \to \left[\begin{array}{ccc|c} 1 & -4 & 2 & -3 \\ 5 & -2 & -3 & 8 \\ 2 & 3 & 5 & 17 \end{array}\right] \quad [R_1 \to (1/2)R_1]$$

Now, we seek zeros in the first column below the leading 1 as shown in the next two steps.

$$\left[\begin{array}{ccc|c} 1 & -4 & 2 & -3 \\ 5 & -2 & -3 & 8 \\ 2 & 3 & 5 & 17 \end{array}\right] \to \left[\begin{array}{ccc|c} 1 & -4 & 2 & -3 \\ 0 & 18 & -13 & 23 \\ 2 & 3 & 5 & 17 \end{array}\right] \quad [R_2 \to R_2 - 5R_1]$$

⟶

$$\begin{bmatrix} 1 & -4 & 2 & | & -3 \\ 0 & 18 & -13 & | & 23 \\ 2 & 3 & 5 & | & 17 \end{bmatrix} \rightarrow \begin{bmatrix} 1 & -4 & 2 & | & -3 \\ 0 & 18 & -13 & | & 23 \\ 0 & 11 & 1 & | & 23 \end{bmatrix} \quad [R_3 \rightarrow R_3 - 2R_1]$$

(Some of the matrix rewriting will not be needed after a little practice.)

Now, the first row and column are shaded to reveal a submatrix. Its lead entry, 18, is to become a leading 1. The entry beneath the leading 1, an 11 in the next row, becomes zero using appropriate row operations.

$$\begin{bmatrix} 1 & -4 & 2 & | & -3 \\ 0 & 18 & -13 & | & 23 \\ 0 & 11 & 1 & | & 23 \end{bmatrix} \rightarrow \begin{bmatrix} 1 & -4 & 2 & | & -3 \\ 0 & 1 & -13/18 & | & 23/18 \\ 0 & 11 & 1 & | & 23 \end{bmatrix} \quad [R_2 \rightarrow (1/18)R_2]$$

$$\begin{bmatrix} 1 & -4 & 2 & | & -3 \\ 0 & 1 & -13/18 & | & 23/18 \\ 0 & 11 & 1 & | & 23 \end{bmatrix} \rightarrow \begin{bmatrix} 1 & -4 & 2 & | & -3 \\ 0 & 1 & -13/18 & | & 23/18 \\ 0 & 0 & 161/18 & | & 161/18 \end{bmatrix} \quad [R_3 \rightarrow R_3 - 11R_2]$$

Now, a leading 1 is needed in the bottom row.

$$\begin{bmatrix} 1 & -4 & 2 & | & -3 \\ 0 & 1 & -13/18 & | & 23/18 \\ 0 & 0 & 161/18 & | & 161/18 \end{bmatrix} \rightarrow \begin{bmatrix} 1 & -4 & 2 & | & -3 \\ 0 & 1 & -13/18 & | & 23/18 \\ 0 & 0 & 1 & | & 1 \end{bmatrix} \quad [R_3 \rightarrow (18/161)R_3]$$

The last matrix is in row echelon form. It is to be altered to a reduced row echelon form by eliminating nonzero entries above the main diagonal.

$$\begin{bmatrix} 1 & -4 & 2 & | & -3 \\ 0 & 1 & -13/18 & | & 23/18 \\ 0 & 0 & 1 & | & 1 \end{bmatrix} \rightarrow \begin{bmatrix} 1 & -4 & 2 & | & -3 \\ 0 & 1 & 0 & | & 2 \\ 0 & 0 & 1 & | & 1 \end{bmatrix} \quad [R_2 \rightarrow R_2 + (13/18)R_3]$$

$$\begin{bmatrix} 1 & -4 & 2 & | & -3 \\ 0 & 1 & 0 & | & 2 \\ 0 & 0 & 1 & | & 1 \end{bmatrix} \rightarrow \begin{bmatrix} 1 & -4 & 0 & | & -5 \\ 0 & 1 & 0 & | & 2 \\ 0 & 0 & 1 & | & 1 \end{bmatrix} \quad [R_1 \rightarrow R_1 - 2R_3]$$

$$\begin{bmatrix} 1 & -4 & 0 & | & -5 \\ 0 & 1 & 0 & | & 2 \\ 0 & 0 & 1 & | & 1 \end{bmatrix} \rightarrow \begin{bmatrix} 1 & 0 & 0 & | & 3 \\ 0 & 1 & 0 & | & 2 \\ 0 & 0 & 1 & | & 1 \end{bmatrix} \quad [R_1 \rightarrow R_1 + 4R_2]$$

The right-hand column in the final reduced row echelon matrix yields the solution $x_1 = 3$, $x_2 = 2$, *and* $x_3 = 1$.

It is wise to check that the solution (3, 2, 1) is correct by substitution into the original system.

Example 3.3.6 Gauss–Jordan Elimination (Word Problem)

A scouting merit badge in stamp collecting requires at least 250 different stamps from at least 15 countries. A scout displays a favorite stamp from each of 15 countries. An observer noted that only 10-, 5-, or 1-cent stamps were displayed having a total face value of 95 cents. Also, there are twice as many 10- and 5-cent stamps combined as 1-cent stamps. How many are there of each denomination? (use Gauss–Jordan elimination.)

Solution:

Let x_1 = number of 10-cent stamps, x_2 = number of 5-cent stamps, and x_3 = number of 1-cent stamps

Three equations determine the value of the variables. First, since there are 15 stamps, we have

$$x_1 + x_2 + x_3 = 15$$

The total of the stamp denominations is the second equation

$$10x_1 + 5x_2 + x_3 = 95$$

The third equation comes from the relationship among the number of stamps: The number of 1-cent stamps is doubled to equal the number of the others combined.

$x_1 + x_2 = 2x_3$ or $x_1 + x_2 - 2x_3 = 0$

The system to solve is

$$\begin{aligned} x_1 + x_2 + x_3 &= 15 \\ 10x_1 + 5x_2 + x_3 &= 95 \\ x_1 + x_2 - 2x_3 &= 0 \end{aligned}$$

The augmented matrix format is

$$\left[\begin{array}{ccc|c} 1 & 1 & 1 & 15 \\ 10 & 5 & 1 & 95 \\ 1 & 1 & -2 & 0 \end{array}\right]$$

There is a leading 1 in the top row so zeros must appear in the first column beneath the leading 1 (as in the next two steps).

$$\left[\begin{array}{ccc|c} 1 & 1 & 1 & 15 \\ 10 & 5 & 1 & 95 \\ 1 & 1 & -2 & 0 \end{array}\right] \to \left[\begin{array}{ccc|c} 1 & 1 & 1 & 15 \\ 0 & -5 & -9 & -55 \\ 1 & 1 & -2 & 0 \end{array}\right] \quad [R_2 \to R_2 - 10R_1]$$

$$\left[\begin{array}{ccc|c} 1 & 1 & 1 & 15 \\ 0 & -5 & -9 & -55 \\ 1 & 1 & -2 & 0 \end{array}\right] \to \left[\begin{array}{ccc|c} 1 & 1 & 1 & 15 \\ 0 & -5 & -9 & -55 \\ 0 & 0 & -3 & -15 \end{array}\right] \quad [R_3 \to R_3 - R_1]$$

The bottom row requires a leading 1.

$$\left[\begin{array}{ccc|c} 1 & 1 & 1 & 15 \\ 0 & -5 & -9 & -55 \\ 0 & 0 & -3 & -15 \end{array}\right] \to \left[\begin{array}{ccc|c} 1 & 1 & 1 & 15 \\ 0 & -5 & -9 & -55 \\ 0 & 0 & 1 & 5 \end{array}\right] \quad [R_3 \to (-1/3)R_3]$$

$\longrightarrow$

The bottom row now indicates five 1-cent stamps. Continuing, form the reduced row echelon matrix and solutions as:

$$\left[\begin{array}{ccc|c} 1 & 1 & 1 & 15 \\ 0 & -5 & -9 & -55 \\ 0 & 0 & 1 & 5 \end{array}\right] \to \left[\begin{array}{ccc|c} 1 & 1 & 1 & 15 \\ 0 & -5 & 0 & -10 \\ 0 & 0 & 1 & 5 \end{array}\right] \quad [R_2 \to R_2 + 9R_3]$$

$$\left[\begin{array}{ccc|c} 1 & 1 & 1 & 15 \\ 0 & -5 & 0 & -10 \\ 0 & 0 & 1 & 5 \end{array}\right] \to \left[\begin{array}{ccc|c} 1 & 1 & 1 & 15 \\ 0 & 1 & 0 & 2 \\ 0 & 0 & 1 & 5 \end{array}\right] \quad [R_2 \to (-1/5)R_2]$$

$$\left[\begin{array}{ccc|c} 1 & 1 & 1 & 15 \\ 0 & 1 & 0 & 2 \\ 0 & 0 & 1 & 5 \end{array}\right] \to \left[\begin{array}{ccc|c} 1 & 1 & 0 & 10 \\ 0 & 1 & 0 & 2 \\ 0 & 0 & 1 & 5 \end{array}\right] \quad [R_1 \to R_1 - R_3]$$

$$\left[\begin{array}{ccc|c} 1 & 1 & 0 & 10 \\ 0 & 1 & 0 & 2 \\ 0 & 0 & 1 & 5 \end{array}\right] \to \left[\begin{array}{ccc|c} 1 & 0 & 0 & 8 \\ 0 & 1 & 0 & 2 \\ 0 & 0 & 1 & 5 \end{array}\right] \quad [R_1 \to R_1 - R_2]$$

There are eight 10-cent stamps, two 5-cent stamps, and five 1-cent stamps in the display.

Recall from Chapter 1 that the solution to a system of two linear equations in two variables (x_1 and x_2) is in one of these three categories:

1. The system is consistent and independent (lines intersect at a single point).
2. The system is consistent and dependent (lines are collinear and there are an infinite number of solutions that lie on the line).
3. The system is inconsistent (the lines are parallel) and there is no solution.

When solving these systems, using Gauss–Jordan elimination the final matrices may have one of these formats:

1. $\left[\begin{array}{cc|c} 1 & 0 & r \\ 0 & 1 & s \end{array}\right]$ when (r, s) is the solution.
2. $\left[\begin{array}{cc|c} 1 & p & q \\ 0 & 0 & 0 \end{array}\right]$ when $x_1 + px_2 = q$ so x_1 depends on the choice of x_2.
3. $\left[\begin{array}{cc|c} 1 & p & q \\ 0 & 0 & r \end{array}\right]$. A row such as $[0 \quad 0 \cdots 0|r]$ indicates that the system has no solution provided $r \neq 0$, since $0x_1 + 0x_2 + \cdots + 0x_n = r$ cannot be satisfied.

These are illustrated in Examples 3.3.7 and 3.3.8.

Example 3.3.7 A System with No Solution

Use Gauss–Jordan elimination to show the following system of equations has no solution.

$$\begin{aligned} x_1 + \ \ x_2 + \ \ x_3 &= 5 \\ x_1 + 2x_2 + 3x_3 &= 4 \\ 3x_1 + 2x_2 + \ \ x_3 &= 8 \end{aligned}$$

Solution:

First, write the augmented matrix. Next, perform the following operations $R_2 \to R_2 - R_1$ and $R_3 \to R_3 - 3R_1$ to yield

$$\left[\begin{array}{ccc|c} 1 & 1 & 1 & 5 \\ 1 & 2 & 3 & 4 \\ 3 & 2 & 1 & 8 \end{array}\right] \to \left[\begin{array}{ccc|c} 1 & 1 & 1 & 5 \\ 0 & 1 & 2 & -1 \\ 0 & -1 & -2 & -7 \end{array}\right]$$

Next, perform the following operation $R_3 \to R_3 + R_2$

$$\left[\begin{array}{ccc|c} 1 & 1 & 1 & 5 \\ 0 & 1 & 2 & -1 \\ 0 & 0 & 0 & -8 \end{array}\right]$$

The bottom row has $0 = -8$; impossible! The system has no solution.

Example 3.3.8 A System with Infinite Solutions

Use Gauss–Jordan elimination to show the following system of equations has an infinity of solutions.

$$\begin{aligned} x_1 + x_2 - \ \ x_3 + \ \ x_4 &= \ \ 6 \\ x_2 + \ \ x_3 + \ \ x_4 &= \ \ 3 \\ x_1 \qquad - 2x_3 + 2x_4 &= 13 \\ 2x_1 - x_2 - 5x_3 - \ \ x_4 &= \ \ 3 \end{aligned}$$

Solution:

First, write the augmented matrix. Next, perform the following operations $R_3 \to R_3 - R_1$ and $R_4 \to R_4 - 2R_1$ to yield

$$\left[\begin{array}{cccc|c} 1 & 1 & -1 & 1 & 6 \\ 0 & 1 & 1 & 1 & 3 \\ 1 & 0 & -2 & 2 & 13 \\ 2 & -1 & -5 & -1 & 3 \end{array}\right] \to \left[\begin{array}{cccc|c} 1 & 1 & -1 & 1 & 6 \\ 0 & 1 & 1 & 1 & 3 \\ 0 & -1 & -1 & 1 & 7 \\ 0 & -3 & -3 & -3 & -9 \end{array}\right]$$

Next, perform the following operations $R_3 \to R_3 + R_2$ and $R_4 \to R_4 + 3R_2$ to yield

$\longrightarrow$

$$\left[\begin{array}{cccc|c} 1 & 1 & -1 & 1 & 6 \\ 0 & 1 & 1 & 1 & 3 \\ 0 & 0 & 0 & 2 & 10 \\ 0 & 0 & 0 & 0 & 0 \end{array}\right]$$

Since there are three rows that are not all zeros and there are four unknowns, there are an infinite number of solutions to this system (more variables than usable equations). Continuing with row operations yields

$$\left[\begin{array}{cccc|c} 1 & 1 & -1 & 1 & 6 \\ 0 & 1 & 1 & 1 & 3 \\ 0 & 0 & 0 & 1 & 5 \\ 0 & 0 & 0 & 0 & 0 \end{array}\right] \rightarrow \left[\begin{array}{cccc|c} 1 & 1 & -1 & 0 & 1 \\ 0 & 1 & 1 & 0 & -2 \\ 0 & 0 & 0 & 1 & 5 \\ 0 & 0 & 0 & 0 & 0 \end{array}\right]$$

$$\rightarrow \left[\begin{array}{cccc|c} 1 & 0 & -2 & 0 & 3 \\ 0 & 1 & 1 & 0 & -2 \\ 0 & 0 & 0 & 1 & 5 \\ 0 & 0 & 0 & 0 & 0 \end{array}\right]$$

The solutions to the system must satisfy $x_4 = 5$, $x_1 - 2x_3 = 3$, *and* $x_2 + x_3 = -2$, *and* x_3 *can be any real number. Setting* x_3 *to an arbitrary number gives a possible solution to the system. For example, if* $x_3 = 0$ *then* $(3, -2, 0, 5)$ *is a solution, and if* $x_3 = 4$ *then* $(11, -6, 4, 5)$ *is another solution, and so on.*

EXERCISES 3.3

1. Solve the following homogeneous systems using Gauss–Jordan elimination.

a) $\begin{aligned} 2x_1 + 3x_2 &= 0 \\ x_1 + 4x_2 &= 0 \end{aligned}$

b) $\begin{aligned} x_1 + 3x_2 &= 0 \\ 2x_1 + 6x_2 &= 0 \end{aligned}$

2. Repeat Exercise 3.3.1 for

a) $\begin{aligned} x_1 + 2x_2 + 3x_3 &= 0 \\ 2x_1 + x_2 \qquad &= 0 \\ x_2 + 3x_3 &= 0 \end{aligned}$

b) $\begin{aligned} x_1 + x_2 + x_3 + 2x_4 &= 0 \\ 5x_1 + 3x_2 + 5x_3 - x_4 &= 0 \end{aligned}$

3. Which of these are elementary matrices?

a) $\begin{bmatrix} 1 & 0 & 0 \\ 0 & 9 & 0 \\ 0 & 0 & 1 \end{bmatrix}$ b) $\begin{bmatrix} 1 & 0 & 0 \\ 5 & 1 & 0 \\ 0 & 0 & 2 \end{bmatrix}$ c) $\begin{bmatrix} 1 & 0 & 0 & 0 \\ 0 & 1 & 3 & 0 \\ 0 & 0 & 1 & 0 \\ 0 & 0 & 0 & 1 \end{bmatrix}$ d) $\begin{bmatrix} 1 & 0 & 0 \\ 0 & 1 & 0 \\ 0 & -8 & 1 \end{bmatrix}$ e) $\begin{bmatrix} 0 & 1 \\ 1 & 0 \end{bmatrix}$

4. Perform the indicated elementary row operations on the identity matrices.

a) On $\mathbf{I}_4$, $R_3 \to 4R_3$ b) On $\mathbf{I}_3$, $R_2 \leftrightarrow R_1$ c) On $\mathbf{I}_5$, $R_2 \to R_2 + 6R_4$

5. If $\mathbf{A} = \begin{bmatrix} 10 & 3 & 7 & 5 \\ 12 & 8 & 1 & 0 \\ 7 & -4 & 2 & 6 \end{bmatrix}$, determine submatrices that result by eliminating these rows and/or columns.

a) Eliminate the third row.

b) Eliminate the second and fourth columns.

c) Eliminate the second row and second column.

d) Eliminate the first and third columns.

6. Identify matrices in reduced row echelon form.

a) $\begin{bmatrix} 1 & 0 & 0 & 0 \\ 0 & 1 & 0 & 0 \\ 0 & 0 & 1 & 0 \\ 0 & 0 & 0 & 0 \end{bmatrix}$ b) $\begin{bmatrix} 1 & 0 & 2 & 0 \\ 0 & 1 & 0 & 0 \\ 0 & 0 & 0 & 0 \\ 0 & 0 & 0 & 0 \end{bmatrix}$ c) $\begin{bmatrix} 1 & 0 & 0 & 2 \\ 0 & 0 & 1 & 1 \\ 0 & 1 & 0 & 4 \end{bmatrix}$ d) $\begin{bmatrix} 1 & 0 & 0 \\ 0 & 2 & 0 \\ 0 & 0 & 1 \end{bmatrix}$

7. Identify matrices in row echelon form.

a) $\begin{bmatrix} 1 & 5 & 0 \\ 0 & 1 & 0 \\ 0 & 0 & 0 \end{bmatrix}$ b) $\begin{bmatrix} 1 & 3 & 4 \\ 0 & 1 & 0 \\ 0 & 0 & 1 \end{bmatrix}$ c) $\begin{bmatrix} 1 & 0 & 0 & 3 \\ 0 & 0 & 1 & 2 \\ 0 & 1 & 0 & 5 \end{bmatrix}$ d) $\begin{bmatrix} 1 & 0 & 0 \\ 0 & 0 & 0 \\ 0 & 0 & 1 \end{bmatrix}$

In Exercises 8–23, solve using Gauss–Jordan elimination

8. $$\begin{aligned} x_1 - x_2 &= 2 \\ 2x_1 + 3x_2 &= 19 \end{aligned}$$

9. $$\begin{aligned} 2x_1 + x_2 &= 7 \\ 4x_1 + x_2 &= 9 \end{aligned}$$

10. $$\begin{aligned} x_1 + 2x_2 &= 4 \\ -x_1 + x_2 &= -1 \end{aligned}$$

11. $$\begin{aligned} x_1 + 2x_2 &= 8 \\ 2x_1 + 4x_2 &= 14 \end{aligned}$$

12. $$\begin{aligned} x_1 + x_2 + 2x_3 &= 9 \\ x_1 + 2x_2 + 2x_3 &= 11 \\ -x_1 - x_2 + x_3 &= 0 \end{aligned}$$

13. $$\begin{aligned} x_1 - x_2 - 2x_3 &= -10 \\ 3x_1 + 2x_2 - x_3 &= 0 \\ 5x_1 + 3x_2 + x_3 &= 16 \end{aligned}$$

14. $$\begin{aligned} x_1 + x_2 + x_3 &= 6 \\ x_1 - x_2 + x_3 &= 2 \\ 2x_1 + x_2 - 3x_3 &= -5 \end{aligned}$$

15. $$\begin{aligned} x_1 + 2x_2 - 3x_3 &= 5 \\ 3x_1 - x_2 + x_3 &= 0 \\ 2x_1 - x_2 + 3x_3 &= 3 \end{aligned}$$

16. $$\begin{aligned} x_1 + x_2 + x_3 + x_4 &= 4 \\ 2x_1 + x_2 - x_3 + x_4 &= 6 \\ x_1 - 2x_2 + 3x_3 + 2x_4 &= -1 \\ x_1 \qquad\qquad + x_4 &= 3 \end{aligned}$$

17. $$\begin{aligned} x_1 + x_2 + x_3 + x_4 &= 1 \\ x_1 - x_2 - 2x_3 &= 0 \\ 3x_1 + x_2 + 3x_3 - x_4 &= 3 \\ 2x_1 - x_2 + 2x_3 + 2x_4 &= 8 \end{aligned}$$

18. An investor has \$25,000 in two stocks with respective 4% and 5% yearly returns. If the investments gained \$1100 for the year, how much was invested in each stock?

19. Advance show tickets for children are \$15 and adult tickets are \$20. Tickets sold at the door are \$25 each for adults or children. The office reports that 1000 tickets were sold and \$18,500 was realized. The number of tickets sold at the door was 1/4 the number of advance sale children's tickets. How many adult tickets were sold in advance?

20. The sum of three numbers is 10. The sum of the smallest and largest is two more than the third number. The largest number is five times the smallest. Find the numbers.

21. A local theater sells children's matinee tickets for \$4 and adult tickets for \$5. The theater sold 2400 tickets during the weekend and realized revenue of \$10,500. How many of each type ticket was sold?

22. A person has 12 coins; quarters, halves, and dollars that total \$6.50. There are two more 50-cent pieces than dollars and one-half as many quarters as 50-cent pieces and dollars combined. How many of each type of coin are there?

23. The table lists calories, carbohydrates, and fats per ounce for three foods. How many ounces of each food should be combined to provide 525 calories, 133 grams of carbohydrates, and 9 grams of fat?

	Calories	Carbohydrates	Fat
Food A	50	20	0
Food B	100	25	2
Food C	125	18	3

3.4 MATRIX INVERSION AND INPUT–OUTPUT ANALYSIS

Any square matrix, **A**, is said to have an **inverse** if there is another square matrix, **B**, such that **AB** = **BA** = **I**, an identity matrix. The inverse matrix **B** having this property is written as $\mathbf{A}^{-1}$.

Example 3.4.1 *Inverse of a Matrix*

*Decide whether **A** and **B** are inverses by computing **AB** and **BA**.*

$$\mathbf{A} = \begin{bmatrix} 2 & 0 & 1 \\ 3 & 1 & -1 \\ 2 & -2 & 4 \end{bmatrix} \quad \text{and } \mathbf{B} = \begin{bmatrix} -1/2 & 1/2 & 1/4 \\ 7/2 & -3/2 & -5/4 \\ 2 & -1 & -1/2 \end{bmatrix}$$

Solution:

*If **B** is the inverse of **A**, we must have **AB** = **BA** = **I**. Computing,*

$$\mathbf{AB} = \begin{bmatrix} 2 & 0 & 1 \\ 3 & 1 & -1 \\ 2 & -2 & 4 \end{bmatrix} \begin{bmatrix} -1/2 & 1/2 & 1/4 \\ 7/2 & -3/2 & -5/4 \\ 2 & -1 & -1/2 \end{bmatrix} = \begin{bmatrix} 1 & 0 & 0 \\ 0 & 1 & 0 \\ 0 & 0 & 1 \end{bmatrix} = \mathbf{I}$$

→

Alternatively, we could have verified that $\mathbf{BA} = \mathbf{I}$. *That is,*

$$\mathbf{BA} = \begin{bmatrix} -1/2 & 1/2 & 1/4 \\ -7/2 & -3/2 & -5/4 \\ 2 & -1 & -1/2 \end{bmatrix} \begin{bmatrix} 2 & 0 & 1 \\ 3 & 1 & -1 \\ 2 & -2 & 4 \end{bmatrix} = \begin{bmatrix} 1 & 0 & 0 \\ 0 & 1 & 0 \\ 0 & 0 & 1 \end{bmatrix} = \mathbf{I}$$

Since $\mathbf{AB}$ *and* $\mathbf{BA}$ *both equal* $\mathbf{I}$, *we conclude that* $\mathbf{B} = \mathbf{A}^{-1}$ *(or* $\mathbf{A} = \mathbf{B}^{-1}$*).*

To find the inverse of a square matrix, **A**, first form the augmented matrix [**A**|**I**] with **A** on the left and an identity matrix, **I**, of the same dimension, on the right. A vertical line to aid in identification of operations performed on the individual matrices separates the two matrices. The vertical line is not intended to replace equal signs but merely for convenience of having the two relevant matrices at hand.

The same types of operations prescribed for Gauss–Jordan elimination are used to transform a matrix [**A**|**I**] into the matrix [**I**|**B**] with $\mathbf{B} = \mathbf{A}^{-1}$, the desired inverse of the matrix **A**.

Example 3.4.2 Matrix Inversion

Find the inverse of $\mathbf{A} = \begin{bmatrix} 1 & 2 & 4 \\ 1 & 3 & 6 \\ 2 & 6 & 10 \end{bmatrix}$, *if it exists.*

Solution:

The starting augmented matrix is

$$\left[\begin{array}{ccc|ccc} 1 & 2 & 4 & 1 & 0 & 0 \\ 1 & 3 & 6 & 0 & 1 & 0 \\ 2 & 6 & 10 & 0 & 0 & 1 \end{array}\right]$$

Next, transform the first column so zeros appear below the leading "1."

$$\left[\begin{array}{ccc|ccc} 1 & 2 & 4 & 1 & 0 & 0 \\ 0 & 1 & 2 & -1 & 1 & 0 \\ 2 & 6 & 10 & 0 & 0 & 1 \end{array}\right] \qquad [R_2 \rightarrow R_2 - R_1]$$

$$\left[\begin{array}{ccc|ccc} 1 & 2 & 4 & 1 & 0 & 0 \\ 0 & 1 & 2 & -1 & 1 & 0 \\ 0 & 2 & 2 & -2 & 0 & 1 \end{array}\right] \qquad [R_3 \rightarrow R_3 - 2R_1]$$

Next, adjust the second column to produce a zero below the "1" on the main diagonal. Beginning with the bottom row

$$\left[\begin{array}{ccc|ccc} 1 & 2 & 4 & 1 & 0 & 0 \\ 0 & 1 & 2 & -1 & 1 & 0 \\ 0 & 0 & -2 & 0 & -2 & 1 \end{array}\right] \qquad [R_3 \rightarrow R_3 - 2R_2]$$

$\longrightarrow$

A leading 1 is introduced in the third row and third column by multiplying by (−1/2)

$$\left[\begin{array}{ccc|ccc} 1 & 2 & 4 & 1 & 0 & 0 \\ 0 & 1 & 2 & -1 & 1 & 0 \\ 0 & 0 & 1 & 0 & 1 & -1/2 \end{array}\right] \quad [R_3 \rightarrow (-1/2)R_3]$$

To have this row echelon form become a reduced row echelon form, zeros are needed above each "1" on the main diagonal.

$$\left[\begin{array}{ccc|ccc} 1 & 2 & 4 & 1 & 0 & 0 \\ 0 & 1 & 0 & -1 & -1 & 1 \\ 0 & 0 & 1 & 0 & 1 & -1/2 \end{array}\right] \quad [R_2 \rightarrow R_2 - 2R_3]$$

$$\left[\begin{array}{ccc|ccc} 1 & 2 & 0 & 1 & -4 & 2 \\ 0 & 1 & 0 & -1 & -1 & 1 \\ 0 & 0 & 1 & 0 & 1 & -1/2 \end{array}\right] \quad [R_1 \rightarrow R_1 - 4R_3]$$

$$\left[\begin{array}{ccc|ccc} 1 & 0 & 0 & 3 & -2 & 0 \\ 0 & 1 & 0 & -1 & -1 & 1 \\ 0 & 0 & 1 & 0 & 1 & -1/2 \end{array}\right] \quad [R_1 \rightarrow R_1 - 2R_2]$$

Therefore, we have determined that $\mathbf{B} = \mathbf{A}^{-1} = \begin{bmatrix} 3 & -2 & 0 \\ -1 & -1 & 1 \\ 0 & 1 & -1/2 \end{bmatrix}$

(Note: if a row of zeros results on the left-side matrix, then an inverse does not exist.)

It is possible that A lacks an inverse and, therefore, is not **invertible**. A noninvertible matrix is called **singular**.

Determinants

A square matrix, **A**, has a **determinant, det A**, to represent its numeric value. The determinant is evaluated by summing all **signed elementary products** from **A**, as illustrated below.

The value of a determinant of a 2 × 2 matrix is

$$\det \begin{bmatrix} a_{11} & a_{12} \\ a_{21} & a_{22} \end{bmatrix} = a_{11}a_{22} - a_{12}a_{21}$$

Note the "rule" for its evaluation as a 2 × 2 matrix: multiply diagonal element pairs from left to right. That is, starting at the "one–one" position, multiply a_{11} by a_{22}. Then, subtract the product of the other diagonal elements.

For a 3 × 3 matrix, its determinant is evaluated as:

$$\det \begin{bmatrix} a_{11} & a_{12} & a_{13} \\ a_{21} & a_{22} & a_{23} \\ a_{31} & a_{32} & a_{33} \end{bmatrix}$$

$$= a_{11}a_{22}a_{33} + a_{12}a_{23}a_{31} + a_{13}a_{21}a_{32} - a_{13}a_{22}a_{31} - a_{12}a_{21}a_{33} - a_{11}a_{23}a_{32}$$

This determinant is evaluated using the signed elementary products of the matrix. A diagram is useful. Note that the matrix is repeated side by side in the diagram to perform the operations, and the sign at the arrow head denotes whether the product is added or subtracted in valuing the determinant.

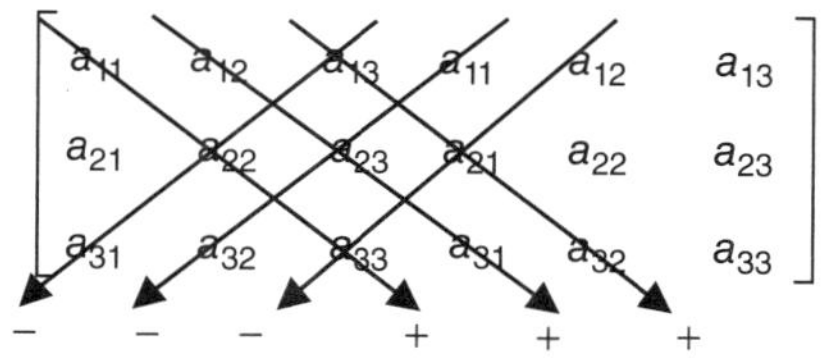

Here are some rules for determinants:

- If a matrix contains a row or column of zeros, the value of the determinant is zero.
- If a matrix has two equal rows or columns, the value of the determinant is also zero.
- If one row or column is a multiple of the other, then the value of the determinant is zero (since they are made equivalent by a single arithmetic operation).
- If the determinant of a matrix is zero, it has no inverse.

A useful formula for the inverse of a 2 × 2 matrix uses a determinant as:

Inverse of A 2 × 2 Matrix

The inverse of a 2 × 2 matrix $\mathbf{A} = \begin{bmatrix} a_{11} & a_{12} \\ a_{21} & a_{22} \end{bmatrix}$ **is given by**

$$\mathbf{A}^{-1} = \frac{1}{\det \mathbf{A}} \begin{bmatrix} a_{22} & -a_{12} \\ -a_{21} & a_{11} \end{bmatrix} = \frac{1}{a_{11}a_{22} - a_{12}a_{21}} \begin{bmatrix} a_{22} & -a_{12} \\ -a_{21} & a_{11} \end{bmatrix}$$

The inverse of a matrix is useful in solving systems of equations. The solution to a system of equations $\mathbf{AX} = \mathbf{B}$ is given symbolically by $\mathbf{X} = \mathbf{A}^{-1}\mathbf{B}$.

Example 3.4.3 Ticket Sales

Suppose that circus tickets cost \$25 for adults and \$10 for children. In one hourly period, 90 tickets sold for \$1650. Determine how many adult tickets and children's tickets were sold by using a matrix inverse.

Solution:
The equations involved are

$$\begin{aligned} x_1 + \quad x_2 &= 90 \qquad \textit{(90 tickets sold)} \\ 25x_1 + 10x_2 &= 1650 \qquad \textit{(tickets sold for \$1650)} \end{aligned}$$

In matrix format, this system is

$$\begin{bmatrix} 1 & 1 \\ 25 & 10 \end{bmatrix} \begin{bmatrix} x_1 \\ x_2 \end{bmatrix} = \begin{bmatrix} 90 \\ 1650 \end{bmatrix}$$

Therefore,

$$\begin{bmatrix} x_1 \\ x_2 \end{bmatrix} = \begin{bmatrix} 1 & 1 \\ 25 & 10 \end{bmatrix}^{-1} \begin{bmatrix} 90 \\ 1650 \end{bmatrix}$$

Using the formula for finding the inverse of a 2×2 *matrix,*

$$\begin{bmatrix} 1 & 1 \\ 25 & 10 \end{bmatrix}^{-1} = \frac{1}{1(10) - 1(25)} \begin{bmatrix} 10 & -1 \\ -25 & 1 \end{bmatrix} = \begin{bmatrix} -2/3 & 1/15 \\ 5/3 & -1/15 \end{bmatrix}$$

Then, $\begin{bmatrix} x_1 \\ x_2 \end{bmatrix} = \begin{bmatrix} -2/3 & 1/15 \\ 5/3 & -1/15 \end{bmatrix} \begin{bmatrix} 90 \\ 1650 \end{bmatrix} = \begin{bmatrix} 50 \\ 40 \end{bmatrix}$

Fifty adult tickets and 40 children tickets were sold in the hour.

Cryptology, the science of secret codes, depends on matrices. For example, in one simple code of letters and punctuation, A = 1, B = 2, … , Z = 26, space = 27, period = 28, exclamation point = 29, and a question mark = 30. Example 3.4.4 illustrates the fundamentals of encryption (and decryption) of secret codes. They depend on an *encoding matrix* or *code key*.

Example 3.4.4 Cryptology

Encode the message: CONGRATULATIONS! The encoding matrix is

$$\mathbf{A} = \begin{bmatrix} 2 & 5 \\ 1 & 3 \end{bmatrix}$$

Solution:
Since **A** *is a* (2×2) *matrix, first enter the message in the columns of a two-row matrix* **B**. *This ensures that* **A** *and* **B** *are conformable and that the multiplication* **AB** *is possible.*

$$\mathbf{B} = \begin{bmatrix} C & N & R & T & L & T & O & S \\ O & G & A & U & A & I & N & ! \end{bmatrix} = \begin{bmatrix} 03 & 14 & 18 & 20 & 12 & 20 & 15 & 19 \\ 15 & 07 & 01 & 21 & 01 & 09 & 14 & 29 \end{bmatrix}$$

→

Next, form the matrix $\boldsymbol{AB}$.

$$\boldsymbol{AB} = \begin{bmatrix} 2 & 5 \\ 1 & 3 \end{bmatrix} \begin{bmatrix} 03 & 14 & 18 & 20 & 12 & 20 & 15 & 19 \\ 15 & 07 & 01 & 21 & 01 & 09 & 14 & 29 \end{bmatrix}$$

$$= \begin{bmatrix} 81 & 63 & 41 & 145 & 29 & 85 & 100 & 183 \\ 48 & 35 & 21 & 83 & 15 & 47 & 57 & 106 \end{bmatrix}$$

The secret code for "CONGRATULATIONS!" is

81 48 63 35 41 21 145 83 29 15 85 47 100 57 183 106

The message is decoded by multiplying the inverse matrix, $\mathbf{A}^{-1}$, by the coded matrix, **AB**. This retrieves **B**, which is used with the number and punctuation codes to decode the message. In matrix notation,

$$\mathbf{A}^{-1}(\mathbf{AB}) = (\mathbf{A}^{-1}\mathbf{A})\mathbf{B} = \mathbf{IB} = \mathbf{B}$$

Example 3.4.5 Decoding a Message

Decode the following message using the simple code of this section.

84 31 152 57 102 39 177 68 18 7 84 33 119 45 81 30 97 35 35 13

Use the code key $\boldsymbol{A}$.

$$\boldsymbol{A} = \begin{bmatrix} 3 & 5 \\ 1 & 2 \end{bmatrix}$$

Solution:
The inverse of $\boldsymbol{A}$ *can be found using the formula given in the text.*

$$\boldsymbol{A}^{-1} = \frac{1}{3(2) - 5(1)} \begin{bmatrix} 2 & -5 \\ -1 & 3 \end{bmatrix} = \begin{bmatrix} 2 & -5 \\ -1 & 3 \end{bmatrix}$$

To continue the decoding, first multiply this inverse matrix by the coded message in matrix format.

$$\begin{bmatrix} 2 & -5 \\ -1 & 3 \end{bmatrix} \begin{bmatrix} 84 & 152 & 102 & 177 & 18 & 84 & 119 & 81 & 97 & 35 \\ 31 & 57 & 39 & 68 & 7 & 33 & 45 & 30 & 35 & 13 \end{bmatrix}$$

$$= \begin{bmatrix} 13 & 19 & 9 & 14 & 1 & 3 & 13 & 12 & 19 & 5 \\ 9 & 19 & 15 & 27 & 3 & 15 & 16 & 9 & 8 & 4 \end{bmatrix}$$

$\longrightarrow$

Next, substitute the appropriate letters and punctuation to yield

$$\begin{bmatrix} M & S & I & N & A & C & M & L & S & E \\ I & S & O & & C & O & P & I & H & D \end{bmatrix} = \textit{MISSION ACCOMPLISHED}$$

Input–Output Analysis

Economic entities such as a corporation, a community, even nations, use various resources to produce a variety of products. In turn, these products become resources for other entities, and so on.

An automobile manufacturer, for example, uses sheet metal, plastics, wire, fabrics, machines, tires, water, electricity, fuel, labor, and other products as resources to produce vehicles – some of them used by employees for deliveries, transportation, and so on and by others to produce other resources.

The central idea of input–output analysis is the complex interactions of *sectors* of an economic entity. Large or small, some sectors become resources for other sectors that contribute to other products.

Matrices are the main mathematical means for the study of input–output models.

Let $x_1, \ldots, x_n$ represent the number of units of output of n sectors and a_{ij} the number of units of input i to produce a single unit of output j; $i, j = 1, \ldots, n$. A matrix of the a_{ij}, **A**, is known variously as "exchange table," "transaction table," "technology coefficient matrix," "technology matrix," "consumption matrix," and probably others. For example, $a_{ij}x_j$ is the number of units of sector i (input) required to produce x_j units of sector j (outputs). When summed over all sectors, $\sum_i a_{ij}x_j$ is the total of sector outputs to produce x_j units of output j. This is to be balanced against the demand, j, for sector j's output.

We can write for sector j's output $x_j - (a_{1j}x_j - a_{2j}x_j - \cdots - a_{nj}x_j) = d_j$. Here, d_j is the demand for sector j. That is, the gross production of sector j is reduced by the amount of j consumed internally. Such an equation applies to each of the n sectors. In matrix notation, $\mathbf{X} - \mathbf{AX} = \mathbf{D}$, where **A** is the technology matrix, **X** is the output or production schedule vector, and **D** is the $n \times 1$ demand vector. This can also be written as $(\mathbf{I} - \mathbf{A})\mathbf{X} = \mathbf{D}$. Solve for **X** by multiplying both sides of the equation by $(\mathbf{I} - \mathbf{A})^{-1}$ to yield

$$\mathbf{X} = (\mathbf{I} - \mathbf{A})^{-1}\mathbf{D}$$

If all entries of $(\mathbf{I} - \mathbf{A})^{-1}$ are nonnegative, then a nonnegative production vector can be found for any nonnegative demand vector and the economy is said to be *productive*.

If a column of the technology or consumption matrix, **A**, sums to less than unity, then the corresponding industry consumes less than \$1 in order to produce \$1 of output and the industry is said to be *profitable*.

Example 3.4.6 A Single Good Economy

Suppose that there is an economy with a single good, x, and that in order to produce x some of the good is consumed in the process (waste). If 1/3 unit of x is consumed in order to produce a unit of x, how many units of x must be produced to fill an order for 100 units of x.

Solution:
Since 1/3 unit of x is considered waste, the following equation is appropriate:

$$x - 1/3x = 100$$

$$2/3x = 100$$

$$x = 150$$

Therefore, 150 units of x are required to produce 100 units of x.

This concept can be extended to an economy with two industries in which each produce a single good. Each industry may use some of its own goods and the others' goods in the production process. This is illustrated in Example 3.4.7.

Example 3.4.7 Input–Output Analysis

Imagine two independent industries (utilities and metal fabrication, agriculture and transportation, etc.) in which each sells portions of its product outputs (A and B) to each other as well as to others. Besides, each industry consumes a portion of its products for the consumption of its outputs.

The production of one unit of A requires 0.1 unit of A and 0.25 unit of B.

The production of one unit of B requires 0.3 unit of A and 0.2 unit of B. In addition, industry A sells 120 units to others, while industry B sells 250 units to others. Express the relation in equations and matrices.

Solution:

Let a_{11} = *the amount of good 1 used to produce one unit of good 1*
Let a_{12} = *the amount of good 1 used to produce one unit of good 2*
Let a_{21} = *the amount of good 2 used to produce one unit of good 1*
Let a_{22} = *the amount of good 2 used to produce one unit of good 2*
Let d_1 = *the external demand for good 1*
Let d_2 = *the external demand for good 2*
Let x_1 = *the output of industry A*
Let x_2 = *the output of industry B.*

Industry A uses 0.10 unit of A to produce 1 unit of A
Industry B uses 0.30 unit of A to produce 1 unit of B

⟶

Industry A uses 0.25 unit of B to produce 1 unit of A
Industry B uses 0.20 unit of B to produce 1 unit of B
Outside demand for A is 120 and for B is 250

The equations are

$$x_1 = 0.10x_1 + 0.30x_2 + 120$$

$$x_2 = 0.25x_1 + 0.20x_2 + 250$$

or, using matrices,

$$\begin{bmatrix} x_1 \\ x_2 \end{bmatrix} = \begin{bmatrix} 0.10 & 0.30 \\ 0.25 & 0.20 \end{bmatrix} \begin{bmatrix} x_1 \\ x_2 \end{bmatrix} + \begin{bmatrix} 120 \\ 250 \end{bmatrix}$$

This imaginative characterization of an economic entity, usually known as an **input–output model**, is attributed to the Nobel Prize winning economist *Wassily Leontief*. It is a mainstay of modern economic theory.

Example 3.4.8 *Three Industry Input–Output Model*

Suppose there is an economy consisting of three industries A, B, and C. The following table has the information for the consumption matrix.

		Outputs	
Inputs	*A*	*B*	*C*
A	*0.60*	*0.10*	*0.20*
B	*0.20*	*0.50*	*0.10*
C	*0.10*	*0.30*	*0.40*

a) How many units of B and C are required to produce 100 units of A?
b) Production of which commodity is most dependent on the other two?
c) An increase in cost of B affects which industry the least?
d) Which industry is most dependent on its own goods for its operation?
e) What is the gross output to have a surplus of 68 units of A, 24 units of B, and 6 units of C?

Solution:

a) Each unit of A requires 0.20 units of B and 0.10 units of C, which is seen from the A column. Therefore, 20 units of B and 10 units of C are required to produce 100 units of A.

⟶

b) Each unit of A requires 0.30 units from B and C, each unit of B requires 0.40 units from A and C, and each unit of C requires 0.30 units from A and B as determined from the columns in the table. Therefore, B is most dependent on the other two industries.

c) An increase in the cost of B least affects industry C as determined by reading across row B.

d) Industry A is most dependent on its own goods for operation as determined by reading the values of the main diagonal of the table.

e) First, **I** − **A** *must be found, where* ***I*** *is an identity matrix and* ***A*** *the consumption matrix.*

$$\begin{bmatrix} 1 & 0 & 0 \\ 0 & 1 & 0 \\ 0 & 0 & 1 \end{bmatrix} - \begin{bmatrix} 0.60 & 0.10 & 0.20 \\ 0.20 & 0.50 & 0.10 \\ 0.10 & 0.30 & 0.40 \end{bmatrix} = \begin{bmatrix} 0.40 & -0.10 & -0.20 \\ -0.20 & 0.50 & -0.10 \\ -0.10 & -0.30 & 0.60 \end{bmatrix}$$

Recall that $(\boldsymbol{I}-\boldsymbol{A})^{-1}\ \boldsymbol{D}=\boldsymbol{X}$.
Therefore,

$$\begin{bmatrix} 0.40 & -0.10 & -0.20 \\ -0.20 & 0.50 & -0.10 \\ -0.10 & -0.30 & 0.60 \end{bmatrix}^{-1} \begin{bmatrix} 68 \\ 24 \\ 6 \end{bmatrix} = \begin{bmatrix} 300 \\ 200 \\ 160 \end{bmatrix}$$

So, 300 units of A, 200 units of B, and 160 units of C are required.

The basic input–output problem can be stated as:

Given the internal demands for the output of each industry, determine the output levels for the various industries that will meet a given final (external) level of demand as well as the internal demand.

EXERCISES 3.4

1. If $\mathbf{A} = \begin{bmatrix} 3 & 1 \\ 5 & 2 \end{bmatrix}$ and $\mathbf{B} = \begin{bmatrix} 2 & -1 \\ -5 & 3 \end{bmatrix}$, determine whether **A** and **B** are inverses by computing both **AB** and **BA**.

2. If $\mathbf{A} = \begin{bmatrix} 2 & 5 \\ 1 & 3 \end{bmatrix}$ and $\mathbf{B} = \begin{bmatrix} 3 & -5 \\ -1 & 2 \end{bmatrix}$, determine whether **A** and **B** are inverses by computing both **AB** and **BA**.

3. If $\mathbf{A} = \begin{bmatrix} 3 & 5 & 9 \\ 1 & 3 & 6 \\ 0 & 1 & 0 \end{bmatrix}$ and $\mathbf{B} = \begin{bmatrix} 3 & 0 & -5 \\ -1 & 2 & 4 \\ 1 & 0 & 0 \end{bmatrix}$, determine whether **A** and **B** are inverses by computing both **AB** and **BA**.

4. If $\mathbf{A} = \begin{bmatrix} 3 & 3 & -1 \\ -2 & -2 & 1 \\ -4 & 5 & -2 \end{bmatrix}$ and $\mathbf{B} = \begin{bmatrix} 1 & -1 & 1 \\ 0 & 2 & -1 \\ 2 & 3 & 1 \end{bmatrix}$, determine whether **A** and **B** are inverses by computing both **AB** and **BA**.

5. Why do each of these square matrices lack an inverse?

a) $\begin{bmatrix} 3 & 4 \\ 6 & 8 \end{bmatrix}$ b) $\begin{bmatrix} 2 & 3 & 5 \\ 0 & 2 & 3 \\ 4 & 6 & 10 \end{bmatrix}$ c) $\begin{bmatrix} 1 & 0 & 6 & 5 \\ 2 & 3 & 3 & 4 \\ 0 & 0 & 0 & 0 \\ 1 & 1 & 7 & 2 \end{bmatrix}$ d) $\begin{bmatrix} 1 & 1 & 3 \\ 2 & 4 & 6 \\ 3 & 0 & 9 \end{bmatrix}$

6. Find the inverses of these matrices. Use both the formula and the computational technique.

a) $\begin{bmatrix} 2 & 7 \\ 1 & 4 \end{bmatrix}$ b) $\begin{bmatrix} 6 & 6 \\ 5 & 2 \end{bmatrix}$ c) $\begin{bmatrix} 2 & 8 \\ 6 & 4 \end{bmatrix}$ d) $\begin{bmatrix} 3 & -4 \\ -2 & 3 \end{bmatrix}$

7. Let $\mathbf{A} = \begin{bmatrix} 1 & 0 & -1/2 \\ 0 & 1 & -1/2 \\ -1/3 & -1/3 & 1 \end{bmatrix}$. Verify $\mathbf{A}^{-1} = \begin{bmatrix} 5/4 & 1/4 & 3/4 \\ 1/4 & 5/4 & 3/4 \\ 1/2 & 1/2 & 3/2 \end{bmatrix}$

a) By showing $\mathbf{A}^{-1}\,\mathbf{A} = \mathbf{I}$ or $\mathbf{A}\,\mathbf{A}^{-1} = \mathbf{I}$.
b) By using the computational technique.

8. Find the inverse of the matrix if it exists:

$$\mathbf{A} = \begin{bmatrix} 1 & 3 & 3 \\ 0 & 1 & 2 \\ 1 & 3 & 6 \end{bmatrix}$$

9. Find the inverse of the matrix if it exists:

$$\mathbf{A} = \begin{bmatrix} 3 & 3 & -1 \\ -2 & -2 & 1 \\ -4 & -5 & 2 \end{bmatrix}$$

10. Find the inverse of the matrix if it exists:

$$\mathbf{A} = \begin{bmatrix} 2 & 1 & 0 \\ 1 & 3 & 1 \\ 3 & 4 & 1 \end{bmatrix}$$

11. Find the inverse of the matrix if it exists:

$$\mathbf{A} = \begin{bmatrix} 1 & 0 & 2 \\ 3 & 1 & 0 \\ 0 & 1 & 4 \end{bmatrix}$$

12. Find the inverse of the matrix if it exists:

$$\mathbf{A} = \begin{bmatrix} 2 & 0 & 4 & 6 \\ 0 & 1 & 8 & 1 \\ 3 & 0 & 5 & 2 \\ 0 & 0 & 1 & 4 \end{bmatrix}$$

13. A change purse contains 14 coins (nickels, dimes, and quarters) with a value of \$1.60. The number of dimes is two fewer than that of nickels and quarters combined.
    a) Write a system of equations for this situation.
    b) Express the system in matrix format.
    c) Using $\mathbf{X} = \mathbf{A}^{-1}\mathbf{B}$, determine the number of each coin denomination.

14. A local elementary school is to have a fund-raiser. Students are to sell boxes of cookies for \$4 and candies for \$5. The students sold 2400 boxes realizing \$10,500. How many boxes of each were sold?

    Write equations for this fund-raiser and use $\mathbf{X} = \mathbf{A}^{-1}\mathbf{B}$ to determine the number of boxes sold of each confection.

15. An investment portfolio has \$18,000 in a mix of stocks, bonds, and money market funds. The amount in stocks is twice that in money market funds. In addition, \$2000 more is placed in bonds than in money market funds. Determine the amounts placed in each by using $\mathbf{X} = \mathbf{A}^{-1}\mathbf{B}$.

16. Encode the message **HAPPY BIRTHDAY!** using the encoding matrix $\mathbf{A} = \begin{bmatrix} 2 & 9 \\ 1 & 4 \end{bmatrix}$.

17. Encode the message **FOUR SCORE AND SEVEN YEARS AGO** using the encoding matrix $\mathbf{A} = \begin{bmatrix} 3 & 4 & 1 \\ 1 & 1 & 0 \\ 2 & 3 & 2 \end{bmatrix}$.

18. Decode your message from Exercise 3.4.16 as a check.

19. Decode your message from Exercise 3.4.17 as a check.

20. Given the technology matrix $\mathbf{A} = \begin{bmatrix} 0.2 & 0.3 \\ 0.6 & 0.1 \end{bmatrix}$, compute the production schedule for the demand vector $\mathbf{D} = \begin{bmatrix} 100 \\ 50 \end{bmatrix}$

21. An economy has three goods – coal, lumber, and steel – and a technology matrix $\mathbf{A}$ and a production schedule $\mathbf{X}$

$$\mathbf{A} = \begin{bmatrix} 0.2 & 0.5 & 0.1 \\ 0.0 & 0.3 & 0.2 \\ 0.5 & 0.0 & 0.4 \end{bmatrix} \quad \mathbf{X} = \begin{bmatrix} 30 \\ 20 \\ 50 \end{bmatrix}.$$

    Find the external demand, $\mathbf{D}$, to be met by the production schedule.

22. Imagine a two-industry economy, A and B, where output is measured in dollars. Each dollar's worth of A's output requires 0.40 units from A and 0.20 units from B. Each dollar's worth of B's output requires 0.50 units from A and 0.30 units from B. If the final demand is for \$60,000 from A and \$40,000 from B, how many units from each industry should be produced to meet the demand?

23. An economy has three goods A, B, C. Each \$1 output from A requires 10 cents from A, 30 cents from B, and 20 cents from C; \$1 from B requires 40 cents from A, 25 cents from B, and 15 cents from C; \$1 from C requires 35 cents from A, 15 cents from B, and 35 cents from C. Demand is \$25,000 from A, \$10,000 from B, and \$15,000 from C. Determine the production schedule.

HISTORICAL NOTES

In "*Matrices and Society*" by Ian Bradley and Ronald L. Meek, a readable Princeton University Press book (1986), an interesting case is made of the importance of matrices to social scientists. They contrast the mathematics of several variables that has been vital to the development of the physical sciences with the huge number of variables required in the social sciences and, in consequence, the different mathematics (mainly matrix theory) required there.

Benjamin Franklin called one of his unusual magic square creations "the most magically magical of any square ever made by any magician". An excellent account appears in Paul C. Pasles' *Benjamin Franklin's Numbers: An Unsung Mathematical Odyssey.*

◆ *Contradancing* can be thought of as matrices in motion. Contradancing, a dance form whose roots can be traced to English country dance. As practiced in the United States, it is similar to square dancing.

The music for the dance is highly structured. Dancers line up in groups of four for each block of two couples to form a 2×2 matrix with dancers as the elements. As a "caller" issues periodic instructions, rows of the matrix are interchanged.

$$\begin{bmatrix} w_1 & m_1 \\ m_2 & w_2 \end{bmatrix}$$

www.cdss.org or Bernie Scanlon, a mathematician and contradancer.

Arthur Cayley (1821–1895) – A brilliant English mathematician known for his work in algebra. He also contributed to theoretical dynamics and astronomy and was one of the first mathematicians to consider a geometry of more than three dimensions. Cayley was a leader of the British school of pure mathematics that emerged during the nineteenth century. During the 14 years he practiced law, he wrote over 250 mathematical papers. In 1863, he became a Cambridge University mathematics professor, and there he developed the algebra of matrices in addition to work on higher dimension geometries. Cayley is one of the early mathematicians to study matrices. A notation for determinants is credited to him.

Cayley's work on matrices provided a foundation for Werner Heisenberg's development of quantum mechanics. Matrices have since become the foundation for the computational operations on large databases as well as for theoretical developments in the social sciences.

Karl Friedrich Gauss (1777–1855) – German born, was notable as a mathematician, physicist, and astronomer. Known as "The Prince of Mathematicians," Gauss made profound contributions to number theory, the theory of functions, probability, statistics, electromagnetic theory, and much more.

Gauss studied at the University of Göttingen from 1795 to 1798. His doctoral dissertation was a proof that every algebraic equation has at least one root (solution). This theorem, which had challenged mathematicians for centuries, is called "the fundamental theorem of algebra." Gauss insisted on a complete proof of any result before publication. Consequently, many discoveries were not credited to him. However, his published works were enough to establish his reputation as perhaps the greatest mathematician of all time.

Camille Jordan (1838–1921) – born in France as Marie-Ennemond Camille Jordan and trained as an engineer. While working as an engineer, he wrote 120 mathematical research papers. He published papers in practically all branches of the mathematics of his time. Known primarily as an algebraist, he made useful refinements to an elimination method for solving systems of linear equations now known as the Gauss–Jordan elimination method.

Wassily Leontief (1906–1999) – the son of an economics professor, was born in St. Petersburg, Russia. He studied economics, philosophy, and sociology at the University of Leningrad, moved to Berlin in 1925, and later earned a Ph.D. at the University of Berlin. In 1931, Leontief joined the National Bureau of Economic Research in the United States, at the time a center of empirical economic research. A year later he joined the Harvard faculty.

Early on, he became interested in how changes in one economic sector affected other economic sectors. His work, even its earlier stages, had an important role in national war production planning during World War II. His Input–Output economic analysis was used to show the extensive process by which inputs in one industry produce outputs for consumption or for input into another industry. It helps to understand trade within and between countries. The matrix format devised by Leontief is often used to show the effect of a change in production of a final good on the demand for inputs. He was the 1973 Nobel Prize winner in economics.

CHAPTER 3 SUPPLEMENTARY EXERCISES

1. Suppose **A** is an $(m \times n)$ matrix and **B** is a $(p \times r)$ matrix. What must be true about m, n, p, and r so these matrices are defined?

   a) **AB**   b) **A** + **B**   c) Both **AB** and **BA**

2. Find **AB** if $\mathbf{A} = \begin{bmatrix} 3 & 2 & 1 \\ 2 & 0 & 1 \\ 6 & 4 & 3 \\ 1 & 7 & 6 \end{bmatrix}$ and $\mathbf{B} = \begin{bmatrix} 1 & -4 \\ 5 & -2 \\ 0 & 4 \end{bmatrix}$.

3. Find **AB** if $\mathbf{A} = \begin{bmatrix} 1 & 3 & 2 & 1 \\ 9 & 2 & 0 & 1 \\ 5 & 6 & 8 & 3 \\ 4 & 2 & 7 & 6 \end{bmatrix}$ and $\mathbf{B} = \begin{bmatrix} 0 & 1 & 3 & -3 \\ 5 & 4 & 9 & -1 \\ 3 & 0 & 7 & -2 \\ 2 & 1 & 0 & 4 \end{bmatrix}$.

4. Compute $\begin{bmatrix} 2 & 3 & 1 & 0 \\ 0 & 1 & -1 & 0 \\ 3 & 2 & 1 & 1 \\ 1 & 0 & 4 & 5 \end{bmatrix} \begin{bmatrix} 6 & 2 \\ 9 & 1 \\ 0 & -3 \\ -1 & -1 \end{bmatrix}$.

5. Find $\mathbf{A}^2$ if $\mathbf{A} = \begin{bmatrix} 2 & 3 & 1 \\ 1 & 4 & 0 \\ 6 & 0 & 5 \end{bmatrix}$.

6. Find $\mathbf{A}^2$ if $\mathbf{A} = \begin{bmatrix} 1 & 2 & -1 & 4 \\ 3 & 0 & 4 & 3 \\ 1 & 6 & 9 & 1 \\ 5 & 1 & 0 & 2 \end{bmatrix}$.

7. Solve the system of equations:

$$\begin{aligned} 3x_1 + 4x_2 &= 7 \\ x_1 - 2x_2 &= -1 \\ 2x_1 + 3x_2 &= 5 \end{aligned}$$

8. Solve the system of equations:

$$\begin{aligned} x_1 + 5x_2 + 3x_3 &= 13 \\ 2x_1 - x_2 + x_3 &= 12 \\ -x_1 + x_2 + x_3 &= 1 \end{aligned}$$

9. Solve the system of equations:

$$\begin{aligned} 2x_1 + 3x_2 + x_4 &= 18 \\ -2x_2 + 3x_3 - 2x_4 &= -5 \\ x_2 + x_3 &= 18 \\ 2x_1 + 3x_3 - x_4 &= 10 \end{aligned}$$

10. Many matrices commute, that is, $\mathbf{AB} = \mathbf{BA}$. Find three such matrices for

$$\mathbf{A} = \begin{bmatrix} 1 & 3 \\ 2 & 4 \end{bmatrix}$$

11. Let $\mathbf{A} = \begin{bmatrix} 4 & 0 \\ 3 & 2 \end{bmatrix}$ and $\mathbf{B} = \begin{bmatrix} 6 & 1 \\ 2 & -1 \end{bmatrix}$. Find constants p and q so the following is true:

$$p\mathbf{A} + q\mathbf{B} = \begin{bmatrix} 16 & 2 \\ 7 & 0 \end{bmatrix}$$

12. Let $\mathbf{A} = \begin{bmatrix} 4 & 1 & 1 \\ 3 & 2 & 0 \\ 1 & 3 & 2 \end{bmatrix}$ and $\mathbf{B} = \begin{bmatrix} 5 & 1 & 3 \\ 3 & 1 & 2 \\ 1 & 2 & 4 \end{bmatrix}$. Find constants p and q so the following is true:

$$p\mathbf{A} + q\mathbf{B} = \begin{bmatrix} 17 & 4 & 6 \\ 12 & 7 & 2 \\ 4 & 11 & 10 \end{bmatrix}$$

13. Find the augmented matrix for the following system:

$$\begin{aligned} 4x_1 - 11x_2 + 5x_3 + 3x_4 + 3x_5 &= 5 \\ x_1 - x_2 + x_3 - 2x_4 + 4x_5 &= 10 \\ 3x_1 + x_2 + 2x_3 - 3x_4 + x_5 &= 4 \\ x_1 + x_2 + 3x_3 + 2x_4 + 2x_5 &= 10 \\ x_1 + 5x_3 + 3x_5 &= 7 \end{aligned}$$

14. Let $\mathbf{A} = \begin{bmatrix} 1 & 2 \\ 3 & 4 \\ 1 & 5 \end{bmatrix}$ and $\mathbf{B} = \begin{bmatrix} 4 & 1 \\ 0 & 2 \end{bmatrix}$. Verify $(\mathbf{AB})^t = \mathbf{B}^t\mathbf{A}^t$.

15. A state fair has tickets for two weekend performances by Clay Aiken from *American Idol.* Ticket prices are \$25, \$15, and \$12. The first performance sold 2000, 3000, and 5000 tickets at the respective prices. The sales for the second performance were 2500, 3500, and 4000 tickets, respectively. Use matrix methods to determine the receipts for each performance. What were the total receipts?

16. A professor weights quizzes as 10%; homework, 15%; tests, 50%; and the final exam, 25%. Use a matrix product to find the grades for three students whose four respective grades are

    Student 1: 90, 70, 80, 85
    Student 2: 95, 80, 90, 75
    Student 3: 75, 95, 85, 90

17. Three integers sum to 6. The second largest is one less than the largest. The largest is equal to the difference between the second largest and the smallest integer. Use Gauss–Jordan methods to determine the integers.

18. Solve $\mathbf{AX} = \mathbf{B}$ using $\mathbf{X} = \mathbf{A}^{-1}\mathbf{B}$ when $\mathbf{A} = \begin{bmatrix} 1 & 1 & 1 \\ 3 & 2 & -1 \\ -2 & 1 & 3 \end{bmatrix}$, $\mathbf{X} = \begin{bmatrix} x_1 \\ x_2 \\ x_3 \end{bmatrix}$ and $\mathbf{B} = \begin{bmatrix} 9 \\ 4 \\ 16 \end{bmatrix}$.

19. A technology matrix, **A**, and demand matrix, **D**, are

$$\mathbf{A} = \begin{bmatrix} 0.1 & 0.5 & 0.2 \\ 0.3 & 0.3 & 0.3 \\ 0.6 & 0.0 & 0.1 \end{bmatrix} \text{ and } \mathbf{D} = \begin{bmatrix} 5 \\ 14 \\ 3 \end{bmatrix}.$$

    Determine the production schedule.

4 Linear Programming – Geometric Solutions

INTRODUCTION

Linear programming is the mathematics of seeking a maximum (or a minimum) to a linear function (the objective) and subject to linear constraints. In the absence of constraints, the problem is simple! Simply assign values to each variable independently. However, the presence of constraints is much more challenging since arbitrary values cannot be assigned to variables. In fact, it was not until the late 1940s that a reliable scheme for an iterative numerical solution developed.

There is much one can write about this fascinating subject and its impact on people's everyday lives. It is the mathematics of allocating scarce resources, which, after all, is everyone's everyday challenge! The use of linear programming in industrial and government operations and planning is widespread.

This chapter solves (small) two variable linear programs (LPs) geometrically. While the method is impractical for larger problems, it does provide valuable insights.

Finite Mathematics: Models and Applications, First Edition. Carla C. Morris and Robert M. Stark.

Companion Website: http://www.wiley.com/go/morris/finitemathematics

4.1 GRAPHING LINEAR INEQUALITIES

The **strict inequalities** studied here are of two types: less than (<) and greater than (>). Inequalities resemble equations except that an inequality sign replaces the equal sign. Inequalities that are not strict allow for the possibility of equality as well as inequality (denoted as $\leq$ and $\geq$).

Example 4.1.1 Identifying Strict Inequalities

Which of these are strict inequalities?

a) $2x + 3y \leq 1$ *b)* $x_1 + 5x_2 > 10$ *c)* $7x + 2y = 6$ *d)* $5x_1 + 3x_2 + x_3 > 15$

Solution:

a) Not strict since the two sides could be equal.
b) Strict.
c) This is an equality.
d) Strict.

Before introducing systems of inequalities, focus on graphing a single inequality in two variables. The graph of an inequality encompasses a region known as a **half-plane**. Inequalities can be represented graphically as in Example 4.1.2.

Example 4.1.2 Graphing a Linear Inequality

Represent $x + y \leq 5$ *graphically.*

Solution:
First, plot the line $x + y = 5$*. Use at least three points that are on the line (recall that while two points suffice, a third point is a useful check). The line separates the xy-plane into two regions: the half-planes:* $x + y < 5$ *and* $x + y > 5$*.*

As the inequality is not strict, the line $x + y = 5$ *is shown as a solid line. (Strict inequalities appear as a dashed line.) The correct half-plane is determined by choosing a* ***test point*** *(test points cannot be on the line). The origin is an easy-to-use test point. Substituting* $x = y = 0$ *yields* $0 + 0 \leq 5$*. This is true! The half-plane that includes the origin has all of the feasible solutions and is shaded. (If the test point had not satisfied the inequality, the other half-plane would have been shaded.)*

$\longrightarrow$

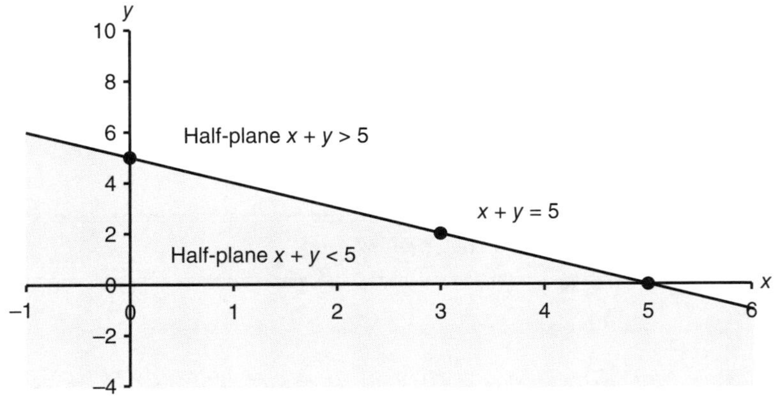

To graph an inequality, follow these steps:

Graphing a Linear Inequality

1. **Consider the inequality as an equality and plot. (Use three points.)**
2. **Use a solid line for the boundary of nonstrict inequalities ($\geq$ and $\leq$) as the line itself is a part of the solution region. Use a dashed line for strict inequalities, as the line itself is not part of the solution.**
3. **Choose a test point (not on the line). If the inequality holds for the test point, shade that half-plane. If not, shade the other (complementary) half-plane. The shaded half-plane has solutions to the LP and is known as the "feasible region."**

Example 4.1.3 More on Graphing Linear Inequalities

Represent $2x + y > 6$ graphically.

Solution:
The inequality being strict, the line $2x + y = 6$ is shown as a dashed line. Again, use the origin as a test point. (It is not on the line and is easy to use.) Here, $2(0) + 0 > 6$ is false! So the half-plane that does not include the origin is the feasible region.

⟶

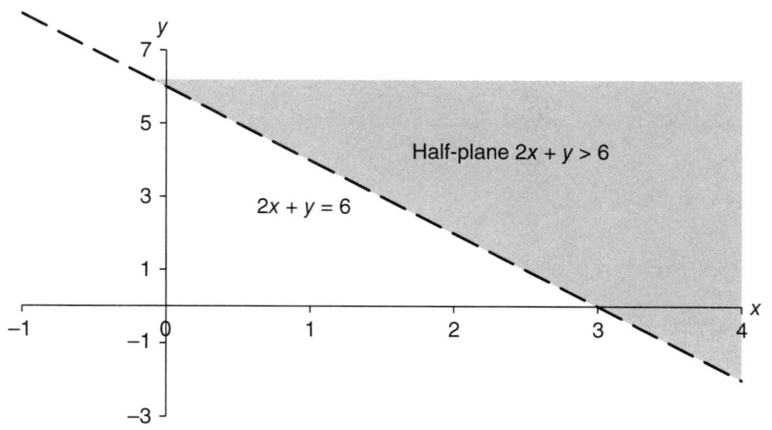

Now, having learned to graph inequalities, can the process be reversed? That is, determining the inequality from its graph. The answer is affirmative! Recall that the equation of a line is determined by two points (although we have used three points as a check). It can be useful to determine the x- and y-intercepts for the line as becomes apparent later in this chapter. The solid or dashed line indicates whether the inequality is strict or not, and a test point in the shaded region provides the direction of the inequality. This is shown in Example 4.1.4.

Example 4.1.4 Determining a Linear Inequality

Determine the inequality depicted in the following graph:

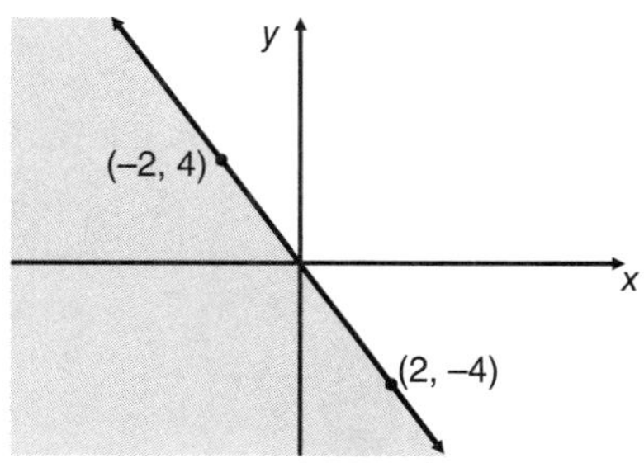

Solution:

We use two boundary points (see graph) to determine the slope of the line. Here, $m = (-4 - 4)/(2 - -2) = -8/4 = -2$. *The origin is a point on the line, so its*

⟶

y-intercept is zero. The equation of the line is $y = -2x$ or $2x + y = 0$. The solid line indicates that the inequality is either $\leq$ or $\geq$. An arbitrary test point $(-2, 0)$ is in the shaded region. Now, $2(-2) + 0 \leq 0$ is true. The inequality depicted is $2x + y \leq 0$.

EXERCISES 4.1

1. Determine which of these are strict inequalities.

   a) $5x + 2y > 10$ b) $x + 25y \leq 50$ c) $x_1 + 3x_2 < 9$

2. Determine which of these are strict inequalities.

   a) $5x + 2y + 3z < 30$ b) $x_1 + 2x_2 + 3x_3 < 18$ c) $4x_1 + x_2 - 2x_3 \geq 8$

In Exercises 3–18, graph the inequalities.

3. $x + 5 < 4$
4. $3x - y \leq 6$
5. $2x + 3y > 12$
6. $x \geq y$
7. $x < 3$
8. $y > 2$
9. $3x \leq 2y$
10. $x > 0$
11. $2x + y \leq 8$
12. $3x - 2y \geq 6$
13. $2x + 5y \geq 10$
14. $x > y$
15. $y \geq 2x + 3$
16. $y < 3x - 1$
17. $(1/3)x + (1/2)y \leq 2$
18. $3x + 5y \geq 15$

19. Determine the appropriate inequality.

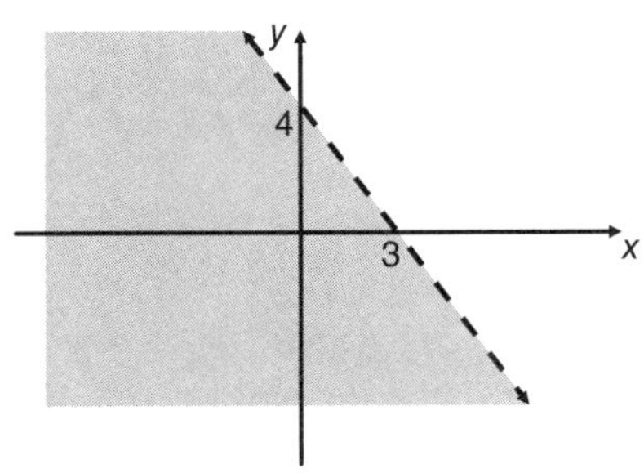

20. Determine the appropriate inequality.

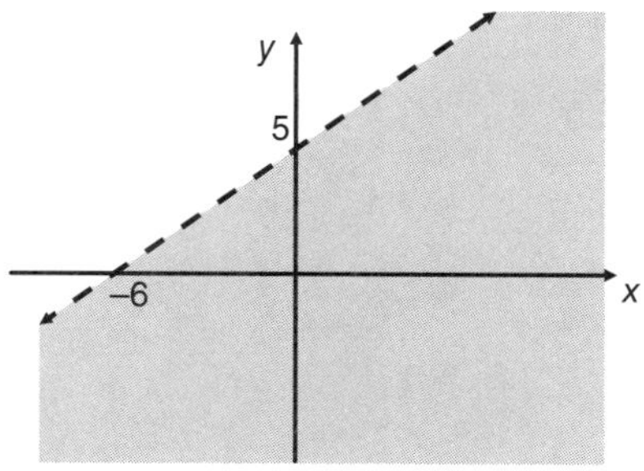

21. The sum of two numbers, x and y, is at most 10. Represent the possibilities graphically.

22. Product A costs \$3 per unit and Product B costs \$4 per unit. Sixty dollars is available for purchases. Depict the possibilities in the first quadrant, assuming fractional units can be purchased.

23. Food A has 5 grams and Food B has 4 grams of protein per serving. Graphically show which combination of foods assures that a person eats at least 40 grams of protein daily.

4.2 GRAPHING SYSTEMS OF LINEAR INEQUALITIES

Systems of inequalities, which arise in many applications, are graphed on common axes. A region whose points satisfy all of the inequalities is the **feasible region**. Each of the system's inequalities is satisfied in the feasible region. Hence, the problem's solution must lie in the feasible region.

A feasible region may be bounded and **closed** or **unbounded**. In either case, the feasible region is **convex**. A convex region has the property that any line segment joining any two points lies entirely within the region. Every point on the line segment also lies in the convex region (see figures).

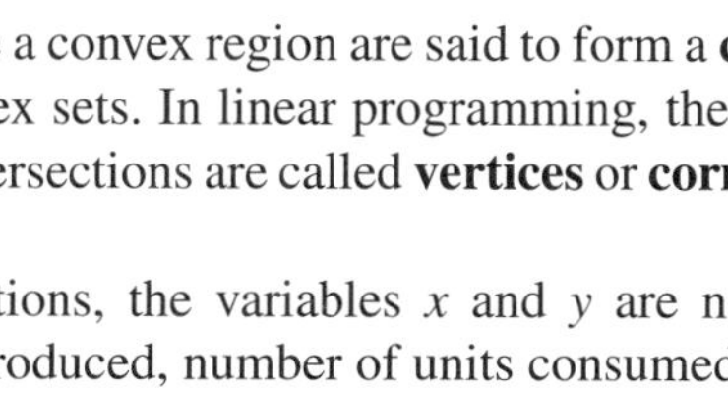

The points that comprise a convex region are said to form a **convex set**. Linear programming deals only with convex sets. In linear programming, the boundaries of a convex set are line segments. Their intersections are called **vertices** or **corner points** and are of central importance.

In many actual applications, the variables x and y are naturally nonnegative (time, demand, number of units produced, number of units consumed, etc.). That is, $x \geq 0$, $y \geq 0$ is a convex region in the first quadrant. The origin is often a corner point. Frequently, intercepts of the inequalities are other corner points. With this in mind, it is good graphing practice to determine the intercepts of each inequality.

Example 4.2.1 Unrestricted Variables

Graphically determine the feasible region that satisfies the system of inequalities:

$$x + 2y \leq 10$$
$$2x + y \leq 14$$

Solution:

Plotting the lines as equalities, the half-planes satisfying the individual inequalities define the convex region satisfying both inequalities.

The origin is included in the half-planes satisfying the individual inequalities. The half-planes that include the origin belong to the ***unbounded*** *feasible region (shaded). The corner point (6, 2) is the intersection of the two lines and is determined by solving the system of equations.*

$$\begin{array}{r} 2x + y = 14 \\ \underline{-2x - 4y = -20} \\ -3y = -6 \\ y = 2 \end{array}$$

Therefore,

$$\begin{array}{r} 2x + 2 = 14 \\ 2x = 12 \\ x = 6 \end{array}$$

The corner point, as noted, is at (6, 2) as shown. Note: since x and y are unrestricted, the feasible region is unbounded and extends into the four quadrants.

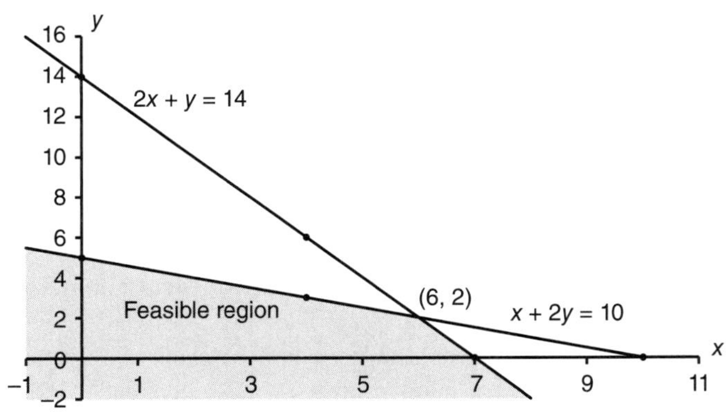

Here is another example!

Example 4.2.2 Bounded Feasible Region

Graph the system of inequalities:

$$x + y \leq 7$$

$$2x + y \leq 9$$

$$x \geq 0, \quad y \geq 0$$

Solution:

The inequalities are plotted as solid lines. The nonnegative variables result in a feasible region within the first quadrant. (Recall $x = 0$ is the y-axis and $y = 0$ is the x-axis.)
Using the origin as the test point, the shaded region is feasible. The feasible region, therefore, consists of the region bounded by the four points (0, 0), (0, 7), (9/2, 0) and (2, 5), where $x + y = 7$ intersects $2x + y = 9$.

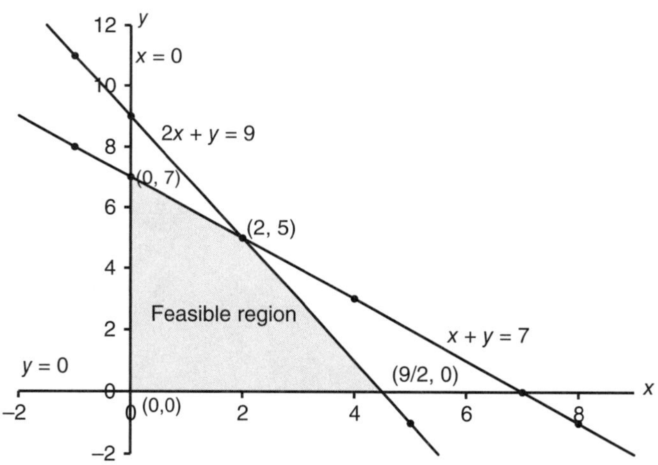

Example 4.2.3 Unbounded Feasible Region (Restricted Variables)

Graph the system of inequalities:

$$x + y \geq 7$$

$$2x + y \geq 9$$

$$x \geq 0, \quad y \geq 0$$

Solution:

Note that in this system inequalities are of the "greater than" type. Also, the nonnegativity of the variables defines a feasible region in the first quadrant. The unbounded feasible

⟶

region contains three corner points (0, 9), the y-intercept of the second inequality; (7, 0), the x-intercept of the first inequality; and (2, 5), the intersection point of the first two inequalities.

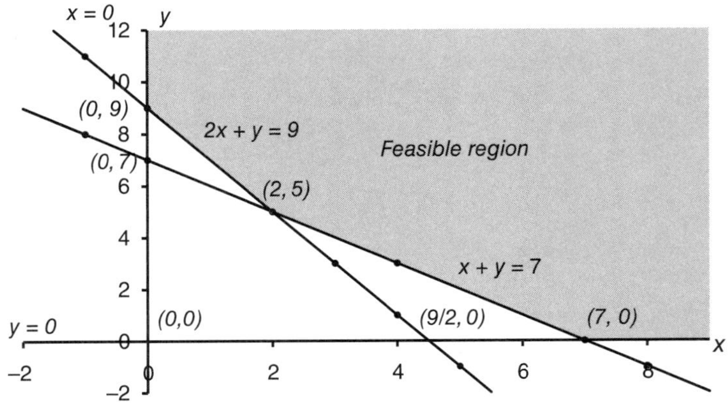

EXERCISES 4.2

Graph each system of inequalities, label corner points, and shade the feasible regions.

1. $x \geq 1, \quad y \leq 3$
2. $x \geq -2, \quad y \leq 5$
3. $-2 \leq x \leq 5, \quad 1 \leq y \leq 4$
4. $-1 \leq x \leq 3, \quad -2 \leq y \leq 6$
5. $x + y \leq 4, \quad 2x - y \leq 1$
6. $x + 2y \leq 8, \; x \geq 0, \; y \geq 0$
7. $2x + y \leq 7, \; x + y \leq 5, \quad x \geq 0, \quad y \geq 0$
8. $4x + 5y - 20 \leq 0, \quad x \geq 0, \quad y \geq 0$
9. $x \geq y, \quad x \geq 0, \quad y \geq 0$
10. $x + y \geq 1, \quad x + y \leq 7$
11. $3x + 2y \geq 6, \quad 3x + 2y \leq 12, \quad x \leq 2, \quad y \geq -3$
12. $y \geq 2x - 3, \quad y \leq -x + 6, \; x \geq 0, \quad y \geq 0$
13. $2x + y \leq 7, \; x + y \leq 4, \; x \geq 0, \quad y \geq 0$
14. $x + y \leq 8, \; x \geq 2, \; x \leq 3y$
15. $x + 2y \leq 8, \; 3x - 2y \leq 0$

16. $2x + 3y \leq 6, \; x \geq 0, \; y \geq 0$

17. $x + y \leq 10, \; x + y \geq 6, \; x \geq 0, \; y \geq 0$

18. $x + y \leq 7, \; 2x + y \leq 10, \; x \geq 0, \; y \geq 0$

19. $3x + 2y \leq 60, \; x + 2y \leq 40, \; x \geq 0, \; y \geq 0$

20. $x + y \leq 7, \; y \leq 2x - 5, \; x \geq 0, \; y \geq 0$

21. $x + y \leq 5, \; 2x - y \leq 2, \; x \geq 0, \; y \geq 0$

22. $-x + y \leq 2, \; x + 2y \leq 10, \; 3x + y \leq 15, \; x \geq 0, \; y \geq 0$

23. $x + 2y \leq 10, \; 3x + 4y \leq 24, \; 3x + 2y \leq 21, \; x \geq 0, \; y \geq 0$

24. $x + 2y \leq 15, \; 3x + 5y \leq 40, \; x + 5y \leq 20, \; x \geq 0, \; y \geq 0$

25. $-x + y \leq 1, \; x + 3y \leq 15, \; 6x - 2y \leq 30, \; x \geq 0, \; y \geq 0$

4.3 GEOMETRIC SOLUTIONS TO LINEAR PROGRAMS

Many practical problems in business, economics, and manufacturing have complicated relationships among requirements for capital, workforce, and raw material. The problems are usually maximization of profits or revenue or, perhaps the minimization of costs or input resources. Optimizations are usually subject to constraints or restrictions on management choices.

If an **objective** to be maximized (or minimized) is linear and constraints can be expressed as linear inequalities, the problem is called a **linear program (LP)**. The applications of LPs span many fields including business, military, agriculture, transportation, the behavioral and social sciences, and much more. The practical importance of LPs developed with the advent of automated computation in complex scheduling and resource allocations by the 1950s.

The examples in this chapter have only two variables, so they can be graphed in two dimensions. In Chapter 5, we introduce methods to solve problems with many more variables.

Example 4.3.1 Formulating a Linear Program

A manufacturer produces two products, A and B, which yield profits of \$2 and \$3, respectively. Resources are sufficient to produce up to 13 items. To produce each unit of product A requires 2 hours, while a unit of product B requires only 1 hour. The manufacturer has 17 hours available for production. How many of each product yields maximum profit?

Solution:

Let x and y represent the number of the respective products to produce and Z the total profit.

Respective profits of \$2 and \$3 are expressed by the objective function, $Z = 2x + 3y$.

The resource constraint is $x + y \leq 13$.

⟶

The time constraint is: $2x + y \leq 17$.
Since negative units of A or B are not sensible, constraints $x \geq 0$, $y \geq 0$ *are included.*
The LP is written as:

$$\begin{aligned} &\textit{Maximize}: Z = 2x + 3y && (\textit{objective/profit function}) \\ &\textit{Subject to}: x + y \leq 13 && (\textit{resource constraint}) \\ &2x + y \leq 17 && (\textit{time constraint}) \\ &x \geq 0,\ \ y \geq 0 && (\textit{nonnegativity constraints}) \end{aligned}$$

Decision variables, as the name implies, are variables whose values are decided by a decision maker. In Example 4.3.1, x and y, used to denote the numbers to produce, are the decision variables. The objective function is the quantity to be maximized (or minimized) and is expressed in terms of decision variables. Here, the profit function, $2x + 3y$, denoted by Z, is the objective function. The four inequalities are the constraints; the variable restrictions $x \geq 0$ and $y \geq 0$ are known as **nonnegativity conditions**.

The maximum or minimum value of the objective function, optimal solutions, of a LP (with a closed feasible solution set) always occurs at corner points (vertices). This can be seen by studying the **level curves**, parallel lines that increase the value of the function from one curve to another in the same direction. The final (last) level curve to intersect the feasible region in the direction of increased objective function values intersects the feasible region at a corner point. This is illustrated in Example 4.3.2 by the level curves $2x + 3y$ (dashed lines) on the graph.

Example 4.3.2 LP Solution – Geometric Solution

Solve the LP graphically:

$$\begin{aligned} &\textit{Maximize}: Z = 2x + 3y\ (\textit{objective function}) \\ &\textit{Subject to}: x + y \leq 13 \\ &2x + y \leq 17 \\ &x \geq 0,\ \ y \geq 0 \end{aligned}$$

Solution:
First, plot the inequalities as equalities to determine the feasible region.

The corner points of the bounded feasible region are (0, 0), (0, 13), (4, 9), and (8.5, 0).

The solution to the system can be found by evaluating Z at each of the four corner points and choosing the largest value since a maximum is desired. Note that while every point in the feasible region is a possible solution, the quest for the optimal solution(s) can be safely limited to the corner points.

⟶

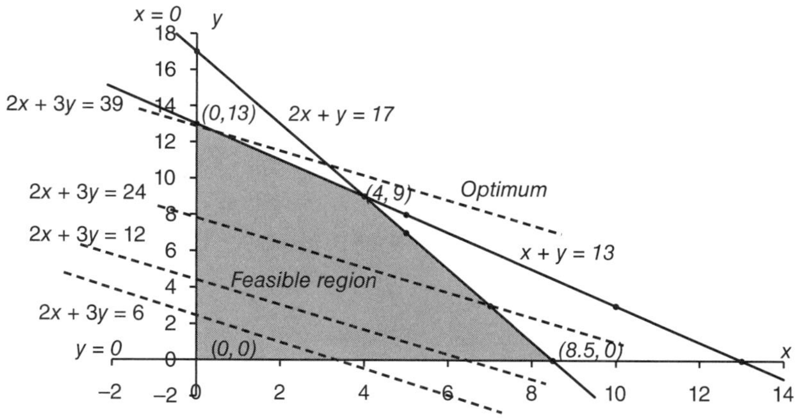

Evaluating Z at each of the four corner points gives

$$Z(0,0) = 2(0) + 3(0) = 0$$

$$Z(0,13) = 2(0) + 3(13) = 39$$

$$Z(4,9) = 2(4) + 3(9) = 35$$

$$Z(8.5,0) = 2(8.5) + 3(0) = 17$$

The maximum value is 39 when $x = 0$ and $y = 13$.

Note: the last of the level curves $2x + 3y = C$ to intersect the feasible region is for $C = 39$. It intersects the feasible region at (0, 13), which is the optimal solution.

Before continuing with the examples, we organize LP characteristics in three shaded boxes.

Properties of Linear Programs

1. **When the feasible region is bounded (finite), the objective function has a finite optimum value.**
2. **When the feasible region is unbounded, the objective function may not have a finite optimum value.**
3. **If the LP has a solution, then the optimum value of the objective function occurs at a corner point of the feasible region.**

Here is a result from the most important theorem of linear programming.

Fundamental Theorem of Linear Programming

Optimal solutions occur at one or more of the vertices of a convex set.
Since a feasible region formed by the constraints is a convex set, the theorem guarantees that at least one optimal solution is at a corner point.

Here are aids to solving LPs geometrically:

Geometric Solutions to a Linear Program

1. **Formulate the LP with its objective function and constraints.**
2. **Graph the constraints and identify the feasible region.**
3. **Locate all corner points of the feasible region.**
4. **Evaluate the objective function at each corner point of the feasible region.**
5. **Corner points having the maximum (or minimum) value of the objective function are optimal solutions. If two or more corner points have the same optimal value for the objective function, all points lying on the segment(s) joining them are optimal solutions.**
6. **If the feasible region is unbounded, there may not be an optimal solution.**

Further examples:

Example 4.3.3 LP Solution (Revisited)

Maximize $Z = 3x + 2y$ subject to the constraints of Example 4.3.2.

Solution:
The corner points of the bounded feasible region are (0, 0), (0, 13), (4, 9), and (8.5, 0). Evaluate Z at the four corner points for its largest value. At least one of them must be optimal as the feasible region is bounded.

$$Z(0,0) = 3(0) + 2(0) = 0$$

$$Z(0,13) = 3(0) + 2(13) = 26$$

$$Z(4,9) = 3(4) + 2(9) = 30$$

$$Z(8.5,0) = 3(8.5) + 2(0) = 25.5$$

$\longrightarrow$

The maximum is $Z = 30$ when $x = 4$ and $y = 9$. Here, the optimal value is unique. Although the feasible region is the same as in Example 4.3.2, the objective is different. The optimum occurs at a different corner point.

Note that the fundamental theorem does not guarantee a solution. For instance, there may be no feasible region or the objective function may be **unbounded**. It does not also guarantee a unique solution since it is possible that the optimal value occurs along a boundary of the feasible region.

A constraint that has no effect on the feasible region is called **redundant**. This means the solution is independent of the redundant constraint. Such a constraint can, but need not, be omitted from the LP.

While LPs in two variables are usually not practical, the ideas from the geometric approach aid in understanding a more general technique such as the **simplex algorithm** in Chapter 5.

These steps are useful when formulating LPs:

Formulation of a Linear Program

1. **It is important to properly identify all relevant constraints and the objective.**
2. **Identify decision variables and whether any could be negative valued.**
3. **Express the objective function in terms of the decision variables.**
4. **Express the constraints as inequalities.**
5. **Solve the resulting LP.**

You can apply these steps in Example 4.3.4.

Example 4.3.4 Jack and Jill

Jack and Jill make large pails and small pails to fetch water from a hill. Jack fabricates the pails and Jill applies designs. Jill places flowers on the small pails and forest scenes on the large pails.

Large pails take Jack 1.5 hours and Jill 1 hour to complete. Each small pail requires 1 hour for Jack and 2 hours for Jill. Jack works up to 16 hours and Jill up to 24 hours per week. They sell the large pails for \$9 and small ones for \$7.

Formulate an LP to determine how many of each type of pail they should make to maximize their revenue.

Solution:

Revenue equals price multiplied by the number of pails sold (demand). The demand for the large pails, x, and for the small, y, are the decision variables.

The revenue, $9x + 7y$, is the objective to be maximized. However, there are constraints. Jack spends 1.5 hours on each large pail and 1 hours on each small pail. His fabrication

⟶

time is 1.5x + y. Jill spends 1 hour on each large pail and 2 hours on each small pail, so her design time is x + 2y. The LP is

$$\begin{aligned} &\textit{Maximize}: Z = 9x + 7y \\ &\textit{Subject to}: 1.5x + y \le 16 \quad (\textit{Fabrication time}) \\ &\qquad x + 2y \le 24 \quad (\textit{Design time}) \\ &\qquad x \ge 0, \ y \ge 0 \end{aligned}$$

Example 4.3.5 Jack and Jill (Revisited)

Solve the LP formulated in Example 4.3.4 graphically.

Solution:

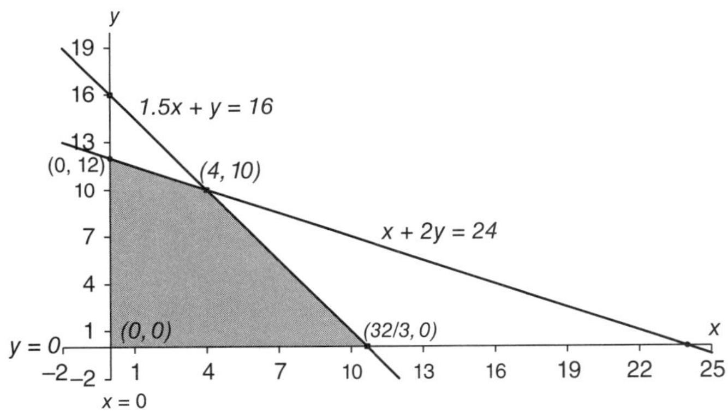

The bounded feasible region has four corner points (0, 0), (4, 10), (0, 12), and (32/3, 0). The values of the objective function at these four points are 0, 106, 84, and 96. Therefore, Jack and Jill should produce 4 large pails and 10 small pails, respectively, for a maximum revenue of $106.

♦ Bank officials regularly decide on the "mix" of investments needed to make good use of funds on deposit. Such investments may include bonds, home mortgages, commercial real estate, and reserves. An excellent tutorial article, intended for bank managers, with examples and graphic illustrations, appeared in The Monthly Review of the Federal Reserve Bank of Richmond (Virginia) by Alfred Broaddus in May 1972.

EXERCISES 4.3

In Exercises 1–5, find the maximum value of each objective function for the feasible region shown.

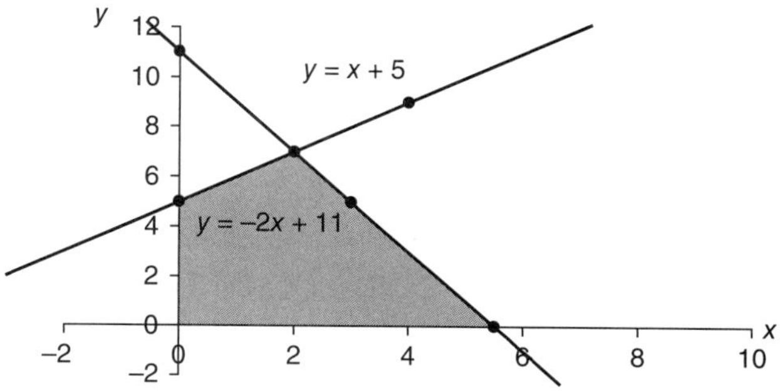

1. $Z = 2x + 3y$
2. $Z = x + 2y$
3. $Z = 10x + 4y$
4. $Z = 2x + y$
5. $Z = 5x - 2y$

In Exercises 6–10, find the minimum value of each objective function for the feasible region shown.

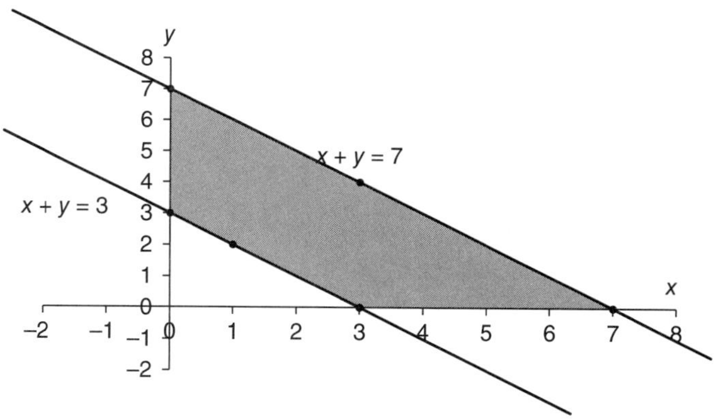

6. $Z = 3x + y$
7. $Z = x + 3y$
8. $Z = 3x + 2y$
9. $Z = 2x + 5y$
10. $Z = 4x - y$

In Exercises 11–16, use the geometric approach to solve each of the LPs.

11.
$$\begin{aligned} \text{Maximize: } & Z = 2x_1 + 3x_2 \\ \text{Subject to: } & 2x_1 + x_2 \leq 8 \\ & x_1 + 2x_2 \leq 7 \\ & x_1,\ x_2 \geq 0 \end{aligned}$$

12.
$$\begin{aligned} \text{Minimize: } & Z = x_1 + 5x_2 \\ \text{Subject to: } & 2x_1 + x_2 \geq 8 \\ & x_1 + 2x_2 \geq 7 \\ & x_1,\ x_2 \geq 0 \end{aligned}$$

13.
$$\begin{aligned} \text{Maximize: } & Z = 5x_1 + 4x_2 \\ \text{Subject to: } & 5x_1 + x_2 \leq 16 \\ & x_1 + 2x_2 \leq 7 \\ & x_1,\ x_2 \geq 0 \end{aligned}$$

14.
$$\begin{aligned} \text{Maximize: } & Z = 4x + 7y \\ \text{Subject to: } & x + y \leq 60 \\ & x + 2y \leq 80 \\ & 4x + y \leq 120 \\ & x,\ y \geq 0 \end{aligned}$$

15.
$$\begin{aligned} \text{Maximize: } & Z = 5x + 2y \\ \text{Subject to: } & -x + y \leq 2 \\ & x + 2y \leq 10 \\ & 3x + y \leq 15 \\ & x,\ y \geq 0 \end{aligned}$$

16.
$$\begin{aligned} \text{Maximize: } & Z = 6x + 5y \\ \text{Subject to: } & x + y \leq 8 \\ & 2x + y \leq 14 \\ & y \leq 6 \\ & x,\ y \geq 0 \end{aligned}$$

In Exercises 17–19, formulate an LP for each described situation. Be sure to identify the decision variables, constraints, and the objective function.

17. A manufacturer produces two products A and B. It takes 15 minutes to make a unit of A and 10 minutes for each unit of B. Forty hours of labor are available for production. Furthermore, it takes 3 ounces of paint to produce an item of A and 4 ounces of paint for an item of B. There are 600 ounces of paint available. Product A sells for \$16 and product B for \$20 each. How should production be scheduled to maximize revenue?

18. An office furniture manufacturer has 10 tons of sheet steel available to produce two-drawer and five-drawer filing cabinets. It requires 50 pounds of steel and 3 hours of labor to produce a two-drawer filing cabinet and 75 pounds of steel and 4 hours of labor to produce a five-drawer filing cabinet. There are 800 hours of labor available, and no more than 200 of the five-drawer filing cabinets are required. The profit for producing a two-drawer filing cabinet is \$25 and \$35 for a five-drawer filing cabinet. How many of each type of cabinet should be made to maximize profit?

19. The ACME truck rental company rents two types of trucks; Type A has 100 cubic feet of refrigerated available space and 200 cubic feet of nonrefrigerated space. Type B has 150 cubic feet of each type of space. The Roadrunner needs to ship at least 4500 cubic feet of refrigerated product and at least 6000 cubic feet of nonrefrigerated product. How many trucks of each type should be rented if Type A rents for \$3 per mile and Type B for \$4 per mile?

20. Solve the ACME truck rental problem of Exercise 19 graphically.

HISTORICAL NOTES

The basics of linear programming can be learned from the simple two variable models depicted in the last section. The next chapter extends the concepts and uses an algorithmic approach to solving systems with more variables. While graphing can also be used for three variables, three-dimensional graphs are impractical. More information on LP follows in the next chapters and their Historical Notes.

CHAPTER 4 SUPPLEMENTARY EXERCISES

1. Graph $x + 2y < 20$
2. Graph $3x + 5y \geq 150$
3. Graph $40x + 30y \leq 120$
4. Graph $y > 2x - 5$
5. A student can work up to 30 hours at two part-time jobs. Depict the possibilities for hours worked at each job graphically.
6. An investor has $50,000 to invest in two stocks and must invest at least $10,000 in each one.
   Depict the investor's options graphically.
7. Graph the system of inequalities and determine the corner points:
$$x + 2y \leq 20, \quad x \geq 0, \quad y \geq 0$$
8. Graph the system of inequalities and determine the corner points:
$$x + y \leq 6, \quad x - y \leq 2, \quad x \geq 0, \quad y \geq 0$$
9. Graph the system of inequalities and determine the corner points:
$$x + 2y \leq 120, \quad 4x - y \leq 30, \quad x \geq 0, \quad y \geq 0$$
10. Graph the system of inequalities and determine the corner points:
$$3x + y \leq 6, \quad 2x - y \leq -1, \quad x \geq 0, \quad y \geq 0$$
11. Graph the system of inequalities and determine the corner points:
$$2x + y \leq 100, \quad 3x - y \leq 100, \quad x \geq 0, \quad y \geq 0$$

12. Maximize: $2x + 5y$

 Subject to: $x + 2y \le 10$

 $x \ge 0, \quad y \ge 0$

13. Maximize: $5x + 2y$

 Subject to: $x + 2y \le 10$

 $x \ge 0, \quad y \ge 0$

14. Maximize: $7x + 10y$

 Subject to: $3x + 2y \le 30$

 $x \ge 0, \quad y \ge 0$

15. Maximize: $3x + 5y$

 Subject to: $3x + 2y \le 30$

 $x \ge 0, \quad y \ge 0$

16. Maximize: $x + 3y$

 Subject to: $x + y \le 5$

 $4x + 3y \le 18$

 $x \ge 0, \quad y \ge 0$

17. Maximize: $20x + 10y$

 Subject to: $x + y \le 6$

 $x - y \le 2$

 $x \ge 0, \quad y \ge 0$

18. Maximize: $5x + y$

 Subject to: $x + y \le 30$

 $2x + y \le 40$

 $x + 2y \le 35$

 $x \ge 0, \quad y \ge 0$

19. Maximize: $3x + 2y$

 Subject to: $x + y \le 30$

 $2x + y \le 40$

 $x + 2y \le 35$

 $x \ge 0, \quad y \ge 0$

20. Maximize: $10x + 8y$

Subject to: $x + y \leq 30$

$2x + y \leq 40$

$x + 2y \leq 35$

$x \geq 0, \quad y \geq 0$

21. Maximize: $3x + 5y$

Subject to: $x + 2y \leq 10$

$3x + 4y \leq 24$

$3x + 2y \leq 21$

$x \geq 0, \quad y \geq 0$

22. Maximize: $4x + 3y$

Subject to: $x + 2y \leq 15$

$3x + 5y \leq 40$

$x + 5y \leq 20$

$x \geq 0, \quad y \geq 0$

23. Maximize: $3x - 2y$

Subject to: $-x + y \leq 1$

$6x - 2y \leq 30$

$x + 3y \leq 15$

$x \geq 0, \quad y \geq 0$

5 Linear Programming – Simplex Method

The limitations of graphical solutions to **linear programs** (Chapter 4) led to searches for alternatives. The best known of these is an iterative numerical scheme known as the **Simplex Method** credited to *George Dantzig*. Its development beginning in the late 1940s along with

Finite Mathematics: Models and Applications, First Edition. Carla C. Morris and Robert M. Stark.

Companion Website: http://www.wiley.com/go/morris/finitemathematics

advances in automated computation and computers has led to **Linear Programming (LP)** being widely used in everyday industrial operations and government planning. Advantages of learning to use the Simplex Method on programs for a few variables give insights into the character of solutions and to computational procedures. Linear programs, having many thousands of variables and constraints are solved routinely in practical contexts.

◆ Google has over 14 million entries related to LP.

5.1 THE STANDARD MAXIMIZATION PROBLEM (SMP)

Consider the linear program:

$$\begin{aligned}
\text{Maximize:} \quad & Z = c_1x_1 + c_2x_2 + \cdots + c_nx_n \\
\text{Subject to:} \quad & a_{11}x_1 + a_{12}x_2 + \cdots + a_{1n}x_n \leq b_1 \\
& a_{21}x_1 + a_{22}x_2 + \cdots + a_{2n}x_n \leq b_2 \\
& \vdots \\
& a_{k1}x_1 + a_{k2}x_2 + \cdots + a_{kn}x_n \leq b_k \\
& x_1,\ x_2,\ \ldots,\ x_n \geq 0
\end{aligned}$$

The $c_1,\ c_2,\ \ldots,\ c_n$ are **objective (or cost) coefficients**, the $a_{11},\ \ldots,\ a_{kn}$ are **constraint coefficients**, and the $b_1,\ \ldots,\ b_k$ are the **resource constants**.

When $b_1,\ b_2,\ \ldots,\ b_k$ are nonnegative, this linear program is a **Standard Maximization Problem** (SMP). It is a linear program as the objectives and constraints are linear.

Example 5.1.1 Identifying a SMP

Why are the following linear programs not SMPs?

a) $$\begin{aligned}
\text{Minimize:} \quad & Z = 5x_1 + 2x_2 \\
\text{Subject to:} \quad & x_1 + 3x_2 \leq 6 \\
& 5x_1 + x_2 \leq 10 \\
& x_1,\ x_2 \geq 0
\end{aligned}$$

b) $$\begin{aligned}
\text{Maximize:} \quad & Z = 3x_1 + 7x_2 \\
\text{Subject to:} \quad & 2x_1 + 3x_2 \leq 12 \\
& 3x_1 + 2x_2 \geq 10 \\
& x_1,\ x_2 \geq 0
\end{aligned}$$

c) $$\begin{aligned}
\text{Maximize:} \quad & Z = 3x_1 + 7x_2 \\
\text{Subject to:} \quad & 2x_1 + x_2 \leq 4 \\
& x_1 + 3x_2 \leq -3 \\
& x_1, x_2 \geq 0
\end{aligned}$$

d) $$\begin{aligned}
\text{Maximize:} \quad & Z = 3x_1 + 2x_2 + 3x_3 \\
\text{Subject to:} \quad & 2x_1 + 3x_2 \leq 12 \\
& 6x_2 + 3x_3 \leq 7 \\
& x_1 + x_2 + 2x_3 \leq 10 \\
& x_1, x_2 \geq 0
\end{aligned}$$

Solution:

a) Seeks minimization, not maximization.

b) The second constraint is not of the "less than" type.

⟶

c) The second constraint has a negative right-hand side *(RHS).*

d) x_3 is not cited as nonnegative valued (and assumed to be unrestricted in sign).

Standard Maximization Problem (SMP)

The standard maximization of a linear program (SMP) requires:

1. **Maximize the objective function.**
2. **All constraints are of the form $a_{j1}x_1 + a_{j2}x_2 + \cdots + a_{jn}x_n \le b_j$, $j = 1, \ldots, k$.**
3. **The constants b_j, $j = 1, \ldots, k$ are nonnegative.**
4. **All $x_i \ge 0, i = 1, \ldots, n$.**

♦ Every linear program can be placed into the SMP form as shown later in this chapter.

Consider this standard maximization problem:

$$\begin{aligned} \text{Maximize:} \quad & Z = 6x_1 + 13x_2 + 20x_3 \\ \text{Subject to:} \quad & 5x_1 + 7x_2 + 10x_3 \le 90{,}000 \\ & x_1 + 3x_2 + 4x_3 \le 30{,}000 \\ & x_1 + x_2 + x_3 \le 9000 \\ & x_1, x_2, x_3 \ge 0 \end{aligned}$$

Clearly, this linear program has the four requirements:

- It is a maximization problem.
- All constraints are of the less than or equal type.
- The RHS constants are nonnegative.
- All variables are nonnegative.

A solution to an SMP is feasible if values of the variable are nonnegative and satisfy all constraints. Such a solution of values $(x_1, x_2, \ldots, x_n)$ is called a **feasible vector**. A feasible vector is a feasible solution.

Example 5.1.2 Feasible Vectors

Are these feasible vectors (solutions) for the SMP above?

a) [1000 2000 3000]

b) [0 3000 6000]

c) [−1000 4000 3000]

⟶

Solution:

a) *The values $x_1 = 1000$, $x_2 = 2000$, and $x_3 = 3000$ satisfy the constraints and constitute a feasible solution (or feasible vector).*

b) *The values $x_1 = 0$, $x_2 = 3000$, and $x_3 = 6000$ is not a feasible solution as it violates the second constraint since $3(3000) + 4(6000) = 31{,}000 > 30{,}000$.*

c) *The values $x_1 = -1000$, $x_2 = 4000$, and $x_3 = 3000$ cannot be feasible because x_1 is negative.*

"Less than or equal to" inequality constraints become equality constraints by adding a new variable, aptly called a **slack variable**. It accounts for the difference between the left-hand side and the **right-hand side (RHS)** of the inequality. For example, $x_1 + 2x_2 \leq 3$ can be written as the equality $x_1 + 2x_2 + s_1 = 3$ by addition of s_1 as a nonnegative slack variable.

Each less than inequality requires its unique slack variable.

The objective and constraints after introducing three necessary slack variables are

$$\begin{aligned}
\text{Maximize:}\quad & Z - 6x_1 - 13x_2 - 20x_3 = 0 \\
\text{Subject to:}\quad & 5x_1 + 7x_2 + 10x_3 + s_1 = 90{,}000 \\
& x_1 + 3x_2 + 4x_3 + s_2 = 30{,}000 \\
& x_1 + x_2 + x_3 + s_3 = 9000 \\
& x_1, x_2, x_3, s_1, s_2, s_3 \geq 0
\end{aligned}$$

It is customary to express the objective function with the variable terms on the left side. This form is called an SMP in **standard form**.

Example 5.1.3 Introducing Slack Variables

Express this linear program in standard form.

$$\begin{aligned}
\textit{Maximize:}\quad & Z = 4x_1 + 3x_2 + 2x_3 \\
\textit{Subject to:}\quad & 3x_1 + 5x_2 + 2x_3 \leq 19 \\
& x_1 + 2x_2 + 3x_3 \leq 30 \\
& x_1 + x_2 + x_3 \leq 6 \\
& x_1, x_2, x_3 \geq 0
\end{aligned}$$

Solution:

The objective is written as $Z - 4x_1 - 3x_2 - 2x_3 = 0$.

The three constraints require slack variables, s_1, s_2, and s_3.

The linear program in standard form becomes:

$$\begin{aligned}
\textit{Maximize:}\quad & Z - 4x_1 - 3x_2 - 2x_3 = 0 \\
\textit{Subject to:}\quad & 3x_1 + 5x_2 + 2x_3 + s_1 = 19 \\
& x_1 + 2x_2 + 3x_3 + s_2 = 30 \\
& x_1 + x_2 + x_3 + s_3 = 6 \\
& x_1, x_2, x_3, s_1, s_2, s_3 \geq 0
\end{aligned}$$

EXERCISES 5.1

For Exercises 1–4, identify SMP linear programs. Explain any non SMPs.

1. Maximize: $Z = 5x_1 + 3x_2 + 2x_3$

   Subject to:
$$\begin{aligned} 3x_1 + x_2 + x_3 &\le 90 \\ x_1 - 3x_2 + 2x_3 &\le 40 \\ x_1 + x_2 - x_3 &\le 10 \\ x_1, x_2, x_3 &\ge 0 \end{aligned}$$

2. Maximize: $Z = x_1 + x_2$

   Subject to:
$$\begin{aligned} 2x_1 + 3x_2 &\ge 9 \\ x_1 + x_2 &\le 3 \\ x_1, x_2 &\ge 0 \end{aligned}$$

3. Maximize: $Z = 4x_1 + 3x_2 + 2x_3 + x_4$

   Subject to:
$$\begin{aligned} x_1 + x_2 + 5x_3 + 2x_4 &\le 17 \\ x_1 + 3x_2 - 4x_3 + 3x_4 &\le 21 \\ x_1 - 5x_2 + 3x_3 + x_4 &\le 13 \\ x_1 + 4x_2 + x_3 + x_4 &\le 10 \\ x_1, x_2, x_3, x_4 &\ge 0 \end{aligned}$$

4. Maximize: $Z = 5x_1 + 3x_2 + 2x_3$

   Subject to:
$$\begin{aligned} x_1 + x_2 + 5x_3 &\le 2 \\ 2x_1 + 3x_2 + 2x_3 &\le -8 \\ 5x_1 - x_2 + x_3 &\le 10 \\ x_1, x_2, x_3 &\ge 0 \end{aligned}$$

5. Which of these vectors are feasible for this SMP?

   Maximize: $Z = 3x + 2y$

   Subject to:
$$\begin{aligned} x + y &\le 7 \\ 2x + y &\le 9 \\ x, y &\ge 0 \end{aligned}$$

   a) [1 1] b) [0 5] c) [5 2] d) [3 3]

6. Which of these vectors are feasible for this SMP?

   Maximize: $Z = x_1 + x_2$

   Subject to:
$$\begin{aligned} 2x_1 + 3x_2 &\le 9 \\ x_1 + x_2 &\le 3 \\ x_1, x_2 &\ge 0 \end{aligned}$$

   a) [2 1] b) [0 3] c) [5 −2] d) [4 0]

7. Which of these vectors are feasible for this SMP?

$$\begin{aligned}
\text{Maximize:}\quad & P = 5x + y + z \\
\text{Subject to:}\quad & x + 2y + z \leq 12 \\
& 2x + y + 3z \leq 18 \\
& x + 3y + z \leq 20 \\
& x,\ y,\ z \geq 0
\end{aligned}$$

a) [1 1 5] b) [0 6 0] c) [2 4 3] d) [4 3 5]

8. Which of these vectors are feasible for this SMP?

$$\begin{aligned}
\text{Maximize:}\quad & P = 3x + 4y + 2z \\
\text{Subject to:}\quad & 5x + y + z \leq 1500 \\
& 4x + 2y + z \leq 2000 \\
& 2x + 2y + z \leq 2500 \\
& x,\ y,\ z \geq 0
\end{aligned}$$

a) [300 0 0] b) [100 500 500] c) [0 1000 400]

9. Show that the vectors [4 3 5] and [3 0 7] are feasible for the following SMP.

$$\begin{aligned}
\text{Maximize:}\quad & Z = 5x_1 + 3x_2 + 2x_3 \\
\text{Subject to:}\quad & x_1 + x_2 + 2x_3 \leq 20 \\
& 2x_1 + x_2 + 2x_3 \leq 25 \\
& 5x_1 + x_2 + x_3 \leq 30 \\
& x_1,\ x_2,\ x_3 \geq 0
\end{aligned}$$

In Exercises 10–14, place the SMP into standard form.

10. The SMP from Exercise 5.1.5.

11. The SMP from Exercise 5.1.6.

12. The SMP from Exercise 5.1.7.

13. The SMP from Exercise 5.1.8.

14. The SMP from Exercise 5.1.9.

15. A manufacturer produces two products, A and B. It takes 15 minutes to make a unit of A and 10 minutes for a unit of B. Forty hours of labor is available to produce them. Furthermore, it takes three ounces of paint to produce a unit of A and 4 ounces for a unit of B. Six hundred ounces of paint are available. The products sell for \$16 and \$20 per unit of A and B, respectively. Formulate a linear program to maximize revenue.

16. An office furniture manufacturer has 10 tons of sheet steel available to produce two-drawer and five-drawer filing cabinets. It requires 50 pounds of steel and 3 hours of labor to produce a two-drawer filing cabinet and 75 pounds of steel and 4 hours of labor to produce a five-drawer filing cabinet. There are 800 hours of labor available and no more than 200 of the five-drawer filing cabinets are required. The profit for producing the two-drawer filing cabinet is \$25 and \$35 for the five-drawer filing cabinet. How many of each type of cabinet should be made to maximize profit?

5.2 TABLEAUS AND PIVOT OPERATIONS

In this section, we develop a format and procedure for solving linear programs.

Recall the linear program rewritten with slack variables (Section 5.1) as:

$$\begin{aligned}
&\text{Maximize:} && Z - 6x_1 - 13x_2 - 20x_3 = 0\\
&\text{Subject to:} && 5x_1 + 7x_2 + 10x_3 + s_1 = 90{,}000\\
& && x_1 + 3x_2 + 4x_3 + s_2 = 30{,}000\\
& && x_1 + x_2 + x_3 + s_3 = 9000\\
& && x_1,\ x_2,\ x_3,\ s_1,\ s_2,\ s_3 \geq 0
\end{aligned}$$

For convenience, organize the program in a **tableau** format as

Row	x_1	x_2	x_3	s_1	s_2	s_3	RHS
1	5	7	10	1	0	0	90,000
2	1	3	4	0	1	0	30,000
3	1	1	1	0	0	1	9000
Objective	−6	−13	−20	0	0	0	$0 = Z$

Note that the bottom row has the objective coefficients and Z moved to the RHS with a current value of zero. If a column of the tableau has a single "1" and zeros elsewhere in that column, then the corresponding variable (at the top of that column) is called a **basic variable**. At the current iteration, above, s_1, s_2, and s_3 are the basic variables.

Nonbasic variables (x_1, x_2, and x_3, here) have the value of zero. As the solution proceeds, basic and nonbasic variables can change status. However, there is always exactly one basic variable in each constraint.

In the initial tableau (the preceding one) with the three slack variables as basic their current values at the RHS are $s_1 = 90{,}000$, $s_2 = 30{,}000$, and $s_3 = 9000$. Geometrically, setting nonbasic variables equal to zero amounts to evaluating the basic variables at corner points of the feasible region.

The solution at this stage or **iteration** is read from the tableau as $x_1 = x_2 = x_3 = 0$, $s_1 = 90{,}000$, $s_2 = 30{,}000$, $s_3 = 9000$, and the objective $Z = 0$.

♦ A partial check for arithmetic errors in any tableau is to substitute current values of basic variables to ensure they satisfy constraints. This is tedious, so work carefully with your calculations.

Having a convenient format for the SMP, we next seek a means for changing from one feasible solution to another and, hopefully, closer to the optimal solution.

One feasible solution is changed into another feasible solution by the exchange of exactly one basic variable for exactly one nonbasic variable. That is, a current basic variable becomes nonbasic while a current nonbasic variable becomes basic. This "one-for-one" exchange maintains the number of basic variables equal to the number of constraint equations, all other variables being nonbasic.

That still leaves open the question of which nonbasic variable to elevate to basic status and, consequently, which basic variable becomes nonbasic. A **pivot element** is the key to the desired change.

To select a pivot element, choose a column with a negative-valued objective coefficient. As a convention, when there is more than one negative objective coefficient, choose the column with the smallest value (most negative). The nonbasic variable for this column is the incoming basic variable. Recall the objective $Z - 6x_1 - 13x_2 - 20x_3 = 0$ in the example introduced at the beginning of this section. If we choose a negative coefficient, a nonbasic variable becomes basic, and therefore increases from zero, the value of Z is increased, progressing toward the maximum. The choice of a nonbasic variable with a positive coefficient will decrease the value of Z and is not useful.

Next, study the ratios of the "b to a" (RHS to constraint coefficients) in the column for the nonbasic variable that is slated to enter the basis. That is, the ratio of the constants of the RHS to the corresponding positive-valued constraint coefficients. This ratio is the amount by which the nonbasic variable can be increased from its current zero value without violating that constraint. If a constraint coefficient happens to be negative, then the new basic variable can be increased to any value without violating that constraint, clearly not a useful choice. The new basic variable can only be increased by the smallest of these positive-valued b/a ratios. The constraint in which that occurs identifies the pivot element (if the smallest b/a ratio is not selected, a constraint is violated).

Example 5.2.1 illustrates the preceding discussion.

Example 5.2.1 The Pivot Element

Determine the pivot element for

Row	x_1	x_2	x_3	s_1	s_2	s_3	*RHS*
1	*5*	*7*	*10*	*1*	*0*	*0*	*90,000*
2	*1*	*3*	*4*	*0*	*1*	*0*	*30,000*
3	*1*	*1*	*1*	*0*	*0*	*1*	*9000*
Objective	*−6*	*−13*	*−20*	*0*	*0*	*0*	*0*

Solution:

There are three negative objective coefficients. The smallest or "most negative" is "−20." The x_3 column becomes the pivot column and x_3 is the incoming basic variable. The constraint coefficients in the x_3 column are positive and the RHS column ratios are, respectively, $\frac{90{,}000}{10} = 9000$, $\frac{30{,}000}{4} = 7500$, and $\frac{9000}{1} = 9000$. The smallest of these, 7500, occurs in the second constraint. Its row becomes the pivot row and the

⟶

coefficient 4 in that row is the pivot element. These are indicated in the next tableau. (Note that s_2, the current basic variable, will leave the basis to make room for x_3.)

Row	x_1	x_2	x_3	s_1	s_2	s_3	RHS
1	5	7	10	1	0	0	90,000
2	1	3	(4)	0	1	0	[30,000]
3	1	1	1	0	0	1	9,000
Objective	−6	−13	[−20]	0	0	0	0

The following are steps for selecting a pivot element.

Selecting a Pivot Element

1. **Among negative coefficients in the objective row, select the smallest. Its column is the pivot column. (In a tie, use either column.)**
2. **Divide each constraint RHS by the positive coefficient in the pivot column of that row, *b/a*. (Skip constraints in which the coefficient is zero or negative valued.) The row in which the smallest of these positive ratios occurs is the pivot row (If all the constraint coefficients in that column are negative, there is no finite solution to the linear program.)**
3. **The pivot element is at the intersection of the pivot column and the pivot row.**

Note: a nonpositive number can never be a pivot element.

If all entries in the column of a negative objective row coefficient are either negative or zero, then the maximum value of Z is infinite. Since this cannot be an answer to any practical problem, probably some boundary constraint(s) was omitted in the formulation, or perhaps an error was made in transcription or arithmetic. For a practical situation, the cause must be remedied.

Once the pivot element is identified, a **pivot operation** is the next step. The pivot element must be a "1" and all other coefficients in its column must be zero. To make the pivot element unity, divide the row by its value. Then add or subtract suitable multiples of the pivot row to the other rows (Gauss–Jordan elimination scheme of Chapter 3) until the required zeros are obtained. Example 5.2.2 illustrates a pivot operation.

Example 5.2.2 The Pivot Operation

Carry out the pivot operation indicated in Example 5.2.1.

Row	x_1	x_2	x_3	s_1	s_2	s_3	RHS
1	5	7	10	1	0	0	90,000
2	1	3	(4)	0	1	0	[30,000]
3	1	1	1	0	0	1	9,000
Objective	−6	−13	[−20]	0	0	0	0

→

Solution:

The pivot element, 4, in the x_3 column is to become a "1" as the coefficient of a new basic variable. Divide row 2 by 4 and place the result in a revised tableau (iteration) before proceeding. Note that the entire row including the RHS value is divided by 4.

x_1	x_2	x_3	s_1	s_2	s_3	RHS
5	7	10	1	0	0	90,000
1/4	3/4	1	0	1/4	0	7500
1	1	1	0	0	1	9000
−6	−13	−20	0	0	0	0

Next, all other entries above and below the pivot element are to become zeros by adding or subtracting suitable multiples of the current pivot row to the other rows (including the objective row). Therefore, multiply row 2 by −10 and add to row 1, that is, $R_1 \to R_1 - 10R_2$, $R_3 \to R_3 - R_2$, $R_{obj} \to R_{obj} + 20R_2$. Completing the pivot operation yields

x_1	x_2	x_3	s_1	s_2	s_3	RHS
5/2	−1/2	0	1	−5/2	0	15,000
1/4	3/4	1	0	1/4	0	7,500
3/4	1/4	0	0	−1/4	1	1,500
−1	2	0	0	5	0	150,000

Note that x_3 has replaced s_2 in the basis. Now, the new solution is $x_1 = 0, x_2 = 0, x_3 = 7500, s_1 = 15,000, s_2 = 0, s_3 = 1500$ and the objective value is $Z = 150,000$, an improvement over the prior feasible solution. However, we do not have an optimal solution since a negative objective coefficient remains ("−1"). The movement from one tableau to the next is called an ***iteration****.*

Example 5.2.3 begins with an SMP and continues through to the first pivot operation.

Example 5.2.3 Readjusting the SMP Tableau

Returning to Jack and Jill Examples 4.3.4 and 4.3.5 (geometric solutions to LPs), we had the SMP:

$$\begin{aligned} \textit{Maximize:} \quad & Z = 9x + 7y \\ \textit{Subject to:} \quad & 1.5x + y \le 16 \\ & x + 2y \le 24 \\ & x \ge 0,\ y \ge 0 \end{aligned}$$

Place this LP in a tableau format and complete an iteration.

⟶

Solution:

First, the SMP in standard form with the objective equal to zero and appropriate slack variables is

$$Z - 9x - 7y = 0$$
$$1.5x + y + s_1 \qquad = 16$$
$$x + 2y \qquad + s_2 = 24$$
$$x \geq 0,\ y \geq 0,\ s_1 \geq 0,\ s_2 \geq 0$$

Next, place the linear program into a tableau format for convenience:

x	y	s_1	s_2	RHS
1.5	1	1	0	16
1	2	0	1	24
−9	−7	0	0	0

Next, locate the pivot column and pivot element!

The most negative objective row coefficient is in the column (x) for the first pivot. The −9 has been boxed as a guide in the next tableau. When the RHS values are divided by the corresponding positive coefficients in the pivot column, there are two ratios: $16/1.5 = 32/3$ and $24/1 = 24$. The minimum ratio is 32/3, so 1.5 is the pivot element (circled below). The first pivot operation occurs on the value 1.5 in the x column.

x	y	s_1	s_2	RHS
(1.5)	1	1	0	16
1	2	0	1	24
[−9]	−7	0	0	0

For the pivot element to become "1," multiply the pivot row by its reciprocal, $1/1.5 = 2/3$.

The resulting numbers for the row are then placed in the revised tableau.

x	y	s_1	s_2	RHS
1	2/3	2/3	0	32/3
1	2	0	1	24
−9	−7	0	0	0

Next, row operations are performed in the x column, so other coefficients in the column become zero (except the pivot element).

This yields the readjusted tableau:

x	y	s_1	s_2	RHS
1	2/3	2/3	0	32/3
0	4/3	−2/3	1	40/3
0	−1	6	0	96

⟶

The new basic variable is x and s_1 becomes a nonbasic variable.

The current solution is $x = 32/3$, $y = 0$, $s_1 = 0$, $s_2 = 40/3$, and an objective is 96.

As mentioned earlier, it may be useful (and optional) to check this solution in the original SMP to help protect against an arithmetic error. First, from the objective, $(32/3) + 7(0) = 96$. It checks! Likewise, from the two constraints $1.5(32/3) + 1(0) + (0) = 16$ and $1(32/3) + 2(0) + (40/3) = 24$. Both check!

In the next section, the tableau iteration is carried to an optimal solution.

EXERCISES 5.2

1. Write the initial tableau for

$$\begin{aligned} \text{Maximize: } & 5x + 2y \\ \text{Subject to: } & x + 3y \le 10 \\ & 2x + y \le 8 \\ & x, y \ge 0 \end{aligned}$$

2. Write the initial tableau for

$$\begin{aligned} \text{Maximize: } & 2x + 5y + z \\ \text{Subject to: } & 2x + 2y + 3z \le 19 \\ & 3x + 4y + 5z \le 26 \\ & x + y + z \le 6 \\ & x, y, z \ge 0 \end{aligned}$$

3. Write the initial tableau for

$$\begin{aligned} \text{Maximize: } & 5x + y + z \\ \text{Subject to: } & x + 2y + z \le 12 \\ & 2x + y + 3z \le 18 \\ & x + 3y + z \le 20 \\ & x, y, z \ge 0 \end{aligned}$$

4. Write the SMP for

x	y	s_1	s_2	RHS
2	4	1	0	20
1	3	0	1	14
−3	−2	0	0	0

5. Write the SMP for

x	y	s_1	s_2	RHS
30	20	1	0	500
5	10	0	1	150
−4	−3	0	0	0

6. Write the SMP for

x_1	x_2	x_3	s_1	s_2	s_3	RHS
2	4	1	1	0	0	13
3	1	2	0	1	0	11
1	6	3	0	0	1	22
−3	−2	0	0	0	0	0

7. Write the SMP for

x_1	x_2	x_3	s_1	s_2	s_3	RHS
3	5	2	1	0	0	19
5	2	4	0	1	0	21
1	7	3	0	0	1	22
−2	−3	0	0	0	0	0

8. Explain why pivoting cannot occur for the following tableau:

x_1	x_2	x_3	s_1	s_2	s_3	RHS
1	−2	1	1	0	0	13
4	−5	1	0	1	0	11
3	0	5	0	0	1	22
3	−2	0	0	0	0	0

9. Find the initial pivot element for the tableau in

a) Exercise 5.2.1.
b) Exercise 5.2.2.
c) Exercise 5.2.3.
d) Exercise 5.2.4.

10. Find the initial pivot element for the tableau in

a) Exercise 5.2.5.
b) Exercise 5.2.6.
c) Exercise 5.2.7.

In Exercises 11–17, perform the pivot for these exercises.

11. Exercise 5.2.1.

12. Exercise 5.2.2.

13. Exercise 5.2.3.

14. Exercise 5.2.4.

15. Exercise 5.2.5.

16. Exercise 5.2.6.

17. Exercise 5.2.7.

18. Write the current solution.

x_1	x_2	x_3	s_1	s_2	s_3	RHS
3/2	2	1	0	0	3	12
3/5	3	0	0	1	−1	1
3/2	4	0	1	0	2	5
1	−2	0	0	0	4	26

19. Write the current solution.

x_1	x_2	x_3	x_4	s_1	s_2	s_3	s_4	RHS
10	−1	0	0	1	0	2	3	15
2	4	1	0	0	0	3	5	20
1	3	0	1	0	0	4	3	75
3	1	0	0	0	1	9	2	10
−1	2	0	0	0	0	4	1	230

20. Write the current solution.

x_1	x_2	x_3	x_4	s_1	s_2	s_3	s_4	RHS
7	0	0	0	1	2	3	2	25
4	0	0	1	0	1	2	6	10
2	0	1	0	0	3	5	1	15
5	1	0	0	0	1	8	2	30
−2	0	0	0	0	4	3	2	90

5.3 OPTIMAL SOLUTIONS AND THE SIMPLEX METHOD

Thus far, we have learned:

- to form an initial tableau;
- to determine a pivot column and element;
- to perform a pivot.

Geometrically, the pivot operation starts at the origin and moves to an adjacent corner point. While this may not prove to be the shortest route to the optimal corner point, it provides a discipline for an easily programmed algorithm.

After the first pivot, the objective row is again surveyed for negative coefficients. As long as there are negative coefficients, the objective value can be increased by continued iteration.

Example 5.3.1 Jack and Jill (continued)

Recall, the Jack and Jill example after the first pivot Example 5.2.3 yielded the tableau

x	y	s_1	s_2	*RHS*
1	2/3	2/3	0	32/3
0	4/3	−2/3	1	40/3
0	−1	6	0	96

Continue to iterate.

Solution:

Additional pivoting is necessary because of the negative value (boxed) in the objective row.

x	y	s_1	s_2	*RHS*
1	2/3	2/3	0	32/3
0	(4/3)	−2/3	1	40/3
0	[−1]	6	0	96

The "b/a" ratios (previous section) identify the pivot element. The pivot column is "y" because of the negative coefficient in the objective row. The minimum of the ratios $(32/3)/(2/3) = 16$ *and* $(40/3)/(4/3) = 10$ *indicates that 4/3 is the pivot element (circled). Performing this pivot yields the next tableau:*

x	y	s_1	s_2	*RHS*
1	0	1	−1/2	4
0	1	−1/2	3/4	10
0	0	11/2	3/4	106

The next solution is $x^* = 4$, $y^* = 10$, $s_1 = 0$, $s_2 = 0$, *and the objective value is 106. Since the objective row no longer contains negative coefficients,* Z^* *cannot increase further. This is an optimal solution (indicated by the asterisks).*

As a check, use the solution values in the constraints (with slack) and objective as described in the last section. This yields the following:

$1.5(4) + 1(10) + 0 = 16$, $(4) + 2(10) + 0 = 24$, *and* $Z^* = 9(4) + 7(10) = 106$. *Checks!*

The steps for using the Simplex Method to solve linear programs are:

The Simplex Method

1. **Among negative-valued objective coefficients, choose the smallest (for clarity, "box it").**
2. **Each positive entry in the column above the boxed coefficient becomes the divisor for the corresponding constant in the right-hand column to form a "*b*/*a*" ratio (RHS/positive column entry). Circle the positive entry for the smallest of these ratios.**
3. **Divide the pivot row (including the constant in the right-hand column) by the value of the pivot element to obtain a unit coefficient. Record the results in a new tableau.**
4. **All other entries in the pivot column are reduced to zero by adding (or subtracting) suitable multiples of the new pivot row to (or from) each row in the preceding (complete) tableau. Use the revised rows to update the new tableau (as in the Gauss–Jordan elimination scheme).**
5. **Repeat Steps 1 through 4 until no negative coefficients remain in the objective row.**
6. **The maximum value of Z is the extreme lower right entry in the final tableau. The optimal value of each basic variable appears in the right-hand column of the row corresponding to its unit coefficient. All other variables are nonbasic and equal zero.**

Correct results with the simplex algorithm depend on accurate arithmetic. Therefore, in hand calculations, fractions are preferable to decimal approximations.

If the optimal value of the objective function in a linear program exists, then that value must occur at one (or more) of the corner points of the feasible solution set or region, known as a **convex set**. This was shown in the geometric solutions to linear programs in the previous chapter. Furthermore, if the feasible region is bounded, then the maximum (or minimum) value of the objective function always exists. If the feasible region for an SMP is unbounded and the coefficients of the objective function are positive, then the maximum value of the objective function is unbounded.

A flow chart summarizes the Simplex Method:

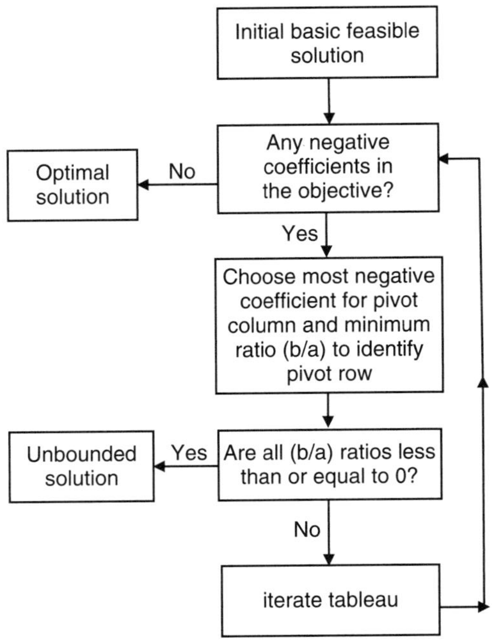

Flow Chart For Simplex Iterations

Example 5.3.2 The Simplex Method

Recall from Example 5.2.2 that after a single pivot the new tableau became:

x_1	x_2	x_3	s_1	s_2	s_3	RHS
5/2	–1/2	0	1	–5/2	0	15,000
1/4	3/4	1	0	1/4	0	7500
(3/4)	1/4	0	0	–1/4	1	1500
[–1]	2	0	0	5	0	150,000

Find the optimal solution.

Solution:

The current solution is not optimal as there is a negative (boxed) coefficient in the objective row. The minimum of the ratios $15{,}000/(5/2) = 6000$, $7500/(1/4) = 30{,}000$, *and* $1500/(3/4) = 2000$ *indicates the pivot element is 3/4 (circled) in the pivot column*

→

(above the boxed number). Multiplying the third row by 4/3 (for the coefficient to become "1") and using the result in a new tableau yields:

x_1	x_2	x_3	s_1	s_2	s_3	RHS
5/2	−1/2	0	1	−5/2	0	15,000
1/4	3/4	1	0	1/4	0	7500
1	1/3	0	0	−1/3	4/3	2000
−1	2	0	0	5	0	150,000

To complete the pivot, use these operations: $R_1 \to R_1 - (5/2)R_3$, $R_2 \to R_2 - (1/4)R_3$, $R_{obj} \to R_{obj} + R_3$. *Completing the pivot operation yields*

x_1	x_2	x_3	s_1	s_2	s_3	RHS
0	−4/3	0	1	−5/3	−10/3	10,000
0	2/3	1	0	1/3	−1/3	7000
1	1/3	0	0	−1/3	4/3	2000
0	7/3	0	0	14/3	4/3	152,000

This indicates the new solution is $x_1^* = 2000$, $x_2^* = 0$, $x_3^* = 7000$, $s_1 = 10{,}000$, $s_2 = 0$, $s_3 = 0$, *and the objective value is* $Z^* = 152{,}000$. *There are no negative numbers in the objective row and this solution is optimal (as indicated by the asterisks).*

Example 5.3.3 The Simplex Method Revisited

Use the Simplex Method to solve the following SMP:

$$\begin{aligned}
\text{Maximize:} \quad & Z = 5x_1 - 2x_2 \\
\text{Subject to:} \quad & 2x_1 + x_2 \le 9 \\
& 2x_1 - 4x_2 \le 4 \\
& -3x_1 + 2x_2 \le 3 \\
& x_1, x_2 \ge 0
\end{aligned}$$

Solution:

Start by rewriting the SMP in a tableau format as:

x_1	x_2	s_1	s_2	s_3	RHS
2	1	1	0	0	9
2	−4	0	1	0	4
−3	2	0	0	1	3
−5	2	0	0	0	0

⟶

Since there is only one negative coefficient (−5) in the objective row, it is in the pivot column. Place a box on the coefficient for clarity.

x_1	x_2	s_1	s_2	s_3	RHS
2	1	1	0	0	9
2	−4	0	1	0	4
−3	2	0	0	1	3
$\boxed{-5}$	2	0	0	0	0

Next, each positive coefficient above the boxed coefficient is divided into the corresponding coefficient for the RHS to give 9/2 and 4/2 = 2. The minimum of these ratios determines the pivot element (circled).

x_1	x_2	s_1	s_2	s_3	RHS
2	1	1	0	0	9
(2)	−4	0	1	0	4
−3	2	0	0	1	3
$\boxed{-5}$	2	0	0	0	0

Now, each entry of the pivot row, including the RHS coefficient, is divided by the value of the pivot element to change it to "1." The results are placed in a new tableau.

x_1	x_2	s_1	s_2	s_3	RHS
2	1	1	0	0	9
1	−2	0	1/2	0	2
−3	2	0	0	1	3
−5	2	0	0	0	0

The remaining coefficients of the pivot column need to be zero. This is accomplished by adding or subtracting suitable multiples of the new pivot row to or from each row of the tableau. The operations that are performed are given as:

$R_1 \rightarrow R_1 - 2R_2$, $R_3 \rightarrow R_3 + 3R_2$, and $R_{obj} \rightarrow R_{obj} + 5R_2$. The results are placed in a new tableau.

x_1	x_2	s_1	s_2	s_3	RHS
0	5	1	−1	0	5
1	−2	0	1/2	0	2
0	−4	0	3/2	1	9
0	−8	0	5/2	0	10

⟶

As there is still a negative coefficient (−8) in the objective row, pivoting continues. The pivot column and element coefficients are boxed and circled as before to give

x_1	x_2	s_1	s_2	s_3	RHS
0	(5)	1	−1	0	5
1	−2	0	1/2	0	2
0	−4	0	3/2	1	9
0	[−8]	0	5/2	0	10

The tableau resulting from making the new pivot element "1" is

x_1	x_2	s_1	s_2	s_3	RHS
0	1	1/5	−1/5	0	1
1	−2	0	1/2	0	2
0	−4	0	3/2	1	9
0	−8	0	5/2	0	10

The row operations needed to complete the pivot operation are $R_2 \rightarrow R_2 + 2R_1$, $R_3 \rightarrow R_3 + 4R_1$, *and* $R_{obj} \rightarrow R_{obj} + 8R_1$. *The tableau that results is*

x_1	x_2	s_1	s_2	s_3	RHS
0	1	1/5	−1/5	0	1
1	0	2/5	1/10	0	4
0	0	4/5	7/10	1	13
0	0	8/5	9/10	0	18

Now, no negative coefficients remain in the objective row indicating an optimal solution (and an end to the pivoting).

This tableau indicates that the optimal solution is that $x_1^* = 4$, $x_2^* = 1$, $s_1 = s_2 = 0$, $s_3 = 13$, *and the maximum objective value is* $Z^* = 18$.

(You may wish to check these values in the original constraint equations and objective to assure that each is satisfied.)

In Example 5.3.3, s_1 and s_2 are nonbasic variables and set to zero. If the objective coefficients (in the objective row) had been zero for either or both (instead of 8/5 and 9/10), it would signal that the optimum is not unique, that is, there are other optimal solutions having the same value for Z. Example 5.3.4 illustrates this.

Example 5.3.4 The Simplex Method Nonunique Optimum

Use the Simplex Method to solve the following SMP and show the optimum is not unique

$$\begin{aligned} \text{Maximize:} \quad & Z = x_1 + x_2 \\ \text{Subject to:} \quad & x_1 + 2x_2 \leq 80 \\ & 4x_1 + x_2 \leq 140 \\ & x_1 + x_2 \leq 50 \\ & x_1, x_2 \geq 0 \end{aligned}$$

⟶

Solution:
Start by rewriting the SMP in a tableau format as:

x_1	x_2	s_1	s_2	s_3	RHS
1	2	1	0	0	80
4	1	0	1	0	140
1	1	0	0	1	50
−1	−1	0	0	0	0

Since there is a tie for the negative coefficient (−1) in the objective row, either is used to determine the pivot column. A box is placed around the coefficient for clarity.

x_1	x_2	s_1	s_2	s_3	RHS
1	2	1	0	0	80
4	1	0	1	0	140
1	1	0	0	1	50
−1	$\boxed{-1}$	0	0	0	0

Next, each positive coefficient above the boxed coefficient is divided into the corresponding coefficient for the RHS to give 80/2, 180/1, and 50/1. The minimum of these ratios determines the pivot element (circled).

x_1	x_2	s_1	s_2	s_3	RHS
1	(2)	1	0	0	80
4	1	0	1	0	140
1	1	0	0	1	50
−1	$\boxed{-1}$	0	0	0	0

Now, each entry of the pivot row, including the RHS coefficient, is divided by the value of the pivot element to change it to "1." The results are placed in a new tableau.

x_1	x_2	s_1	s_2	s_3	RHS
1/2	1	1/2	0	0	40
4	1	0	1	0	140
1	1	0	0	1	50
−1	−1	0	0	0	0

The remaining coefficients of the pivot column need to be zero. This is accomplished by adding or subtracting suitable multiples of the new pivot row to or from each row of the tableau to yield

⟶

x_1	x_2	s_1	s_2	s_3	RHS
1/2	1	1/2	0	0	40
7/2	0	−1/2	1	0	100
1/2	0	−1/2	0	1	10
−1/2	0	1/2	0	0	40

As there is still a negative coefficient (−1/2) in the objective row, pivoting continues. The pivot column and element coefficients are boxed and circled as before to give

x_1	x_2	s_1	s_2	s_3	RHS
1/2	1	1/2	0	0	40
7/2	0	−1/2	1	0	100
(1/2)	0	−1/2	0	1	10
[−1/2]	0	1/2	0	0	40

The tableau that results after pivoting is

x_1	x_2	s_1	s_2	s_3	RHS
0	1	1	0	−1	30
0	0	3	1	−7	30
1	0	−1	0	2	20
0	0	0	0	1	50

Now, there are no negative coefficients in the objective row; indicating an end to the pivoting and an optimal solution.

This tableau has the optimal solution as $x_1^* = 20$, $x_2^* = 30$, $s_1 = s_3 = 0$, $s_2 = 30$, and the objective value is 50. However, the "0" in the bottom row of the nonbasic variable s_1 column of this final tableau means that the solution is not unique. There are other choices for x_1 and x_2 that will also yield an optimum value of 50. In fact, values of x_1 and x_2 on the segment connecting (20, 30) and (30, 20) yield a value of 50 and satisfy the constraints. ***Hint: a clue to a possible nonunique optimum is that the objective is a multiple of any of the constraints. Here, the objective $x_1 + x_2$ and the third constraint $x_1 + x_2 \leq 50$.***

Matrix Format

Linear programs are often expressed in a matrix format. The matrix **A** denotes the constraint coefficients; **B**, the constants of the RHS; and **C**, the objective coefficients. The variables matrix is denoted by **X**. Matrices **B**, **C**, and **X** are column matrices.

An SMP in matrix notation is written

$$\text{Maximize: } \mathbf{Z} = \mathbf{C}^{\mathrm{T}}\mathbf{X}$$

$$\text{Subject to: } \mathbf{AX} \leq \mathbf{B}$$

$$\mathbf{X} \geq \mathbf{0}$$

where $\mathbf{C}^{\mathrm{T}}$ denotes the transpose of **C**.

Actually, computers that solve linear programs use matrix operations even when the data input is in a user-friendly algebraic form as ours. Example 5.3.5 is an illustration.

Example 5.3.5 The Matrix Format

Express the SMP in Example 5.3.3 in matrix format

Solution:
Recall,

$$\begin{aligned} \text{Maximize: } & Z = 5x_1 - 2x_2 \\ \text{Subject to: } & 2x_1 + x_2 \le 9 \\ & 2x_1 - 4x_2 \le 4 \\ & -3x_1 + 2x_2 \le 3 \\ & x_1, x_2 \ge 0 \end{aligned}$$

In matrix format,

$$\begin{aligned} \text{Maximize: } & \mathbf{Z} = \mathbf{C}^T\mathbf{X} \\ \text{Subject to: } & \mathbf{AX} \le \mathbf{B} \\ & \mathbf{X} \ge \mathbf{0} \end{aligned}$$

$$\text{where } \mathbf{A} = \begin{bmatrix} 2 & 1 \\ 2 & -4 \\ -3 & 2 \end{bmatrix}; \mathbf{B} = \begin{bmatrix} 9 \\ 4 \\ 3 \end{bmatrix}; \mathbf{C} = \begin{bmatrix} 5 \\ -2 \end{bmatrix}; \text{ and } \mathbf{X} = \begin{bmatrix} x_1 \\ x_2 \end{bmatrix}.$$

Note that these matrices are conformable:
*$\mathbf{C}^T$, the transpose of **C**, is a (1×2) while **X** is a (2×1) so Z is a $(1 \times 1)(= (1 \times 2)(2 \times 1))$ matrix. Similarly, **A** is a (3×2) while **X**, being (2×1), results in a (3×1) $(= (3 \times 2)(2 \times 1))$ for **B**.*

The computational importance of the matrix format derives from the incredible speed and size of matrices routinely manipulated by automated computation. The practical importance of linear programs derives from this extraordinary computational facility.

Our discussion has centered on the SMP linear program. Recall our earlier mention that every linear program can be placed in an SMP standard form. The Simplex Method here can easily be extended to situations requiring minimization, having "greater than" constraints, and negative-valued variables. That is the subject of the next section.

EXERCISES 5.3

Use the Simplex Method to solve these exercises.

1. $$\begin{aligned} \text{Maximize: } & Z = 2x_1 + 3x_2 \\ \text{Subject to: } & 2x_1 + x_2 \le 8 \\ & x_1 + 2x_2 \le 7 \\ & x_1, x_2 \ge 0 \end{aligned}$$

2. Maximize: $Z = 4x_1 + 5x_2$
Subject to: $x_1 + 2x_2 \leq 8$
$2x_1 + x_2 \leq 10$
$x_1, x_2 \geq 0$

3. Maximize: $Z = 3x_1 + 2x_2$
Subject to: $x_1 + x_2 \leq 24$
$2x_1 + x_2 \leq 32$
$x_1, x_2 \geq 0$

4. Maximize: $Z = 5x_1 + 4x_2$
Subject to: $5x_1 + x_2 \leq 16$
$x_1 + 2x_2 \leq 7$
$x_1, x_2 \geq 0$

5. Maximize: $Z = x_1 + 3x_2$
Subject to: $x_1 + x_2 \leq 10$
$2x_1 + x_2 \leq 14$
$x_1, x_2 \geq 0$

6. Maximize: $Z = 4x_1 + 7x_2$
Subject to: $x_1 + x_2 \leq 60$
$x_1 + 2x_2 \leq 80$
$4x_1 + x_2 \leq 120$
$x_1, x_2 \geq 0$

7. Maximize: $Z = 5x_1 + x_2 + x_3$
Subject to: $2x_1 + x_2 + x_3 \leq 50$
$x_1 + 3x_2 \leq 20$
$x_2 + x_3 \leq 15$
$x_1, x_2, x_3 \geq 0$

8. Maximize: $Z = 3x_1 + 4x_2 + x_3$
Subject to: $3x_1 + 5x_2 + 4x_3 \leq 30$
$3x_1 + 2x_2 \leq 4$
$x_1 + 2x_2 \leq 8$
$x_1, x_2, x_3 \geq 0$

9. Maximize: $Z = 2x_1 + 3x_2 + 4x_3$
   Subject to: $2x_1 + x_2 + x_3 \leq 4$
   $x_1 + 2x_2 \leq 10$
   $x_2 + 3x_3 \leq 8$
   $x_1, x_2, x_3 \geq 0$

10. A manufacturer produces two products X and Y. It takes 15 minutes to make a unit of X and 10 minutes for each unit of Y. There are 40 hours of labor available to produce the units. Furthermore, it takes 3 ounces of paint to produce a unit of X and 4 ounces for a unit of Y. There are 600 ounces of paint available to produce the products. Product X is priced at \$16 and product Y at \$20 each. How should production be scheduled to maximize revenue?

11. An office furniture manufacturer has 10 tons of sheet steel available to produce two-drawer and five-drawer filing cabinets. It requires 50 pounds of steel and 3 hours of labor to produce a two-drawer cabinet and 75 pounds of steel and 4 hours of labor to produce a five-drawer cabinet. There are 800 hours of labor available, and no more than 200 of the five-drawer cabinets are required. The profit for producing a two-drawer cabinet is \$25 and \$35 for the five-drawer cabinet. How many of each cabinet type are needed to maximize profit?

12. Write the SMP in matrix format:

$$\begin{aligned} \text{Maximize:} \quad & Z = 3x_1 + 2x_2 \\ \text{Subject to:} \quad & 2x_1 + x_2 \leq 50 \\ & x_1 + 3x_2 \leq 75 \\ & x_1 \geq 0,\ x_2 \geq 0 \end{aligned}$$

13. Write the SMP in matrix format:

$$\begin{aligned} \text{Maximize:} \quad & Z = 5x_1 + 3x_2 + 2x_3 \\ \text{Subject to:} \quad & x_1 + 3x_2 \leq 30 \\ & 2x_1 + x_2 + x_3 \leq 60 \\ & x_1 + 2x_2 + 4x_3 \leq 125 \\ & x_1 \geq 0,\ x_2 \geq 0,\ x_3 \geq 0 \end{aligned}$$

14. Write the SMP in matrix format:

$$\begin{aligned} \text{Maximize:} \quad & Z = 9x_1 + 2x_2 + 3x_3 \\ \text{Subject to:} \quad & x_1 + 5x_3 \leq 50 \\ & 6x_1 + 4x_2 + 3x_3 \leq 100 \\ & 3x_1 + 2x_2 \leq 250 \\ & x_1 \geq 0,\ x_2 \geq 0,\ x_3 \geq 0 \end{aligned}$$

5.4 DUAL PROGRAMS

Objective functions of linear programs are to be optimized, that is, a maximization or a minimization. To here SMPs were solved with the Simplex Method. It is a remarkable fact that every linear program has associated with it another linear program. One of them is designated as a **primal** program and a complement, derived from it, is a **dual** program.

The pairing is illustrated in Example 5.4.1.

Example 5.4.1 The Dual Program

For the (primal) linear program

$$\begin{aligned} \text{Maximize: } \quad & Z = 7x_1 - 4x_2 + x_3 \\ \text{Subject to: } \quad & 2x_1 + 3x_2 + 5x_3 \leq 1000 \\ & x_1 + 2x_2 + 3x_3 \leq 600 \\ & 5x_1 + \ x_2 + 6x_3 \leq 1200 \\ & x_1 \geq 0,\ x_2 \geq 0,\ x_3 \geq 0 \end{aligned}$$

write the dual program.

Solution:

*To formulate a dual program begin by identifying new variables, one for each constraint. Call them the **dual variables** y_1, y_2, and y_3, corresponding to the three constraints.*

*The **dual objective** function, V, is formed from the primal constraint RHSs, and, since the primal objective requires a maximization, the dual objective is a minimization. That is,*

$$\text{Minimize: } V = 1000y_1 + 600y_2 + 1200y_3$$

In this case, the dual constraints are written
Subject to:

$$\begin{aligned} 2y_1 + \ y_2 + 5y_3 &\geq 7 \\ 3y_1 + 2y_2 + \ y_3 &\geq -4 \\ 5y_1 + 3y_2 + 6y_3 &\geq 1 \\ y_1, y_2, y_3 &\geq 0 \end{aligned}$$

Note that the primal objective coefficients are the RHSs of the dual constraints.

Also, dual constraints are formed from the transposed primal constraint coefficients, and "less than or equal to" constraints (≤) in the primal become "greater than or equal to" constraints (≥) in the dual.

Finally, note that a negative-valued RHS (not permissible in the SMP) is permitted in the dual.

A comparison is useful:

Maximize: $Z = 7x_1 - 4x_2 + x_3$	*Minimize:* $V = 1000y_1 + 600y_2 + 1200y_3$
Subject to:	*Subject to:*
$2x_1 + 3x_2 + 5x_3 \leq 1000$	$2y_1 + y_2 + 5y_3 \geq 7$
$x_1 + 2x_2 + 3x_3 \leq 600$	$3y_1 + 2y_2 + y_3 \geq -4$
$5x_1 + x_2 + 6x_3 \leq 1200$	$5y_1 + 3y_2 + 6y_3 \geq 1$
$x_1, x_2, x_3 \geq 0$	$y_1, y_2, y_3 \geq 0$

The primal-dual relationship has an important interpretation. The primal objective, here, is a maximization of output (profit, say) for fixed input or available resources (the constraint RHSs).

The dual is a complement with inputs (resources) to be minimized to meet fixed output requirements.

An economic interpretation of the dual is to maximize profits or minimize costs when revenues are fixed as

$$\text{Profit} = \text{Revenue} - \text{Cost}$$

Economists often refer to dual variables as "shadow prices". They measure the marginal value of the resource or the rate at which the objective, Z, increases with (slight) increase in resources.

Clearly, maximizing profits and minimizing costs, when revenues are fixed, are "two sides of the same coin."

It is a mathematical fact that the dual of the dual returns the primal problem. Furthermore, not only are the optimal objective values of dual and primal identical, having the solution to one of them, dual or primal, yields the solution to the other.

Fundamental Principle of Duality

A primal problem has a solution if, and only if, its dual problem has a solution. If a solution exists, then the optimal objective values of the primal and dual problems are identical.

Example 5.4.2 The Dual of the Dual

Show that the dual of the dual problem is the primal problem for

Primal: Maximize: $Z = c_1x_1 + c_2x_2$

Subject to:

$$\begin{aligned} a_{11}x_1 + a_{12}x_2 &\le b_1 \\ a_{21}x_1 + a_{22}x_2 &\le b_2 \\ a_{31}x_1 + a_{32}x_2 &\le b_3 \\ x_1, x_2 &\ge 0 \end{aligned}$$

Solution:

Use y_1, y_2, *and* y_3 *as the three dual variables corresponding to the three primal constraints. (Note that the primal has two variables* x_1, *and* x_2 *and two objective coefficients* c_1, *and* c_2.*)*

The dual problem is

Minimize: $V = b_1y_1 + b_2y_2 + b_3y_3$

Subject to:

$$\begin{aligned} a_{11}y_1 + a_{21}y_2 + a_{31}y_3 &\ge c_1 \\ a_{12}y_1 + a_{22}y_2 + a_{32}y_3 &\ge c_2 \\ y_1, y_2, y_3 &\ge 0 \end{aligned}$$

⟶

Next, to form the dual of this dual, begin by assigning two variables, say, w_1 and w_2, corresponding to the two constraints. Next, replace the minimization by a maximization and the objective, V, becomes

Maximize: $V = c_1 w_1 + c_2 w_2$

There are three constraints since there are three objective coefficients in the dual problem. The transpose of the earlier dual constraint coefficients reverses their greater than signs

$$a_{11}w_1 + a_{12}w_2 \le b_1$$
$$a_{21}w_1 + a_{22}w_2 \le b_2$$
$$a_{31}w_1 + a_{32}w_2 \le b_3$$
$$w_1,\ w_2,\ \ge 0$$

This is the original primal problem with w_1 and w_2 replacing x_1 and x_2, respectively, and the maximization objective denoted by Z.

Note that duality permits changing a linear program having a minimization with greater than constraints to its dual as an SMP. An SMP can be solved by the Simplex Method described earlier in this chapter.

Example 5.4.3 Dual Solution

Convert the primal linear program to an SMP and solve.

$$\begin{aligned} \text{Minimize:}\quad & V = 4x_1 + 3x_2 + 85x_5 + 93x_6 \\ \text{Subject to:}\quad & 2x_1 + x_2 - x_3 + x_5 \ge 6 \\ & x_1 + 2x_2 - x_4 + x_6 \ge 5 \\ & x_1,\ x_2,\ x_3,\ x_4,\ x_5,\ x_6 \ge 0 \end{aligned}$$

Solution:

Since there are only two constraints, there are only two dual variables, y_1 and y_2, associated with them. Also, since the objective is to be minimized and the constraints are $\ge$, the dual problem will be a maximization with $\le$ constraints, that is, an SMP. The six primal objective coefficients are translated into six dual constraints as:

$$\begin{aligned} 2y_1 + y_2 &\le 4 \\ y_1 + 2y_2 &\le 3 \\ -y_1 \quad &\le 0 \\ -y_2 &\le 0 \\ y_1 \quad &\le 85 \\ y_2 &\le 93 \\ y_1, y_2 &\ge 0 \end{aligned}$$

and the dual objective is

→

Maximize: $V = 6y_1 + 5y_2$

Note that the third and fourth constraints simply imply that y_1 *and* y_2 *are* ≥ 0. *They can be omitted.*

Solving yields $y_1{}^* = 5/3$, $y_2{}^* = 2/3$ *and the final tableau*

y_1	y_2	s_1	s_2	s_3	s_4	RHS
1	0	2/3	−1/3	0	0	5/3
0	1	−1/3	2/3	0	0	2/3
0	0	−2/3	1/3	1	0	250/3
0	0	1/3	−2/3	0	1	277/3
0	0	7/3	4/3	0	0	40/3

Example 5.4.4 Dual Variables in the Simplex Tableau

Use the Simplex Method to solve the SMP, then formulate its dual, and verify that the values of the dual in the simplex tableau are correct.

$$\begin{aligned} \text{Maximize:}\quad & Z = 5x_1 + 3x_2 \\ \text{Subject to:}\quad & x_1 + x_2 \leq 5 \\ & -3x_1 + x_2 \leq 1 \\ & x_1, x_2 \geq 0 \end{aligned}$$

Solution:

x_1	x_2	s_1	s_2	RHS
1	1	1	0	5
−3	1	0	1	1
−5	−3	0	0	0

x_1	x_2	s_1	s_2	RHS
1	1	1	0	5
0	4	3	1	16
0	2	(5)	(0)	25

→

The solution to the primal is $x_1^* = 5$, $x_2^* = 0$, *and* $Z^* = 25$.
Now, formulating the dual problem using y_1 *and* y_2 *as dual variables*

$$\begin{aligned} \textit{Minimize:} \quad & V = 5y_1 + y_2 \\ \textit{Subject to:} \quad & y_1 - 3y_2 \geq 5 \\ & y_1 + \ \ y_2 \geq 3 \\ & y_1,\ y_2 \geq 0 \end{aligned}$$

The theory of Linear Programming asserts that y_1^* *and* y_2^*, *the optimal values of the two dual variables, are the objective coefficients (circled in the tableau) of the slack variables in the final tableau of the simplex problem. That is,* $y_1^* = 5$ *and* $y_2^* = 0$.
Evaluating $V^* = 5(5) + 1(0) = 25 = Z^*$, *as expected,* $V^* = Z^*$ *at optimality.*
(You may check that y_1^* *and* y_2^* *satisfy the dual constraints.)*

The theory of primal and dual linear programs includes other interesting facets beyond our scope.

EXERCISES 5.4

Write the dual for Exercises 1 and 2

1. $$\begin{aligned} \text{Maximize:} \quad & Z = 5x_1 + x_2 + x_3 \\ \text{Subject to:} \quad & x_1 + 2x_2 + 3x_3 \leq 12 \\ & 2x_1 + \ \ x_2 + 3x_3 \leq 18 \\ & x_1 + 3x_2 + \ \ x_3 \leq 20 \\ & x_1,\ x_2,\ x_3 \geq 0 \end{aligned}$$

2. $$\begin{aligned} \text{Maximize:} \quad & Z = 6x_1 + 13x_2 + 20x_3 \\ \text{Subject to:} \quad & 5x_1 + 7x_2 + 10x_3 \leq 90{,}000 \\ & x_1 + 3x_2 + \ \ 4x_3 \leq 30{,}000 \\ & x_1 + \ \ x_2 + \ \ \ \ x_3 \leq 9000 \\ & x_1,\ x_2,\ x_3 \geq 0 \end{aligned}$$

3. Solve the primal linear program by converting to a dual and solving as an SMP.

$$\begin{aligned} \text{Minimize:} \quad & Z = 11x_1 + 7x_2 \\ \text{Subject to:} \quad & x_1 + 2x_2 \geq 10 \\ & 3x_1 + \ \ x_2 \geq 15 \\ & x_1,\ x_2 \geq 0 \end{aligned}$$

4. Solve the primal linear program by converting to a dual and solving as an SMP.

$$\begin{aligned} \text{Minimize:} \quad & Z = 12x_1 + 18x_2 \\ \text{Subject to:} \quad & x_1 + 2x_2 \geq 8 \\ & x_1 + x_2 \geq 6 \\ & x_1,\ x_2 \geq 0 \end{aligned}$$

5. Solve the primal linear program by converting to a dual and solving as an SMP.

$$\begin{aligned} \text{Minimize:} \quad & Z = 3x_1 + 2x_2 \\ \text{Subject to:} \quad & 3x_1 + 4x_2 \geq 46 \\ & 5x_1 + 3x_2 \geq 40 \\ & x_1,\ x_2 \geq 0 \end{aligned}$$

6. Solve the primal linear program by converting to a dual and solving as an SMP.

$$\begin{aligned} \text{Minimize:} \quad & Z = 12x_1 + 18x_2 + 20x_3 \\ \text{Subject to:} \quad & x_1 + 2x_2 + x_3 \geq 5 \\ & 2x_1 + x_2 + 3x_3 \geq 1 \\ & x_1 + 3x_2 + x_3 \geq 3 \\ & x_1,\ x_2,\ x_3 \geq 0 \end{aligned}$$

7. Write the dual linear program and then solve both the primal and the dual using the Simplex Method.

$$\begin{aligned} \text{Minimize:} \quad & Z = 2x_1 + 5x_2 \\ \text{Subject to:} \quad & 4x_1 + x_2 \geq 12 \\ & x_1 + x_2 \geq 9 \\ & x_1 + x_2 \geq 15 \\ & x_1,\ x_2 \geq 0 \end{aligned}$$

8. Write the dual linear program and then solve both the primal and the dual using the Simplex Method.

$$\begin{aligned} \text{Minimize:} \quad & Z = 12x_1 + 48x_2 + 8x_3 \\ \text{Subject to:} \quad & x_1 + 3x_2 \geq 1 \\ & 4x_1 + 6x_2 + x_3 \geq 3 \\ & 4x_2 + x_3 \geq 1 \\ & x_1,\ x_2,\ x_3 \geq 0 \end{aligned}$$

5.5 NON-SMP LINEAR PROGRAMS

The Simplex Method for the standard maximization problem, SMP, can be used to solve every linear program of the form

$$\begin{aligned}\text{Maximize:}\quad & Z = c_1x_1 + c_2x_2 + \cdots + c_nx_n \\ \text{Subject to:}\quad & a_{11}x_1 + a_{12}x_2 + \cdots + a_{1n}x_n \le b_1 \\ & a_{21}x_1 + a_{22}x_2 + \cdots + a_{2n}x_n \le b_2 \\ & \vdots \\ & a_{k1}x_1 + a_{k2}x_2 + \cdots + a_{kn}x_n \le b_k \\ & x_1,\ x_2, \cdots,\ x_n \ge 0\end{aligned}$$

However, recall our earlier assertion that every linear program can be cast into an SMP. Let us examine every possibility!

Minimization Objective

If an objective calls for minimization as in

$$\text{Minimize:}\quad Z' = c_1x_1 + c_2x_2 + \cdots + c_nx_n$$

Simply multiply Z' by -1 so

$$-Z' = -c_1x_1 - c_2x_2 - \cdots - c_nx_n$$

and maximize $-Z' = Z$.

The maximization of a positive linear quantity is equivalent to the minimization of its negative! Thus, any minimization objective can be placed in SMP format by multiplying by -1.

For example, the objective

$$\text{Minimize:}\quad Z' = -6x_1 + 2x_2 + 5x_3$$

is replaced by

$$\text{Maximize:}\quad Z = -Z' = 6x_1 - 2x_2 - 5x_3.$$

Example 5.5.1 Minimization

Convert this linear program into an SMP format.

$$\begin{aligned}\textit{Minimize:}\quad & Z = x_1 + 2x_2 - x_3 + x_4 \\ \textit{Subject to:}\quad & x_1 - x_2 + x_3 \le 4 \\ & x_1 + 2x_2 + 2x_4 \le 6 \\ & x_2 - 2x_3 + x_4 \le 2 \\ & -x_1 + 2x_2 + x_3 \le 2 \\ & x_1,\ x_2,\ x_3,\ x_4 \ge 0\end{aligned}$$

⟶

Solution:
The SMP format requires maximization. Multiplying the objective by −1 and adding slack variables yields

$$\begin{array}{llll} \textit{Maximize:} & -Z = -x_1 - 2x_2 + x_3 - x_4 & & \\ \textit{Subject to:} & x_1 - x_2 + x_3 + s_1 & = 4 \\ & x_1 + x_2 + 2x_4 + s_2 & = 6 \\ & x_2 - 2x_3 + x_4 + s_3 & = 2 \\ & -x_1 + 2x_2 + x_3 + s_4 & = 2 \\ & x_1,\ x_2,\ x_3,\ x_4,\ s_1,\ s_2,\ s_3,\ s_4 \geq 0 & \end{array}$$

An SMP format!

Variables Unrestricted in Sign

By an **unrestricted variable**, we mean that it may assume negative and positive values within the problem context. Recall, only nonnegative-valued variables appear in an SMP, that is, $x_1,\ x_2,\ \ldots,\ x_n \geq 0$.

Suppose that a linear program has variables that are unrestricted in sign. Specifically, let x be a variable in the program that could conceivably be negative valued. Such instances do not arise often in practice. Exceptions arise in applications to curve-fitting, deteriorating or spoiled inventory, and modeling of chemical processes, as examples. Simply replace x by the difference of two nonnegative variables, say $x^+ - x^-$, where x^+ and x^- are each ≥ 0. Every negative number can be expressed as the difference of two positive numbers. For example, −3 can be written as (+3) – (+6). Every unrestricted variable is treated in this way. For example, consider the "less than or equal to" constraint.

$$\begin{gathered} 2x_1 + x_2 + 3x_3 \leq 6 \\ x_1,\ x_2 \geq 0,\ x_3 \text{ unrestricted} \end{gathered}$$

The unrestricted variable, x_3, is replaced by the nonnegative variables x_3^+ and x_3^- as

$$\begin{gathered} 2x_1 + x_2 + 3x_3^+ - 3x_3^- \leq 6 \\ x_1,\ x_2,\ x_3^-,\ x_3^+ \geq 0 \end{gathered}$$

You may not be familiar with this simple means of replacing an unrestricted variable by two positive ones probably because it increases the number of variables. Small numbers of variables is far more important in other areas of mathematical applications than in LP where the number of variables is virtually no impediment to automated computation.

Example 5.5.2 Unrestricted Variables

Rewrite this linear program as an SMP format.

$$\begin{array}{ll} \textit{Maximize:} & Z = -x_1 - 2x_2 + x_3 - x_4 \\ \textit{Subject to:} & x_1 - x_2 + x_3 + s_1 = 4 \\ & x_1 + x_2 + 2x_4 + s_2 = 6 \\ & x_2 - 2x_3 + x_4 + s_3 = 2 \\ & -x_1 + 2x_2 + x_3 + s_4 = 2 \\ & x_1,\ x_4,\ s_1,\ s_2,\ s_3,\ s_4 \geq 0 \textit{ and } x_2,\ x_3 \textit{ unrestricted in sign.} \end{array}$$

⟶

Solution:
Replacing x_2 and x_3 by nonnegative variables $x_2^+ - x_2^-$ and $x_3^+ - x_3^-$, respectively, yields

$$\begin{aligned}
\textit{Maximize:}\quad & Z = -x_1 - 2x_2^+ + 2x_2^- + x_3^+ - x_3^- - x_4 \\
\textit{Subject to:}\quad & x_1 - x_2^+ + x_2^- + x_3^+ - x_3^- + s_1 = 4 \\
& x_1 + x_2^+ - x_2^- + 2x_4 + s_2 = 6 \\
& x_2^+ - x_2^- - 2x_3^+ + 2x_3^- + x_4 + s_3 = 2 \\
& -x_1 + 2x_2^+ - 2x_2^- + x_3^+ - x_3^- + s_4 = 2 \\
& x_1,\ x_2^-,\ x_2^+,\ x_3^-,\ x_3^+,\ x_4,\ s_1,\ s_2,\ s_3,\ s_4 \geq 0
\end{aligned}$$

An SMP format!

Greater Than or Equality Constraints

Recall, that the inequality constraints in an SMP are of the "less than or equal to" type. This means that the addition of a slack variable serves two purposes: the inequality becomes an equality and the slack variable becomes a basic variable. For example, the addition of a nonnegative slack variable, s, alters

$$2x_1 + x_2 + 3x_3 \leq 6$$
$$x_1,\ x_2,\ x_3 \geq 0$$

to

$$2x_1 + x_2 + 3x_3 + s = 6$$
$$x_1,\ x_2,\ x_3,\ s \geq 0$$

in accord with the SMP format.

When the constraint is a "greater than or equal to" as

$$2x_1 + x_2 + 3x_3 \geq 6$$

one can subtract an **excess** or **surplus variable**, s.

That is, the constraint becomes

$$2x_1 + x_2 + 3x_3 - s = 6$$
$$x_1,\ x_2,\ x_3,\ s \geq 0$$

The surplus variable results in a constraint equation. However, unlike the less than case, s is not a basic variable since its coefficient is not unity. Multiplying by -1 does not help here if the RHS becomes negative. Recall that an SMP requires nonnegative RHSs for constraints.

What to do?

The next example demonstrates three ways to create a basic variable in a "greater than or equal to" constraint: "by hand," or automated Phase I, or an alternate method. You only need to learn one (any one!) of the three ways to solve, in principle at least, every linear program by converting it into an SMP.

"By Hand" Method

Example 5.5.3 Toward an SMP

Rewrite as an SMP format.

$$\begin{aligned} \text{Maximize:} \quad & Z = -x_1 - 3x_2 \\ \text{Subject to:} \quad & x_1 + 4x_2 \geq 48 \\ & 5x_1 + x_2 \geq 50 \\ & x_1,\ x_2 \geq 0 \end{aligned}$$

Solution:

First, introduce nonnegative surplus variables x_3 and x_4 (for the greater than constraints) to yield

$$\begin{aligned} x_1 + 4x_2 - x_3 \quad &= 48 \\ 5x_1 + x_2 \quad - x_4 &= 50 \\ x_1,\ x_2,\ x_3,\ x_4 &\geq 0 \end{aligned}$$

Clearly, x_1 and x_2 must become basic variables to achieve an SMP format (as the negative coefficients of x_3 and x_4 preclude their solution). Algebraic manipulation of the two constraints yield

$$\begin{aligned} x_1 \quad + \frac{1}{19}x_3 - \frac{4}{19}x_4 &= 8 \\ x_2 - \frac{5}{19}x_3 + \frac{1}{19}x_4 &= 10 \end{aligned}$$

(Hint: use three steps in the manipulation $R_2 \to R_2 - 5R_1,\ R_2 \to (-1/19)R_2,\ R_1 \to R_1 - 4R_2$)

Next, x_1 and x_2 must be replaced in the objective since they are now basic variables. To do this, solve for x_1 and x_2 in the constraints and substitute into the objective as

$$\text{Maximize:} -\left(8 - \frac{1}{19}x_3 + \frac{4}{19}x_4\right) - 3\left(10 + \frac{5}{19}x_3 - \frac{1}{19}x_4\right) = -38 - \frac{4}{19}x_3 - \frac{1}{19}x_4$$

The linear program in SMP format becomes

$$\begin{aligned} \text{Maximize:} \quad & -38 - \frac{4}{19}x_3 - \frac{1}{19}x_4 \\ \text{Subject to:} \quad & x_1 \quad + \frac{1}{19}x_3 - \frac{4}{19}x_4 = 8 \\ & x_2 - \frac{5}{19}x_3 + \frac{1}{19}x_4 = 10 \\ & x_1,\ x_2,\ x_3,\ x_4 \geq 0 \end{aligned}$$

Now, in SMP format, one can iterate to an optimal solution.

By inspection, the maximum is -38 (retaining x_3 and x_4 at zero). This yields $x_1^* = 8$ and $x_2^* = 10$.

This method, although simple, is not useful when there are many constraints, the labor-intensive algebraic manipulation being impractical.

Phase I Method

A second method uses the Simplex Method itself to generate an SMP. Called, a "**Phase I Method,**" another new nonnegative variable is added to each "greater than or equal to" constraint. This is in addition to the surplus variable that has been subtracted to achieve an equality constraint. For example, from the "greater than or equal to" constraint

$$2x_1 + x_2 + x_3 \geq 6$$

a surplus variable s_1 is subtracted and a new variable w_1 added for an equality constraint

$$2x_1 + x_2 + x_3 - s_1 + w_1 = 6$$

The new variable, w_1, known as an **artificial variable**, is a basic variable as it appears with a positive unit coefficient and not elsewhere in the linear program at the moment.

The logic of this **Phase I** procedure is both clever and easy to understand. Having added artificial variables to constraints as needed to form a basis, we now proceed to destroy them – more precisely, to use the Simplex Method itself to bring their values to zero. This is accomplished by temporarily replacing the original objective, Z, by a new objective, W, as the sum of the artificial variables. A new artificial variable is added to every greater than constraint in the linear program.

Having formed a new, albeit temporary, objective function, W, in an SMP format, the Simplex Method is used to drive it to zero. The idea being that with W as zero each artificial variable, being nonnegative, must also be zero. However, the constraints satisfy the SMP format. Next, the original objective, Z, is retrieved. However, it likely requires an "updating" so that it contains none of the current basic variables. This can be easily done algebraically. The newly revised objective and the constraints constitute an SMP. This procedure is illustrated in Example 5.5.4.

Example 5.5.4 SMP Using Phase I

Use a Phase I procedure to solve the linear program in Example 5.5.3.

Solution:

After introducing surplus variables x_3 and x_4, we have

$$\begin{aligned} \text{Maximize:}\quad & Z = -x_1 - 3x_2 \\ \text{Subject to:}\quad & x_1 + 4x_2 - x_3 = 48 \\ & 5x_1 + x_2 - x_4 = 50 \\ & x_1,\ x_2,\ x_3,\ x_4 \geq 0 \end{aligned}$$

Since no basis is apparent and we seek an alternative to the algebraic manipulation of the last example, add an artificial variable to each constraint lacking a convenient basic variable. The constraints become

$$\begin{aligned} & x_1 + 4x_2 - x_3 + w_1 = 48 \\ & 5x_1 + x_2 - x_4 + w_2 = 50 \\ & x_1,\ x_2,\ x_3,\ x_4,\ w_1,\ w_2 \geq 0 \end{aligned}$$

and the "artificial objective" is $W = w_1 + w_2$.

⟶

Use the constraints with w_1 and w_2 to substitute for them in the objective, W, since they are now basic. One obtains

$$W = (48 - x_1 - 4x_2 + x_3) + (50 - 5x_1 - x_2 + x_4) = 98 - 6x_1 - 5x_2 + x_3 + x_4$$

The resulting Phase I SMP linear program is

$$\begin{aligned} \text{Maximize:}\quad & -W = -98 + 6x_1 + 5x_2 - x_3 - x_4 \\ \text{Subject to:}\quad & x_1 + 4x_2 - x_3 + w_1 = 48 \\ & 5x_1 + x_2 - x_4 + w_2 = 50 \\ & x_1,\ x_2,\ x_3,\ x_4,\ w_1,\ w_2 \geq 0 \end{aligned}$$

To solve this linear program using the Simplex Method, begin with the tableau

Row	x_1	x_2	x_3	x_4	w_1	w_2	RHS
1	1	4	−1	0	1	0	48
2	**5**	1	0	−1	0	1	50
Z	1	3	0	0	0	0	0
−W	−6	−5	1	1	0	0	−98

Notice that the original objective is included in the tableau. By keeping it as if a constraint, when W becomes zero Z will be current in nonbasic variables. This avoids having to update Z later, after W has served its purpose and has been discarded. The next iteration illustrates the procedure.

The largest negative in the W row is "−6" and the minimum of ratios of 48/1 and 50/5 indicates that the 5 in the x_1 column is to be the pivot element. After this first pivot, the tableau becomes

Row	x_1	x_2	x_3	x_4	w_1	w_2	RHS
1	0	**19/5**	−1	1/5	1	−1/5	38
2	1	1/5	0	−1/5	0	1/5	10
Z	0	14/5	0	1/5	0	−1/5	−10
−W	0	−19/5	1	−1/5	0	6/5	−38

Another pivot is necessary (since there are still negative coefficients in −W) and the pivot element is shown in bold type. After pivoting, the tableau becomes

Row	x_1	x_2	x_3	x_4	w_1	w_2	RHS
1	0	1	−5/19	1/19	5/19	−1/19	10
2	1	0	1/19	−4/19	−1/19	4/19	8
Z	0	0	14/19	1/19	−14/19	−1/19	−38
−W	0	0	0	0	1	1	0

⟶

Now, W has been driven to zero, its coefficients no longer being negative valued. At this point, discard W and the artificial variables and, for clarity, we prepare for the next iteration using the original objective, Z

Row	x_1	x_2	x_3	x_4	*RHS*
1	*0*	*1*	$-5/19$	*1/19*	*10*
2	*1*	*0*	*1/19*	$-4/19$	*8*
Z	*0*	*0*	*14/19*	*1/19*	*−38*

Actually, we need not have rewritten the tableau (one could have ignored the W row and the w_1 *and* w_2 *columns as their usefulness has ended and they are all at value zero).*

One might ask whether the artificial variables could possibly reenter the basis later. It is a mathematical fact that once W has been driven to zero, each artificial variable is at zero and cannot return to the basis.

At this point, the Simplex Method continues with the original objective, Z, as there are usually negative objective coefficients. In this lucky instance, there are no negative-valued coefficients (ignore any for artificial variables) so no ***Phase II*** *procedure is needed as an optimal solution is at hand (*$x_1^* = 10$*,* $x_2^* = 8$*, and* $Z^* = -38$*). The Phase II procedure, usually required, is the iterative procedure to solve the program with the original objective.*

To the earlier flowchart for the Simplex Method, renamed here as Phase II, a Phase I is added as shown.

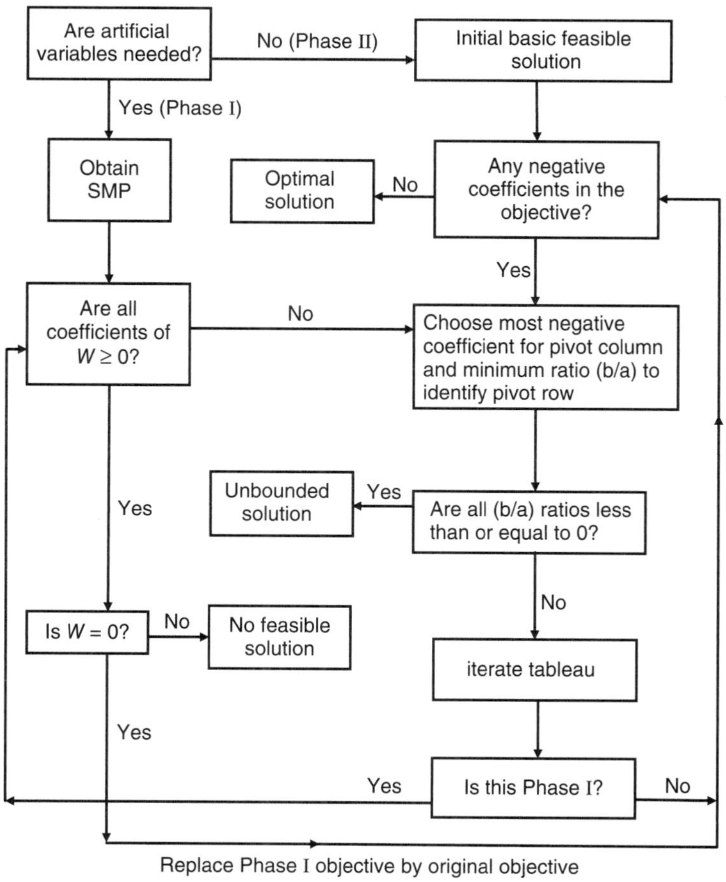

♦ An analogy may be useful. Rockets are boosted into space in stages. The first stage, analogous to an LP Phase I, provides lift off. That achieved, the first stage is jettisoned in favor of the second stage. Similarly, Phase I is jettisoned in favor of Phase II to complete the solution to the linear program.

An "alternate method" to the Phase I procedure is outlined below:

Alternate Method

Alternate Method to Solve Non-SMPs

1. If the objective is to minimize Z, multiply by −1 to maximize −Z.
2. Multiply any "≥" constraints by (−1) to become "≤" constraints
3. Add slack variables to form the simplex tableau.
4. Check the tableau RHS constraint (far right column) for any negative values. If there are no negative constants, apply the Simplex Method. If there are negative constants, go to Step 5.
5. a) When there is a negative value for a RHS constraint (upper portion of the tableau), select any negative coefficient in the same row and use its column as the pivot column.
b) In the pivot column, compute the minimum positive ratio to determine the pivot element.
c) After completing the pivot operations, return to Step 4.

Example 5.5.5 An Alternate Method

Use the alternate method to solve the linear program in Example 5.5.3.

$$\begin{aligned} \text{Minimize:} \quad & Z = x_1 + 3x_2 \\ \text{Subject to:} \quad & x_1 + 4x_2 \geq 48 \\ & 5x_1 + x_2 \geq 50 \end{aligned}$$

Solution:
The objective is rewritten as a maximization and the constraints are multiplied by (−1) to yield:

$$\begin{aligned} \text{Maximize:} \quad & -Z = -x_1 - 3x_2 \\ \text{Subject to:} \quad & -x_1 - 4x_2 + s_1 = -48 \\ & -5x_1 - x_2 + s_2 = -50 \end{aligned}$$

⟶

The initial tableau is:

x_1	x_2	s_1	s_2	RHS
−1	−4	1	0	−48
−5	−1	0	1	−50
1	3	0	0	0

Since there is a negative on RHS, use the first column as pivot: −48/− 1, −50/− 5. This indicates a pivot on the −5 (bold italic)

x_1	x_2	s_1	s_2	RHS
−1	−4	1	0	−48
1	1/5	0	−1/5	10
1	3	0	0	0

Completing the pivot yields:

x_1	x_2	s_1	s_2	RHS
0	***− 19/5***	1	−1/5	−38
1	1/5	0	−1/5	10
0	14/5	0	1/5	−10

As a negative RHS appears, pivot, again, as indicated

x_1	x_2	s_1	s_2	RHS
0	1	−5/19	1/19	10
1	1/5	0	−1/5	10
0	14/5	0	1/5	−10

After completing the pivot, we have:

x_1	x_2	s_1	s_2	RHS
0	1	−5/19	1/19	10
1	0	1/19	−4/19	8
0	0	14/19	1/19	−38

This is an optimal tableau so the optimal solution is $x_1^* = 8$, $x_2^* = 10$, *and minimum 38* ($Z^* = -38$).

EXERCISES 5.5

1. Minimize: $Z = 4x_1 + 3x_2$

   Subject to: $2x_1 + x_2 \geq 25$

   $x_1 + x_2 \geq 20$

   $x_1 - 5x_2 \leq 4$

   $x_1,\ x_2 \geq 0$

2. Maximize: $Z = 2x_1 + x_2$

   Subject to: $3x_1 - 2x_2 \geq 6$

   $x_1 + \ x_2 \leq 13$

   $x_1,\ x_2 \geq 0$

3. Maximize: $Z = 5x_1 + 6x_2$

   Subject to: $4x_1 + 7x_2 \leq 28$

   $x_1 - 3x_2 \geq 2$

   $x_1,\ x_2 \geq 0$

4. Minimize: $Z = 8x_1 + 5x_2$

   Subject to: $2x_1 - \ x_2 \geq 4$

   $x_1 + \ x_2 \leq 80$

   $5x_1 - 4x_2 \leq 50$

   $x_1,\ x_2 \geq 0$

5. Minimize: $Z = x_1 + 2x_2 - x_3 + x_4$

   Subject to: $x_1 - x_2 + x_3 \leq 4$

   $x_1 + x_2 + 2x_4 \leq 6$

   $x_2 - 2x_3 + x_4 \leq 2$

   $-x_1 + 2x_2 + x_3 \leq 2$

   $x_1,\ x_4 \geq 0$

   $x_2,\ x_3$ unrestricted

6. Solve the ACME truck problem of Exercise 4.3.19 using a technique from this section.

HISTORICAL NOTES

Linear Programming History

The identification, solution, and application of linear programs from about 1950 ranks as an outstanding development in a century marked by outstanding scientific achievement. It addresses a fundamental human quest toward the optimal allocation of scarce resources.

However, a reliable solution scheme for linear programs did not appear until 1947 with the "Simplex Method" credited to mathematician George B. Dantzig. In brief, in the Simplex Method the solution passes from vertex to vertex of the feasible convex set, with increasing (or unchanged) objective value until either an optimal solution is reached or it is established that no finite solution exists.

The history of LP is a remarkable convergence of theory, computation, and application. Theoretical development, while continuing, has been sufficient for diverse practical application. The timely solution of linear programs is bound with computational speed. The remarkable development of automated computation has steadily increased the size of linear programs that can be solved inexpensively and quickly. The millions of Google LP listings provide a sense of the diversity of applications. It has been said that LP is the most widely used mathematics outside an academic setting.

CHAPTER 5 SUPPLEMENTARY EXERCISES

1. Explain whether this is an SMP.

$$\begin{aligned}
\text{Maximize:}\quad & Z = 5x_1 + 3x_2 + 2x_3 + 3x_4 \\
\text{Subject to:}\quad & x_1 + x_2 + x_3 + 2x_4 \le 100 \\
& 3x_1 - 2x_2 + 5x_3 - 2x_4 \le 60 \\
& x_1 - x_2 + x_4 \le 30 \\
& 5x_1 + 2x_3 + x_4 \le 150 \\
& x_1,\ x_2,\ x_3 \ge 0
\end{aligned}$$

2. Given the SMP

$$\begin{aligned}
\text{Maximize:}\quad & Z = 4x_1 + 3x_2 \\
\text{Subject to:}\quad & 2x_1 + 3x_2 + 5x_3 \le 23 \\
& 3x_1 + x_2 + x_3 \le 8 \\
& x_1 + 5x_2 + 3x_3 \le 20 \\
& x_1,\ x_2,\ x_3 \ge 0
\end{aligned}$$

Which vectors are feasible?

a) [2 1 0] b) [1 2 3] c) [5 0 4] d) [1 0 −2]

3. Place the SMP from Exercise 2 in standard form.
4. What is the SMP represented by the following standard form?

$$\begin{aligned} \text{Maximize:} \quad & Z = 3x_1 + 5x_2 + 7x_3 + x_4 \\ \text{Subject to:} \quad & x_1 + x_2 + x_3 + 2x_4 + s_1 = 35 \\ & 3x_1 + x_2 + 5x_3 + 3x_4 + s_2 = 95 \\ & x_1 + x_2 + s_3 = 5 \\ & 2x_1 + 2x_3 + x_4 + s_4 = 40 \\ & x_1,\ x_2,\ x_3,\ x_4,\ s_1,\ s_2,\ s_3,\ s_4 \geq 0 \end{aligned}$$

5. Write the initial tableau for the following:

$$\begin{aligned} \text{Maximize:} \quad & Z = x_1 + 2x_2 + 7x_3 \\ \text{Subject to:} \quad & 2x_1 + 5x_2 + x_3 \leq 25 \\ & 3x_1 + x_2 + 3x_3 \leq 21 \\ & x_1 + x_2 \leq 7 \\ & x_1, x_2, x_3 \geq 0 \end{aligned}$$

6. Write the SMP for

x_1	x_2	x_3	x_4	s_1	s_2	s_3	s_4	RHS
3	5	1	2	1	0	0	0	12
2	4	3	6	0	1	0	0	19
1	6	0	2	0	0	1	0	25
2	0	5	1	0	0	0	1	34
−3	−2	−5	−1	0	0	0	0	0

7. Write the SMP for

x_1	x_2	x_3	x_4	s_1	s_2	s_3	s_4	RHS
2	1	3	5	1	0	0	0	25
1	0	2	5	0	1	0	0	34
0	3	1	4	0	0	1	0	59
3	1	2	1	0	0	0	1	17
−4	−1	−3	−2	0	0	0	0	0

8. Determine the pivot element for

x_1	x_2	x_3	s_1	s_2	s_3	RHS
3	1	4	1	0	0	32
2	0	3	0	1	0	20
2	2	5	0	0	1	42
−3	−5	0	0	0	0	0

9. Determine the pivot element for

x_1	x_2	x_3	s_1	s_2	s_3	RHS
5	25	10	1	0	0	100
3	15	1	0	1	0	75
4	6	15	0	0	1	240
−4	−3	0	0	0	0	0

10. Identify the current solution and whether it is optimal.

x_1	x_2	s_1	s_2	RHS
0	1	4/11	−1/11	6
1	0	−1/11	3/11	2
0	0	1	1	34

11. Determine the pivot element and then perform the indicated pivot.

x_1	x_2	s_1	s_2	RHS
1	5	1	0	200
2	3	0	1	134
−3	−7	0	0	0

12. Determine the pivot element and then perform the indicated pivot.

x_1	x_2	x_3	s_1	s_2	s_3	RHS
2	3	4	1	0	0	60
1	4	1	0	1	0	48
5	1	1	0	0	1	134
−2	−4	−7	0	0	0	0

13. Use the Simplex Method to solve:

$$\begin{aligned} \text{Maximize:} \quad & Z = 5x_1 + 4x_2 \\ \text{Subject to:} \quad & 5x_1 + \ \ x_2 \le 16 \\ & x_1 + 2x_2 \le 7 \\ & x_1,\ x_2 \ge 0 \end{aligned}$$

14. Use the Simplex Method to solve:

$$\begin{aligned} \text{Maximize:} \quad & 2x_1 + 5x_2 + 2x_3 \\ \text{Subject to:} \quad & 2x_1 + 7x_2 + 9x_3 \le 100 \\ & 6x_1 + 5x_2 + \ \ x_3 \le 145 \\ & x_1 + 2x_2 + 7x_3 \le 90 \\ & x_1,\ x_2,\ x_3 \ge 0 \end{aligned}$$

15. Use the Simplex Method to solve:

$$\begin{aligned} \text{Maximize:} \quad & Z = 3x_1 + 5x_2 + 4x_3 \\ \text{Subject to:} \quad & 4x_1 + \ \ 2x_2 + 2x_3 \le 40 \\ & 8x_1 + 10x_2 + 5x_3 \le 90 \\ & 4x_1 + \ \ 4x_2 + 3x_3 \le 30 \\ & x_1,\ x_2,\ x_3 \ge 0 \end{aligned}$$

16. Write the dual linear program and then solve both the primal and the dual using the Simplex Method.

$$\begin{aligned} \text{Minimize:} \quad & Z = 12x_1 + 8x_2 + 10x_3 \\ \text{Subject to:} \quad & x_1 + \ \ x_2 \qquad\quad \ge 12 \\ & 2x_1 + \qquad\quad 4x_3 \ge 20 \\ & 2x_1 + 2x_2 + \ \ x_3 \ge 8 \\ & x_1,\ x_2,\ x_3 \ge 0 \end{aligned}$$

17. Write the dual linear program and then solve both the primal and the dual using the Simplex Method.

$$
\begin{aligned}
\text{Minimize:} \quad & Z = 18x_1 + 15x_2 + 10x_3 \\
\text{Subject to:} \quad & 3x_1 + 2x_2 + x_3 \geq 48 \\
& 3x_1 + 2x_2 + 2x_3 \geq 70 \\
& 4x_1 + 3x_2 + 2x_3 \geq 96 \\
& x_1,\ x_2,\ x_3 \geq 0
\end{aligned}
$$

18.

$$
\begin{aligned}
\text{Minimize:} \quad & Z = 2x_1 + 6x_2 + 7x_3 \\
\text{Subject to:} \quad & x_1 + x_2 + x_3 \leq 25 \\
& x_1 + x_2 + 2x_3 \geq 15 \\
& 2x_1 + x_2 + x_3 \leq 18 \\
& x_1,\ x_2,\ x_3 \geq 0
\end{aligned}
$$

6 Linear Programming – Application Models

Two earlier chapters developed mathematical procedures for solving linear programs. Although such understanding is interesting, useful, and desirable, easily available software enables one to usefully apply linear programming with hardly any mathematical knowledge. Applications of linear programming span an immense variety of industrial, governmental, and educational situations seeking an optimal allocation of resources. It has been estimated that linear programming is one of the most practical and widely used mathematical techniques in industry and government. This is less surprising when one realizes that many of life's decisions relate to allocating resources. In addition, relatively inexpensive computation is readily available.

♦ It is said in jest that when resources are scarce, people look to "lean-year" programming.

Modeling linear programs for practical problems is a valuable and rewarding skill. It develops with practice! This chapter seeks to aid you to develop skill by using prototypes

Finite Mathematics: Models and Applications, First Edition. Carla C. Morris and Robert M. Stark.

Companion Website: http://www.wiley.com/go/morris/finitemathematics

to stimulate your imagination for some common and classic models and problem types. The focus is upon modeling linear programs and less upon solving them, although solutions are often included for completeness sake.

In the Linear Programming model prototypes and examples that follow, we have retained the historic names given them as they were identified, mostly in the 1950s. Our focus is on the underlying model, not a particular application.

♦ To gain a perspective on the vastness of linear programming models, browse among more than fourteen million Internet references.

PRODUCT MIX AND FEED MIX PROBLEMS

The *product mix* and *feed mix* prototypes are the simplest of linear program models. In the product mix, an imaginary firm produces several products each of which requires some materials, labor, and processing. Management seeks production quantities that maximize profits. That is, a total profit is to be maximized subject to constraints on available resources.

Example 6.1 Product Mix

A toy manufacturer plans limited editions of two toys using some surplus materials. The first toy is to be priced at \$25 and the second at \$35. The toys use the following materials:

	First Toy	*Second Toy*	*Quantity Available*
Wood strips	*2 feet*	*4 feet*	*300 feet*
Iron struts	*1 foot*	*2 feet*	*75 feet*
Plastic sheet	*3 square feet*	*4 square feet*	*100 square feet*
Plastic cement	*4 ounces*	*3 ounces*	*600 ounces*
Black paint	*1 ounce*	*3 ounces*	*700 ounces*
Blue paint	—	*4 ounces*	*200 ounces*
Red paint	*2 ounces*	*1 ounce*	*500 ounces*

What product mix maximizes revenue?

Solution:

We seek the number of each toy to manufacture to maximize revenue, Z. They are the ***decision variables*** *and are denoted by x_1 and x_2, the respective number of toys to produce. The linear program can be written as:*

$$\text{Maximize: } Z = 25x_1 + 35x_2 \qquad \text{Profit}$$

Subject to:

$$2x_1 + 4x_2 \leq 300 \qquad \text{wood strips}$$

$$x_1 + 2x_2 \leq 75 \qquad \text{iron struts}$$

$$3x_1 + 4x_2 \leq 100 \qquad \text{plastic sheet}$$

⟶

$4x_1 + 3x_2 \leq 600$	*plastic cement*
$x_1 + 3x_2 \leq 700$	*black paint*
$4x_2 \leq 200$	*blue paint*
$2x_1 + x_2 \leq 500$	*red paint*
$x_1, x_2 \geq 0$	*nonnegativity*

This completes the formulation.

The numerical answers for x_1 and x_2 can easily be found by the Simplex Method in this instance or by one of the many software packages such as LINDO. While the answers are of little interest here, for the curious, they are

$$\text{Maximum } Z^* = \$875, \; x_1^* = 0, \text{ and } x_2^* = 25$$

That is, only produce 25 of the second toy for a maximum revenue of \$875.

Note that in the (above) product mix prototype, since revenues or profits are to be maximized, at least some constraints must be of the "less than" variety since it is resources that set the limits on revenue.

In the feed mix problem, total cost is minimized while meeting some minimum resource requirements as, say, nutritional needs. "Feed mix" constraints are mainly of the "greater than" type.

♦ *A note on linear programming software.*

A large variety of software packages are readily available; many free. In practical situations where firms require many hours of computation on specific situations, the choice of software can be cost effective. However, for many everyday uses, the choice is often arbitrary. Google provides more on software comparisons.

Example 6.2 Feed Mix

Several foods help fulfill a certain supplementary diet. Each unit of appropriate foods contributes to the desired dietary result.

The unit costs (in cents) of five foods being considered are:

$$c_1 = 25, \quad c_2 = 22, \quad c_3 = 75, \quad c_4 = 39, \quad \text{and} \quad c_5 = 43$$

Three nutrients comprise the dietary supplement in these minimum amounts

$$N_1 = 200 \text{ units}, \quad N_2 = 250 \text{ units}, \quad \text{and} \quad N_3 = 400 \text{ units}$$

Let a_{ij} be the amount of nutrient i in a unit of food j

$$i = 1, 2, 3 \quad \text{and} \quad j = 1, 2, \ldots, 5$$

⟶

with values

$$a_{11} = 4, \quad a_{12} = 2, \quad a_{13} = 4, \quad a_{14} = 2, \quad a_{15} = 5,$$
$$a_{21} = 2, \quad a_{22} = 3, \quad a_{23} = 2, \quad a_{24} = 4, \quad a_{25} = 2,$$
$$a_{31} = 5, \quad a_{32} = 2, \quad a_{33} = 4, \quad a_{34} = 8, \quad a_{35} = 1$$

What food selection meets the dietary goal at least cost?

Solution:

Let x_j be the amount of food j, j = 1, 2, … , 5, to include in the diet.

The cost Z is

$$\text{Minimize:} \quad Z = c_1x_1 + c_2x_2 + \cdots + c_5x_5$$
$$= 25x_1 + 22x_2 + 75x_3 + 39x_4 + 43x_5$$

$$\text{Subject to:} \quad 4x_1 + 2x_2 + 4x_3 + 2x_4 + 5x_5 \geq 200 \quad \text{first nutrient}$$
$$2x_1 + 3x_2 + 2x_3 + 4x_4 + 2x_5 \geq 250 \quad \text{second nutrient}$$
$$5x_1 + 2x_2 + 4x_3 + 8x_4 + x_5 \geq 400 \quad \text{third nutrient}$$
$$x_1, \ldots, x_5 \geq 0$$

The minimum value is $Z^* = 2330$ cents, $x_1^* = 20$ cents, $x_2^* = 30$ cents, $x_3^* = 0$ cents, $x_4^* = 30$ cents, and $x_5^* = 0$ cents.

Here, the lowest cost diet, at \$23.30, consists only of the first, second, and fourth foods.

In the above example, the constraints are of the "greater than" type. A useful insight is to note that the feed mix prototype seeks to minimize *input* (nutrients) for a fixed *output* (dietary goal), while the product mix prototype is the opposite, to maximize an output (profit) for a given input (available resources).

◆ The scarcity of food and money in the Great Depression of the 1930s prompted the formulation of minimum cost diets that met nutritional requirements of the time. In mathematical format, they came to be known as the "diet problem" of linear programming.

An early diet problem with 9 constraints and 77 variables required 120 man-days for a solution on a desk calculator of the period. The problem was first formulated by J. Stigler ("The Cost of Subsistence", *Journal of Farm Economics 1945;* 27). However, at that time, he had to seek a trial and error solution as mathematicians had not yet solved the linear programming problem. Later, Stigler's diet problem was one of the first linear programs to be solved by the Simplex Method on a computer. Today, this diet problem can be solved in less than a second on a PC.

Similar problems arose during World War II (WWII) in the allocation of scarce resources to wartime production and in military situations such as the allocation of aircraft to targets.

These examples as in the earlier linear programming chapters use relatively few variables for ease of presentation and calculation. Practical problems usually employ hundreds and even many thousands of variables and constraints. The development of relatively inexpensive automated computation is a prime factor in the widespread practical usage of linear programs in the allocation of resources in government and industry.

◆ Linear programs with hundreds of thousands of constraints and even millions of variables are fairly common. The importance of program size can be likened to the number of pixels for a digital camera – more pixels, greater resolution.

FLUID BLENDING PROBLEM

The *fluid blending* problem combines features of both the product mix and feed mix types. An objective may be either minimization of blending costs or maximization of profits. Some constraints may be of the "less than" type for resource limitations, while "greater than" constraints ensure that certain minimum blend specifications are met or exceeded.

Example 6.3 Fluid Blending

A blender will use three basic wines A, B, and C to create three blends Dragon, Quaker, and Tiger. The blends are priced to fill market niches at \$6.80, \$5.75, and \$4.50 per liter, respectively. The basic wines cost the blender \$7.00, \$5.00, and \$4.00 per liter, respectively. The blending rules are:

Dragon must have at least 60% of wine A and no more than 20% of C.
Quaker cannot have more than 60% of wine C and at least 15% of A.
Tiger cannot have more than 75% of wine C.

Finally, the blender has on hand 2400, 3300, and 4000 liters of the respective basic wines. How should they be blended to maximize profit?

Solution:
Let X_{ij} be the amount (in liters) of basic wine i (i = A, B, C) in a liter of blend j (j = D for Dragon, Q for Quaker, and T for Tiger).

$$\text{The amount of Dragon is } X_{ADragon} + X_{BDragon} + X_{CDragon};$$
$$\text{of Quaker, } X_{AQua\,\mathrm{ker}} + X_{BQua\,\mathrm{ker}} + X_{CQua\,\mathrm{ker}};$$
$$\text{and of Tiger, } X_{ATiger} + X_{BTiger} + X_{CTiger}.$$

The revenue is the selling price per liter multiplied by the number of liters:

$$6.80(X_{AD} + X_{BD} + X_{CD}) + 5.75(X_{AQ} + X_{BQ} + X_{CQ}) + 4.50(X_{AT} + X_{BT} + X_{CT})$$

⟶

The cost is proportional to the amounts of the basic wines.

$$X_{AD} + X_{AQ} + X_{AT} \qquad \text{Wine A}$$
$$X_{BD} + X_{BQ} + X_{BT} \qquad \text{Wine B}$$
$$X_{CD} + X_{CQ} + X_{CT} \qquad \text{Wine C}$$

So, the cost of the basic wines is

$$7.00(X_{AD} + X_{AQ} + X_{AT}) + 5.00(X_{BD} + X_{BQ} + X_{BT}) + 4.00(X_{CD} + X_{CQ} + X_{CT})$$

The profit is

$$Z = REVENUE - COST$$

Now, for the resource constraints,

$$X_{AD} + X_{AQ} + X_{AT} \le 2400$$
$$X_{BD} + X_{BQ} + X_{BT} \le 3300 \qquad \text{BASIC WINE LIMITS}$$
$$X_{CD} + X_{CQ} + X_{CT} \le 4000$$

And the blend constraints:
for Dragon,

$$\frac{X_{AD}}{X_{AD} + X_{BD} + X_{CD}} \ge 0.60; \quad \frac{X_{CD}}{X_{AD} + X_{BD} + X_{CD}} \le 0.20$$

for Quaker,

$$\frac{X_{CQ}}{X_{AQ} + X_{BQ} + X_{CQ}} \le 0.60; \quad \frac{X_{AQ}}{X_{AQ} + X_{BQ} + X_{CQ}} \ge 0.15$$

for Tiger,

$$\frac{X_{CT}}{X_{AT} + X_{BT} + X_{CT}} \le 0.75$$

$$\text{All } X_{ij} \ge 0 \qquad i = A, B, C \qquad j = D, Q, T$$

At first sight, the proportionality of the blend constraints appears to be nonlinear. As written, they are. However, in this instance, elementary algebra removes the nonlinearity. For example, the last proportion can be written as

$$X_{CT} \le 0.75(X_{AT} + X_{BT} + X_{CT})$$

And, simplifying,

$$0.75X_{AT} + 0.75X_{BT} - 0.25X_{CT} \ge 0,$$

Clearly, a linear constraint! A similar procedure for the other constraints removes their nonlinearity.

⟶

Again, you may have an interest in carrying out the computation with LINDO or similar software. An optimal solution is

$$Maximum\ Profit = \$8680,$$

$$X_{AD}^* = 1260, \quad X_{BD}^* = 420, \quad X_{CD}^* = 420,$$

$$X_{AQ}^* = 1140, \quad X_{BQ}^* = 2880, \quad X_{CQ}^* = 3580,$$

$$X_{AT}^* = 0, \quad X_{BT}^* = 0, \quad X_{CT}^* = 0.$$

Blend 2100 liters of Dragon, 7600 liters of Quaker, and do not blend Tiger.

You will encounter the use of double subscripts, as X_{AD} above. In linear programming usage, they are a convenience in modeling "from, to" situations. For example, x_{ij} for "from factory i to warehouse j." So x_{23}, for example, suggests at a glance the number of units to move from, say, factory #2 to warehouse #3.

◆ Some of the earliest applications of linear programming arose in refining petroleum. Various types of crude oil are processed in a multitude of products called *streams*. They, in turn, are blended into other products (from gasoline to Vaseline). For the early work, see A. Charnes, W.W. Cooper, and B. Mellon, "Blending aviation gasolines – a study in programming interdependent activities in an integrated oil company". Econometrica, 1952; 20: 135–159. A Web search yields more recent work; however, many results are proprietary.

ASSIGNMENT PROBLEMS

Imagine a number of workers and a like number of tasks to which they are to be assigned. Each worker has an efficiency relative to each task. The problem is to assign tasks to workers in a way that maximizes total efficiency. The complementary problem of a cost or time for each worker relative to each task results in a minimization problem. The principles are the same in either case.

Example 6.4 An Assignment Problem

A supervisor is to assign four people A, B, C, and D to carry out four jobs I, II, III, and IV.

The table indicates the supervisor's estimates of the time for each person to complete each job. For example, C is estimated to require 15 hours to complete job III and so on.

	Job			
Personnel	*I*	*II*	*III*	*IV*
A	*7*	*18*	*23*	*14*
B	*19*	*27*	*33*	*26*
C	*18*	*24*	*15*	*17*
D	*23*	*19*	*14*	*21*

⟶

How should assignments be made to minimize the total time required to complete the four jobs?

Solution:

Here, let x_{ij} represent the assignment of worker i (i = A, B, C, D) to job j (j = I, II, III, IV).

The total completion time Z is

$$\text{Minimize: } Z = 7x_{A1} + 18x_{A2} + 23x_{A3} + 14x_{A4}$$
$$+ 19x_{B1} + 27x_{B2} + 33x_{B3} + 26x_{B4}$$
$$+ 18x_{C1} + 24x_{C2} + 15x_{C3} + 17x_{C4}$$
$$+ 23x_{D1} + 19x_{D2} + 14x_{D3} + 21x_{D4}$$

Subject to:

$$x_{A1} + x_{B1} + x_{C1} + x_{D1} = 1$$
$$x_{A2} + x_{B2} + x_{C2} + x_{D2} = 1$$
$$x_{A3} + x_{B3} + x_{C3} + x_{D3} = 1 \qquad \textit{assignments to jobs}$$
$$x_{A4} + x_{B4} + x_{C4} + x_{D1} = 1$$

$$x_{A1} + x_{A2} + x_{A3} + x_{A4} = 1$$
$$x_{B1} + x_{B2} + x_{B3} + x_{B4} = 1$$
$$x_{C1} + x_{C2} + x_{C3} + x_{C4} = 1 \qquad \textit{assignments to workers}$$
$$x_{D1} + x_{D2} + x_{D3} + x_{D4} = 1$$
$$x_{ij} \geq 0$$

These constraints ensure that no job is assigned too many workers and workers do not exceed "full time."

The minimum value is $Z^* = 65$ when

$$x_{A1}^* = 1,\ x_{A2}^* = 0,\ x_{A3}^* = 0,\ x_{A4}^* = 0,$$
$$x_{B1}^* = 0,\ x_{B2}^* = 1,\ x_{B3}^* = 0,\ x_{B4}^* = 0,$$
$$x_{C1}^* = 0,\ x_{C2}^* = 0,\ x_{C3}^* = 0,\ x_{C4}^* = 1,$$
$$x_{D1}^* = 0,\ x_{D2}^* = 0,\ x_{D3}^* = 1,\ x_{D4}^* = 0.$$

While the numbers of jobs and workers are the same in the example, they need not. One can think that workers might divide some jobs so that the x_{ij} may not simply be 0 or 1. A theorem in the theory of linear programming (not quoted here) ensures that since the right-hand sides of the constraints are unity, the optimal values of x is unity or zero. Carrying out the numerical computation by hand can clarify the character of the solution.

♦ The versatility of linear program models has spawned a vast spectrum of applications. They are described in thousands of published papers. Forests are cut and lumber mills operate from solutions of linear programs. Hospital menus, automobile production lines, structural designs, oil refining, water resources, shipping schedules, and much more are routine industrial and government uses of linear programs.

TRANSPORTATION PROBLEM

The designation of a transportation type problem, a generalization of the assignment problem earlier, suggests the context in which it arose. Imagine a number of factories (which manufacture products) and a number of warehouses (which store them). Different factories and warehouses generally produce and store differing quantities.

Costs of shipping may vary from each factory to each warehouse. The problem: What quantities should be shipped from each factory to each warehouse that meets their needs so that total shipping costs are a minimum?

Example 6.5 A Trans-shipment Problem

A company has three warehouses to supply six stores. The warehouses store the following quantities of product:

Warehouse 1	*2000*
2	*1800*
3	*2400*
	6200 units

The stores require these quantities of product:

Store 1	*1000*
2	*2200*
3	*1500*
4	*800*
5	*1200*
6	*200*
	6900 units

Note that the warehouses cannot fully meet the store requirements here.
Unit trans-shipment costs of moving product from warehouse to store are:

	Store					
Warehouse	*1*	*2*	*3*	*4*	*5*	*6*
1	*1*	*3*	*1*	*2*	*5*	*1*
2	*3*	*1*	*2*	*6*	*1*	*1*
3	*2*	*3*	*1*	*4*	*5*	*2*

For example, to move 1 unit of product from Warehouse 2 to Store 4 costs 6 units of money (or time).

How shall allocations be made from warehouses to stores to minimize trans-shipment costs?

⟶

Solution:

Let x_{ij} be the amount to ship from warehouse i (i = 1, 2, 3) to store j (j = 1, 2, … , 6).

$$\begin{aligned} \text{Minimize:} \quad Z = \; & x_{11} + 3x_{12} + x_{13} + 2x_{14} + 5x_{15} + x_{16} && \text{total} \\ & + 3x_{21} + x_{22} + 2x_{23} + 6x_{24} + x_{25} + x_{26} && \text{transshipment} \\ & + 2x_{31} + 3x_{32} + x_{33} + 4x_{34} + 5x_{35} + 2x_{36} && \text{cost} \end{aligned}$$

$$\begin{aligned} \text{Subject to:} \quad & x_{11} + x_{12} + x_{13} + x_{14} + x_{15} + x_{16} \leq 2000 \\ & x_{21} + x_{22} + x_{23} + x_{24} + x_{25} + x_{26} \leq 1800 && \text{warehouse stocks} \\ & x_{31} + x_{32} + x_{33} + x_{34} + x_{35} + x_{36} \leq 2400 \end{aligned}$$

$$\begin{aligned} & x_{11} + x_{21} + x_{31} \leq 1000 \\ & x_{12} + x_{22} + x_{32} \leq 2200 \\ & x_{13} + x_{23} + x_{33} \leq 1500 \\ & x_{14} + x_{24} + x_{34} \leq 800 && \text{store needs} \\ & x_{15} + x_{25} + x_{35} \leq 1200 \\ & x_{16} + x_{26} + x_{36} \leq 200 \end{aligned}$$

$$\begin{aligned} & x_{11} + x_{21} + x_{31} + x_{12} + x_{22} + x_{32} + x_{13} + x_{23} + x_{33} + && \text{total warehouse} \\ & x_{14} + x_{24} + x_{34} + x_{15} + x_{25} + x_{35} + x_{16} + x_{26} + x_{36} = 6200 && \text{stock} \\ & x_{ij} \geq 0 \qquad \text{all } i \text{ and } j \end{aligned}$$

Since all store needs cannot be satisfied, "less than" constraints are used for the store needs. Therefore, to ensure that all available stock is shipped, the last constraint has been added.

At the minimum, $Z^* = 8800$ and

$$\begin{array}{llllll} x_{11}^* = 1000, & x_{12}^* = 0, & x_{13}^* = 0, & x_{14}^* = 800, & x_{15}^* = 0, & x_{16}^* = 200, \\ x_{21}^* = 0, & x_{22}^* = 600, & x_{23}^* = 0, & x_{24}^* = 0, & x_{25}^* = 1200, & x_{26}^* = 0, \\ x_{31}^* = 0, & x_{32}^* = 900, & x_{33}^* = 1500, & x_{34}^* = 0, & x_{35}^* = 0, & x_{36}^* = 0. \end{array}$$

Note also the necessity of the last (equality) constraint! Without it, the minimum cost solution would be to ship nothing!

You have probably noticed again the usefulness of double subscripted variables in the assignment and transportation problems. Multiple subscripts are common in mathematical notation to indicate dimensionality. That is not the case in linear programming, where double subscript notation often indicates "from–to" relations. In more complicated situations, three or more subscripts are useful, perhaps to reflect the movement of material or parts through successive production stages. Equally satisfactory notation, and clearly less convenient, could number variables sequentially. However, that likely entails frequent reference to a "directory" to identify variables.

THE KNAPSACK PROBLEM

In the classic knapsack problem, a variety of items of known weight and value (or utility) and in various quantities are available to fill a figurative knapsack subject to a total weight limit. The objective is to maximize the value of the knapsack contents.

Example 6.6 Packing a Knapsack

The following items, their weights and values, are proposed for an expedition's members.

Item #	*Items*	*Weight per Unit, Pounds*	*Value per Unit*	*Maximum # Units Desired*	*Minimum # Units Desired*
1	*Meat*	*0.5*	*6*	*9*	*3*
2	*Carbohydrate*	*1.0*	*5*	*15*	*12*
3	*Liquids*	*1.5*	*8*	*12*	*6*
4	*First aid*	*0.5*	*4*	*6*	*3*
5	*Tools*	*1.0*	*6*	*6*	*3*

(values may be in units of dollars, utility, and so on).

What mix maximizes the total value within a total weight constraint of 30 pounds?

Solution:

Let x_j be the units of item j ($j = 1, 2, \ldots, 5$) to include in a "knapsack". Denoting the value of the jth item by v_j and its weight by w_j, the value of the knapsack to be maximized is

$$\text{Maximize}: \quad Z = \sum_{j=1}^{5} v_j x_j$$

The constraint on weight is

$$\sum_{j=1}^{5} w_j x_j \leq 30 \qquad \textit{weight constraint}$$

The constraints on the unit requirements are

$$\begin{aligned} 3 &\leq x_1 \leq 9 \\ 12 &\leq x_2 \leq 15 \\ 6 &\leq x_3 \leq 12 \qquad \textit{requirements} \\ 3 &\leq x_4 \leq 6 \\ 3 &\leq x_5 \leq 6 \\ x_j &\geq 0, \quad j = 1, \ldots, 5 \qquad \textit{nonnegativity} \end{aligned}$$

Again, the solution obtained by LINDO software is: $Z^* = 192$ when $x_1^* = 9$, $x_2^* = 12$, $x_3^* = 6$, $x_4^* = 3$, and $x_5^* = 3$.

THE TRIM PROBLEM

The trim problem is another classic linear program. Film, paper, fabric, and so on are manufactured in huge master rolls that are cut into smaller sizes for commercial uses. The objective is to minimize trim waste during cutting while meeting commercial needs.

Example 6.7 Trim Waste

Master rolls of fixed length and 50 inch width are cut into smaller widths of 8.5, 12, and 24 inch rolls. The respective sales requirements are 120, 80, and 150 rolls. Formulate a linear program to minimize total waste. Rolls in excess of requirements are also considered waste.

Solution:

There are a number of combinations of cuts or cutting strategies to meet demand. They are enumerated in the table.

Strategy Number	8.5 Inch	12 Inch	24 Inch	Trim Waste, Inch
1	5	0	0	7.5
2	0	4	0	2.0
3	0	0	2	2.0
4	4	1	0	4.0
5	3	2	0	0.5
6	3	0	1	0.5
7	0	2	1	2.0
8	1	3	0	5.5
9	1	1	1	5.5

Let x_i be the number of master rolls that are cut using the ith strategy, $i = 1, \ldots, 9$. The trim waste is

$$7.5x_1 + 2x_2 + 2x_3 + 4x_4 + 0.5x_5 + 0.5x_6 + 2x_7 + 5.5x_8 + 5.5x_9$$

Rolls that are cut in excess of demand are considered waste. Such waste is the difference between the numbers of rolls produced and the numbers sold.

$$\begin{aligned} &8.5(5x_1 + 4x_4 + 3x_5 + 3x_6 + x_8 + x_9 - 120) \\ &+ 12(4x_2 + x_4 + 2x_5 + 2x_7 + 3x_8 + x_9 - 80) \\ &+ 24(2x_3 + x_6 + x_7 + x_9 - 150) \qquad \textit{Product waste} \end{aligned}$$

The objective is to minimize the total waste, which is the sum of the two previous expressions. The constraints ensure that demand is met. They are

$$\begin{aligned} 5x_1 + 4x_4 + 3x_5 + 3x_6 + x_8 + x_9 &\geq 120 \\ 4x_2 + x_4 + 2x_5 + 2x_7 + 3x_8 + x_9 &\geq 80 \qquad \textit{Demand constraints} \\ 2x_3 + x_6 + x_7 + x_9 &\geq 150 \end{aligned}$$

→

Here, the minimum trim waste, $Z^* = 170$ inches, $x_1^* = 0$, $x_2^* = 0$, $x_3^* = 75$, $x_4^* = 0$, $x_5^* = 40$, $x_6^* = 0$, $x_7^* = 0$, $x_8^* = 0$, and $x_9^* = 0$.

Example 6.8 Trim Waste (Revisited)

Formulate the trim problem in symbols.

Solution:

Let d_j be the demand for the jth product whose width is a_j, $j = 1, \ldots, n$.

Again, master rolls of width W are cut into n products by m cutting strategies. The ith strategy produces k_{ij} rolls of width a_j and trim waste w_i,

$$i = 1, \ldots, m \text{ and } j = 1, \ldots, n$$

Forming the objective as the sum of wastes from each possible strategy, we have

$$\text{Minimize: } \sum_{i=1}^{m}[w_i x_i + \sum_{j=1}^{n} a_j(k_{ij} x_j - d_j)] \qquad \text{trim + product waste}$$

where x_i is the number of times the ith cutting strategy is used.

Note that there is a single summation of index i in the first term and double summation of indices i and j in the second term. You may find it helpful to compare the formulation with the numerical example above.

The constraints are

$$\sum_{i=1}^{m} a_j(k_{ij} x_i) \geq d_j, \qquad j = 1, \ldots, n \qquad \text{demand constraints}$$

$$x_i \geq 0, \qquad i = 1, \ldots, m$$

You may question the use of a "greater than" sign in the requirements constraints when only the precise quantities d_i, $j = 1, \ldots, n$ are needed. The answer is partly one of general advice – give the optimization technique all the latitude that can be spared. This aids the mathematical search for the best possible allocation. An unnecessary equality constraint might result in overlooking a superior alternative.

♦ Some of the names given to models in this chapter are apt, such as "assignment" and "transportation" problems. Some are not, as in the "caterer" problem below. We have retained the descriptive names given by others to the models here and as they have become known. This enables readers to more easily search for additional information.

A CATERER PROBLEM

The Caterer Problem, another classic in the literature of linear programming, requires increased formulation skills. Here, a mythical caterer requires r_j napkins at the start of the jth day, $j = 1, \ldots, n$. All soiled napkins are to be laundered at the end of each day.

The caterer has three options:

1. Purchase w_j new napkins for use on j^{th} day at a dollars per napkin.
2. Use normal (full day) laundering of x_j soiled napkins at the end of the jth day at b dollars per napkin and ready for use on day $j + 2$.
3. Use rapid (overnight) service of y_j napkins at the end of the jth day at c dollars per napkin and ready for use on day $j + 1$.

The total cost to be minimized for the next n days is

$$\text{Minimize: } Z = \sum_{j=1}^{n}(aw_j + bx_j + cy_j)$$

The napkin requirement constraint is

$$w_j + x_{j-2} + y_{j-1} + k_j \geq r_j, \quad j = 3, 4, \ldots, n \qquad \text{napkin needs}$$

where k_j is the number of clean napkins left over from the previous day that can be used on the jth day.

A constraint is needed to define the leftover napkins that were not used earlier.

Assume that we begin with newly purchased napkins.

$$\text{On the first day: } w_1 - r_1 = k_1$$

For the second day, add napkins that were laundered overnight as well as yesterday's leftovers.

$$\text{Second day: } w_2 + y_1 + k_1 - r_2 = k_2$$

For the third day, add, in addition, the napkins that were sent for normal laundering on the first day.

$$\text{Third day: } w_3 + y_2 + k_2 + x_1 - r_3 = k_3$$

Updating for the fourth day,

$$\text{Fourth day: } w_4 + y_3 + k_3 + x_2 - r_4 = k_4$$

More generally,

$$j^{\text{th}} \text{ day} \qquad w_j + y_{j-1} + k_{j-1} + x_{j-2} - k_j = r_j \quad j = 3, 4, \ldots, n$$

$$w_j, x_j, y_j \geq 0, \qquad \text{all } j = 1, 2, \ldots, n$$

Also, we add a set of constraints to ensure that all soiled napkins each day are placed in either normal or overnight laundering. The following are used:

First day $x_1 + y_1 + k_1 - w_1 = 0$

Second day $x_1 + x_2 + y_2 + k_2 - w_1 - w_2 = 0$

Third day $x_2 + x_3 + y_3 + k_3 - w_1 - w_2 - w_3 = 0$

and so on until the jth day

j^{th} day $x_{j-1} + x_j + y_j + k_j - w_1 - w_2 - w_3 - \cdots - w_j = 0, \quad j = 1, \ldots, n-1$

The objective and constraints comprise a linear program.

One could add "termination constraints" for napkin usage. By omitting them, we are in effect assuming that another planning horizon is used and begin with k_n napkins.

In formulating the caterer problem, numerical quantities were replaced by symbols, such as r_j for the number of napkins required in the jth day, and various prices a, b, and c. Feel free to choose some numerical values and solve the program by computer as illustrated in the next example.

Example 6.9 Next Week's Napkin Needs

Using the information about the caterer's problem suppose that the following quantities of napkins are needed next week

Monday 32, Tuesday 45, Wednesday 62, Thursday 71, Friday 92, Saturday 107, and Sunday 87.

The costs are as: purchase $a = \$2.00$, full day launder $b = \$0.25$, and rapid service $c = \$0.50$. Formulate a linear program for an ordering strategy that minimizes costs for the next week.

Solution:

Minimize:

$$2w_1 + 2w_2 + 2w_3 + 2w_4 + 2w_5 + 2w_6 + 2w_7 + 0.25x_1 + 0.25x_2 + 0.25x_3 + 0.25x_4$$
$$+ 0.25x_5 + 0.25x_6 + 0.25x_7 + 0.5y_1 + 0.5y_2 + 0.5y_3 + 0.5y_4 + 0.5y_5 + 0.5y_6 + 0.5y_7$$

Subject to:

$$w_1 - k_1 = 32$$
$$w_2 + y_1 + k_1 - k_2 = 45$$
$$w_3 + y_2 + k_2 + x_1 - k_3 = 62$$
$$w_4 + y_3 + k_3 + x_2 - k_4 = 71$$
$$w_5 + y_4 + k_4 + x_3 - k_5 = 92$$
$$w_6 + y_5 + k_5 + x_4 - k_6 = 107$$

⟶

$$w_7 + y_6 + k_6 + x_5 - k_7 = 87$$
$$x_1 + y_1 + k_1 - w_1 = 0$$
$$x_1 + x_2 + y_2 + k_2 - w_1 - w_2 = 0$$
$$x_2 + x_3 + y_3 + k_3 - w_1 - w_2 - w_3 = 0$$
$$x_3 + x_4 + y_4 + k_4 - w_1 - w_2 - w_3 - w_4 = 0$$
$$x_4 + x_5 + y_5 + k_5 - w_1 - w_2 - w_3 - w_4 - w_5 = 0$$
$$x_5 + x_6 + y_6 + k_6 - w_1 - w_2 - w_3 - w_4 - w_5 - w_6 = 0$$

The minimum cost is $381.50 when

$$w_1^* = 77, \; w_3^* = 30, \; w_2^* = w_4^* = w_5^* = w_6^* = w_7^* = 0$$
$$x_1^* = 32, \; x_2^* = 45, \; x_3^* = 36, \; x_4^* = 15, \; x_5^* = 0, \; x_6^* = 20, \; x_7^* = 0$$
$$y_1^* = y_2^* = 0, \; y_3^* = 26, \; y_4^* = 36, \; y_5^* = 92, \; y_6^* = 87, \; y_7^* = 0$$
$$k_1^* = 45, \; k_2^* = k_3^* = k_4^* = k_5^* = k_6^* = k_7^* = 0$$

♦ The model for the caterer problem arose in determining the number of spare aircraft and engines needed to maintain specified operational levels for a fleet of airplanes. The original paper by W. W. Jacobs titled "The Caterer Problem" (by which it has been known since) appeared in *Naval Research and Logistics Quarterly*, 1: pp 154–165 (1954).

Similar problems arose during WWII in the allocation of scarce resources to wartime production and military situations.

ANOTHER PLANNING HORIZON

Sometimes, allocations are required over a planning horizon as in the caterer problem. In such instances, the decision for an earlier period can affect later periods. The next example is another illustration.

Example 6.10 Yearly Oil Purchases

Oil refiners process crude oil into a variety of products depending on seasonal demands. While the petroleum industry has traditionally been one of the largest users of linear programming in the planning of refinery operations, this oversimplified example does not do justice to the size and complexity of actual problems.

Nevertheless, suppose that we are to plan for crude oil purchases for the coming year. The table indicates the prices at which crude can be bought and sold in advance by contract.

⟶

	Cost, Dollars per Barrel	*Sales Price, Sales Price,*	*Estimated Demand, Barrels*
Summer	*28*	*32*	*800*
Fall	*33*	*36*	*1300*
Winter	*39*	*41*	*2100*
Spring	*34*	*35*	*1100*

The refinery purchases crude oil in barrels. The sales price is the sum received for the products derived from a single barrel of crude. The number of barrels needed to produce the products expresses the estimated customer demand. Other material and processing costs are neglected.

Products can be stored from one season to the next at a cost equivalent of $3 per barrel. There is a maximum storage capacity of 2500 barrels.
How should managerial decisions be made to maximize profit?

Solution:
At each season, there are as many as three managerial decisions: amount to purchase, amount to sell, and amount to store.

Let w_i and x_i be the number of barrels to sell and to buy, respectively, for the ith season, $i = 1, \ldots, 4$. Let y_{ij} be the number of barrels to purchase in the ith season for sale in the jth season, $j = 2, 3, 4$.

First, for revenues, the total is $32w_1 + 36w_2 + 41w_3 + 35w_4$.

The basic cost of raw materials is $-28x_1 - 33x_2 - 39x_3 - 34x_4$.

In addition there is a $3 charge per barrel for each season it is stored.

Therefore, $-3y_{12} - 6y_{13} - 9y_{14}$ is the storage cost for crude purchased in the summer ($i = 1$).

Similarly, the total storage cost for the remaining seasons is $-3y_{23} - 6y_{24} - 3y_{34}$.

Gathering parts, the profit to be maximized is

$$Z = 32w_1 + 36w_2 + 41w_3 + 35w_4 - 28x_1 - 33x_2 - 39x_3 - 34x_4 - 3y_{12}$$
$$- 6y_{13} - 9y_{14} - 3y_{23} - 6y_{24} - 3y_{34}$$

Now, consider the constraints! An immediate group of constraints ensures that no more is sold than estimated demand. Note that the amount sold may be less than the demand but not more. So,

$$w_1 \leq 800, \quad w_2 \leq 1300, \quad w_3 \leq 2100, \quad \text{and} \quad w_4 \leq 1100$$

Another group of constraints ensures that there is no waste: all that is purchased is sold. That is, the supply purchased for the first season x_1 is either sold in that season or in subsequent seasons.

$$x_1 - w_1 - y_{12} - y_{13} - y_{14} = 0$$

Similarly,

$$x_2 - w_2 + y_{12} - y_{23} - y_{24} = 0$$
$$x_3 - w_3 + y_{13} + y_{23} - y_{34} = 0 \qquad \textit{No Waste Constraint}$$
$$x_4 - w_4 + y_{14} + y_{24} + y_{34} = 0$$

⟶

To ensure that tank capacity of 2500 barrels is not exceeded, use the storage constraints

$$
\begin{aligned}
x_1 &\le 2500 \\
x_2 + y_{12} + y_{13} + y_{14} &\le 2500 \\
x_3 + y_{13} + y_{14} + y_{23} + y_{24} &\le 2500 \qquad \textit{Storage Capacity} \\
x_4 + y_{14} + y_{24} + y_{34} &\le 2500
\end{aligned}
$$

Another constraint, easily overlooked, is the necessity to ensure that sales in any season must be at least as large as the amount stored for sale in that season. That is,

$$
\begin{aligned}
w_2 &\ge y_{12} \\
w_3 &\ge y_{13} + y_{23} \qquad \textit{Sales Constraints} \\
w_4 &\ge y_{14} + y_{24} + y_{34}
\end{aligned}
$$

Collecting the results,
Maximize:

$$
\begin{aligned}
Z = {} & 32w_1 + 36w_2 + 41w_3 + 35w_4 \\
& - 28x_1 - 33x_2 - 39x_3 - 34x_4 \\
& - 3y_{12} - 6y_{13} - 9y_{14} - 3y_{23} - 6y_{24} - 3y_{34}
\end{aligned}
$$

Subject to:

$$
\begin{aligned}
w_1 &\le 800 \\
w_2 &\le 1300 \\
w_3 &\le 2100 \qquad \textit{Demand} \\
w_4 &\le 1100
\end{aligned}
$$

$$
\begin{aligned}
x_1 - w_1 - y_{12} - y_{13} - y_{14} &= 0 \\
x_2 - w_2 + y_{12} - y_{23} - y_{24} &= 0 \\
x_3 - w_3 + y_{13} + y_{23} - y_{34} &= 0 \qquad \textit{Waste} \\
x_4 - w_4 + y_{14} + y_{24} + y_{34} &= 0
\end{aligned}
$$

$$
\begin{aligned}
x_1 &\le 2500 \\
x_2 + y_{12} + y_{13} + y_{14} &\le 2500 \qquad \textit{Storage} \\
x_3 + y_{13} + y_{14} + y_{23} + y_{24} &\le 2500 \\
x_4 + y_{14} + y_{24} + y_{34} &\le 2500
\end{aligned}
$$

$$
\begin{aligned}
w_2 - y_{12} &\ge 0 \\
w_3 - y_{13} - y_{23} &\ge 0 \qquad \textit{Sales} \\
w_4 - y_{14} - y_{24} - y_{34} &\ge 0
\end{aligned}
$$

$$w_i \ge 0, \quad x_i \ge 0 \quad y_{ij} \ge 0, \qquad i = 1, 2, 3, 4 \text{ and } j = 2, 3, 4$$

The solution is:

$$
\begin{aligned}
& \text{Maximum profit } Z^* = 19{,}400 \\
& w_1^* = 800, \quad w_2^* = 1300, \quad w_3^* = 2100, \quad w_4^* = 1100 \\
& x_1^* = 2500, \quad x_2^* = 800, \quad x_3^* = 900, \quad x_4^* = 1100 \\
& y_{12}^* = 500, \quad y_{13}^* = 1200 \\
& y_{14}^* = y_{23}^* = y_{24}^* = y_{34}^* = 0
\end{aligned}
$$

HISTORICAL NOTES

Interested students can explore the remarkably diverse applications of linear programming to seemingly every area of human endeavor. A few more examples are suggestive.

The Nobel Prize in economics was awarded in 1975 to the mathematician Leonid Kantorovich (USSR) and the economist Tjalling Koopmans (USA) for their contributions to the theory of optimal allocation of resources, in which linear programming has a key role.

Among military applications "Defending the Roman Empire: A Classical Problem in Military Strategy" (C.S. ReVelle and K. Rosing in *American Mathematical Monthly*, 107: 585–594, Aug–Sept 2000) is of interest.

A somewhat whimsical sociological study used linear programming to conclude that "monogamy is best" or, perhaps, "worst" (Halmos, P. and Vaughan, H. The marriage problem. American Journal of Mathematics Jan 1950; 72(1): 214–215).

Dantzig's "Reminiscences About the Origins of Linear Programming" (*Memoirs of the American Mathematical Society*, 1984) is illuminating.

♦ The "traveling salesman problem" is among the most famous of the last half-century. A "salesman" is required to visit a number of cities, each exactly once, that is, no backtracking! What itinerary will minimize the total distance (time) traveled?

While the problem in its entirety remains unsolved, many inroads have been made in the form of algorithms (rules) based on linear programming,

It is instructive to review the challenges of the problem and the commercial importance of its solution.

If there are only a few sales stops, an optimal route is easy to choose by enumeration. As the number of stops increases to, say, 50, the number of possible routes increases exponentially. Even automated computation to enumerate them is often not practical despite some algorithms that compute up to thousands of times faster than just a few years ago.

The traveling salesman problem has important applications in scheduling cable TV service trucks, UPS routes, NFL games, and much more.

A corporation, Waste Management Inc. (WMI), reduced its fleet by over 700 trucks and still met trash haul requirements at a savings of $91 million in operating costs according to the *Wall Street Journal*. Also Google has interesting references to current work.

EXERCISES 6.1

1. a) A company has four warehouses to supply six stores. The stores require a total of 22 units as follows:

Store #	Requirements
1	6
2	4
3	4
4	2
5	4
6	2

The warehouses have a combined surplus of 22 units of production as:

Warehouse #	Surplus
1	9
2	2
3	6
4	5

The shipping costs for a single unit from warehouse to store are

		Store					
		1	2	3	4	5	6
	1	9	7	9	9	6	10
Warehouse	2	12	6	9	7	11	5
	3	3	6	7	5	10	11
	4	5	8	2	2	11	3

Find the best shipping schedule.

b) Actually, the stores require only 20 units since store #1 had two cancelled sales. Find the best shipping schedule in this case.

2. Solve the trim problem when master rolls are 75 inches wide and cuts of 12, 24, and 36 inches are required in respective quantities of 50, 120, and 75 rolls.

3. A whiskey blender imports three whiskeys A, B, and C and mixes them according to recipes, which specify maximum and minimum percentages of A and C in each blend. The table of recipes is

Blend	Specifications	Price/Fifth, $
Blue Hen	Not less than 40% of A Not more than 30% of C	8.50
Old College	Not less than 25% of A Not more than 40% of C	7.00
Modeler's Choice	Not more than 60% of C	5.50

Supplies of the basic whiskeys and their costs are:

Whiskey	Maximum Quantity Available	Cost per Fifth, $
A	2100	7
B	800	6
C	1700	5

Formulate and solve for a blending strategy that maximizes profit.

4. Larger cargo ships accept shipments consistent with capacity limits in their sections as:

Forward	8 tons	40 cubic feet
Center	12 tons	50 cubic feet
Rear	6 tons	30 cubic feet

The cargoes available to the truckers, which can be accepted in whole or in part.

Commodity	Amount, tons	Volume, cubic feet per ton	Profit, dollars per ton
A	14	6	3
B	15	3	2
C	9	4	4

To maintain balanced loads, the weight in each section must be proportional to the capacity in tons.

Formulate a linear program to determine how to load a ship to maximize profit.

5. A caterer estimates next week's napkin needs as

Sunday	250
Monday	75
Tuesday	95
Wednesday	120
Thursday	70
Friday	300
Saturday	500

New napkins cost \$5.00 each. Normal laundering cost \$1.00 each and overnight service is \$1.50 each. What is the least cost way for the caterer to minimize napkin costs? All soiled napkins each day are to be laundered.

6. Reconsider Example 6.10 using a \$5 per barrel per season holding cost.

7 Set and Probability Relationships

Finite Mathematics: Models and Applications, First Edition. Carla C. Morris and Robert M. Stark.

Companion Website: http://www.wiley.com/go/morris/finitemathematics

Probability theory is the branch of mathematics that deals with chance events. While historically linked to games of chance, probability concepts are increasingly encountered in business, social sciences, biology, medicine, physical sciences, and engineering. From assessment of investment risks to bridge safety, behavior, spread of disease, efficacy of medication, and spam filters, mathematical probability has a role.

A study of mathematical probability in this chapter and in Chapter 8 requires some knowledge of **set theory**, which explains its presence as the next topic.

7.1 SETS

Loosely said, sets are collections or groups of things. For example, the group of digits $\{1, 3, 2, 6\}$ constitute a set and can be designated as $\mathbf{A}$. The members of a set are called its **elements**. To indicate that an element a belongs to a set $\mathbf{A}$, one writes $a \in \mathbf{A}$. If a is not an element of $\mathbf{A}$, write $a \notin \mathbf{A}$. The totality of elements under consideration is the **universal** set $\mathbf{U}$.

Sometimes, a rule defines the elements of a set. Some examples are "faces on a die," "vowels," or "days of a week." These are indicated as set

$$\mathbf{A} = \{x/x \text{ is a number on a die}\}$$

and so on, where the forward slash (/) is read as "such that."

A set without elements is called an **empty** or **null** set. The empty set is denoted by { } or $\varnothing$, but not both. Either notation for an empty set is appropriate. Their combination as $\{\varnothing\}$ is not an empty set. It is a set containing the symbol, $\varnothing$.

Example 7.1.1 Set Relations

Let $\mathbf{A} = \{x/x \text{ is a vowel}\}$
Which of these set statements are true?

a) $u \in \mathbf{A}$ *b)* $n \in \mathbf{A}$ *c)* $f \notin \mathbf{A}$

Solution:

a) True, u is a vowel.
b) False, n is not a vowel.
c) True, f is not a vowel.

If each element of a set **A** is also an element of set **B,** then we say **A** is a **subset** of **B**, written $\mathbf{A} \subset \mathbf{B}$ or "**A is contained in B**." Its negation is written $\mathbf{A} \not\subset \mathbf{B}$, or "**A is not in B.**" If the two sets **A** and **B** have exactly the same elements, they are said to be **equal** or **identical** and written as $\mathbf{A} = \mathbf{B}$. Its negation, "**A** not equal to **B**", is $\mathbf{A} \neq \mathbf{B}$.

Example 7.1.2 Set Relations (continued)

Let $\mathbf{A} = \{a, e, i, o, u\}$ *and* $\mathbf{B} = \{a, b, c, d, e, f, g, h\}$
Which of these set statements are correct?

a) $\{a, e, u\} \subset \mathbf{A}$ *b)* $\{d, f, o\} \subset \mathbf{B}$ *c)* $\mathbf{A} = \mathbf{B}$

Solution:

a) True! The elements a, e, and u are a subset of set ***A****.*
b) False! While d and f are elements of set ***B****, o is not.*
c) False! Sets ***A*** *and* ***B*** *do not have exactly the same elements.*

Composite Sets

A **union** of two or more sets is another set that includes the elements of each of them. For two sets **A** and **B**, their union set $\mathbf{A} \cup \mathbf{B}$ consists of all elements that are "in **A** and/or **B**." For three sets **A**, **B**, and **C** the union, $\mathbf{A} \cup \mathbf{B} \cup \mathbf{C}$, consists of every element that occurs in at least one of the sets, and so on.

Within a union set, there is an **intersection** set. It corresponds to the elements common to sets comprising the union set. The intersection set of two sets **A** and **B** is denoted by $\mathbf{A} \cap \mathbf{B}$. Only elements that appear in both sets form the intersection set. Two sets are called **disjoint** or **mutually exclusive** if they have no elements in common; that is, $\mathbf{A} \cap \mathbf{B} = \varnothing$.

The intersection of three sets is denoted by $\mathbf{A} \cap \mathbf{B} \cap \mathbf{C}$. Only elements that appear in all three sets can form the intersection set.

The **complement** of a set **A** (relative to the universal set **U**) is its negation. The complement of a set **A** is the set "**not A**" denoted by $\mathbf{A^c}$. Note that **c** is not an exponent here as sets are not exponentiated. To avoid confusion, some writers use $\mathbf{A'}$ or $\mathbf{\bar{A}}$ although these notations also have disadvantages. The complement $\mathbf{A^c}$ consists of the elements of the universal set **U** which do not belong to the set **A**. The union $\mathbf{A} \cup \mathbf{A^c}$ is the universal set while $\mathbf{A} \cap \mathbf{A^c}$ is an empty set.

Example 7.1.3 Composite Sets

Let $U = \{1, 2, 3, 4, 5, 6, 7, 8, 9, 10\}$, $A = \{1, 2, 3, 4, 5\}$, *and* $B = \{2, 4, 6, 8\}$ *Identify*

a) $A \cap B$ *b)* $A \cup B$ *c)* A^c *d)* $A^c \cap B$ *e)* $(A \cup B)^c$

Solution:

a) The intersection set contains elements common to both sets. Therefore,

$$A \cap B = \{2, 4\}$$

b) The union contains elements belonging to either of the two sets. Therefore,

$$A \cup B = \{1, 2, 3, 4, 5, 6, 8\}$$

*c) The complement to set **A** is any element from the universal set that is not an element of **A**. Therefore,*

$$A^c = \{6, 7, 8, 9, 10\}.$$

d) Using c), $A^c \cap B = \{6, 8\}$.

e) Using b), $(A \cup B)^c = \{7, 9, 10\}$.

The **commutative, associative**, and **distributive** laws of algebra have counterparts in the algebra of sets. They are summarized as:

Commutative Law

$$\mathbf{A} \cup \mathbf{B} = \mathbf{B} \cup \mathbf{A}$$

$$\mathbf{A} \cap \mathbf{B} = \mathbf{B} \cap \mathbf{A}$$

Associative Law

$$\mathbf{A} \cup (\mathbf{B} \cup \mathbf{C}) = (\mathbf{A} \cup \mathbf{B}) \cup \mathbf{C}$$

$$\mathbf{A} \cap (\mathbf{B} \cap \mathbf{C}) = (\mathbf{A} \cap \mathbf{B}) \cap \mathbf{C}$$

Distributive Law

$$\mathbf{A} \cup (\mathbf{B} \cap \mathbf{C}) = (\mathbf{A} \cup \mathbf{B}) \cap (\mathbf{A} \cup \mathbf{C})$$

$$\mathbf{A} \cap (\mathbf{B} \cup \mathbf{C}) = (\mathbf{A} \cap \mathbf{B}) \cup (\mathbf{A} \cap \mathbf{C})$$

The size of a set **A**, measured by the number of its elements, is denoted by $n(\mathbf{A})$. For example, if the elements of **A** are the alphabet's consonants, then $n(\mathbf{A}) = 21$, since the other five letters are vowels.

The number of elements in a set when added to the number in its complement is the size of the universal set. That is, $n(\mathbf{A}) + n(\mathbf{A}^c) = n(\mathbf{U})$. Note that while sets do not add arithmetically, numbers do!

Similarly, the size of a union set is the sum of the sizes of its component sets after subtracting duplicate elements. In symbols, for two sets **A** and **B**,

$$n(\mathbf{A} \cup \mathbf{B}) = n(\mathbf{A}) + n(\mathbf{B}) - n(\mathbf{A} \cap \mathbf{B})$$

Set Size

$$n(\mathbf{A}) + n(\mathbf{A}^c) = n(\mathbf{U})$$

$$n(\mathbf{A} \cup \mathbf{B}) = n(\mathbf{A}) + n(\mathbf{B}) - n(\mathbf{A} \cap \mathbf{B})$$

Example 7.1.4 Analyzing a Survey

In a survey of 100 businesses, 80 were profitable. Fifty-five of the 60 that planned to expand within a year were profitable. How many businesses were neither profitable nor planning expansions?

Solution:

Let $\boldsymbol{A} = \{\textit{profitable business}\}$ *and* $\boldsymbol{B} = \{\textit{business planning expansion}\}$.

Here, $n(\boldsymbol{U}) = 100$, $n(\boldsymbol{A}) = 80$, $n(\boldsymbol{B}) = 60$, *and* $n(\boldsymbol{A} \cap \boldsymbol{B}) = 55$.

The event "neither profitable nor expanding" can be expressed as $(\boldsymbol{A} \cup \boldsymbol{B})^c$. *Therefore, we seek* $n(\boldsymbol{A} \cup \boldsymbol{B})^c$.

⟶

It is easier to calculate $n(A \cup B) = 80 + 60 - 55 = 85$ and find its complement. So $n(A \cup B)^c = n(U) - n(A \cup B) = 100 - 85 = 15$. Therefore, 15 businesses were neither profitable nor planning expansions.

If **A** and **B** are any two sets, then the set of all possible pairs (a, b) such that $a \in \mathbf{A}$ and $b \in \mathbf{B}$ is the **direct product** of **A** and **B** and is denoted by $\mathbf{A} \times \mathbf{B}$. It can also be called the **cross product** or the **Cartesian product** of the two sets.

Number of Elements in A Direct Product

$$n(\mathbf{A} \times \mathbf{B}) = n(\mathbf{A}) \cdot n(\mathbf{B})$$

Example 7.1.5 Location Planning

---

*Planning for a new gasoline station identified six possible locations (**A**) and a choice among four brands of gasoline (**B**). How many choices of location and brand are there?*

Solution:
Here $n(A) = 6$ and $n(B) = 4$. Therefore, $n(A \times B) = n(A) \cdot n(B) = 6 \cdot 4 = 24$.

EXERCISES 7.1

1. If $\mathbf{A} = \{1, 2, 3, 4, 5\}$, which statements are true?

   a) $7 \notin \mathbf{A}$
   b) $\{2, 3\} \subset \mathbf{A}$
   c) $\{1\} \subset \mathbf{A}$
   d) $\{2, 3, 4\} \not\subset \mathbf{A}$
   e) $4 \notin \mathbf{A}$

2. If $\mathbf{A} = (1, 2, 3, 4, 5, 6\}$ and $\mathbf{B} = (4, 5\}$, which statements are true?

   a) $\mathbf{A} \subset \mathbf{B}$
   b) $4 \in \mathbf{A}$
   c) $\{4\} \subset \mathbf{B}$
   d) $\mathbf{A} \cap \mathbf{B} = \varnothing$
   e) $\mathbf{A} \cup \mathbf{B} = \{1, 2, 3, 6\}$

3. If $\mathbf{U} = \{1, 2, 3, 4, 5, 6, 7, 8, 9\}$, $\mathbf{A} = \{2, 3, 5, 7\}$, and $\mathbf{B} = \{1, 2, 3, 4, 5\}$, find

   a) $n(\mathbf{A} \cap \mathbf{B})$
   b) $n(\mathbf{A} \cup \mathbf{B})$
   c) $n(\mathbf{A}^c)$
   d) $n(\mathbf{A}^c \cup \mathbf{B})^c$
   e) $n(\mathbf{A} \cap \mathbf{B}^c)$

4. If $\mathbf{A} = \{1, 3, 5, 7\}$ and $\mathbf{B} = \{2, 4, 6, 8\}$, find

   a) $n(\mathbf{A} \cup \mathbf{B})$
   b) $n(\mathbf{A} \cap \mathbf{B})$

5. If $\mathbf{U} = \{1, 2, 3, \ldots, 10\}$, $\mathbf{A} = \{1, 3, 5, 7\}$, and $\mathbf{B} = \{2, 3, 4, 5, 6, 7\}$, find

a) $\mathbf{A} \cap \mathbf{B}$ b) $\mathbf{A} \cup \mathbf{B}$ c) $\mathbf{A}^c$ d) $\mathbf{A}^c \cup \mathbf{B}^c$ e) $(\mathbf{A} \cap \mathbf{B}^c)$

6. If $\mathbf{U} = \{1, 2, 3, \ldots, 10\}$, $\mathbf{A} = \{1, 3, 5, 7\}$, $\mathbf{B} = \{5, 6, 7\}$, and $\boldsymbol{c} = \{1, 4, 8\}$, find

a) $(\mathbf{A} \cup \mathbf{B}) \cap \mathbf{C}$ b) $\mathbf{C} \cap \mathbf{B}$ c) $(\mathbf{A} \cap \mathbf{C})^c$ d) $\mathbf{A} \cup \mathbf{B} \cup \mathbf{C}$

7. Let **A** = {months with 31 days}, **B** = {February}, and **C** = {months with exactly 30 days}.

a) Find $n(\mathbf{A})$ b) Find $n(\mathbf{C})$ c) Find $n(\mathbf{B} \cup \mathbf{C})$

8. List the elements in the set {letters in "DELAWARE"}.

9. Verify $(\mathbf{A} \cup \mathbf{C}) \cap (\mathbf{B} \cup \mathbf{C}) = (\mathbf{A} \cap \mathbf{B}) \cup \mathbf{C}$ by creating three sets **A**, **B**, and **C** and finding the elements in the corresponding sets.

10. Let **A** = {vowels) and **B** = {consonants}.

a) What is $\mathbf{A} \cup \mathbf{B}$? b) What is $\mathbf{A} \cap \mathbf{B}$?

11. Let **A** = {Natural numbers less than 50}, **B** = {perfect squares}, and **C** = {prime numbers}.

a) Find $\mathbf{A} \cap \mathbf{B}$ b) $\mathbf{A} \cap \mathbf{C}$

12. a) What is $\varnothing^c$? b) $\mathbf{U}^c$?

13. A set is partitioned into three subsets x_1, x_2, and x_3. The number of elements in x_1 is four-thirds the number in x_2 while the number in x_3 is three times the number in x_2. Find $n(x_1)$, $n(x_2)$, and $n(x_3)$ when $n(x) = 80$.

14. Find $n(\mathbf{A} \cup \mathbf{B})$, given $n(\mathbf{A}) = 28$, $n(\mathbf{B}) = 32$, and $n(\mathbf{A} \cap \mathbf{B}) = 7$.

15. An English professor has assigned 10 readings. Call them the Universal set. **U** = {Tom Sawyer, The Black Cat, Huckleberry Finn, Cask of Amontillado, The Fall of the House of Usher, Pit and the Pendulum, Moby Dick, Billy Budd, Murders in the Rue Morgue, A Connecticut Yankee in King Arthur's Court}.

Let **A** = {author Mark Twain}, **B** = {author Edgar Allen Poe}, and **C** = {author Herman Melville}. Determine the numbers

a) $n(\mathbf{A})$ b) $n(\mathbf{B})$ c) $n(\mathbf{C})$.

16. Among these personages: Hubert Humphrey, Spiro Agnew, Nelson Rockefeller, Alexander Hamilton, Walter Mondale, George Washington, Benjamin Franklin, Al Gore, Dan Quayle, Joseph Biden, and Richard Cheney, let **A** = {USA Vice Presidents}. List the elements of **A**.

17. In a survey of 150 town residents, 50 subscribed to *Time*, 90 to *US News & World Report*, and 30 to both magazines. Determine the numbers that subscribe to

    a) *Time* alone   b) neither magazine.

18. List the set of three-letter "words" that can be formed from the letters in the word PEA. Identify the subset of English words.

19. List the set of four-letter "words" from the letters in the word STOP. Identify the subset of English words.

20. Let $\mathbf{A} = \{\text{colors of the rainbow}\}$, $\mathbf{B} = \{\text{primary colors}\}$, and $\mathbf{C} = \{\text{blue, black, white}\}$. List the elements of

    a) $\mathbf{A} \cup \mathbf{C}$ b) $\mathbf{A} \cap \mathbf{B}$ c) $\mathbf{A} \cap \mathbf{B} \cap \mathbf{C}$

21. A coin and a die are tossed. List the possible outcomes. Compare with the number of outcomes of a direct product.

22. A business can be located at one of five possible locations and can choose one of three franchise names. How many choices of location and name are there?

23. A student can conduct a survey on weekdays by telephone, e-mail, or in person. How many possibilities of mode and weekday are there?

24. (Exercise 9 revisited) Verify $(\mathbf{A} \cup \mathbf{C}) \cap (\mathbf{B} \cup \mathbf{C}) = (\mathbf{A} \cap \mathbf{B}) \cup \mathbf{C}$ using the laws of sets.

25. Determine a formula for $n(\mathbf{A} \cup \mathbf{B} \cup \mathbf{C})$ where **A**, **B**, and **C** are any subsets of a universal set **U**. Check your result with an example.

7.2 VENN DIAGRAMS

It is useful to represent set relationships diagrammatically. Those known as **Venn diagrams** are widely used. Typically, Venn diagrams represent the universal set as an arbitrary rectangle inside which elements of a set are represented by any enclosed area; usually a circle. Shaded areas illustrate basic set relationships in the figures.

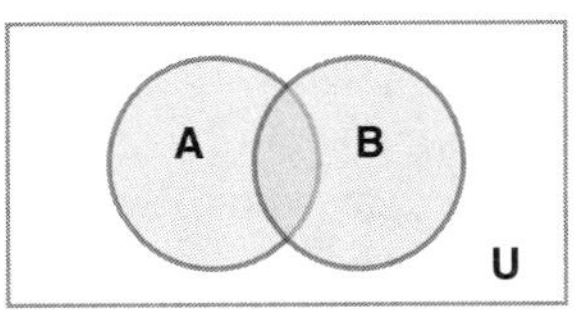

Venn diagram for $\mathbf{A} \cup \mathbf{B}$

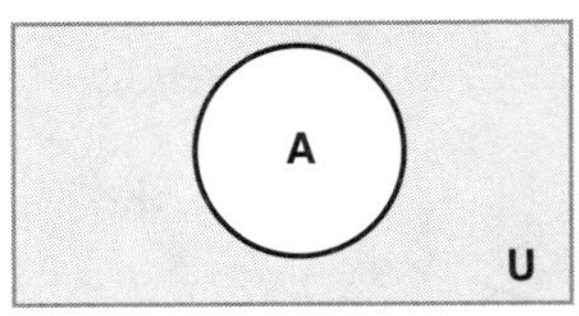

Venn diagram for $\mathbf{A}^{\mathbf{C}}$

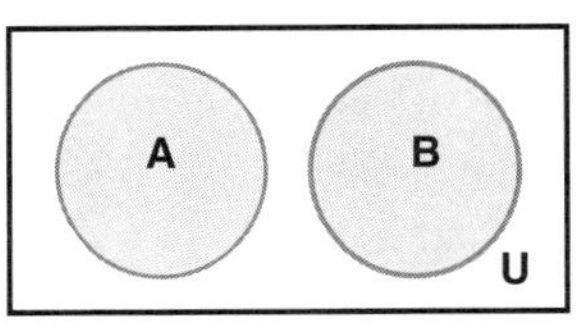

A and **B** are mutually exclusive

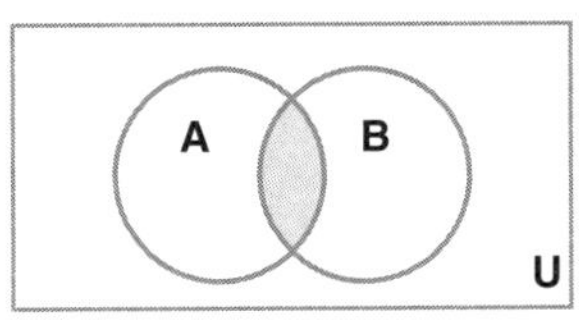

Venn diagram for $\mathbf{A} \cap \mathbf{B}$

Sets obey commutative, associative, and distributive laws, which can be verified from their definitions or with Venn diagrams.

Example 7.2.1 Set Check

Use a Venn diagram to verify that $(\boldsymbol{A}^c \cup \boldsymbol{B})^c = \boldsymbol{A} \cap \boldsymbol{B}^c$.

Solution:

The numerals I, II, III, and IV denote regions on the Venn diagram.
Here, $\boldsymbol{A} = \{I, II\}, \boldsymbol{A}^c = \{III, IV\}, B = \{I, III\}$, *and* $\boldsymbol{B}^c = \{II, IV\}$.
From, $\boldsymbol{A}^c \cup \boldsymbol{B} = \{III, IV\} \cup \{I, III\} = \{I, III, IV\}$. *Its complement clearly is region II.*
Now, $\boldsymbol{A} \cap \boldsymbol{B}^c = \{I, II\} \cap \{II, IV\} = \{II\}$. *The set relation is verified.*

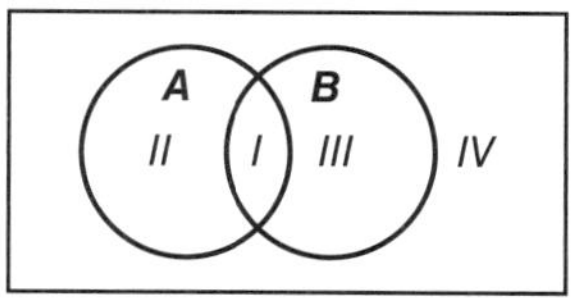

Example 7.2.2 Class Composition

In a survey of 50 students, 40 were in a math class, 30 were in a biology class, and 25 were in both classes. Use a Venn diagram to determine how many students were not in either class.

Solution:

*Circles represent math (**M**) and biology (**B**) sets. It is useful to start with the intersection set of 25 students who are in both sets. Since 40 students are in math, then 15 are in math only. Similarly, there are five students in biology only.*

Finally, because there are 50 students in the universal set, the 5 remaining students are in neither set and depicted outside the two course sets.

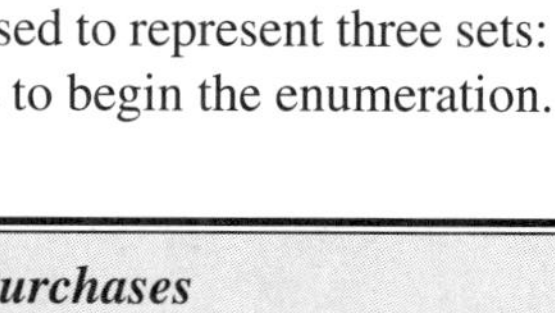

Venn diagrams can also be used to represent three sets: **A**, **B**, and **C**. Their mutual intersection is usually a useful place to begin the enumeration. An example follows.

Example 7.2.3 Product Purchases

A consumer survey queried 100 people about preferences for three new products (A, B, C).
The results were:

→

4 selected none of the three products
17 selected all three products
30 selected products A and B
27 selected products A and C
22 selected products B and C
56 selected product A
55 selected product B
15 selected product C but neither A nor B

Use a Venn diagram to determine how many prefer at least two of the three new products.

Solution:
With three overlapping circles it is useful to start enumeration in the center, $(\mathbf{A} \cap \mathbf{B} \cap \mathbf{C})$, *or outside the three sets,* $(\mathbf{A} \cup \mathbf{B} \cup \mathbf{C})^c$. *The information is given for both regions of the Venn diagram so the 4 and 17 (bold) are placed first.*

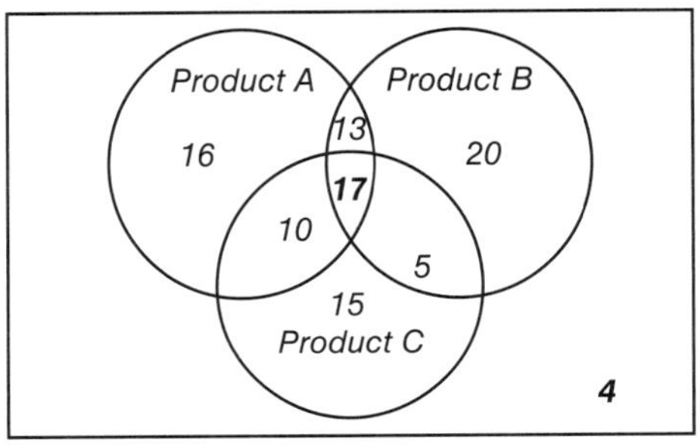

Next, regions where two circles overlap are useful to consider remembering that 17 of these people have already been counted. This means that there are 13 additional people who prefer products A and B, 10 additional people who prefer products A and C, and 5 additional people who prefer products B and C.

The next logical steps are to consider individual products. Fifteen people prefer product C only; place this number appropriately.

The remaining numbers for products A and B fill the missing numbers.

Recall, there are 100 people. As a check, add the numbers in the eight regions to verify that they sum to 100.

Now, to answer the question, 45 (= 17 + 10 + 13 + 5) people prefer at least two of the three products.

De Morgan's Laws

Two useful set equalities are **De Morgan's laws**. One has the complement of a union of sets as the intersection of the sets' complements. In symbols,
$(\mathbf{A} \cup \mathbf{B})^c = \mathbf{A}^c \cap \mathbf{B}^c$.

The second set is obtained by replacing **A** by $\mathbf{A}^c$ and **B** by $\mathbf{B}^c$. That is,
$(\mathbf{A}^c \cup \mathbf{B}^c)^c = \mathbf{A} \cap \mathbf{B}$, since the complement of a complement is the set itself. Now, taking the complement of each side yields $\mathbf{A}^c \cup \mathbf{B}^c = (\mathbf{A} \cap \mathbf{B})^c$.

De Morgan's Laws

$$(\mathbf{A} \cup \mathbf{B})^c = \mathbf{A}^c \cap \mathbf{B}^c$$

$$\mathbf{A}^c \cup \mathbf{B}^c = (\mathbf{A} \cap \mathbf{B})^c$$

Example 7.2.4 Pipeline Operations

The diagram depicts a power plant that receives fuel from two sources along pipelines A and B through a pipeline C.

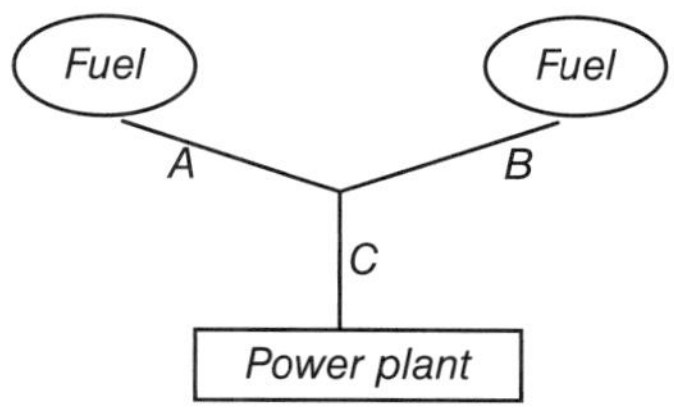

If C fails, or if both A and B fail, the power plant cannot operate. However, if either A or B fail (but not both), and C does not fail, the power plant can operate.

Express the event "{Power plant does not operate}" as a set relation. Then find an alternate using De Morgan's laws.

Solution:

For convenience, let ***A****,* ***B****, and* ***C*** *denote the events that the respective pipelines are operating. Then, it is easy to express the event*

$$\{\textit{Power plant operates}\} \text{ as } (\boldsymbol{A} \cup \boldsymbol{B}) \cap \boldsymbol{C}.$$

The complementary event

{Power plant does not operate} is $[(\boldsymbol{A} \cup \boldsymbol{B}) \cap \boldsymbol{C}]^c = (\boldsymbol{A} \cup \boldsymbol{B})^c \cup \boldsymbol{C}^c$ *(using the second of De Morgan's laws regarding* $(A \cup B)$ *as a single entity).*

Using the first of De Morgan's laws yields $(\boldsymbol{A}^c \cap \boldsymbol{B}^c) \cup \boldsymbol{C}^c$.

♦ Now it is clearer that this chapter on probability began with a mini-course in set theory and Venn diagrams.

As written language depends on grammar and syntax, so translating real-life chance events and chance circumstances into a mathematical probability requires some basic principles of sets.

EXERCISES 7.2

1. Use Venn diagrams to verify the commutative, associative, and distributive laws of sets.
2. For two sets **A** and **B**, use a Venn diagram to show

   a) $(\mathbf{A} \cap \mathbf{B})^c$ b) $\mathbf{A}^c \cap \mathbf{B}$ c) $\mathbf{A}^c \cup \mathbf{B}^c$

3. For two sets A and B, use a Venn diagram to show the situations below

   a) $\mathbf{B} \subset \mathbf{A}$ b) $\mathbf{A} \cap \mathbf{B} = \varnothing$

4. For three sets **A**, **B**, and **C**, use a Venn diagram to show

   a) $(\mathbf{A} \cup \mathbf{B}) \cap \mathbf{C}$. b) $(\mathbf{A} \cap \mathbf{B} \cap \mathbf{C})^c$. c) $\mathbf{A}^c \cup \mathbf{B}^c \cup \mathbf{C}^c$.

5. Use a Venn diagram to show that $(\mathbf{A} \cap \mathbf{B})^c$ is not the same as $\mathbf{A}^c \cap \mathbf{B}^c$.
6. Suppose $n(\mathbf{U}) = 100$, $n(\mathbf{A}) = 35$, $n(\mathbf{B}^c) = 55$, and $n(\mathbf{A} \cup \mathbf{B})^c = 30$. Use a Venn diagram to determine

   a) $n(\mathbf{A} \cap \mathbf{B})$. b) $n(\mathbf{B} \cap \mathbf{A}^c)$.

7. In the Venn diagram below, determine $n(\mathbf{A}^c \cup \mathbf{B}^c)$.

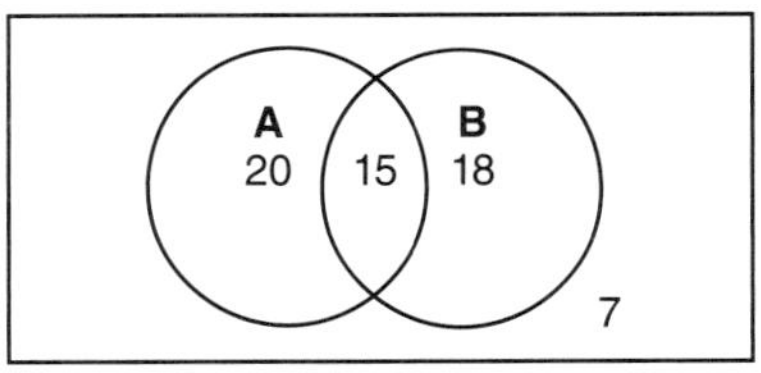

8. A quality control analyst for WXYZ Corporation is reviewing the performance of 50 different products. Compared with the previous year, the analyst finds that 32 have improved reliability and 27 have improved durability. Only six products have not improved at all. How many products have improved in both reliability and durability?
9. In a survey of 150 town residents, 50 subscribed to *Time*, 90 to *US News & World Report*, and 30 to both news magazines. How many residents subscribe to

   a) *Time* alone? b) Neither magazine?

10. There are 150 people at a picnic. Each drank at least one of three beverages: iced tea, soda, and lemonade. There were 21 who drank each beverage, 32 drank lemonade and tea, 33 drank iced tea and soda, and 44 drank soda and lemonade. Eighty drank iced tea and 18 drank soda only.
    a) How many drank only lemonade?
    b) How many drank soda?
    (Hint: use a Venn diagram with overlapping circles, one for each beverage. Start at the central intersection set).

11. In a math class with 250 students, 100 are also in an accounting class, 150 in economics, and 200 in English. Of those students in the math class who are in an economics class, 20 are in neither accounting nor English, 75 are in both accounting and English, and 30 are in English but not in accounting. How many math students are not in any of the other three classes?

12. A survey of 150 job applicants revealed that 69 were college graduates, 82 had some prior job experience, 90 were men, and of these 45 had prior job experience. Fifteen of the female college graduates had prior job experience. Twenty of the male college graduates had some job experience. In addition, there were five women with neither job experience nor college background. Use a Venn diagram to determine how many of the applicants were female college graduates.

13. In a restaurant survey of 200 patrons, 90 had no significant complaints. Three main complaints dealt with issues of cleanliness, quality of food, and quality of service. Seventy-six patrons complained of the service, 6 complained of all three, and 10 complained only about food quality. Fifteen patrons complained about the cleanliness and of food quality; 26 about the cleanliness and service; and 46 complained about the food quality and service.

    Use a Venn diagram to determine the number of patrons that had a complaint about only one of the three issues.

14. If $n(\mathbf{U}) = 65$ and x is 10 more than y, determine the values of x and y.

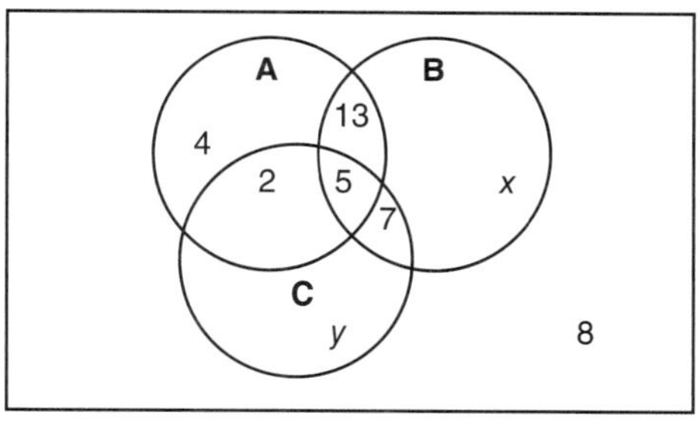

15. Blood types are determined by the presence or absence of three antigens: A antigen, B antigen, and an antigen called the Rh factor. The resulting blood types are

A – A antigen present	AB – both A and B antigen are present
B – B antigen present	O – neither A nor B antigen is present

    These blood types are further classified as Rh-positive or Rh-negative based on the presence or absence of the Rh antigen.

    a) Use a Venn diagram to depict the eight possible blood types.
    b) What percentage of the US population has each of the four blood types A, B, AB, and O? (Hint: use an Internet search)
    c) Examine blood types of the world's population. Are there differences among geographic areas? (For instance, South America and United States.)

16. In a market survey of 100 randomly chosen households, the following three questions were asked.
    1. Do you own a computer?
    2. Do you own a DVD player?
    3. Do you own a cell phone?

    Some results of the survey were that 70 owned a computer, 25 a DVD, and 48 a cell phone. There were three people who owned all three items and four owned none of them. Eleven owned a computer and DVD but not a cell phone, 15 a cell phone only, and 32 a computer only.

    Determine the number surveyed who owned

    a) a DVD only? b) exactly two of the three items? c) a cell phone or DVD?

7.3 TREE DIAGRAMS

A **tree diagram** is an example of a *directed graph* whose branches emanate from a starting point. The starting point of a "tree" is its **root**. A **path** indicates travel from the root to a **terminal point** without backtracking. Properly formed trees simplify counting problems and calculating probabilities. They are a popular means of diagramming complex decisions (called **decision trees**) such as in large-scale construction and design projects as in the space program and business planning.

Example 7.3.1 Class Schedule Tree

Mathematics (M), English (E), and History (H) classes are each scheduled at 9:00, 10:00, and 11:00 AM. Use a tree diagram to determine the number of possible schedules that include each subject.

Solution:
A tree diagram of choices follows:

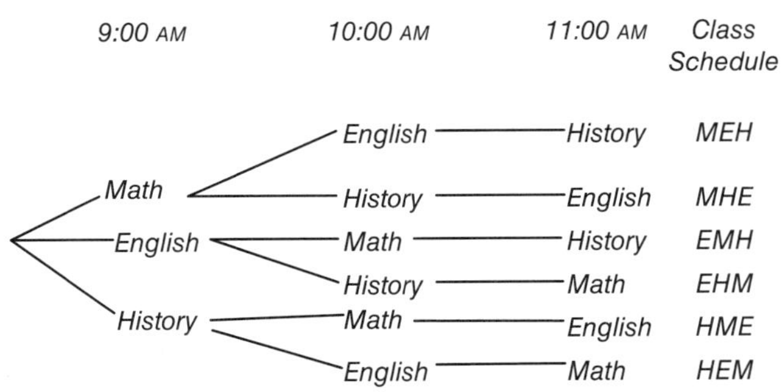

There are six possible schedules!

Example 7.3.2 Venus versus Serena

Venus and Serena are to play a "best of three" tennis match. The first to win two sets is the winner. Display outcomes on a tree diagram .

Solution:

There are six possible outcomes. In two cases, the winner is determined after two sets. In the other four outcomes, three sets determine the winner.

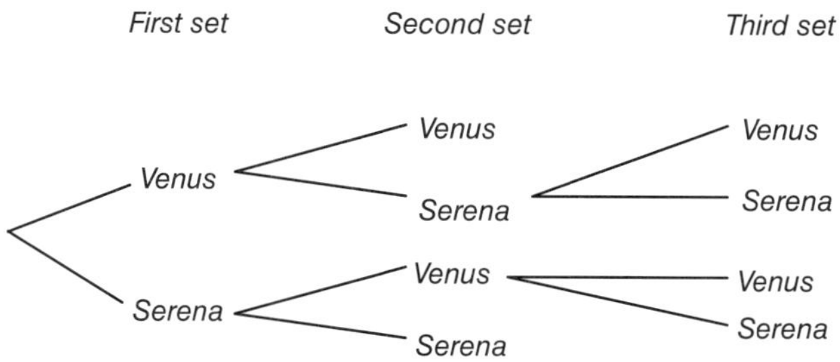

The number of subsets of a set with n elements is 2^n, including the null set and the set itself. This is illustrated in the next example.

Example 7.3.3 Subset Choices

Find all possible subsets of $\{A, B, C\}$.

Solution:

A tree diagram can determine the subsets. Alternatively, a systematic listing of the sets with none of the elements, one of the elements, two of the elements, and all three elements determines the solution.

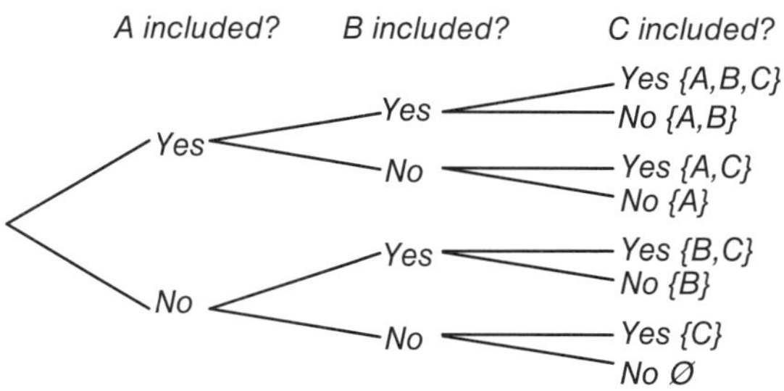

There are 2^3 *or 8 possible subsets. The subsets are*
$\{A\}$, $\{B\}$, $\{C\}$, $\{A, B\}$, $\{A, C\}$, $\{B, C\}$, $\{A, B, C\}$, *and* $\varnothing$.
Note the difference from Example 7.3.1, where the three items had to be included.

Example 7.3.4 Possible Outcomes

How many different four-digit numbers greater than 4900 can be formed from the digits {3, 4, 5, 9}? Use a tree diagram. Digits can be repeated.

Solution:
To exceed 4900, the first digit cannot be 3. Nor can it be 4 unless the second digit is 9. The third and fourth digits are unrestricted. If the first digit is 5 or 9, the remaining three digits are unrestricted.

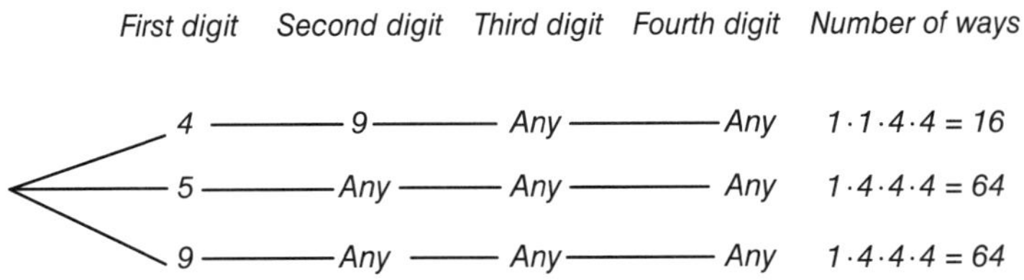

There are 16 + 64 + 64 = 144 possibilities.

Sometimes, tree diagrams are easier to display than Venn diagrams. The next example revisits the Venn diagram from Example 7.2.3.

Example 7.3.5 Product Purchases Revisited

A consumer survey queried 100 people about preferences for three new products (A, B, C).

The results are outlined as:

4 selected none of the three products
17 selected all three products
30 selected products A and B
27 selected products A and C
22 selected products B and C
56 selected product A
55 selected product B
15 selected product C but neither A nor B

Use a tree diagram to determine the number who prefer at least two of the three new products.

Solution:
Previously (in Example 7.2.3), three circles of a Venn diagram depicted products A, B, and C. To form a tree diagram, ask in turn whether or not each product is preferred.

⟶

We start knowing 56 of the 100 people prefer A; hence, 44 people do not. These are the numbers on the initial branches of the tree. The top and bottom branches represent "all or nothing" situations and are often easily determined. Here, these numbers are 17 and 4 for the example. Pairs of products are examined, especially the A and B split of 30 (leaving 26 for the branch below). The 30 people are divided into a group of 17 (already placed), and 13 others. Products A and C are represented by the YYY and YNY branches. The 27 people are divided into groups of 17 and of 10 for the YNY branch.

Since the YN branch has 26 people and YNY has 10, there is a 26 on the YNN branch. The YYY and NYY represent those who preferred products B and C. The 22 are divided into groups of 17 and 5.

These analyses appear as:

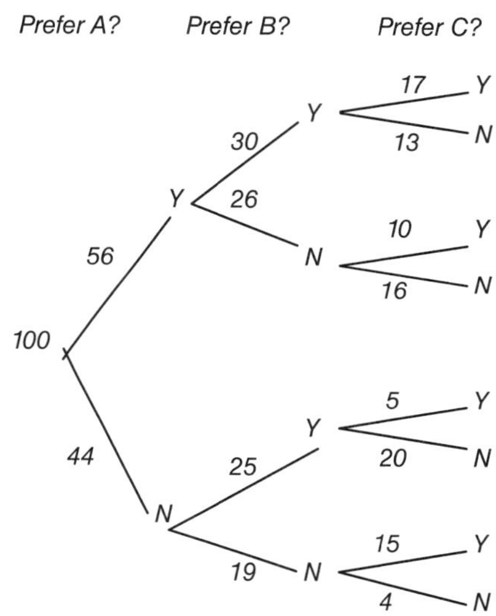

The people prefer at least two of the three product numbers 17 + 13 + 10 + 5 = 45 in agreement with the Venn diagram of Example 7.2.3.

♦ While tree diagrams can be popular, visual representations do not rely solely on them. For larger numbers in practical problems, tree diagrams can become impractical.

Still, they have a role (sometimes, wall sized) in planning large construction projects. In the space program and other large programs, they are adapted to coordinate a multitude of activities.

EXERCISES 7.3

1. Find all subsets of **A** = {X, Y, Z}.
2. Find all subsets of **A** = {1, 2, 3, 4}.

3. An experiment consists of drawing a ball from an urn that contains three yellow, two white, one red, and four green balls. After the color is noted, a second ball is drawn (without replacement) and the color is again noted. Use a tree diagram to determine the number of outcomes.

4. An inspector rates four items as "excellent," as "satisfactory," or as "unacceptable." How many possible reports can the inspector make?

5. In a psychology experiment, five rats are placed in a cage. Two rats are trained and three untrained. A rat is selected and labeled **T** if it is trained and **U** if not. The rat is placed in another cage and cannot be selected again. Use a tree diagram to describe the experiment when three rats are selected.

6. How many different four digit numbers less than 6500 can be formed from the digits $\{3, 6, 8, 9\}$

   a) if digits can be used repeatedly? b) if digits cannot repeat?

Answer Exercises 7–11 using tree diagrams.

7. Exercise 7.2.8.

8. Exercise 7.2.9.

9. Exercise 7.2.10.

10. Exercise 7.2.11.

11. Exercise 7.2.12.

12. A coin is flipped until either of two heads appears or four flips have occurred. Use a tree diagram to list the possible outcomes.

13. A die is tossed and its face is noted. If it is an odd number, a coin is flipped. If it is an even number, a letter (a or b or c) is randomly selected. Use a tree diagram to depict possible outcomes.

14. A quality control inspector tests the items produced on an assembly line and rates them as G (good) or D (defective). An experiment consists of rating items in sequence until either a defective is found or four items are rated. After each rating the item is set aside. Use a tree diagram to list all possible outcomes.

15. Suppose the quality control inspector in Exercise 14 rates items until either two defectives appear or five items are rated. Use a tree diagram to list all possible outcomes.

16. A game charges \$1 per die toss. If a 4, 5, or 6 appears, a dollar is won in addition to the dollar charge. Otherwise, the dollar charge is forfeited. A player starts with \$1 and can play up to three times if funds are available.

    Use a tree diagram to display outcomes.

17. In a class of 100 students, 57 are women. The class has 50 freshmen of whom 23 are male. Use a tree diagram to determine how many freshmen are women.

7.4 COMBINATORICS

With n_1 elements of one kind and n_2 elements of a second kind, there are exactly $n_1 \cdot n_2$ pairs of elements, a pair consisting of an element of each kind. This is the direct product considered earlier in this chapter.

With n_1, n_2, and n_3 elements of the first, second, and third kinds, respectively, there are $n_1 \cdot n_2 \cdot n_3$ triads, known as **3-tuples**, consisting of an element of each kind, and so on.

The result generalizes as the **Multiplication Principle** and is sometimes called the **Fundamental Principle of Counting**.

Multiplication Principle

With n_1 elements of a first kind, n_2 of a second kind, ..., n_r of an rth kind, there are exactly $n_1 \cdot n_2 \cdots n_r$ r-tuples consisting of an element of each kind.

Example 7.4.1 Table Seating

How many arrangements of eight people are possible at a banquet head table?

Solution:

Using the multiplication principle:

$$8 \cdot 7 \cdot 6 \cdot 5 \cdot 4 \cdot 3 \cdot 2 \cdot 1 = 40,320$$

Example 7.4.2 License Plates

The first digits on Wyoming license plates are coded to one of its 23 counties followed by a bucking bronco and then either three or four digits are followed by two letters. How many distinct license plates are possible for

a) Laramie County (2)?

b) the entire state?

Solution:

a) If three digits and two letters follow the bronco, $(10)(10)(10)(26)(26) = 676,000$ possibilities.
If four digits and two letters follow, there are $(10)(10)(10)(10)(26)(26) = 6,760,000$ possibilities.
Adding on, there are $676,000 + 6,760,000 = 7,436,000$ possible license plates for Laramie County.

⟶

b) Since each of the 23 counties has the same number of possible license plates as Laramie county (part a), there are $(23)\ (7,436,000) = 171,028,000$ *possible license plates for the entire state.*

A product of successive natural numbers is known as a **factorial**. For instance "five factorial," denoted by 5!, is the product $5 \cdot 4 \cdot 3 \cdot 2 \cdot 1 = 120$. By definition, $0! = 1$.

More generally,

$$n! = n(n-1)(n-2)\cdots(3)(2)(1).$$

Example 7.4.3 Factorials

Evaluate

a) 10!/8! *b) 1500!/1499!*

Solution:

a) $$\frac{10!}{8!} = \frac{10 \cdot 9 \cdot 8 \cdot 7 \cdot 6 \cdot 5 \cdot 4 \cdot 3 \cdot 2 \cdot 1}{8 \cdot 7 \cdot 6 \cdot 5 \cdot 4 \cdot 3 \cdot 2 \cdot 1} = 10 \cdot 9 = 90.$$

b) $$\frac{1500!}{1499!} = \frac{1500 \cdot 1499 \cdot 1498 \cdots 2 \cdot 1}{1499 \cdot 1498 \cdots 2 \cdot 1} = 1500.$$

◆ Most nongraphing calculators calculate to 69!. Graphing calculators reach much higher numbers. As technology advances, expect current limits to be exceeded.

Combinatorics

How many three-number combinations are possible for a combination lock? How many symbols (dots and dashes) are required for Morse code? How many connections are needed in a large telecommunications network or for a large population to be sorted into several categories?

These are a few examples associated with a branch of mathematics called **combinatorics**. It is "super counting," often dealing with numbers well beyond everyday experience and well beyond one's fingers and toes. Combinatorics has attracted more interest as computer software has developed. Some uses include cryptography and other security codes or devices, network designs, and software programs.

Our main use of combinatorics is in probabilistic and statistical analyses to calculate probabilities. With equally likely outcomes, the probability of a particular event is the ratio of the number of ways that event can materialize to the total number of outcomes.

Consider the question:

In how many ways can a sample of r items be selected from a population of n elements?

There are two ambiguities in this query!

The first is whether an item selected from the population is replaced before the next selection is made. If so, then repeated selections are effectively from the original population. If not, the population is altered after each selection and prior to the next one.

The second is whether items are distinguishable from one another. If not, the order in which the items are selected is irrelevant. If items are distinguishable from each other, an order of selection becomes relevant.

The following tree diagram shows the four possible outcomes according to whether outcomes are chosen **without replacement** or **with replacement** and with regard to order.

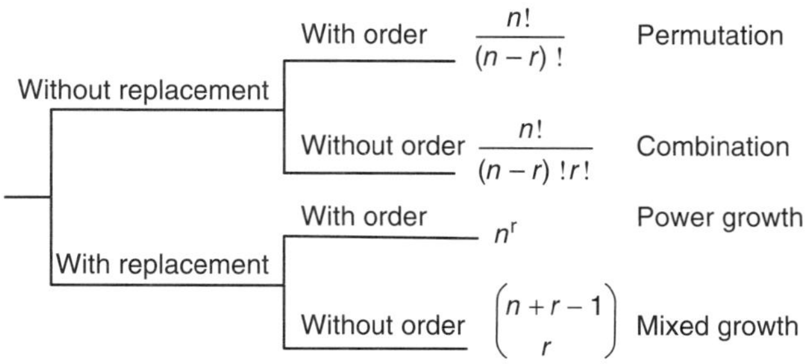

Combinatoric Tree

We consider them in turn!

Permutation

Drawing without replacement and with regard to order is known as a **permutation**. It is easily calculated using the multiplication principle. The first element can clearly be chosen in one of n ways. Since it is without replacement, the next draw is from among $n - 1$ elements, and so on, as $n(n - 1)\cdots(n - r + 1)$. [This is easily seen to be the product of the r numbers $(n - 0)(n - 1)\cdots(n - (r - 1))$]. It can also be written in factorial format as $n!/(n - r)!$

Example 7.4.4 Seating

In how many ways can five people be seated in three seats, each to a seat?

Solution:

Any one of the five people can be seated first. Any of the remaining four people can be seated next for a subtotal of $5 \times 4 = 20$ ways. The third seat can be filled by any of the three remaining people in $5 \times 4 \times 3 = 60$ ways.

This is the same as (5!/2!); a permutation with $n = 5$ and $r = 3$.

A permutation takes account of order. For example, Jane seated before Joe is different than Joe seated first and Jane next.

Combination

In many instances, order is probably not relevant, as in the above instance, had Jane and Joe been a couple. For convenience, let y represent the number of ways r elements can be selected from among n elements without replacement and without regard to order. Now, to restore order, each of the y ways must be multiplied by $r!$, the number of arrangements (permutations) of r elements. Then $(y)(r!)$ becomes the permutation $n!/(n-r)!$. That is, $(y)(r!) = n!/(n-r)!$. Solving for y yields $\frac{n!}{(n-r)!r!}$ and is called a **combination** and denoted by $\binom{n}{r}$. That is, $\binom{n}{r} = \frac{n!}{(n-r)!r!}$.

Sometimes, the combination $\binom{n}{r}$ is written as ${}_nC_r$ and the permutation as $\frac{n!}{(n-r)!} = {}_nP_r$. It can be shown that $(r!)_nC_r = {}_nP_r$. There are usually more permutations than there are combinations, unless of course $r = 0$ or 1. It should be also noted that $\binom{n}{r} = \binom{n}{n-r}$. This occurs as the ways to include r of n elements is the same as excluding $n - r$ of n elements.

For instance, a billiard set is of 15 ($n = 15$) numbered balls. The number of ways three ($r = 3$) balls can be selected is the permutation $15 \cdot 14 \cdot 13 (= 15!/12!)$. Suppose the actual balls selected are (arbitrarily) numbered "1", "8", and "15". The permutation $15 \cdot 14 \cdot 13$ includes the 6 $(= 3 \cdot 2 \cdot 1)$ $(r!)$ ways in which the three selected balls can be arranged ("1", "8", "15"), ("1", "15", "8"), ("8", "1", "15"), ("8", "15", "1"), ("15", "1", "8"), and ("15", "8", "1").

Now, suppose the order in which the "1", "8", and "15" balls is of no interest; that is, the selected order is irrelevant. The number without order (denoted as y) can be multiplied by $3!(= r!)$ to restore the order. However, this is the permutation above, that is, $3!y = 15!/12!$ so $y = \frac{15!}{3!12!}$; the combination $\binom{15}{3} = \binom{15}{12}$.

Example 7.4.5 **Blocks**

Three blocks are labeled A, B, and C. Two blocks are to be selected without replacement and without regard to order. Display the possible outcomes.

Solution:
First, select blocks with regard to order:

AB AC BA BC CA CB

This follows from the permutation $3!/(3-2)! = 3! = 6$. Since AB and BA are the permutation of the letters A and B, they are a single pair of the two letters. Now, to remove order either AB or BA, AC or CA, BC or CB are deleted so AB AC BC remain.

Without order, this is the combination $\binom{3}{2} = \frac{3!}{(3-2)!(2!)} = 3.$

While playing with blocks seems childish, they are a useful way to build your intuitive grasp of combinatoric outcomes.

Power Growth and Mixed Growth

Recall the combinatoric tree that appeared earlier in this section. Between the two lower branches (with replacement), the upper one, power growth, follows immediately from the multiplication principle. Since the population is restored to its original configuration (i.e., with replacement), each repeated draw is from among n elements or n^r.

The lowest branch, mixed growth, has the number $\binom{n+r-1}{r}$ for the number of ways r elements can be selected from among n elements with replacement and without regard to order. We illustrate this with throws of a die.

Imagine a die thrown five times to yield 2, 6, 4, 2, and 1. This outcome can be displayed as

0 | 00 | | 0 | | 0

using a 0 to designate an outcome and a vertical divider to separate die faces. These can be written symbolically as 0100110110 where the vertical dividers are replaced by "ones" (for convenience). The number of dividers or "ones" needed to separate the categories is $n-1$. The total number of 1s and 0s is $n+r-1$, $(6+5-1=10$ here). Among this total, these r outcomes are to be placed; $r=5$ here. The number of ways that five outcomes can be selected from among six elements with replacement and without regard to order is the combination $\binom{n+r-1}{r}$; here $\binom{6+5-1}{5}=\binom{10}{5}=252$.

♦ During the tenth century, Bishop Wibold of Cambrai invented a moral dice game in which players rolled three dice, a virtue being associated with each distinct outcome. The problem of calculating and enumerating possible outcomes became the subject of a thirteenth-century poem "De Vetula." The result is the mixed growth number.

If there is replacement and regard to order, then there would be $n^r = 6^3 = 216$ possible outcomes for a throw of 3 dice as $r=3$. However, here the order in which dice are thrown does not matter. Using $n=6$ and $r=3$ there are $\binom{6+3-1}{3}=56$ possible outcomes. Therefore, when dice are thrown for divination, in this instance, the number of possible throws of three dice is 56.

♦ Some years ago, a popular faster food chain used TV ads to show that diners could select three sides from among 6 offered claiming that there were 216 meal choices. It was quickly pointed out to the chain that the number was actually 56. How did it happen? Well $216 = 6^3$ includes replacement as well as order. While replacement is possible, order is not. "Peas and carrots" is the same as "carrots and peas!" The correct calculation is the Wibold number $\binom{6+3-1}{3}=56$ since $n=6$ and $r=3$ here.

♦ A book by Florence David and an article by Kreith and Kysh (Historical Notes) provide interesting background on Bishop Wibold's role in combinatoric probability.

Products of combinations are helpful in determining the ways a series of events can occur. Many events can be decomposed into steps or stages. Analyzing each stage separately, they can be combined for the total. The number of elements in the set of outcomes is often determined using the multiplication principle as in the next examples.

Example 7.4.6 Species–Island Pairings

Charles Darwin observed 13 species of finches among the 17 Galapagos Islands. Ecologists study the presence and absence of species on the various islands seeking pairs of species that never appear together. How many pairs of species are there? How many species pairs and island pair combinations are theoretically possible?

Solution:

The number of possible species pairings is $\binom{13}{2} = 78$.

The number of island pairings is $\binom{17}{2} = 136$.

We consider both species pairs and island pairs to find that there are exactly $\binom{13}{2}\binom{17}{2} = 10{,}608$ *possible species–island pairings.*

Example 7.4.7 Subcommittees

Suppose that five undergraduates, three graduates, and four faculty members currently serve on a committee. A subcommittee is to consist of one from each group. How many possible subcommittees can be formed?

Solution:

Using combinations, one member from each of the three groups is chosen to yield the desired number $\binom{5}{1}\binom{3}{1}\binom{4}{1} = 5 \cdot 4 \cdot 3 = 60$. *(The multiplication principle applies as well).*

♦ Blaise Pascal's abilities as a foremost seventeenth-century mathematician, physicist, and religious scholar were recognized at an early age. Unfortunately, his frail health contributed to his premature death at 39 years. The famous Pascal's Triangle, named in his honor, did not originate with him although he seems to be the first to have recognized its many curious properties. Origins of the Pascal Triangle date to the Chinese in the fourteenth century and earlier. From China, it reached the Arabian Peninsula before reaching Europe. It is known by various names. A Google search is rewarding.

♦ Pascal's Triangle

The Pascal Triangle begins as:

$$
\begin{array}{ccccccccc}
 & & & & 1 & & & & \\
 & & & 1 & & 1 & & & \\
 & & 1 & & 2 & & 1 & & \\
 & 1 & & 3 & & 3 & & 1 & \\
1 & & 4 & & 6 & & 4 & & 1
\end{array}
\qquad
\begin{array}{l}
\\
n = 1 \\
n = 2 \\
n = 3 \\
n = 4
\end{array}
$$

Rows begin and end with a "1." The other entries are the sum of successive pairs of entries from the row above. Pascal's Triangle evolves in this manner for the positive integers.

The entries have the remarkable property of being the coefficients in a binomial expansion. Triangle entries are also combinations. For example, the row for $n = 4$ has the respective combinations $\binom{4}{0}\binom{4}{1}\binom{4}{2}\binom{4}{3}\binom{4}{4}$ corresponding to $C(n, r)$ where $n = 4$ and $r = 0, 1, 2, 3, 4$.

Special cases arise when arrangements are circular or when there are several groups of distinct objects. The number of permutations of n distinct objects arranged in a circle, such as keys on a ring, is $(n - 1)!$

The number of permutations of n distinct objects of which n_1 are of one kind, n_2 are of a second kind, …, and n_r are of the rth kind is

$$\frac{n!}{n_1!n_2!\cdots n_r!}, \quad \text{where } n_1 + n_2 + \cdots + n_r = n$$

This is usually denoted by the symbol

$$\binom{n}{n_1 \quad n_2 \cdots n_r}$$

and called a **multinomial**.

Example 7.4.8 Group Outing

Find the number of ways that a group of 20 senior citizens may be divided into three groups on a cruise so that one group of 10 has tango lessons, 6 play bingo, and the other 4 do shopping.

Solution:

There are three groups (or labels) with $n_1 = 10$, $n_2 = 6$, and $n_3 = 4$. The problem can be viewed as three stages in which 10 of the 20 have tango lessons, 6 of the remaining 10 play bingo, and, finally, the other 4 seniors are shopping.

$$\binom{20}{10}\binom{10}{6}\binom{4}{4} = \frac{20!}{10!10!}\,\frac{10!}{6!4!}(1) = \frac{20!}{10!6!4!} = \binom{20}{10 \quad 6 \quad 4}$$

♦ Students have understandable difficulty with combinatoric problems. Unfortunately, there is not much help available.

Two reasons for the difficulty are, first, one usually lacks the experience to have a useful intuition for an approximate answer. For example, few people can estimate the number of possible poker hands or the number of possible routes a delivery truck might choose about a city. A second source of difficulty stems from not having a simple way to obtain a "ball park" answer in the same way that one might estimate the number of patrons in a well-filled auditorium by counting them in one section and multiplying by the number of sections.

Still, being able to "super count" is a valuable skill. Using combinatorics, one can estimate the number of operations to run a computer program. This estimate combined with the computer speed yields an estimate of the computational time (and cost!) to execute the program. How extensive a market survey a company might afford depends on the cost of sampling or of a sorting among a large list of addresses, and so on.

While, exact answers are needed for your homework and exams, for many practical purposes a "ball park" or "order of magnitude" estimate is often adequate.

EXERCISES 7.4

1. How many odd three-digit numbers can be formed using the digits $\{2, 3, 6, 7, 8\}$ if no digit can be used more than once?
2. A quiz consists of five multiple-choice questions with four possible options for each. In how many ways can the quiz be completed?
3. A computer company has four types of monitors, two types of printers, and three types of hard drives. How many computer configurations consisting of a monitor, printer, and hard drive are possible?
4. A familiar rhyme begins:

   As I was going to St. Ives,
   I met a man with seven wives.
   Each wife had seven sacks,
   Each sack had seven cats,
   Each cat had seven kits. …

   a) How many sacks are there? b) kits?

♦ Curiously, this Mother Goose nursery rhyme is strikingly similar to one found in a 3500-year-old Egyptian Rhind papyrus.

5. Evaluate

a) $7!$ b) $8!/6!$ c) $9!/0!$ d) $10!/10!$ e) $52!/(48!4!)$

6. A die is tossed three times. How many possible outcomes are there?

7. How many five-card poker hands are possible?

8. Evaluate

a) ${}_5P_3$ b) ${}_6P_5$ c) ${}_{20}P_{20}$ d) ${}_8P_0$

9. Evaluate

a) ${}_7C_2$ b) ${}_6C_5$ c) $\binom{9}{6}$ d) $\binom{4}{1}$

10. A committee has five members. In how many ways can three members be chosen to be secretary, treasurer, and president?

11. A map is to have Delaware, Maryland, New Jersey, Virginia, and Pennsylvania in different colors. If there are seven color choices, in how many ways can the map be colored?

12. A committee consists of four members; Al, Barbara, Charles, and Diana. They wish to form a subcommittee for their next event. One member is responsible for ticket sales, and the other for refreshments. In how many ways can this be done?

13. How many three-digit numbers can be formed from 3, 5, 6, 7, 8, and 9 if no digit can be used more than once?

14. A horse race has 10 entries. In how many ways can the horses finish in win, place, and show? Neglect a dead heat.

15. A coin is tossed four times. How many outcomes have three heads?

16. A bag contains six balls numbered from 1 to 6. A ball is drawn, its number noted, and then replaced. If the draw is repeated, how many outcomes are possible?

17. Verify that $(r!)_nC_r = {}_nP_r$.

18. A Shakespearean sonnet typically consists of three quatrains (four lines) and a couplet (2 lines). The sonnets are written in iambic pentameter so a line consists of 10 syllables: five stressed and five unstressed.

    a) How many stressed syllables are in the 154 Shakespearean sonnets?
    b) How many syllables are in 50 sonnets?

19. From a group of 10 finalists on American Idol, you are asked to randomly choose three. In how many ways can you correctly choose at least two of the show's three finalists?

20. In a lottery, a win is a correct choice of 5 of the 59 white balls and the red Powerball number from among 35 possibilities. Verify the published odds of 1:175,223,510 that a single ticket wins the Powerball lottery.

21. Three cars are used to carry 14 students on a field trip. Two cars can each seat five students while the third can only seat four students. How many different seating arrangements are possible?

22. Three coins are to be chosen from among six coins: three nickels, two dimes, and a quarter. In how many ways can the selection be made so the value of the coins is at least 30 cents?

23. How many distinct license plates can have two letters followed by four digits?

24. At a US Olympic track and field trial, eight men were in the final heat of the 400-m hurdles vying for three spots on the Olympic team. In how many ways can this be done assuming no ties?

25. A bank has several identity theft TV commercials. What is the probability that someone using your ATM card successfully guesses your four-digit pin in a single attempt? Suppose a card issues an alert when the wrong PIN is entered for the fourth time. What is the probability someone trying to use your ATM card would set off an alert (assume no PIN is repeated)?

26. In Morse code, dots and dashes represent letters of the alphabet and the 10 digits (e.g., A is ·- , B is -· · ·, etc.). How many characters are needed to code the alphabet?
    The alphabet and the digits?

◆ While Morse code is largely obsolete, a web search tells of its amazing 150-years.

27. In their earlier years, three-digit Area Codes were restricted by
    i. first digit cannot be a zero
    ii. second digit can only be a 0 or 1
    a) How many possible seven-digit phone numbers are possible in an Area Code?
    b) What is the largest number of possible phone numbers using the rules cited?

◆ By 1995, the boom in fax machines, cell phones, and computers compelled alteration of telephone company computers to accept any number as a middle digit for an area code – a change from the earlier use of 0 or 1.

The system of codes began in 1947 with direct dial long-distance service. Designed when rotary dial phones were in use, low-numbered digits were assigned to areas receiving high call volumes to reduce dialing time: Washington, DC, 202; New York, 212; Los Angeles, 213; and so on.

Area codes have actually become a sociological topic as phone numbers have become personal, contrasting an era when phone calls that began with "hello, may I speak to so-and-so" rather than "Hey" or "What's up?." www.areacode-info.com and "Area Codes, Now Divorced from Their Areas," *New York Times*.

28. Postal service equipment "reads" zip codes at a rapid rate whether hand written or not and regardless of envelope orientation. Some letters are troubling as the optical reader can mistake a zip code such as 16988 when read upside down as 88691.
    a) How many possible five-digit zip codes are there?
    b) How many zip codes are potentially subject to being misread?
    c) How many zip codes read the same regardless of envelope orientation?
    d) One example of a zip code that reads the same in either direction is 18,081. To what city or town does this zip code belong?
    e) Three cities that have troubling zip codes for an optical reader are Wheatland, PA 16161; Media, PA 19091; and Wilmington, DE 19899. Where would mail to these zip codes be sent if read upside down?

29. A DJ has a collection of 10 R&B CDs, 15 country CDs, and 25 pop CDs. In how many ways can the DJ select four different CDs to play so that each of the three groups is represented?

30. Evaluate the multinomial coefficient $\binom{6}{4\quad 1\quad 1}$.

31. In how many ways can an apple, cherry, plum, and pear tree be planted in a circle?

32. In a game, the starting lineup of the Philadelphia Phillies hit six home runs. In how many ways might the six home runs be distributed among the nine players assuming all are equally capable?

33. A mail carrier has 75 pieces of "junk" mail to deliver to 50 homes. In how many ways can they be delivered?

34. In a throw of three dice, there are $6^3 = 216$ possible outcomes. For example, "one-three-six" is distinct from "six-one-three." Without regard to order, this reduces to 56 (as mentioned in the text).
    a) Without regard to order, how many distinct ways can the face total on a roll of three dice be a 9? A 10?
    b) Although the number of ways for a 9 or a 10 is the same, their probabilities are not. Explain! (This is a famous problem reportedly posed to Galileo in the seventeenth century by a Duke who keenly observed that a "10" was more likely than a "9" in equally likely play).

7.5 MATHEMATICAL PROBABILITY

Over centuries, **probabilities** have been understood in various ways as numbers on the unit interval. Earliest, the *geometric* view held that the probability that a coin flip yields "a head" is 1/2. The logic? A coin, assumed two sided, obverse and reverse, one being head, so 1 divided by 2 is 1/2. This early view of probability, still useful today, with the understanding that it assumes all outcomes is equally likely.

By the 1800s, the limitations of the "equally likely" assumption were apparent. A coin or a die could be "loaded" or "skewed" so that one clearly could not assume equally likely outcomes. A *frequency* notion emerged. Probabilities would be determined by experiment, for example, by tossing a coin repeatedly and noting the number of heads. The ratio of the number of heads to the total number of tosses estimates the probability of a head; the presumption being that the greater the number of tosses the better the estimate of the probability. Although one could not know a probability exactly, as it requires an infinite number of tosses, that limitation was balanced by the ability to apply probability concepts to a broader range of phenomena.

♦ Have you ever heard that "lightning never strikes the same place twice"? Well, what about striking the same person seven times? A park ranger named Roy "Dooms" Sullivan cited it in the Guinness Book of World Records for being the most lightning struck person on record. Injured from head to toe – he survived!

Some usage in probability theory requires a brief explanation. We say an **experiment** is an act or process by which an observation or measurement occurs. An **outcome** is a result of the experiment. A **sample space** of an experiment is the complete (and exhaustive) set of its mutually exclusive outcomes. The elements of a sample space, denoted by **S**, are called **sample points**.

Example 7.5.1 *Sample Spaces*

Which of these are sample spaces?

a) Playing cards: {face cards, number cards}
b) Toss of a die: {1, 2, 3, 4, 6}
c) Political party: {Democrat, Republican, not Republican}
d) Playing cards: {Red cards, clubs, spades}

Solution:
As sample spaces, a), b), and c) are faulty. Four aces are necessary for a complete card deck; the five is missing for a die toss; and, since "Democrats" are also "not Republicans," outcomes overlap.
In d), all 52 cards are included and no card belongs to more than one category, so it is a sample space.

Probability relationships stem from the theory of sets. The customary notation for the probability of an event described by a set **A** is $P(\mathbf{A})$. Since probabilities are values between 0 and 1, inclusive, and since the probabilities for a universal set, or sample space, sum to unity, one can write

$$P(\mathbf{A}) + P(\mathbf{A}^c) = 1$$

where the complementary set $\mathbf{A}^c$ means "not **A**" and $P(\mathbf{A}^c)$ its probability. The following are some properties to keep in mind when determining probabilities.

Properties of Probabilities

$$\mathbf{0 \leq P(A) \leq 1}$$

If $P(A) = 0$, then event A is impossible
If $P(A) = 1$, then event A is certain

$$P(A) + P(A^c) = 1$$

$$P(A) = 1 - P(A^c)$$

$$P(A^c) = 1 - P(A)$$

For disjoint events A and B, $P(A \cup B) = P(A) + P(B)$
For any events A and B, $P(A \cup B) = P(A) + P(B) - P(A \cap B)$

For a simple probability, one often assumes each simple event of a sample space **S** is equally likely. A card drawn from a standard deck, a roll of a fair die, and so on are examples. Therefore, probabilities are calculated by the number of elements in the sample space for the specific event of interest as follows:

$$P(\mathrm{A}) = \frac{n(\mathbf{A})}{n(\mathbf{S})} = \frac{\textit{number elements in } \mathbf{A}}{\textit{number elements in } \mathbf{S}}$$

♦ **"What are the odds?"** is an often heard phrase in sports, gambling, and other chance events. If the odds of an event are "5 to 3" (or 5:3), for example, that implies a probability of $5/8$ (= $5/(5 + 3)$). In general, odds ***a:b*** imply an event probability of $a/(a+b)$.

Alternatively, the probability of an event A, expressed as the odds of its occurrence, is

$$\frac{P(\mathrm{A})}{1 - P(\mathrm{A})} = \frac{P(\mathrm{A})}{P(\mathrm{A^c})}$$

For example, if the probability of an event is, say, 2/3, then the odds are $\frac{2/3}{1-(2/3)} = \frac{2}{1}$ or "2 to 1."

Example 7.5.2 Coin Tosses

A fair coin is tossed three times and the results noted. Determine the sample space and the probability of exactly two heads.

Solution:
There are $2^3 = 8$ *sample points in S, the sample space, as follows:*

$$\mathbf{S} = \{HHH, HHT, HTH, HTT, THH, THT, TTH, TTT\}$$

Three sample points have exactly two heads. Therefore, $n(\mathbf{A}) = 3$ *and* $n(\mathbf{S}) = 8$. *The probability of exactly two heads is 3/8.*

Example 7.5.3 Dice Tosses

A pair of dice is tossed. What is the probability of a "seven?"

Solution:
There are $6^2 = 36$ *possible outcomes (1, 1), (1, 2), ..., (1, 6), (2, 1), ..., (6, 5), (6, 6). Therefore,* $\mathbf{n}(\mathbf{S}) = 36$. *The following sample points are a "seven," (1, 6), (2, 5), (3, 4), (4, 3), (5, 2), or (6, 1). Since* $\mathbf{n}(\mathbf{A}) = 6$, *the probability of a "seven" is 6/36 or 1/6.*

◆ How fair are casino dice? Ivars Peterson's MathLand site has an answer. Read his "Fun and Games in Nevada": http://www.maa.org/mathland/mathland_5_5.html.

Example 7.5.4 Card Draws

A card is drawn from a standard deck of 52 cards. Determine the probability of drawing

a) *a black card or an ace.*
b) *a face card or a club.*
c) *a number card or a red ace.*

Solution:

a) *Let* $\mathbf{A} = \{a\ black\ card\ is\ drawn\}$ *and* $\mathbf{B} = \{an\ ace\ is\ drawn\}$
There are two black aces, so $P(\mathbf{A} \cup \mathbf{B}) = \frac{26}{52} + \frac{4}{52} - \frac{2}{52} = \frac{28}{52} = \frac{7}{13}$.

⟶

b) Let $\boldsymbol{A} = \{a\ face\ card\ is\ drawn\}$ *and* $\boldsymbol{B} = \{a\ club\ is\ drawn\}$.

There are three-face card clubs so $P(\boldsymbol{A} \cup \boldsymbol{B}) = \frac{12}{52} + \frac{13}{52} - \frac{3}{52} = \frac{22}{52} = \frac{11}{26}$.

c) Let $\boldsymbol{A} = \{a\ number\ card\ is\ drawn\}$ *and* $\boldsymbol{B} = \{a\ red\ ace\ is\ drawn\}$.

These events are disjoint, so $P(\boldsymbol{A} \cup \boldsymbol{B}) = \frac{36}{52} + \frac{2}{52} = \frac{38}{52} = \frac{19}{26}$.

Example 7.5.5 Sharing Birthdays

Among 30 people, what is the probability that no two share the same birthday? (Assume a 365 day year and that all days are equiprobable.)

Solution:

For no two people to have the same birthday, the first person's birthday can be any of 365 days. The next person's birthday must be one of the 364 other days. The third person's birthday, different from the first two, is one of 363 days and so forth until the thirtieth person's birthday must be one of the 336 remaining days. The intersection event has the probability:

$$\left(\frac{365}{365}\right)\left(\frac{364}{365}\right)\cdots\left(\frac{336}{365}\right) \approx 0.294$$

This seems surprising! It means that among 30 people there is about a $(1 - 0.294)100\% \approx 71\%$ *chance that at least two people share a birthday. For only 23 people, there is about a 50% chance and for 50 people the chance for a shared birthday is about 97%!*

You may have noticed shared birthdays among members of your family tree or in one of your classes or groups. This example illustrates their likelihood in larger families.

Notice that the computation has been organized by multiplication of successive fractions! The permutation $_{365}P_{30}$ *will not "fit" in a calculator.*

♦ Ever had someone tell you something dramatic about a person who was in your thoughts earlier that day? Wondered about your ESP prowess?

A rough sense of the frequency of such a happening might follow along these lines. Suppose there are 100 people in your "every day" circle. With 365 days per year and a US population of about 320 million people, the expected number of people who annually learn that a dramatic and life-altering event has happened to one of their "inner circle" is about 88 million (or (100/365)(320 million)). This translates into an average of 240,000 people nationwide per day. Not too unusual!

The authors of a recent book titled "Debunked!" are convinced that a "lack of knowledge of probability is one of the most pernicious sources of superstition and deception." A lean volume, its 136 pages aids readers to acquire paranormal power themselves as they unmask trickery and fraud in common experiences. ("Debunked!" by Georges Charpak and Henri Broch, Johns Hopkins University Press, 2004.)

Conditional Probability

The **conditional probability** of an event **A** given that an event **B** has occurred is defined by

$$P(\mathbf{A}|\mathbf{B}) = \frac{P(\mathbf{A} \cap \mathbf{B})}{P(\mathbf{B})} \qquad P(\mathbf{B}) \neq 0.$$

That is, the probability of the intersection event divided by the probability of the given event is the conditional probability of event **A** given event **B**.

The calculation of the probability of an intersection event is aided by the conditional probability definition written as

$$P(\mathbf{A} \cap \mathbf{B}) = P(\mathbf{B})P(\mathbf{A}\,|\,\mathbf{B}) \text{ or } P(\mathbf{A} \cap \mathbf{B}) = P(\mathbf{A})P(\mathbf{B}\,|\,\mathbf{A})$$

Example 7.5.6 Colored Balls

A bag contains four red balls numbered 1, 2, 3, 4 and two white balls numbered 5 and 6.

a) An even numbered ball is drawn at random. What is the probability it is white?

b) The ball drawn is not a 2. What is the probability it is red?

Solution:

a) $P(white|even) = \dfrac{P(white\&even)}{P(even)} = \dfrac{1/6}{3/6} = \dfrac{1}{3}.$

b) $P(red|not\ 2) = \dfrac{P(red\&not\ 2)}{P(not\ 2)} = \dfrac{3/6}{5/6} = \dfrac{3}{5}.$

Example 7.5.7 Manufacturer Warranty

Manufacturers have records of customer complaints. A manufacturer noted that 40% of complaints were for electrical failures within the warranty period and 10% afterward. For mechanical failures, 35% were within the warranty period and 15% afterward. Find the probability that an after warranty period complaint was for mechanical failure.

Solution:

*Let **A** = mechanical failure and **B** = product failure after the warranty period. After the warranty period, the probability of failure is the sum of the electrical and mechanical failures. Therefore, **P(B)** = 0.10 + 0.15 = 0.25. The intersection of events **A** and **B** is the event "mechanical failure after the warranty period." That is, **P(A ∩ B)** = 0.15. Now,*

$$P(\mathbf{A}\,|\,\mathbf{B}) = P(A \cap B)\,/\,P(B) = 0.15\,/\,0.25 = 0.60$$

Probabilistic Independence

Events **A** and **B** are considered to be **probabilistically independent,** or simply **independent,** if the probability of event **A** is unaffected by the occurrence (or nonoccurrence) of event **B**. Therefore, events **A** and **B** are independent if, and only if,

$$P(\mathbf{A} \mid \mathbf{B}) = P(\mathbf{A}) \quad \text{(or if } P(\mathbf{B} \mid \mathbf{A}) = P(\mathbf{B})\text{)}.$$

In other words, when the occurrence or nonoccurrence of an event **B** has no influence upon the chance occurrence of **A**, then events **A** and **B** are independent. Events that are not independent are **dependent**. Incidentally, probabilistic independence is also known as *stochastic independence* or *statistical independence*.

> ***Probabilistic Independence***
>
> **Events A and B are probabilistically independent if, and only if**
>
> $$P(\mathbf{A} \mid \mathbf{B}) = P(\mathbf{A}) \quad \textbf{or} \quad P(\mathbf{B} \mid \mathbf{A}) = P(\mathbf{B})$$

♦ Students often confuse "mutually exclusive" and dependence. Mutually exclusive events are always dependent since $P(\mathbf{A} \cap \mathbf{B}) = 0$. However, events that are dependent need not be mutually exclusive.

Example 7.5.8 Die Toss Independence

A fair die is thrown. Let $\mathbf{A} = \{\textit{even number}\}$ *and* $\mathbf{B} = \{\textit{greater than 1}\}$.
Are these independent events?

Solution:
With $\mathbf{A} = \{2, 4, 6\}$ *and* $\mathbf{B} = \{2, 3, 4, 5, 6\}$, $\mathbf{A} \cap \mathbf{B} = \{2, 4, 6\}$.

$$P(A|\mathbf{B}) = \frac{P(A \cap \mathbf{B})}{P(\mathbf{B})} = \frac{3/6}{5/6} = \frac{3}{5}.$$

Here, $P(A \mid \mathbf{B}) = 3/5$ *and* $P(A) = 3/6$ *so* $P(\mathbf{A} \mid \mathbf{B}) \neq P(\mathbf{A})$, *and* **A** *and* **B** *are dependent.*

Example 7.5.9 Assembly Line Inspections

Products on an assembly line must pass two sequential inspections. The probability that the first inspector will miss a defective item is 0.05. If a defective item is missed by the first inspector, the probability that the second inspector will also miss the defect is 0.10. What is the probability that a defective item is missed by both inspectors?

Solution:

$$\begin{aligned} P(\textit{defect missed twice}) &= P(\textit{missed by first and missed by second}) \\ &= P(\textit{missed by first}) \cdot P(\textit{missed by second} \mid \textit{missed by first}) \\ &= (0.05)(0.10) \\ &= 0.005 \end{aligned}$$

In many practical situations, joint or intersection probabilities pose a difficulty since the desired data are often not available. For example, in forecasting the movement of spilled oil at sea, wind speeds and wave heights are primary factors. Data for the joint event should be wind and wave measurements at the same instant and at the same place. While wind and wave measurements are easily made and generally available, they usually are not measured simultaneously. It is of importance to assess the extent of dependence of joint events since they are a clue to the accuracy of the often used assumption that

$$P(\mathbf{A} \cap \mathbf{B}) = P(\mathbf{A} \mid \mathbf{B})P(\mathbf{B}) = P(\mathbf{A})P(\mathbf{B})$$

It is important to understand that probabilistic independence is not something that is generally proved "practically" or "mathematically." It is an assumption about events. Assumptions about independence come from experience and knowledge of them. For example, flips of a coin are often regarded as independent events as are the genders of successive children born to a family. However, these beliefs cannot be "proved mathematically."

♦ Medical researchers, who may base conjectures about the efficacy of medications and therapies on case studies, depend on independent randomized clinical trials for conclusiveness. They refer to such "trials" as the "gold standard."

Example 7.5.10 Two-Link Chain

Consider a two-link chain (shown below) with $\mathbf{A}$ and $\mathbf{B}$ denoting the events "link failure." Let $P(\mathbf{A})$ and $P(\mathbf{B})$ denote the respective probabilities of failure. Assume $P(\mathbf{A}) = P(\mathbf{B}) = 10^{-3}$.

⟶

What is the probability of chain failure, ***F****?*

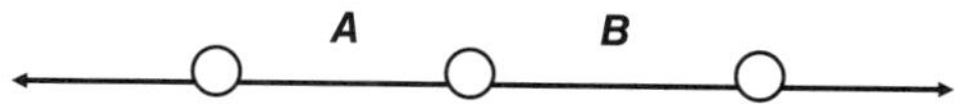

Solution:

Since the chain fails if either or both links fail, the probability of failure of the chain is $P(\boldsymbol{F}) = P(\boldsymbol{A} \cup \boldsymbol{B}) = P(\boldsymbol{A}) + P(\boldsymbol{B}) - P(\boldsymbol{A} \cap \boldsymbol{B})$.

Note the need for the joint probability $P(\boldsymbol{A} \cap \boldsymbol{B})$*, for which data usually are not available. Still it is possible to at least estimate the required probability.*

First, assume that link failures are independent events, so

$$P(\boldsymbol{A} \cap \boldsymbol{B}) = P(\boldsymbol{A})P(\boldsymbol{B}) = (10^{-3})(10^{-3}) = 10^{-6}$$

Then, $P(\boldsymbol{F}) = 10^{-3} + 10^{-3} - 10^{-6} \approx 2(10^{-3})$.

However, if it cannot safely be assumed that failures of the links are independent events, the value above may not be a trustworthy figure.

Now, at the other extreme, suppose that the links are completely dependent so $P(\boldsymbol{A} \mid \boldsymbol{B}) = P(\boldsymbol{B} \mid \boldsymbol{A}) = 1$. *That is, if it is known that if one of the links fails, it is certain that the other has also failed.*

Then, $P(\boldsymbol{F}) = (10^{-3}) + (10^{-3}) - (10^{-3})(1) = (10^{-3})$.

Therefore, even without knowledge of link interaction, one can form the useful bounds

$$10^{-3} \leq P(\boldsymbol{F}) \leq 2(10^{-3})$$

♦ We remarked earlier that knowledge of probabilistic independence must come from other than mathematical considerations. In the last example, suppose that, in practice, the difference between 10^{-3} and $2(10^{-3})$ is not significant to the purpose at hand. Then, one can safely ignore interactive effects of the links.

However, suppose that it is desirable to ascertain whether the true $P(\mathbf{F})$ is closer to one bound or the other. What questions might one ask? One can inquire whether the iron ore, say, used in the links' manufacture came from the same source; came from the same furnace; or came from the same batch. Additional measurements can be persuasive evidence.

♦ There are many English language uses of the word "independent" in mathematics and elsewhere, for example, Independence Day and independent equations. Its use in mathematical probability is distinct and has no relation to its use elsewhere.

EXERCISES 7.5

1. Two dice are rolled and the sum of their faces recorded. What is the sample space?
2. Four coins are tossed. Write the sample space.
3. Explain whether these are sample spaces.
   a) Playing cards {face, ace, number card}
   b) Coin tossed twice {1 head, 2 heads}
   c) Die toss {number less than 6, even number}
4. Find the probability that a card, drawn at random from a standard deck, is
   a) a diamond.
   b) a heart or a number card.
   c) a black number card or a club.
5. A die is tossed. Let event **A** be that an odd number appears and event **B** that the number is more than four. Find $P(\mathbf{A} \cup \mathbf{B})$.
6. Suppose four red, two blue, three green, and one yellow marble are placed in an urn. A marble is randomly drawn. Find the probability that it is either green or not blue.
7. Suppose $P(\mathbf{A}) = 0.2$ and $P(\mathbf{B}^c) = 0.3$ where **A** and **B** are mutually exclusive events. What is $P(\mathbf{A} \cup \mathbf{B})$?
8. If $P(\mathbf{A}) = 0.6$, $P(\mathbf{B}) = 0.3$, and $P(\mathbf{A} \cap \mathbf{B}) = 0.2$, find $P(\mathbf{A} \cup \mathbf{B})$.
9. In one of your larger classes (or groups) what is the probability that at least two people share a birthday? What is the actual result?
10. A pair of dice are tossed and the sum of their faces recorded. Determine the probability
    a) 5 appears.
    b) 7 or 11 appears.
    c) a number that is a multiple of 4 appears.
11. Three dice are tossed and the sum of their face numbers recorded. Determine the probability that the sum is five.
12. A die is tossed and an odd numbered face appears. What is the probability it is 3?
13. A win in the New Jersey Lottery requires the correct choice of six numbers from among the given 49. What is the probability of a win with a single ticket?
14. Find data to support answers to these chances of IRS audits.
    a) Since chances of an audit vary regionally, are individual tax returns from the Mid-Atlantic or Western IRS regions more likely to be audited?
    b) Is an audit more likely to be for an individual or a corporation tax return?

15. The Old Testament consists of 39 books: Pentateuch (5), Historical Books (12), Psalms and Writings (5), and Prophets (17).

The New Testament consists of 27 books: Gospels (4), Historical Narrative (1), Pauline Letters (13), and General Epistles and Revelation (9).

A book of the Bible is chosen at random:

a) What is the probability that it is from the Old Testament?

b) An Old Testament book is chosen at random. What is the probability it is of the Pentateuch?

c) What is the probability the book is named for a woman?

d) A New Testament book is chosen. What is the probability it is one of the four Gospels?

16. Annika Sorenstam, a golfer since age 12, made history as the first woman to play in a PGA tour in 2003. Her scores of 71 and 74, a 145 total, missed the cut by four strokes.

The following probabilities were obtained by using her hole-by-hole results for the two rounds. If she shot a hole at par or better, there was a 11/14 chance that the next hole was also at par or better.

The chance was 3/14 for a hole at par or better and next have a Bogey or worse. When a Bogey or worse was shot, the chance was 6/7 that the next hole scores at par or under and 1/7 for a Bogey or worse. Assuming these are actual probabilities, find the probability of par or under for five consecutive holes on her next PGA outing.

17. At one time 38 states had a death penalty and there were 65 executions in the United States during 2003.

U. S. Executions in 2003

Alabama	3
Arkansas	1
Federal	1
Florida	2
Georgia	3
Indiana	2
Missouri	2
North Carolina	7
Oklahoma	14
Ohio	3
Texas	24
Virginia	3
Total	65

Use the data to determine the probability that an execution at random took place in

a) Texas

b) Delaware

c) one of the first 10 states to ratify the constitution.

18. Suppose P_i is the probability that the ith switch is closed, $i = 1, 2, 3, 4$. If switches function independently, what is the probability that a current flows between A and B for the circuits shown?

a)

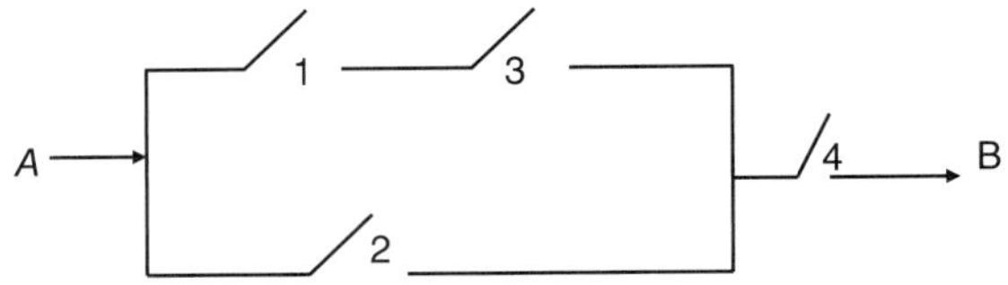

b)

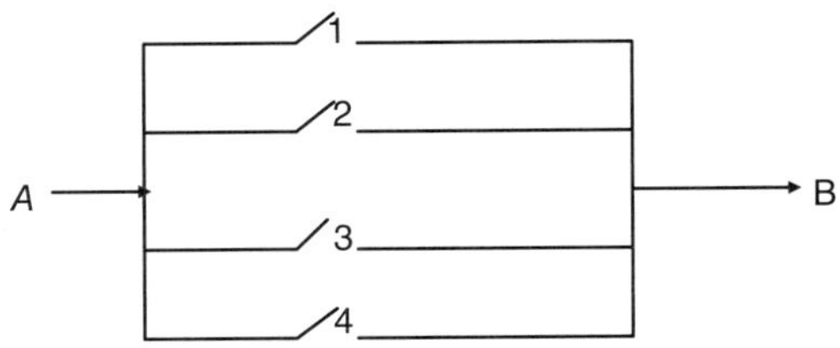

c)

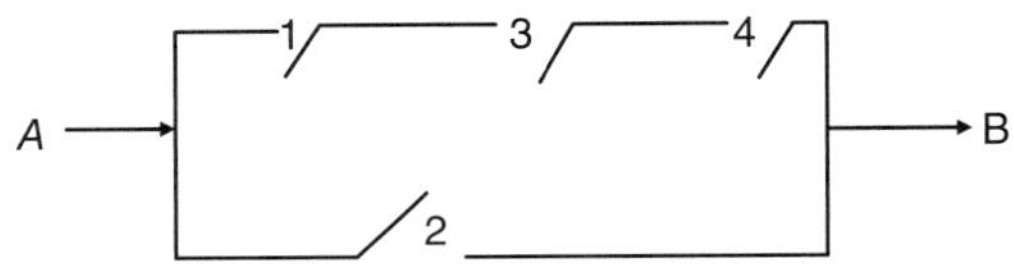

19. While blood types may have some ethnic and racial variations, a distribution of blood types in the US population is:

Blood Type	Percentage
O Rh-positive	38
O Rh-negative	7
A Rh-positive	34
A Rh-negative	6
B Rh-positive	9
B Rh-negative	2
AB Rh-positive	3
AB Rh-negative	1

Using these values:

a) What percentage of the US population is Rh-negative?

b) What is the probability that a person at random has O-type blood, given the person is Rh-positive?

c) What is the probability that a person at random is a "Universal Donor?"

d) What is the probability that a person at random is a "Universal Receiver?"

20. A playing card is drawn at random from a standard deck. Find the probability that the card is
    a) a heart given that it is red
    b) an ace given that it is a face card
    c) black given it is not a heart
    d) not the jack of spades given it is a face card
21. A red die and a white die are tossed. What is the probability that the sum is 7 if it is known to be less than or equal to 8?
22. A die is loaded so that the probability that an odd-numbered face appears is twice as likely as for an even-numbered face. For a toss, what is the probability a perfect square appears if the number is greater than "2?"
23. A department store accepts Visa, MasterCard, Discover, or cash/check. The probabilities that a customer uses these options are 0.2, 0.3, 0.1, and 0.4, respectively. What is the probability that a credit card purchase did not use the MasterCard?
24. In a game, the probability that Jack wins over Jill is 3/5. If three games are played, what is the probability that Jack wins all three if it is known that he wins at least two games?
25. In a public opinion poll, each respondent's age and opinion are recorded. The table summarizes the results:

	Support, %	Oppose, %	Undecided, %
Under 50	32	17	3
50 or over	25	22	1

    Find the probability that a randomly chosen person
    a) under 50 supports the issue.
    b) over 50 does not oppose the issue.
    c) opposes the issue given he or she does not support it.
26. A poll of 500 college students resulted in the following:

	Undergraduate	Graduate
Male	170	80
Female	190	60

    A student is chosen at random. What is the probability that the student is
    a) a graduate?
    b) a male undergraduate?
    c) a female graduate given a female was chosen?
    d) is a female given the student chosen was not a female undergraduate?
27. A committee comprises five Democrats and two Republicans. A subcommittee of two is selected at random. Find the probability that both are Democrats given that at least one is a Democrat.

28. A faculty census showed that 55% were women, 65% had earned a Ph.D., and 10% were males without a Ph.D. Find the probability a faculty member chosen at random is a woman if the person has a Ph.D.

29. Some Shakespearean sonnets are addressed to a young man while others are addressed to a "dark lady". Determine how many Shakespearean sonnets are written to each. Then determine the following probabilities:
    a) a sonnet chosen at random is addressed to a "dark lady".
    b) a sonnet is addressed to a "dark lady" given it is not one of the first 100 sonnets.
    c) a sonnet is addressed to a "dark lady" given the sonnet number is a perfect square.

30. In a die toss, let $\mathbf{A} = \{1, 2, 3\}$ and $\mathbf{B} = \{2, 4, 6\}$. Are events **A** and **B** independent?

31. Are the events "draw an ace" and "draw a black card" independent?

32. Suppose a card is drawn at random from a standard deck of 52 cards. Let **A** be the event that the card is a diamond and **B** be a number card. Are **A** and **B** independent?

33. Late in the 2003 Nascar season, a controversy arose over whether Ryan Newman was the winner. A race was chosen at random from among the 36 Nascar races that season (nascar.com). What was the chance that
    a) Ryan Newman won the race?
    b) Jeff Gordon won the race?
    c) a driver in the pole position at the start was the winner?

34. The human body contains 30,000 to 40,000 *genes*. They carry information that determines characteristics inherited from one's parents. Genes lie on the body's *chromosomes*. Many cells of species, including humans, have 23 pairs of chromosomes. Fruit fly cells have only four pairs, which is why they are often used in genetic experiments. Each gene has two *alleles* that can be categorized as *dominant* or *recessive*. The two alleles of a gene are denoted B-dominant and b-recessive. A *Punnett square* is typically used in genetics to determine the probabilities of "crosses" between parents. Suppose hair color can be determined from a single pair of genes. Brown hair is dominant and blond is recessive. A newborn child receives a hair gene from each parent and the gene is equally likely to be either of the two genes from the parent.
    a) If both parents have brown hair, what is the probability a newborn is a blonde?
    b) Suppose the newborn and both parents have brown hair. A sibling has blonde hair. What is the probability the newborn has a blonde hair gene?
    c) Explain how the human genome project may help calculate probabilities associated with genetics.

35. Skin color in humans is determined primarily by the amount of melanin. Darker skinned individuals tend to produce more melanin than lighter ones. Three genes mainly regulate melanin production. The dominant dark skin alleles are A, B, C and the recessive light skin alleles are a, b, c. There are mainly seven different shades of skin colors (based on up to six capital letters) from very light aabbcc to very dark AABBCC with most individuals having AaBbCc. Determine the probability that two parents with AaBbCc produce an offspring with
    a) darkest skin shade.
    b) lightest skin shade.
    c) skin shade corresponding to one capital letter.

◆ Mathematical probability is central to genetics, the study of random mating of living organisms. Discovered in his experiments with growing garden peas in the mid-1800s by Gregor Mendel, an Austrian friar, its importance was not recognized to the 20th century.

36. Based on "The Lottery" by Shirley Jackson (on the Internet), a lottery is conducted in three stages: heads of families, heads of households, and members of the "winning" household. The village being a patriarchal society, one stage in the process, was eliminated. Bill Hutchison's household is the only household for the "Hutchison" family.

    After reading the story, answer the following:

    a) Given that a member of Bill Hutchison's household wins the lottery, what is the probability that Tess is the winner?
    b) Is each family equally likely to be chosen when slips are drawn alphabetically?
    c) Why are probabilities not equally likely for all of the villagers?
    d) Explain why "gender of a family's married children" and "winning the lottery" are statistically dependent.
    e) How can the lottery be conducted so that each villager has an equal chance of winning the lottery and being stoned?

7.6 BAYES' RULE AND DECISION TREES

The tree diagrams of Section 7.3 display sample spaces of events. In this section, we assign probabilities to the component "branch" events. Sequences of events in which outcomes of some events affect subsequent ones are frequently of interest. It is sometimes helpful to represent them with a tree diagram. The probability on each branch of the tree is a conditional probability that the specified event occurs, given that the events on preceding branches have already occurred. A "tree" with conditional probabilities on its "branches" is called a **probability tree**.

After identifying all terminal points, the sample space probabilities are noted along paths from roots to terminal points. The probability of interest is the sum of the product of branch probabilities. The following example illustrates the technique.

Example 7.6.1 Basketball 3-Point Shots

A basketball player has a 2/5 chance of making a 3-point shot. However, if a shot misses, there is only a 1/5 chance of making the next 3-point attempt. If a shot succeeds, the chance is 3/5 of making the next shot. What is the probability that the player makes the first two 3-point shot attempts?

→

Solution:
Using a tree diagram, the associated probabilities are recorded on the branches as shown below.

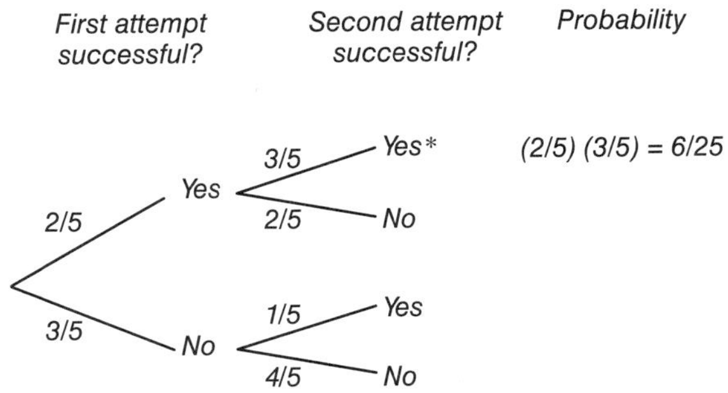

An asterisk denotes the terminal event "both shots are successful"; a probability of 6/25.

Example 7.6.2 Venus versus Serena (Revisited)

Recall the Venus versus Serena tennis match of Example 7.3.2. Suppose the two are equally matched for the first set. However, winning or losing has a psychological effect on Venus. If Venus wins a set, her probability of winning the next set is 3/5; if she loses a set, her probability of winning the next set is 1/3. Find the probability that she wins the match.

Solution:
The tree records the possible outcomes and their probabilities.

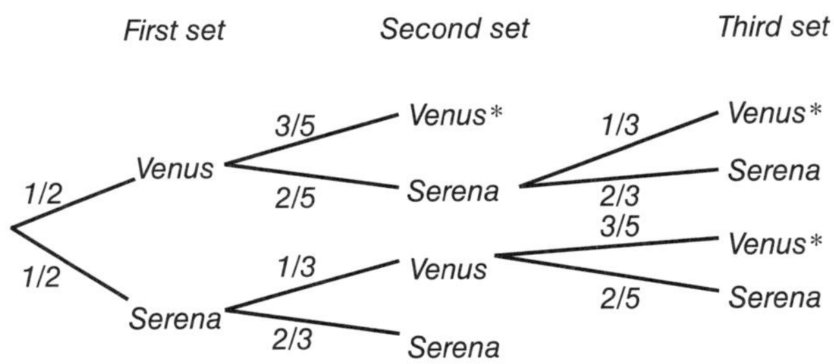

Terminal points with asterisks denote matches won by Venus.
The probability Venus wins is

$$(1/2)(3/5) + (1/2)(2/5)(1/3) + (1/2)(1/3)(3/5)$$

$$= 3/10 + 1/15 + 1/10 = 7/15$$

Total Probability Theorem and Bayes' Rule

Events of interest often depend on the occurrence of other events. For example, if a destination can be reached by any of several routes, then the event "the destination is reached" can only occur in conjunction with travel along exactly one of the routes. Another example: A lost object is in one of several distinct locations. The event "object located" can only occur in connection with one of the several locations.

These examples have implied that the contingent events are *mutually exclusive*; that is, the occurrence of one precludes the occurrence of the others (e.g., if a coin shows head, it cannot show tail). In addition, it is assumed that the contingent events are *collectively exhaustive*; that is, no other (unstated) contingencies are possible. These ideas are expressed in the set relationship of unions and intersections as

$$\mathbf{A} = (\mathbf{A} \cap \mathbf{X}_1) \cup (\mathbf{A} \cap \mathbf{X}_2) \cup \cdots \cup (\mathbf{A} \cap \mathbf{X}_n)$$

where $\mathbf{X}_1, \mathbf{X}_2, \ldots, \mathbf{X}_n$ are the n mutually exclusive and collectively exhaustive contingent events one of which must occur in conjunction with an event **A**.

Example 7.6.3 Defective CDs

Among a group of five CDs, one is known to be defective. Express that event in a set relation.

Solution:
Let **A** *denote the event "defective CD" and* $X_1, \ldots, X_5$ *denote the five CDs, respectively. Then*

$$\mathbf{A} = (\mathbf{A} \cap \mathbf{X}_1) \cup (\mathbf{A} \cap \mathbf{X}_2) \cup \cdots \cup (\mathbf{A} \cap \mathbf{X}_5)$$

It follows that

$$\begin{aligned} P(\mathbf{A}) &= P(\mathbf{A} \cap \mathbf{X}_1) + P(\mathbf{A} \cap \mathbf{X}_2) + \cdots + P(\mathbf{A} \cap \mathbf{X}_n) \\ &= P(\mathbf{A} \mid \mathbf{X}_1)P(\mathbf{X}_1) + P(\mathbf{A} \mid \mathbf{X}_2)P(\mathbf{X}_2) + \cdots + P(\mathbf{A} \mid \mathbf{X}_n)P(\mathbf{X}_n) \end{aligned}$$

All of the joint events are mutually exclusive, so their probabilities vanish. The last expression, known as the **Total Probability Theorem**, is important in theory and in applications. It permits calculation of the probability of an often-elusive event, **A**, using a knowledge of conditional probabilities, which may be recovered from data. Furthermore, the expression properly weights these conditional probabilities by the probabilities of the contingent events.

Example 7.6.4 Search for a Lost Camper

Two airplanes search independently for a lost camper in different sections of a rugged mountainous region. Let $\mathbf{X}_1$ *be the event "first plane reports sighting camper" and* $\mathbf{X}_2$ *the corresponding event for the second plane.*

⟶

*Let **A** be the event, "camper located." Express the probability of **A** using the total probability theorem.*

Solution:

*Since **A** occurs as a consequence of either the first plane* (X_1) *or the second plane* (X_2)*, spotting the lost camper,* $\mathbf{A} = (A \cap X_1) \cup (A \cap X_2)$ *so* $P(A) = P(A \cap X_1) + P(A \cap X_2)$. *The joint event* $P(A \cap X_1 \cap X_2)$ *vanishes since* X_1 *and* X_2 *are mutually exclusive.*

Furthermore, $P(A) = P(A \mid X_1)P(X_1) + P(A \mid X_2)P(X_2)$ *using the definition of conditional probability.*

Bayes' Rule

The total probability theorem leads to an important result known as **Bayes' Rule** (or Bayes' Formula or Bayes' Theorem). The conditional probabilities $P(\mathbf{A}|\mathbf{X}_i)$, $i = 1, 2, \ldots, n$ are-often available from records. For example, that a destination is reached (event **A**) using route $\mathbf{X}_i$; or structural failure (event **A**) results from failure mode $\mathbf{X}_i$; or a blood test (event **A**) is actually false (mode $\mathbf{X}_i$), where $i = 1, \ldots, n$.

However, the practical questions often concern conditional probabilities with the events reversed, that is, $P(\mathbf{X}_i|\mathbf{A})$, $i = 1, \ldots n$. Thus, the catastrophic (or event of interest) event **A** has occurred, the interest is in the cause or associated event.

Clearly, $P(\mathbf{X}_i|\mathbf{A}) = \dfrac{P(\mathbf{X}_i \cap \mathbf{A})}{P(\mathbf{A})} = \dfrac{P(\mathbf{A}|\mathbf{X}_i)P(\mathbf{X}_i)}{P(\mathbf{A})}$, $i = 1, \ldots, n$.

This is Bayes' Rule. It estimates chances for the causal or originating event, given that the main event of interest has occurred. It is a "backward look" of sorts. As another example, "a tornado is reported," a probability estimate of its location is sought, or "a fever is recorded," and the probabilities of its possible causes are sought. The use of such probability estimates of chance is to narrow the search of the actual cause.

Note that $P(\mathbf{X}_i \mid \mathbf{A})$ (called the ***posterior probability***) is expressed in terms of the known conditional probabilities $P(\mathbf{A} \mid \mathbf{X}_i)$ (called the ***likelihood function***) and of the $P(\mathbf{X}_i)$ (called ***the prior probability***), $i = 1, \ldots, n$. $P(\mathbf{A})$, is obtained from these probabilities using the total probability theorem. It is instructive to reflect on why the words posterior, prior, and likelihood have been adopted in application.

Bayes' Rule has a vital role in many applications. Indeed, specialized activities have been formed about Bayes' Rule such as in medical and forensic testing, oil exploration, industrial quality control, fault tree analysis, and artificial intelligence, to name some.

♦ "HyperActive Bob" is the name of a software program in experimental use at several fast-food chain restaurants. Rooftop cameras monitor parking lot and drive-through to assess the type and size of arriving vehicle during peak times. Using a predictive system, based on Bayes' Rule, probabilities are updated with experience.

One McDonald's manager says of his 28-year experience, "It's the most impressive thing I've seen." Over 2 years, he claimed that waste has been halved and wait times at the drive-through reduced by 25 to 40 seconds per customer, "an eternity in the fast-food industry." (As reported in a *New York Times* article, "New Technology Estimates Fast-Food Demand.")

A few examples illustrate Bayes' Rule usage.

Example 7.6.5 Road Contractors

The paving of a road involves three contractors, A, B, and C. Their respective daily productions average 500, 2000, and 1500 yards. Let **F** *denote the event "defective pavement." Prior experience with these contractors produced these estimates:* $P(\mathbf{F} \mid \mathbf{A}) = 0.02$,

$$P(\mathbf{F} \mid \mathbf{B}) = 0.015, \quad P(\mathbf{F} \mid \mathbf{C}) = 0.03.$$

On this day, 4000 yards of road are to be paved, separate portions of road for each contractor. The probability of defective pavement is

$$\begin{aligned} P(\mathbf{F}) &= P(\mathbf{F} \cap \mathbf{A}) + P(\mathbf{F} \cap \mathbf{B}) + P(\mathbf{F} \cap \mathbf{C}) \\ &= P(\mathbf{F} \mid \mathbf{A})P(\mathbf{A}) + P(\mathbf{F} \mid \mathbf{B})P(\mathbf{B}) + P(\mathbf{F} \mid \mathbf{C})P(\mathbf{C}). \end{aligned}$$

$$P(F) = (0.02)\left(\frac{500}{4000}\right) + (0.015)\left(\frac{2000}{4000}\right) + (0.03)\left(\frac{1500}{4000}\right) = 0.02125.$$

A defective pavement has been reported! What is the probability that **A** *paved it?* **B**? **C**?

Solution:
Using Bayes' Rule,

$$P(\mathbf{A}|\mathbf{F}) = \frac{P(\mathbf{F}|\mathbf{A})P(\mathbf{A})}{P(\mathbf{F})} = \frac{(0.02)\left(\frac{500}{4000}\right)}{0.02125} = 0.118$$

and, similarly, $P(\mathbf{B} \mid \mathbf{F}) = 0.353$ *and* $P(\mathbf{C} \mid \mathbf{F}) = 0.529$. *These conditional probabilities properly add to unity.*

Tree diagrams are often used to display and calculate with Bayes' Rule as the next example illustrates.

Example 7.6.6 Tennis Match

In Example 7.6.2, suppose that Venus wins the tennis match. What is the probability that the match went to three sets?

Solution:
In Example 7.6.2, the probability of Venus as winner is 7/15.There are two cases where the match takes three sets and Venus is the winner. The probabilities for these cases are $(1/2)(2/5)(1/3) = 1/15$ *and* $(1/2)(1/3)(3/5) = 1/10$ *(as indicated below). Therefore,*

$$p(\text{three sets}|\text{Venus winner}) = \frac{1/15 + 1/10}{7/15} = \frac{5/30}{14/30} = 5/14$$

There is a 5/14 probability the match goes three sets given that Venus wins.

⟶

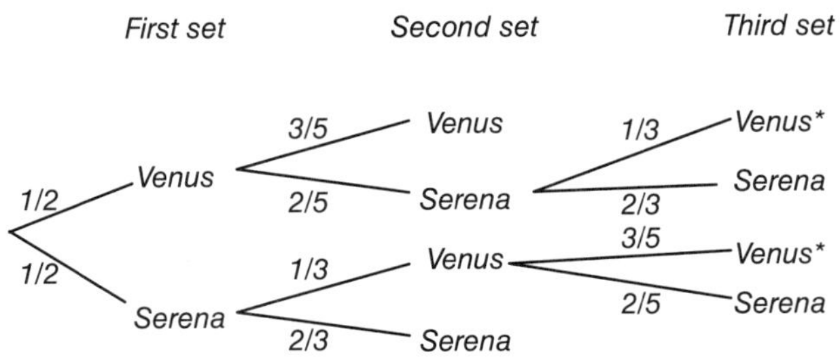

♦ To combat e-mail Spam, filtering programs use Bayes' Rule. A probability is assigned to each word of suspect messages to calculate a score for the likelihood that it is Spam. Higher scores are indicative of Spam.

Example 7.6.7 Factory Effluent

The discharge (effluent) from a factory meets environmental standards 96% of the time. An inspection system correctly identifies toxic effluent 98% of the time. Unfortunately, it also declares satisfactory effluent to be toxic 5% of the time. What is the probability that a discharge declared "OK" is actually "OK?"

Solution:

$$P(actually\ OK|declared\ OK)$$

$$= \frac{\overset{(1-0.05)}{P(dec\ OK|act\ OK)}\overset{(0.96)}{P(act\ OK)}}{\underset{(1-0.05)}{P(dec\ OK|act\ OK)}\underset{(0.96)}{P(act\ OK)} + \underset{(0.02)}{P(dec\ OK|act\ notOK)}\underset{(1-0.96)}{P(act\ not\ OK)}}$$

$$= \frac{(1-0.05)(0.96)}{(1-0.05)(0.96)+(0.02)(0.04)} = \frac{0.912}{0.912+0.008} = 0.9991$$

Note that despite the 96% rate for satisfactory effluent, the result of the reliability testing system (98%) is to boost the chances to 99.9% that the effluent is satisfactory. Here, $P(actually\ not\ OK|declared\ OK) = 1 - 0.9991 = 0.0009$ *or about 0.1%.*

It is useful to note that likelihood functions need not be symmetric. That is, while $P(\mathbf{A}\,|\,\mathbf{B}) = 1 - P(\mathbf{A}^c|\mathbf{B})$, it has no direct relation to $P(\mathbf{A}|\mathbf{B}^c)$. In the last example, P(declared not ok | actually not ok) = 0.98 while P(declared not ok | actually ok) = 0.05. Clearly, there is no relation between these probabilities.

Example 7.6.8 Factory Effluent – A Second Test

Suppose there is a second test of effluent. What is the probability that the sample is actually OK if the second test declares it OK?

Solution:
Using Bayes' Rule again yields,

$$P(actually\ \ OK|declared\ \ OK) = \frac{(1-0.05)(0.9991)}{(1-0.05)(0.9991)+(0.02)(0.0009)} = 0.99998$$

In view of the first test, the P(actually OK) = 0.9991 is used here.

The second successful test has properly increased the chances that the sample is actually OK.

◆ The last example suggests how repeat tests increase (decrease) one's confidence in a conclusion. This explains why medical tests are often repeated (false negatives and false positives). In oil exploration, additional information (geologic, test drilling, etc.) decreases the chances of an erroneous conclusion.

A *New York Times* article ("A Second Test Shows Animal Did Not Have Mad Cow Disease") shows that subsequent tests can differ. A first test being relatively inexpensive (and/or quick) and, if suspicions arise, a second test, perhaps more precise, is used. In this instance, the prior probability changes in subsequent Bayes' Rule calculations.

Example 7.6.9 Evaluating a Blood Pressure Monitor

Periodic comparisons of blood pressure measurements using a home monitor were made against a physician's instrument (assumed accurate). They showed that about 3% of the time the home monitor recorded blood pressures to be in the normal range when actually they were not. About 94% of the time, the home instrument correctly reported that blood pressures were within normal range. Long-term records indicate that the person's blood pressure is normal about 90% of the time. What is the probability that a home measurement of normal is actually in the normal range?

Solution:
One seeks the following:

$$P(act.normal|read\ \ normal) = \frac{P(read\ \ normal|act.norm)P(act.normal)}{P(read\ \ normal)}$$

$$P(read\ \ normal) =$$

$$P((read\ \ normal \cap act.normal) \cup (read\ \ normal \cap not\ \ normal))$$

⟶

The probability the home monitor gives a normal reading is determined from two possibilities

$$P(read\ \ normal|act.normal)P(act.normal) + P(read\ \ normal|not\ \ normal)P(not\ \ normal)$$
$$= (0.94)(0.90) + (0.03)(1 - 0.90) = 0.846 + 0.003 = 0.849$$

Using Bayes' Rule ,

$$P(act.normal|read\ \ normal) = \frac{(0.94)(0.90)}{0.849} = \frac{0.846}{0.849} = 0.996$$

It is nearly certain that the blood pressure is actually normal at this measurement. An alternate solution uses tree diagrams.

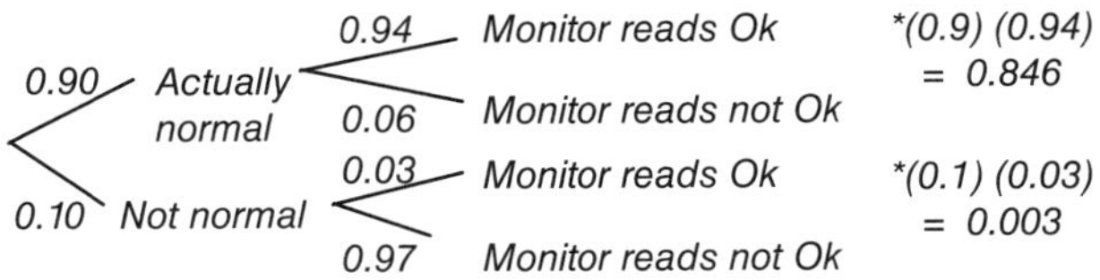

The asterisk denotes the cases relevant to Bayes' Rule.

◆ The above example has relevance as home blood pressure monitoring has become prevalent in the management of hypertension. Dr. John W. Graves at the Mayo clinic notes that "... it is critical that the monitor used is a validated one or the information received from the patient is of little value" (www.dableducational.com).

EXERCISES 7.6

1. A card hand has seven red cards and three black cards. A card is drawn at random and replaced by a card of the opposite color. Then a second card is drawn from the hand. Find the probability that
   a) the second card is red
   b) both cards are black given they are the same color
2. Two balls are drawn without replacement from a bag containing four blue balls and three green balls. What is the probability
   a) both balls are the same color?
   b) the second ball is blue given the first was green?

3. A coin is tossed. If it is a head, then it is tossed again. If it is a tail, then a die is tossed. What is the probability of

   a) exactly one head?

   b) two or more on the die?

4. A coin is tossed three times. Find the probability that

   a) the third toss is a head given the first two tosses are heads.

   b) the first and third tosses show the same face.

5. A 13-card bridge hand contains 5 clubs, 6 spades, and 2 hearts. Two cards are drawn at random without replacement from the hand. Find the probability that

   a) both are spades.

   b) both are spades given first card is a spade.

   c) one is a club given first card is a heart.

6. A box has three red balls and five green balls. Two balls are selected at random without replacement and the color of each is noted. Find the probability that the first ball drawn is green if at least one drawn ball is green.

7. What is the probability a couple will have at least two daughters if they plan to have three children? Assume genders are equally likely.

8. An industrial plant produces items on two machines. The newer machine produces items that are 95% satisfactory, while the older machine produces only 85% satisfactorily. If a machine is chosen at random and two items of its output are checked, find the probability that

   a) both items are satisfactory.

   b) they came from the older machine if both items are unsatisfactory.

9. An inspector at a paint factory found that 12% of cans were overfilled, 13% underfilled, and 75% filled correctly. What is the probability that there is no enough paint in a can, given that it does not contain the correct amount?

10. One of two coins is weighted so the probability of a tail is 3/4. The other coin is fair. A coin is selected at random and tossed twice. The result of each toss is noted. Find the probability that the unfair coin was selected given an outcome of two tails.

11. Two cards are to be drawn at random without replacement and they are noted as faces, aces, or number cards. Find the probability

    a) at least one face card drawn given neither is an ace.

    b) both are number cards given at least one of them is a number card.

12. At one university, 55% of students are underclassmen (freshmen and sophomores), 35% are upperclassmen (juniors and seniors), and 10% are graduate students. Foreign students make up 10% of underclassmen, 25% of upperclassmen, and 60% of graduate students. If a randomly selected student is foreign, what is the probability that he or she is a graduate student?

13. Two independent diagnostic tests are available to detect an elusive disease. The probability that the "A test" detects the disease is only 0.6. For the "B test," the corresponding probability is 0.9. Once the disease is detected in the body, an attempt is made to pinpoint its location. The probability that test A (alone) has detected the disease and correctly pinpoints its location is 0.8. The corresponding probability for test B is 0.5. When both tests detect the disease, its location is pinpointed with certainty.
    a) Derive the probability of detecting the disease.
    b) Calculate the probability that only one of the tests detects the disease.
    c) Derive an expression and calculate the probability of pinpointing location.
    d) A person is identified as having the disease and its location pinpointed. What is the probability that it was detected by test B alone?

14. Parts reaching an assembly line arise from four sources of production that account for 20%, 30%, 25%, and 25% of the total. The respective portions of their production that tend to be defective are 6%, 3%, 1%, and 4%. On the assembly line, a defective part is found. Find the probability that it was produced by the first source.

15. Steel cables for a new bridge are obtained from three sources, A, B, and C, which supply 30%, 25%, and 45%, respectively, of the required number of cables. In the past, on-site testing has found 2%, 3%, and 5%, respectively, of the cables have been defective. A cable chosen at random is found to be defective, what is the probability it came from source C?

16. There are three modes of transporting material from New York to Florida; by land, sea, or air. Also, land transport may be by rail or highway. About half of the materials are transported by land, 30% by sea, and the rest by air. Also 40% of all land transport is by highway and the rest by rail. The percentages of damaged cargo are 10% by highway, 5% by rail, 6% by sea, and 2% by air, respectively.
    a) What percentage of all cargoes may be expected to be damaged?
    b) If a damaged cargo is received, what is the probability that it was shipped by land, by sea, and by air?

17. An instrument that detects flaws has a reliability of 98%. It also indicates flaws on satisfactory items 8% of the time. Experience indicates that 12% of all items tested are flawed. An item has just been tested and the instrument indicates that it is flawed. What is the probability that the item is actually not flawed?

18. In a manufacturing process, defectives occur at a 2% rate for a large machine and 3% for a smaller machine. The smaller machine produces 30% of the output. The outputs are mixed and are indistinguishable. An item is selected at random and found to be defective. What is the probability it was made by the larger machine?

19. Two machines produce 30% and 70% of the silicon chip output and with defective rates of 2% and 3%, respectively. What is the probability that a chip chosen at random, and found to be defective, was made by the first machine?

20. A manufacturer whose product is made on three machines A, B, and C mixes the output into one large bin. The machines manufacture 25%, 45%, and 30%, respectively, of the output. However, a small percentage of each machines output is defective. Suppose that the respective defective rates are 3%, 2%, and 1%. An item is selected at random from the output bin and found to be defective. What is the probability that it was made by machine B?

21. Urn I contains three white and five black marbles. Urn II contains two white and three black marbles. A marble is randomly drawn from Urn I and the color is noted before it is placed into Urn II, then a marble is drawn from Urn II.
    a) What is the probability the marble drawn from Urn II is black?
    b) What is the probability that two white marbles are drawn given that at least one is white?

22. False positives and false negatives arise in the testing of computer antiviral software. Which is the more serious error? Why?

23. About 4% of a firm's e-messages has been found to be virus infected. On rare occasions, its antiviral software signals false positives and false negatives with respective probabilities of 0.01 and 0.0005. What is the probability that an e-message chosen at random and found to be "clean" is actually infected?

24. In a town, winter roads are snow covered 80% of days. On 60% of the days when roads are snow covered, accidents occur. When roads are clear, accidents occur only 15% of days. Find the probability that roads were snow covered on a day when an accident occurred.

25. A car dealership classifies a car buyer's credit rating as E (excellent) 25% of the time, F (fair) 60% of the time, and P (poor) 15% of the time. The probability a buyer's credit is approved is 95% for E, 75% for F, and 25% for P. A new car is purchased. What is the probability the buyer's credit rating was not poor?

26. Suppose a bridge hand has seven hearts, four clubs, and two diamonds. Two cards are drawn without replacement. Find the probability that one card is a heart, given at least one of them was a club.

27. In a large test group the probability that women over the age of 80 will develop breast cancer is about 10% and under the age of 50, about 2%. A sample consists of 70 women over age 80 and 30 women under age 50. If a randomly chosen woman from the sample has breast cancer, what is the probability she is under age 50?

28. WWW The Monty Hall problem (loosely based on the game show, "Let's Make a Deal") is fairly well known in mathematical probability. Conduct an Internet search of the Monty Hall problem and decide whether to switch your curtain numbers.

    Are there certain conditions or insights in the host's thinking that warrant a switch? What would change your solution? What assumptions have you made about whether a switch should be made? Hint: (one answer uses Bayes' Rule).

HISTORICAL NOTES

Mathematical probability has a storied history. Neglected by Western mathematicians well into the twentieth century, it has a fundamental role in modern science.

Its early development in games of chance – cards and dice – probabilities provided some measure of financial support for some who developed mathematical skills to aid gamblers. Mathematicians of the nineteenth and twentieth centuries tended to neglect serious mathematical development of the subject perhaps because of its less savory origins.

The "equally likely" assumption was well suited for gambling applications, and as such there was little motivation for its generalization. The assumption was generalized in the nineteenth century and led to the development of "classical probability," the subject of this and the next chapter.

Thomas Bayes (1702–1761). Mathematician and theologian Bayes is credited with the early use of inductive probability. He established a mathematical basis for the probability of a future event based on its frequency of (past) occurrence. His probability findings were published as "Essay Toward Solving a Problem in the Doctrine of Chances" in 1763. This became the basis of modern Bayesian statistics.

◆ Bayesian models are used in the diagnosis of disease, the study of traffic patterns, and in making search engines more helpful. The paper clip office assistant from Microsoft Office is an application of a Bayesian model. It is an example of an attentive user interface (AUI) that mimics human judgment. Anti-spam software has a similar basis.

Augustus de Morgan (1806–1871). Although a member of the Church of England, his refusal, on principle, to take a theological examination deprived him of the master's degree required for continued advanced studies at Cambridge University. Nonetheless, in 1828, at age 21, he became the first professor of mathematics at the newly formed University College, London. He held that chair for some 38 years except for occasional resignations on matters of principle.

John Venn (1834–1923). Ordained a priest in the Church of England, he later returned to Cambridge University as a lecturer in moral sciences. Interested in logic, he wrote "*Principles of Empirical Logic*" in 1889 and, later, "*Symbolic Logic*." He is best known for the geometrical representation for syllogistic logic, now known as Venn diagrams. Curiously, he was not the first to use them, that honor belonging to Leibnitz. Still, he made notable advances in their development.

CHAPTER 7 SUPPLEMENTARY EXERCISES

1. $\mathbf{A} = \{1, 2, 3, 4, 5\}$, $\mathbf{B} = \{3, 4, 5\}$, and $\mathbf{C} = \{4, 6, 8, 10\}$. Which of the following are true?

   a) $\mathbf{B} \subset \mathbf{A}$ b) $\mathbf{B} \cup \mathbf{C} = \{4\}$ c) $\mathbf{A} \cap \mathbf{B} = \{3, 4, 5\}$

2. List elements of the sets of letters in the name of

   a) the forty-ninth state b) the fiftieth state

3. On a coordinate axes represent "wind speed" and "wind direction" as a sample space in the first quadrant. Identify possible sample values. On the diagram, identify the portions of the sample space where wind speed is between 25 and 50 mph in a direction between 30° and 60°.

4. Verify $(\mathbf{A} \cup \mathbf{C}) \cap (\mathbf{B} \cup \mathbf{C}) = (\mathbf{A} \cap \mathbf{B}) \cup \mathbf{C}$ using the laws of sets.

5. Let $\mathbf{A} = \{\text{multiples of 10 that are less than 500}\}$ and $\mathbf{B} = \{\text{Perfect squares}\}$, find $\mathbf{A} \cap \mathbf{B}$.

6. Display the sets in a Venn diagram.

   a) $(\mathbf{A} \cup \mathbf{B} \cup \mathbf{C})^c$ b) $(\mathbf{B} \cup \mathbf{C}) \cap \mathbf{A}$

7. An unfair coin is tossed three times and the outcome (H or T) noted. Assume $P(\text{H}) = 2/3$. Use a tree diagram to calculate the probability of two or three heads.

8. A bag of marbles contains three red, two blue, and one white. Two marbles are selected at random and without replacement. Use a tree diagram to determine the probability that a white marble is not drawn.

9. The 400-meter Olympic event had eight finalists for a gold medal heat. In how many ways can three medals be awarded? Neglect ties.

10. In how many ways can six balls be drawn from among 36 distinct balls?

11. In how many ways can six people be assigned to single-, double-, and triple-bedded hotel rooms?

12. In how many different ways can six red, four green, and two white bulbs be placed in a string of a dozen Christmas lights?

13. There are nine coins: six nickels and three dimes. In how many ways can three coins be randomly selected to include both denominations?

14. Which is larger ${}_7C_5$ or ${}_8C_6$?

15. A set contains six elements. How many of its subsets contain either one or two elements?

16. In a car lot, 30% of the cars are less than a 1-year old, 45% are 1–3 years old, and the remaining 25% are older than 3 years. The manager of the lot offers a selective rebate. Ninety percent of cars less than a year old are eligible for the rebate, 50% of the cars that are 1–3 years old are eligible, and 10% of the cars older than 3 years are eligible. If a person at random received a rebate, what is the probability that a car 1–3 years old was purchased?

17. An electrical supply buys 30% of its fuses from Company A, 50% from Company B, and 20% from Company C. There may be defective fuses in the lots. The fuses from A are 98% defect free; from B, 95%; and from C, 90%. What is the probability that a fuse chosen at random is not defective? If a randomly chosen fuse is defective, what is the probability that it came from B?

FURTHER READING

Kreith, K. and Kysh, J. "The fourth way to sample k objects from a collection of n". Mathematics Teacher, Feb 1988: 146, et seq.
David, F. N., Games, Gods, and Gambling. Hafner; 1962.
Feller, W., An Introduction to Probability Theory and Its Applications. Vol I, John Wiley and Sons; 1957.

8 Random Variables and Probability Distributions

8.1 RANDOM VARIABLES

In Chapter 7 you studied the elements of probability theory. The probabilities calculated there assumed that all outcomes were equally likely. Besides, each problem required an

---

Finite Mathematics: Models and Applications, First Edition. Carla C. Morris and Robert M. Stark.

Companion Website: http://www.wiley.com/go/morris/finitemathematics

individual or customized solution. In this chapter, we study probabilities for generic situations. They lead to probability formulas called **probability distributions** that apply to classes of situations. The advantage is that when a recognizable situation arises, or similar ones, they can be associated with an already calculated probability distribution. This is generally easier than devising a customized solution.

Recall that outcomes of a chance situation are called *sample points* and their totality, the *sample space*. Sample points can be described by a **random variable**, that is, a variable that takes its values by chance. When the sample points are discrete or countable, such as the numbers of heads in coin tosses or number of auto accidents, the random variable is **discrete valued**. When the sample points are on a continuum, such as rainfall or fuel consumption, the random variable is **continuous valued**.

Example 8.1.1 Classifying Random Variables

Classify these random variables as discrete or continuous valued.

a) The number of auto accidents daily.
b) The volume of soda in a 2-liter bottle.
c) The time to walk a mile.
d) The number of months a sales quota is exceeded.

Solution:
In a) and d), the random values are discrete valued.
In b) and c), the random values are continuous valued.

Probability Distributions

A **probability distribution,** $p(x)$**,** for a discrete random variable, $X = x$, associates a probability with each value of x. It can be a graph, a table, or a mathematical formula. For example, $p(\mathrm{x}) = 1/6, x = 1, 2, \ldots, 6$, represents probabilities associated with the toss of a die.

Probability distributions for discrete random variables have these properties:

(a) All probabilities are nonnegative and bounded by unity.
That is, $0 \le p(x) \le 1$.

(b) The sum of probabilities on a sample space is exactly 1.
That is, $p(x_1) + p(x_2) + \cdots + p(x_n) = 1$, where $x_1, x_2, \ldots, x_n$ are possible values of X.

Example 8.1.2 Probability Distributions

Which of these are probability distributions?

⟶

a)

x	0	2	5
p(x)	1/3	2/3	1/3

b)

x	−1	1	5
p(x)	0.2	0.3	0.5

c)

x	1	2	6	8	10	20
p(x)	0.10	0.25	0.05	0.15	0.30	0.10

Solution:
Here, all numbers are nonnegative and less than unity. However, only in b) do they properly sum to one. Therefore, a) and c) cannot be probability distributions.

Example 8.1.3 Graphing a Probability Distribution

Display this probability distribution graphically.

x	1	2	4	5
p(x)	0.10	0.20	0.40	0.30

Solution:

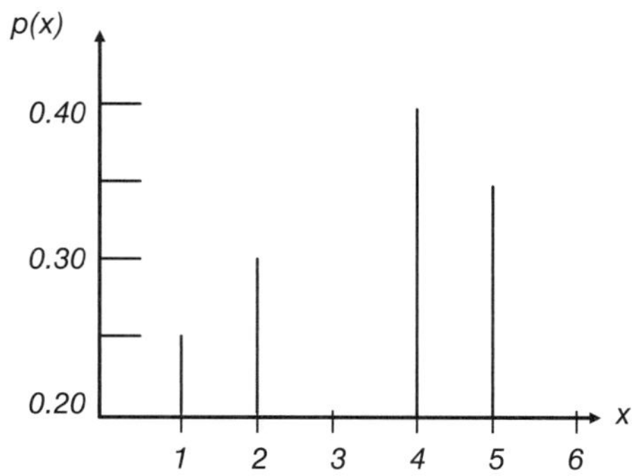

The "line height" indicates the probability value.

Example 8.1.4 Coin Tossing

A fair coin is tossed three times and the number of tails, x, is observed. Display the probability distribution for x in tabular format.

→

Solution:
The eight possible outcomes of three tosses are:

Outcome	*x*	*Outcome*	*x*
HHH	0	*THH*	1
HHT	1	*THT*	2
HTH	1	*TTH*	2
HTT	2	*TTT*	3

The random variable, x, takes values x = 0, 1, 2, and 3 corresponding to the number of tails. The probability of "no tails," p(0), is 1/8, as there is only one corresponding event among the eight equally likely outcomes. Continuing, the probability distribution in tabular format is

x	0	1	2	3
$p(x)$	1/8	3/8	3/8	1/8

Expected Value

Expected value of a random variable, X, is the formal name for its (long-run) **average** or **mean** value. This is an unfortunate name, as an average need neither be "expected" nor even a possible outcome. For example, the (long-run) average number of tails when a fair coin is tossed repeatedly in groups of three tosses is 1.5 tails. Of course, that cannot be an actual outcome.

The expected value of a random variable, X, is universally denoted by $\boldsymbol{E(X)}$. It is obtained by multiplying each value of the random variable by its associated probability and summing over every value of the random variable. For example, $\boldsymbol{E(X)}$ for the toss of three coins (using the probability table in Example 8.1.4) is
$0(1/8) + 1(3/8) + 2(3/8) + 3(1/8) = 12/8 = 3/2$, as expected.

To Calculate an Expected Value

1. **Obtain the probability distribution $p(x)$ for the random variable, $X = x$, whose average is sought.**
2. **Multiply each value of x by its corresponding probability $p(x)$.**
3. **Add the products to yield the expected value of X**

$$E(X) = x_1 p(x_1) + x_2 p(x_2) + \cdots + x_n p(x_n)$$

Not only is the naming of an average as an expected value unfortunate, so is the notation $E(X)$. Usually, such notations are used for functions. An expected value of X certainly cannot be a function of X.

Example 8.1.5 Gambling

A gambler pays \$3 to play in a game in which a fair coin is tossed three times, and \$2 is paid for each tail. Is the game fair?

(In a single play, either \$1 or \$3 is won or lost, respectively.)

Solution:

*A game is called **fair** when the expected gain/loss in long-term play is zero. Now, let the random variable x denote winnings. If no tails are tossed, the gambler loses \$3; if one tail, the loss is \$1; if two tails, the win is \$1; and if three tails are tossed, \$3 is won. The associated probabilities are:*

x	-3	-1	1	3
$p(x)$	1/8	3/8	3/8	1/8

The expected value is $E(x) = (-3)(1/8) + (-1)(3/8) + (1)(3/8) + (3)(3/8) = \$0$.

The game is fair. (Recall earlier mention of the interchangeable usage of X for a random variable and x for values of the random variable for convenience).

EXERCISES 8.1

1. Classify these random variables as discrete or continuous:
   a) The time to build a new home.
   b) The number of misprints on a page.
   c) The number of employees on weekly vacation.
   d) The volume of paint in a gallon can.
   e) The weight of a stem of grapes.
2. Cite at least two examples of discrete random variables of interest to
   a) a boutique owner;
   b) a biologist;
   c) an economist.
3. Cite at least two examples of continuous random variables of interest to economists.
4. Which of these are proper probability distributions?

a)

x	100	200	300	400
$p(x)$	0.13	0.17	0.35	0.45

b)

x	−5	2	3	7
$p(x)$	0.2	0.1	0.3	0.4

c)

x	4	9	11	13	16	20	24
$p(x)$	0.12	0.18	0.15	0.15	0.20	0.25	0.05

5. Complete the tables to depict proper probability distributions.

a)

x	10	20	30	40
$p(x)$	0.20	0.10	—	0.40

b)

x	1	5	9	15	20
$p(x)$	0.05	0.35	0.25	0.20	—

c)

x	−3	5	7	8	10	25
$p(x)$	1/16	1/8	3/16	5/16	—	1/8

6. For this probability distribution calculate:

x	1	2	5	6	8	10
$p(x)$	0.20	0.15	0.05	0.30	0.10	0.20

a) $P(x \geq 3)$
b) $P(x > 5)$
c) $P(x \leq 8)$
d) $P(2 \leq x < 10)$
e) $P(x \leq 7)$
f) $P(x = 9)$

7. A fair die is tossed. Let x denote the face value. Display the probability distribution for x in tabular format.

8. Four fair coins are tossed. Let x be the number of heads. Display the probability distribution for x graphically.

9. Calculate the expected values for the three random variables in Exercise 8.1.5.

10. Calculate the expected value for the random variable in Exercise 8.1.6.

11. A pair of dice is tossed. Let the random variable x denote their sum.
    a) Find the probability distribution for x.
    b) Find the expected value of the random variable x.

12. A gambler pays \$3 to play a game. A dollar amount equal to the face value of the die is won. For instance, if a "one" appears on the die, the gambler wins a dollar, etc. (but has paid \$3 to play). Is the game favorable to the gambler?

13. A farmer expects \$100,000 profit if weather remains favorable. However, there is a 25% chance that unfavorable weather will reduce profit to \$60,000. An insurance company will insure the profit of a favorable crop for \$15,000. Is the insurance cost-effective?

14. A thousand raffle tickets are sold at \$1 each. There are three prizes: a grand prize of \$350, a second prize of \$100, and a third prize of \$50. Three random draws determine the winners. What is the expected value for a single ticket?

15. A lottery offers a \$1.5 million prize for correctly choosing six winning numbers from a group of 39 numbers. A \$500 prize is awarded if five of the winning numbers are correctly chosen and a \$50 prize if four winning numbers are correctly chosen. What is the expected value of a single ticket?

16. An athlete ponders two offers: Team A offers \$100,000 plus 4% of home game ticket sales. Team B offers \$125,000 and 2% of home game ticket sales. Each team estimates ticket sales to home games by this probability distribution.

Ticket sales	\$500,000	\$600,000	\$750,000	\$1,000,000
Probability	0.2	0.4	0.3	0.1

Which is the better offer?

17. A die is loaded to be twice as likely to yield a "6" as any other face. Calculate the expected value of a single toss of the loaded die.

18. An urn contains three nickels, two dimes, four quarters, and a Sacagawea dollar. Find the expected value of a single random draw from the urn.

19. A wallet has six bills displaying Washington, three with Lincoln, four each with Hamilton and Jackson, two with Grant, and one with Franklin. A bill is randomly pulled from the wallet. What is its expected value of a single draw?

8.2 BERNOULLI TRIALS AND THE BINOMIAL DISTRIBUTION

While solutions to problems of chance often require "custom crafting" to individual situations, many analyses can usefully begin with an *idealized* or a *prototype* model. From such prototypes, desired **probability distributions** can be derived.

To begin, imagine a series of repeated trials akin to coin flips, known as **Bernoulli Trials** and have the following characteristics.

Characteristics of Bernoulli Trials

1. **A series of repeated and nominally identical and independent trials.**
2. **A trial result only can be one of two possibilities, say, "success" and "failure."**
3. **The "success" and "failure" trial probabilities remain constant from trial to trial.**

It is important to remember that these are idealized trials. They are the basis for several widely used probability distributions. Coin tossing is the simplest example of Bernoulli Trials. Tossing the same coin repeatedly, without posturing the coin, is a reasonable approximation of the first characteristic. The second considers only two possible outcomes: head or tail. A coin could land on its edge; a third possible outcome. Here, such trials are simply called mistrials and are not counted. The third assumes that the probability of, say, head, is the same for each toss.

Batches of manufactured items can often be regarded as a series of Bernoulli Trials. In fact, deviations from Bernoulli Trials are a clue to a defect in the manufacturing process.

For example, a seeming lack of independence in repeated outcomes may be a clue to an imperfection or flaw in their manufacture. Another example, a change in the frequency of occurrence (trial probability) of defectives may be traced to a worn machine part and so on.

Many instances of Bernoulli Trials, or reasonable approximations, are not always immediately evident. For example, the maximum (or minimum) annual river flow or winds and snow loads, among others. Care is required to ascertain whether successive annual maxima are sensibly Bernoulli Trials since, for example, alterations in the riverbed or a changed pattern of land use can affect the assumptions. Periodic stock market changes or sales fluctuations may be "up" or "down." Tests of concrete cylinders, number of vehicles leaving a freeway, and data transmission may be reasonably described as Bernoulli Trials. Sometimes, the Bernoulli Trials are contrived. For example, to evaluate an anticorrosive naval paint, one side of a group of boats can be treated conventionally while the other has the new treatment. Afterward, each boat is regarded as a Bernoulli Trial with "success" if the new treatment is "useful" and "failure" otherwise.

Typically, the trial probability of success is denoted by p and failure probability by q. Since each outcome results in either a success or a failure, $p + q = 1$.

Example 8.2.1 Die Tossing

A fair die is tossed twice and the outcome noted. What is the probability that a 5 or 6 appeared at each toss? Exactly once? Not at all?

Solution:

The tosses are considered to be two repetitions of a Bernoulli Trial with "success" being "5 or 6" and failure as a "1, 2, 3, or 4." The probability of success is $p = 1/3$ (and failure $q = 2/3$). Using a tree diagram,

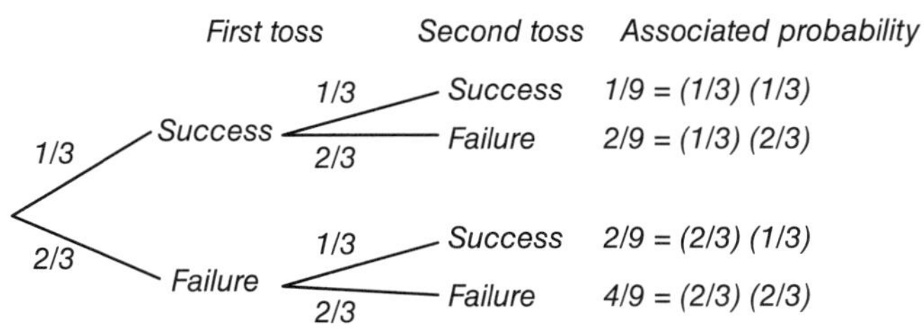

(Note: check that the "tree probabilities" add to unity.)

Using the tree diagram, there is a 1/9 probability of a "5 or 6" for both tosses, a $2/9 + 2/9 = 4/9$ probability of a "5 or 6" for exactly one of the tosses, and a 4/9 probability that a "5 or 6" does not appear at all.

Example 8.2.2 Bernoulli Trials

An experiment consists of three Bernoulli Trials with probability of success, p (and probability of failure, q). Compute the probability of exactly x successes, $x = 0, \ldots, 3$.

Solution:

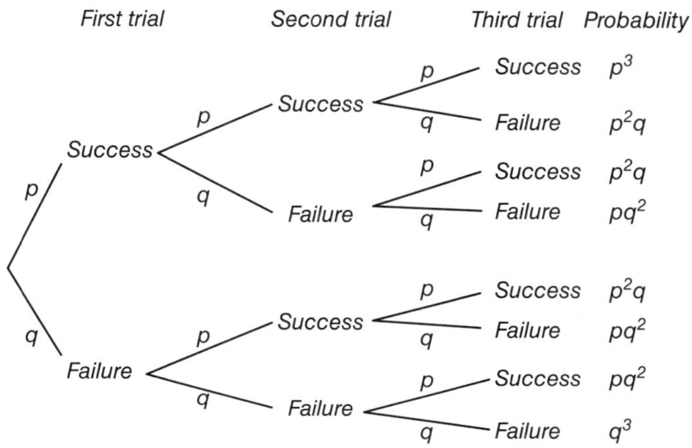

for $x = 0, 1, 2,$ and 3 the probabilities are q^3, $3pq^2$, $3p^2q$, and p^3 respectively.
(Note the sum to unity: $q^3 + 3pq^2 + 3p^2q + p^3 = (q + p)^3 = (1)^3 = 1$).

Binomial Probability Distribution

Consider a series of n Bernoulli Trials, $n = 1, 2, \ldots,$ with probability p for "success" and $1 - p = q$ for "failure." The probability of exactly x "successes" followed by $n - x$ "failures" is

$$p^x q^{n-x}, \quad x = 0, 1, \ldots, n.$$

However, this is the probability for one specific ordering of "successes" and "failures." For example, the event "two successes in five Bernoulli Trials" might be **S S F F F** with probability p^2q^3. Another possibility might be **S F F F S**. Clearly, each possible ordering has the same probability.

Recall from combinatorics (Chapter 7), there are exactly $\binom{5}{2} = 10$ possible orderings of two successes in five Bernoulli Trials. In general, the probability of x "successes" in n Bernoulli Trials is

$$\binom{n}{x} p^x q^{n-x}, \qquad x = 0, 1, \ldots, n.$$

This is the well-known **binomial distribution** for the number of successes in n Bernoulli Trials.

Some use the compact functional notation $b(x;\ n,\ p)$, or simply $b(x)$, to represent this binomial distribution as a shorthand since, at a glance, b indicates the binomial distribution. The random variable is x, the number of trials is n, and the trial success probability is p.

♦ John Allen Paulos, popular author of "*A Mathematician Plays the Stock Market*", observes that sizable stock runs, up or down, are easily mistaken for fundamental market trends when, in fact, they are chance outcomes. Additional keen insights into streaks in sports, stocks, and coin flipping appear on pp. 66–67, for example.

The binomial distribution is associated with many practical problems. These include industrial sampling, drug testing, genetics, epidemics, medical diagnosis, opinion polls, analysis of social phenomena, qualifying tests, hydrology, and more. The exercises are illustrative.

The Binomial Probability Distribution

$$b(x;n,p) = \binom{n}{x} p^x q^{n-x}, \quad x = 0,\ 1, \ldots, n$$

x = Number of successes in n trials

n = Number of trials

p = Probability of success on a single trial

$q = 1 - p$ = Probability of failure on a single trial

where $\binom{n}{x} = \dfrac{n!}{x!(n-x)!}$.

Example 8.2.3 _Binomial Experiment_

In three Bernoulli Trials with p = 0.60, what is the probability of exactly one success?

Solution:
Here, n = 3 and x = 1. The probability

$$b(1;\ 3,\ 0.60) = \binom{3}{1} (0.60)^1(0.40)^2 = 3(0.6)(0.16) = 0.288$$

Examples 8.2.3 and 8.2.4 illustrate the value of prototypes, as mentioned at the beginning of the chapter, as easy computations of probabilities.

Example 8.2.4 Manufacturer Guarantee

A parts manufacturer guarantees that a box of 12 parts will contain at most one defective part. The manufacturer's records show that defective parts average a 5% rate. What is the probability that a randomly chosen box of 12 parts satisfies the guarantee?

Solution:
To satisfy the guarantee, there can be at most one defective part, denoted by $P(x \leq 1)$. So, assuming Bernoulli Trials,

$$P(x \leq 1) = P(x = 0) + P(x = 1) = b(1; 12, 0.05) + b(0; 12, 0.05)$$
$$= \binom{12}{1}(0.05)^1(0.95)^{11} + \binom{12}{0}(0.05)^0(0.95)^{12} = 0.3413 + 0.5404 = 0.8817.$$

Example 8.2.5 Health Care

One contributor to the cost of health care is that many seek medical treatment when no discernible physical basis for their ailments is evident. One physician believes that as many 10% of patients are in this category. Twenty-five patients are randomly selected and six had no evident ailment. What is the probability of observing six or more physically healthy patients in such a group? What might one conclude about the physician's belief?

Solution:
Here, we assume $p = 0.10$. Using binomial tables (or calculating), the probability that at least 6 will not exhibit signs of an ailment is $P(x \geq 6) = 1 - P(x \leq 5) = 1 - 0.967 = 0.033$.

This is a small probability, occurring on average only 33 times in a thousand. While the true rate may be 10%, the experienced outcome suggests it is higher.

The results of random outcomes are sometimes counterintuitive. For example, a machine manufactures computer chips that have a characteristically high rate of defectives, say $p = 0.20$. Suppose that an adjustment is made in an attempt to improve upon the output. For $n = 10$ chips to be examined and none ($x = 0$) is found defective, the probability is $b(0; 10, 0.20) = 0.1074$. If just three more chips are similarly examined and none is found defective (i.e., $b(0; 13, 0.20)$, the probability drops to 0.0550, about half of 0.1074 and provides stronger evidence that the adjustment was effective. If $n = 20$ chips are examined and only one is found defective (i.e., $x = 1$), the associated probability is 0.0576, almost the same as if none had been defective in a group of 13 chips.

The graph of a binomial distribution is asymmetric unless $p = 0.50$. When $p \neq 0.50$ the distribution is skewed. As p approaches 0.50, particularly with larger numbers of trials, the greater the graphical symmetry.

Theoretical or prototype probability distributions (as the binomial) derive from assumptions about the chance situation (e.g., coin tosses). Empirical probability distributions derive from (clinical) observations and measurements of frequencies of occurrence.

The binomial distribution derives its name from the binomial expansion of $(a + b)^r$. Using the **binomial theorem**,

$$(a+b)^r = \binom{r}{r} a^r b^0 + \binom{r}{r-1} a^{r-1} b^1 + \binom{r}{r-2} a^{r-2} b^2 + \cdots + \binom{r}{0} a^0 b^r$$

If a is replaced by p, b by q, and r by n, it is easy to see that the terms of the expansion are the successive values of the binomial distribution $b(x;\ n,\ p)$ as x takes on its values. In effect, we have verified that $b(x;\ n,\ p)$ is a proper probability distribution since $(a + b) = (p + q) = 1$.

Example 8.2.6 Binomial Expansion

Show that the terms of the binomial expansion of $(p + q)^3$ are the successive values of binomial probabilities $b(x;\ 3,\ p)$.

Solution:
The expansion of $(p + q)^3$ is $p^3 + 3p^2q + 3pq^2 + q^3$. Here, p^3 is the probability that three trials yielded three successes; $3p^2q$, the probability of two successes in three trials; and so on. The four terms are indeed the successive values of $b(x;\ 3,\ p)$.

Binomial Trial Models

This last derivation, using the binomial expansion, provides no particular insights nor does it allow further conceptual development. The following "event statement" approach is far more enlightening!

Consider the event:

A = "*exactly x successes in n Bernoulli Trials*" $n = 1,\ 2,\ 3,\ \ldots$ *and* $x = 0,\ 1,\ 2,\ \ldots,\ n$.

The realization of **A** can only occur if one of the following events occurred:

B = "*exactly x successes in* $n - 1$ $(n \neq 0)$ *Bernoulli Trials and a failure on the* n^{th} *trial.*"

or

C = "*exactly* $x - 1$ $(x \neq 0)$ *successes in* $n - 1$ $(n \neq 0)$ *Bernoulli Trials and a success on the* n^{th} *trial.*"

Event **A** is the union of events **B** and **C**. That is,

$$\mathbf{A} = \mathbf{B} \cup \mathbf{C}$$

This can be written as the probability statement

$$\mathrm{P}(\mathbf{A}) = \mathrm{P}(\mathbf{B} \cup \mathbf{C}) = \mathrm{P}(\mathbf{B}) + \mathrm{P}(\mathbf{C})$$

since **B** and **C** are mutually exclusive events.

Let $f(x;\ n)$ represent $P(\mathbf{A})$ since it is a function of two (discrete) variables. Events **B** and **C** themselves are joint events. However, under the assumption of Bernoulli Trials, each is composed of independent events. Letting p be the probability of success at each trial, then $P(\mathbf{B})$ can be expressed as $(1-p)\,[f(x;\ n-1)]$ and $P(\mathbf{C})$ as $p[f(x-1;\ n-1)]$. (We have simplified the discussion for the moment by ignoring the values of x and n.)

Therefore,

$$f(x;\ n) = (1-p)\,[f(x;\ n-1)] + p\,[f(x-1;\ n-1)], \quad x = 0,\ 1,\ \dots,n$$

provided $n \geq 1$ and $x \leq n$.

This is a difference equation in two discrete variables, x and n. The solution of such an equation is not difficult, but may be unfamiliar. Its solution is

$$f(x\,;\,n) = b(x\,;\,n,p) = \binom{n}{x} p^x (1-p)^{n-x}, \quad x = 0,\ 1,\ \dots,n \text{ and } n \geq 1$$

and can easily be verified.

This approach based on development of an event statement has many advantages. First, the event statement at the outset encodes the two possible outcomes. It is then developed into a probability statement, which, in turn, evolves into a functional relationship. The functional format uses the independence assumption for this case. Finally, determining the probability distribution becomes a problem in solving a difference equation.

Note the potentialities for generalization in this modeling approach. In this format, it is easy to extend the Bernoulli Trials to more than two possible outcomes; they are known as Poisson Trials. If the independent trials assumption does not hold, one can cope with the greater complexity of the difference equation and/or seek approximate representations. This is in every way preferable to the mathematical "dead end" of usual elementary derivations. There is a large body of knowledge for the solution of difference equations of various kinds by analytic and/or numerical techniques.

Additionally, situations in which the trial probability, p, is not constant but, for example, may depend on the prior number of successes, that is, a function of x and/or n, can be studied in this format. Finally, and very important, this model approach to probability distributions does not rely on the "equally likely" assumption common to many probability arguments.

◆ Practitioners of therapeutic touch (TT) claim to treat many medical ailments using their hands to intentionally manipulate a "human energy field" that exists above the patient's skin. Of course, the practice is controversial, and claims have been subject to a number of tests.

In one test, a sixth grade student, Emily Rosa, tested 21 TT practitioners to assess whether they could detect her hand when it was placed near theirs. Using a cardboard screen to shield her hand from the practitioners' view, the experiment was repeated for 280 trials.

Using the binomial distribution, as trials and subjects can be regarded as mutually independent, and assuming no particular therapeutic prowess so $p = 1/2$, one expects the average 140 ($= 1/2(280)$) "hits." Actually, there were only 122 trials in which Emily's hand was correctly located – about 44%, which is within a reasonable deviation from 50%.

More detailed information can be found in an article by Rosa, *et al.* in *The Journal of the American Medical Association* 1998; 279: 1005–1010.

EXERCISE 8.2

1. For Bernoulli Trials, compute the probabilities
   a) two successes in three trials with $p = 0.40$;
   b) three successes in four trials with $p = 1/3$;
   c) one success in five trials with $p = 0.30$.
2. For Bernoulli Trials, compute the probabilities
   a) at least two successes in three trials with $p = 0.75$;
   b) no more than three successes in seven trials with $p = 0.50$;
   c) at most two successes in five trials with $q = 0.90$.
3. The probability that a sharpshooter hits a target is 0.90 on a single shot. What is the probability that the sharpshooter will hit the target?
   a) 10 out of 10 shots
   b) 20 out of 20 shots
4. The probability that a machine component will survive a stress test is 4/5. Find the probability that exactly 4 of the next 5 components tested survive.
5. What is the probability of rolling "5" two or three times in four tosses of a fair die?
6. A manufacturing process averages 6% defectives. Ten items are chosen at random. Determine the probability that
   a) only one item is defective;
   b) at most three items are defective;
   c) four or more items are defective.
7. A quiz consists of five multiple-choice questions, each with four possible choices. Determine the probability that a student who randomly chooses an answer to each question has
   a) answered none correctly;
   b) at most, answered three correctly;
   c) answered all five correctly.
8. Each of ten questions on a multiple-choice test has four options. Choosing answers randomly, what is the chance of achieving a passing grade (60% or better)? (Questions are valued equally.)
9. Choosing randomly, what is the probability of receiving a passing grade of 60% or more on a ten-question true or false test?
10. Experience indicates that 30% of patients recover from a certain illness. Twenty patients are randomly chosen from a large group with the illness. Determine the probabilities that
    a) more than six recover;
    b) exactly five recover;
    c) at most three recover.

11. Three blue balls, six red balls, and one white ball are placed in a bag. A ball is drawn at random and replaced after its color is noted. What is the probability of drawing
    a) exactly six red balls in 10 trials?
    b) at least one white ball in seven trials?
    c) four blue balls in four trials?

12. In seven tosses of a fair coin, what is the probability of observing
    a) at least two heads?
    b) both heads and tails?
    c) exactly five heads?
    d) two or three heads?

13. The probability of a weather delay for a NASA space shuttle mission is 15%. What is the probability that at least two of the next five missions are delayed due to weather?

14. Reportedly, some 80% of college students admit to dishonesty. If true, what is the probability that at least 15 of 20 students at random have been dishonest? (Center for Academic Integrity, http://www.plagiarism.org).

15. About 90% of runners completed a recent Boston Marathon. What is the probability that each of ten competitors chosen at random finished this marathon?

16. An article in *The Journal of the American Medical Association* concluded that newly approved drugs are riskier than older ones. More than 10% of such drugs were later removed from the market or required to post new side-effects warnings. Suppose that 10% of US food and Drug Administration (FDA)-approved drugs are later removed from the market. What is the probability that two or more of fifteen new drugs will later be taken off the market?

17. It is estimated that 10% of youth suffer from depression. What is the chance that in an elementary school teacher's class of 30 students
    a) at least five suffer depression?
    b) none suffer depression?

18. How many tosses of a fair coin are needed for a 99% chance of at least two tails?

19. How many tosses of a fair die are needed for a 95% chance that a "5" appears at least once?

20. About 20% of adults smoke. Among 10 randomly chosen adults, what is the chance that at least 3 smoke?

8.3 THE HYPERGEOMETRIC DISTRIBUTION

Imagine a large group of animals, some with a tiny radio "tag" to study migratory habits. A sample of k animals is randomly selected from a large group n $(n \geq k)$, of which r $(r \leq n)$ are tagged, $n = 1, 2, \ldots$. What is the probability that x selected at random are tagged?

To begin, calculate the number of ways the sample can have the indicated composition. Next, divide by the total number of ways a sample of size k can be selected from among n animals. This last number is $\binom{n}{k}$ since there is no replacement and order is immaterial. By similar reasoning, think of the population as made up of two kinds of animals: "tagged" and "not tagged." The number of ways to select x from the r tagged animals is $\binom{r}{x}$, while the number of ways to select $(k - x)$ untagged ones from the other $(n - r)$ untagged ones is $\binom{n-r}{k-x}$.

Using the Fundamental Principle of Counting (Section 7.4), the number of ways two groups of animals can be selected is the product

$$\binom{r}{x}\binom{n-r}{k-x}$$

The required probability is, therefore,

$$\frac{\binom{r}{x}\binom{n-r}{k-x}}{\binom{n}{k}}$$

This is the **hypergeometric probability distribution** and can conveniently be denoted by the functional notation $h(x;\ n,\ r,\ k)$, or simply $h(x)$, in which the random value, x, again, appears before the semicolon and followed by parameters n, r, and k. (This descriptive functional notation is similar to the binomial distribution usage.)

♦ The field use of the hypergeometric distribution in studying migratory habits of animals is widespread. Sometimes referred to as "capture–recapture," it is a common means for estimating sizes of animal populations as described later in this section.

Hypergeometric Distribution

$$h(x;n,r,k) = \frac{\binom{r}{x}\binom{n-r}{k-x}}{\binom{n}{k}}, \quad x = 0, 1, \ldots, r$$

$$r \le n, \qquad n = 1, 2, \ldots, \qquad n \ge k \ge x$$

where x = Number "tagged" in the sample
n = Population size, ("tagged" plus "untagged")
r = Number "tagged" in population
k = Sample size.

It is common to confuse the binomial and hypergeometric distributions as both are concerned with the number of "successes" in a sample of observations. What distinguishes them is the manner in which the data are obtained and/or the size of the population.

For the binomial probability distribution, data are drawn with replacement from a finite population or equivalently, drawn (without replacement) from an infinite population. For example, in repeated tosses, the coin is "restored" after each toss.

Hypergeometric sample data, on the other hand, are drawn without replacement from a finite population. While the trial probability, p, is constant in the binomial case and its outcomes are independent, in a hypergeometric case, the actual probability changes as drawings are made. With each draw, the population composition changes. That is, for example, a "tagged" animal will not be counted twice, and after an animal is selected, the probability for the next draw may change noticeably (unless the population is extremely large).

Example 8.3.1 Poker Hands

What is the probability of exactly three kings in a five-card poker hand?

Solution:

There are 52 cards (the population), of which five (the sample) are chosen without replacement. Therefore, $n = 52$ and $k = 5$. A deck has four kings, of which three are to be chosen. Therefore, $r = 4$ and $x = 3$. The remaining two cards to complete the hand are drawn from the remaining 48 cards that are not kings.

So,

$$h(3; 52, 4, 5) = \frac{\binom{4}{3}\binom{48}{2}}{\binom{52}{5}} = 0.0034$$

♦ Poker, other card games, and games of chance have a storied history in the origins of mathematical probability. Mathematical minds of earlier centuries earned their keep from wealthy patrons by aiding their gambling prowess. A recent resurgence of poker tournaments and sites via satellite TV and the Internet permits large number of people to watch, play, and bet.

While card game examples are still a staple of beginning texts, as this one, mathematical probability long ago achieved an importance well removed from its origins.

Example 8.3.2 Committee Membership

A committee consists of four women and three men. What is the probability that a subcommittee of three chosen at random has two or three women?

Solution:

Either the subcommittee has two of the four women and one of the three men or it has three of the four women and none of the three men.

⟶

The desired probability is the union of these two mutually exclusive events.
Applying the hypergeometric distribution to each event and adding them yields

$$\frac{\binom{4}{2}\binom{3}{1}}{\binom{7}{3}} + \frac{\binom{4}{3}\binom{3}{0}}{\binom{7}{3}} = \frac{18}{35} + \frac{4}{35} = \frac{22}{35}$$

Industrial Quality Control

Quality control inspection is an integral part of manufactured production. Indeed, industrial quality control people have active professional associations. The hypergeometric distribution has a role in estimating the number of defectives in a large production batch.

A quantity produced, n, is apt to be known, while the number of defectives, r, is not. Among ways to decide whether the production run meets standards, assume that r, the number "tagged" in the population, is the largest number of defectives that can be tolerated in the batch of n items. Next, use a sample of size, k, and count the random number of defectives among them, x. The probability of x or more defectives can be calculated using the hypergeometric distribution. If this probability is sufficiently small, it may suggest that the true value of r does not exceed the assumed value.

Population Estimates

In some instances, an estimate is sought for a population size n. For example, estimates are frequently sought for the number of animals in a region. A common practice is to catch a number, r, and mark them in some harmless, but distinguishable way. These marked animals are released and mix with the larger population. When the mixing is judged complete, a second catch, k, is made and the number of marked animals, x, is recorded.

Certainly, there are at least $(k + r - x)$ animals, the numbers caught minus the duplicates. However, it is usually very unlikely that $n = (k + r - x)$. It is also unlikely that n is vastly greater than $(k + r - x)$.

In this instance, one can view the hypergeometric probability distribution $h(x;\ n,\ r,\ k)$ as a function of four variables, of which only three are known: k, r, and x. Therefore, additional information is needed to evaluate the probability. One way to obtain this information is to assume that the experience that resulted in x marked animals was the most probable one. That is, let n be the value that maximizes $h(x;\ n,\ r,\ k)$ for the known values of x, k, and r. Since $h(x;\ n,\ r,\ k)$ is a unimodal function, one way to seek its maximum is to note where the ratio

$$\frac{h(x; n+1, r, k)}{h(x; n, r, k)}$$

changes from less than 1 to greater than 1 (or vice versa). An algebraic argument (not included here) yields the maximizing condition

$$n \approx \frac{kr}{x}$$

which matches the intuitive ratio $\frac{x}{r} = \frac{k}{n}$.

Example 8.3.3 Production Line Inspection

A batch of 10,000 items is just off from a production line. A sample of 100 is inspected, and 10 are defective. What is the likely number of defectives in the batch?

Solution:
By manipulating the earlier approximation for n, r is

$$r \approx \frac{nx}{k} = \frac{10,000(10)}{100} = 1000$$

♦ A splatometer? Actually, it is a grid devised by Britain's Royal Society for the protection of birds. Mounted on the cars of volunteer motorists to aid in counting "splats" – dead bugs. Bugs, being a bird food, estimates of their numbers is relevant. Plausibly, the hypergeometric distribution might be used for such estimates along the lines of the above example.

James Gorman, writing in *The New York Times* some years ago ("Here a Skunk, There a Skunk: A New Kind of Wildlife Census"), suggests (tongue in cheek?) a similar program be devised for drivers to report ungulate road kills from which animal populations might be estimated.

Statistics or Probability?

The preceding discussion provides a good opportunity to clarify a difference between probability as one subject and statistics as another, intertwined though they may be! The derivation of the hypergeometric distribution can be viewed as a "probability problem." However, the quality control and population estimates examples don't deal directly with probability but with the estimation of parameters as r and n. This is a "statistical problem" and large bodies of concepts and techniques have been devised to deal with it. Indeed, our suggestion that n be estimated by maximizing the probability corresponding to the data at hand is the well-known statistical **principle of maximum likelihood**.

Generalizations

The hypergeometric distribution is easily generalized to populations of several types of elements. Our discussion has been limited to two types: "good" and "bad." Specifically, consider a population of size n consisting of r_1 elements of one type, r_2 elements of a second type, …, r_k elements of the kth type from which a sample of size n is drawn. The probability that the sample consists of x_1 of the first type, and so on, is

$$\frac{\binom{r_1}{x_1}\binom{r_2}{x_2}\cdots\binom{r_k}{x_k}}{\binom{n}{k}}$$

where $n = r_1 + r_2 + \cdots + r_k$ and $k = x_1 + x_2 + \cdots + x_k$.

The range of the hypergeometric random variable is limited to the sample size or to the number of successes in the population, whichever is smaller.

EXERCISES 8.3

1. Six red marbles and four black marbles are placed in a bag. Three marbles are drawn at random and simultaneously. What is the probability
   a) all are red?
   b) that both colors are represented?
2. There are five "AA" batteries in a box and three of them are "good." If two are chosen at random for a two-battery Walkman, what is the probability that it operates?
3. In a lotto game, six numbers are drawn without replacement from a set of 36 differently numbered balls. What is the probability of picking
   a) all six numbers correctly?
   b) only five numbers correctly?
4. A bridge hand of 13 cards contains 5 clubs, 6 spades, and 2 hearts. Two cards from the hand are drawn at random without replacement. Find the probability that both cards are spades.
5. What is the probability a five-card poker hand contains
   a) four aces?
   b) a flush (all cards are of the same suit)?
6. If eight cards are drawn without replacement from a standard deck, what is the probability of drawing either one or two kings?
7. An employer plans to hire three people from a group of eight, three of whom are minority candidates. If the employer selects the candidates randomly, what is the probability that
   a) no minority candidate is hired?
   b) all three minority candidates are hired?
8. An investor is to randomly choose three of seven companies as investments. Two of the companies will be bankrupt next year. What is the probability that at least one of the investor's choices will bankrupt next year?
9. A box of 20 items has 6 that are defective. A sample of five is selected at random. Find the probability that
   a) none are defective.
   b) exactly one is defective.
   c) all five are defective.
10. Among 50 components, two are defective. A lot of five is chosen at random. The lot is rejected if at least one defective is found. What is the probability of rejecting the lot?
11. Among 12 individuals in a biological study, 5 have brown hair; 4, black; 1, red; and 2 are blonde. What is the probability that a random sample of six individuals will have three with brown hair, two with black, and one a blonde?

12. Suppose there are 20 female and 80 male United States senators. What is the probability that a five-member subcommittee chosen randomly has exactly one female?
13. Suppose the US Senate consists of 53 Democrats, 45 Republicans, and 2 Independents. A subcommittee of 13 Democrats and 12 Republicans is formed. What is the probability that a subcommittee of Senators chosen at random would have such composition?
14. Suppose the US House of Representatives includes 53 members from California. Of these, 18 are women. If a committee of seven Representatives from California is selected at random, what is the probability that it has three women?
15. If three women and six men serve on the US Supreme Court, what is the probability that a random selection of four justices includes a woman?

8.4 THE POISSON DISTRIBUTION

The **Poisson process** is an important random process that is a foundation for several more complex random processes. Several probability distributions arise naturally from the process including the **Poisson distribution** (Historical Notes).

The Poisson distribution is probably the most useful descriptor of discrete chance events. Some examples include the random daily number of industrial accidents, defects in a production batch, cars crossing a bridge or passing a marker per day, errors on a page of type, cell phone calls received in an hour, mutations in a stretch of irradiated DNA, studies of seat belt use, blood counts, and so on.

Characteristics of a Poisson Random Variable

1. **The number of occurrences of some chance event in a unit of time (or unit area, volume, or weight) is the random variable.**
2. **The probability of a chance event in a unit of time is same for all time units.**
3. **The number of occurrences in a unit of time is independent of the number of occurrences in other time units.**

To illustrate a Poisson random variable, consider events that occur at random on a time line (view hits below). For example,

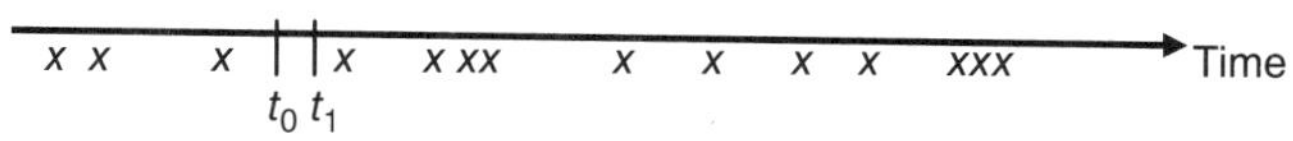

Consider a (small) time interval $(t_1 - t_0)$.The chance of a random occurrence in $(t_1 - t_0)$ is proportional to its duration, not to either t_1 or t_0. That is, the chance of an occurrence in an interval $(t_1 - t_0)$ is the same everywhere along the time line. Also, the probability of two or more occurrences in an arbitrarily small interval is assumed to be negligible. Finally, the number of occurrences in any interval is independent of any other nonoverlapping interval.

Phenomena with the aforementioned characteristics are described by the Poisson distribution as follows:

Poisson Distribution

$$p(x;\ \lambda) = \frac{\lambda^x e^{-\lambda}}{x!}, \quad x = 0, 1, 2, \ldots$$

λ = Mean number of occurrences per unit time
where $e \approx 2.71828$.

Again, we have used the functional notation $p(x;\ \lambda)$, or simply $p(x)$, x being a value the random variable.

The single parameter, λ, in the Poisson distribution is the mean of the number of random occurrences ("hits") in a particular area, volume, unit of time, etc.

Example 8.4.1 Gas Station Arrivals

Cars arrive at a gas station according to a Poisson distribution at a rate of 2.1 cars per minute. What is the probability that in any given minute exactly three cars arrive?

Solution:
For the Poisson distribution, the average (mean) number of cars per minute is $\lambda = 2.1$. The probability that $x = 3$, $p(3; 2.1)$, is

$$p(3; 2.1) = \frac{2.1^3 e^{-2.1}}{3!} = \frac{9.261(0.122)}{6} \approx 0.189$$

Example 8.4.2 eBay Purchases

On average, an eBay sale occurs at a rate of three per minute for a certain item. What is the probability that in a given minute there are two or fewer sales of the item?

Solution:
The mean, λ, is 3. We seek $p(x \leq 2)$.

$$p(x \leq 2) = \frac{3^0 e^{-3}}{0!} + \frac{3^1 e^{-3}}{1!} + \frac{3^2 e^{-3}}{2!}$$
$$= e^{-3} + 3e^{-3} + \frac{9}{2}e^{-3} = 8.5e^{-3} \approx 0.4232$$

In theory, the Poisson random variable takes integer values 0, 1, 2, ... The Poisson distribution is right skewed when λ is small and tends to symmetry for larger values of λ.

A Poisson model provides a reasonable approximation to a binomial distribution when the number of trials is large and the trial probability is small.

Characteristics of the Poisson random variable are usually difficult to verify for practical examples. As with all probability models, the test of adequacy of a Poisson model is whether it provides a reasonable approximation to reality – that is, whether the empirical data support it.

◆ A historic nineteenth-century example of the Poisson distribution is the recorded number of soldiers killed by horse kicks each year in the Prussian cavalry made famous in a book by Ladislaus Bortkiewicz. It has been suggested that the distribution should have been named after him rather than Siméon Poisson.

EXERCISES 8.4

1. A random variable with a Poisson distribution has $\lambda = 2.6$. Calculate

   a) $p(x \leq 3)$ b) $p(x = 1)$ c) $p(x > 2)$

2. A random variable with a Poisson distribution has a mean of 2.0. Find the probabilities

   a) $p(x \leq 1)$ b) $p(x = 2)$ c) $p(x > 3)$

3. Assume that the daily number of employees who call in "sick," x, follows a Poisson distribution with $\lambda = 2.4$.

   a) What is the probability that exactly three people call in sick in a particular day?
   b) What is the probability that at most four people will call in sick?

4. Among small businesses in a city, the annual number of failures averages five. Assume x, the number of failures per year can be characterized by a Poisson distribution.

   a) What is the probability that fewer than three small businesses fail next year?
   b) What is the probability that at least six fail next year?

5. A typist averages one error per page. What is the probability that on a page chosen at random there are

   a) no errors? b) three or more errors?

6. An area of the eastern seaboard averages three major hurricanes in a season. What is the probability a season at random is

   a) not hit by hurricanes? c) hit by exactly three hurricanes?
   b) hit by at least four hurricanes?

7. A production line averages 2.6 breakdowns per day. What is the probability that on a given day

   a) exactly three breakdowns occur?
   b) no more than four breakdowns occur?
   c) at least two breakdowns occur?

8. The number of defects in a square yard of carpet is a random variable. If the average number of defects per square yard of carpet is 1.2, what is the probability the number of defects in a given square yard of carpet will not exceed two?

9. Flaws in a composite material occur at a rate of two per square foot. What is the probability of a flawless square yard of material?

10. Satellite signals arrive according to a Poisson law with $\lambda = 1.5$ signals per hour. An observer takes a 15-minute nap. What is the probability that signals were received during the nap?

11. The number that arrives at a bank counter during a 5-minute period averages three customers and follows a Poisson distribution. What is the probability that no customers arrive in the next 5 minutes and that more than five customers arrive in the next 5 minutes?

12. An estimate of the average number of birth defects near Chernobyl before 1986 was 6 per 1500 births. After the 1986 nuclear accident, many assumed the number would rise. Suppose that defects averaged 10 per 1500 births after 1986. Does this tend to support or refute claims of increased birth defects?

13. For a large fleet of delivery trucks, the average number of daily breakdowns is 2.4 trucks. What is the probability of exactly four breakdowns tomorrow? Less than two breakdowns occur next Thursday?

HISTORICAL NOTES

Jacob Bernoulli (1654–1705) — Swiss mathematician and a founder of probability theory. His brother, Johan Bernoulli, with whom he made important contributions to calculus, and uncle, Daniel Bernoulli, were also well-known mathematicians of their time. Jacob's parents wanted him to study philosophy and theology much to his dislike. While completing his philosophy and theology degrees, he studied mathematics and astronomy.

Siméon Poisson (1781–1840) — Mathematician and physicist, Poisson developed many applications of mathematics in statistics and physics. He authored over 300 manuscripts and books on a variety of mathematical topics.

◆ The Poisson distribution reasonably describes a surprising variety of discrete phenomena. Examples include white blood cells in a blood sample, vehicular traffic, usage of telephone lines, cosmic ray hits, stars in galaxies and more.

CHAPTER 8 SUPPLEMENTARY EXERCISES

1. Determine the expected value of x, whose probability distribution is given. Note that a missing probability is to be determined.

x	10	20	40	60	75	100
$p(x)$	0.20	—	0.05	0.30	0.20	0.10

2. Determine the expected value of x, whose probability distribution is given. Note that missing probabilities are to be determined.

x	20	30	40	50	60	80
$p(x)$	0.15	y	$2y$	$2y$	0.27	0.18

3. Assume that a couple's children are equally likely to be a boy or a girl. If the family plans for five children, find the probability

   a) at least two are girls; b) all five children are of like gender.

4. A fair coin is tossed three times. Find the probability of three heads given that the first two tosses yielded heads.

5. An IRS auditor randomly samples 10 of 100 tax returns. If two or more of these indicate improper deductions, all 100 returns are audited. Determine the probability that the entire group is audited if the long-run percentage of improper returns is

   a) 5% b) 10% c) 25%

6. Among a large number of promising natural gas sites, eight are estimated to have a 40% chance of being viable. What is the probability that at least three of these sites prove viable?

7. A die is loaded so that the probability of a 2, 4, or 6 is twice as likely as a 1, 3, or 5. In one game, a player receives the dollar amount of the face value of a thrown die (for instance, if a 4 is thrown the player receives $4). What should the player be charged to have a fair game?

8. A 10-question multiple-choice test consists of questions with four options. If answers are randomly chosen, what is the probability of at least seven correct answers?

9. A carton contains 20 items, of which three are defective. What is the probability in a sample of five, chosen at random, exactly two are defective?

10. In poker, a "full house" beats a "flush." Show that the probability of a full house is less than that of a flush. (Any five cards of the same suit are a "flush," while a "full house" is three cards of one denomination and two of another denomination.)

11. Tabulate the probability distribution for the number of aces in a five-card poker hand.

12. An experiment consists of randomly drawing three cards from a standard deck without replacement. Let $x =$ be the number of number cards drawn. Find the probability distribution for x.

13. Assume x is a random variable having a Poisson distribution with a mean of 2.4. Calculate

    a) $p(x \leq 4)$ b) $p(x = 2)$ c) $p(x > 1)$

14. A bus service in a city averages 3.8 accidents a week. What is the probability that there are no accidents in a particular week?

15. The number of customers entering a music store was recorded during 15-minute intervals for a 2-hour period. The results were

15-Minute Interval	Number of Customers
1	10
2	8
3	9
4	12
5	10
6	11
7	8
8	12

a) Determine the mean number of customers per 15-minute interval.

b) Determine the probability of exactly nine customers entering the store in a 15-minute interval.

c) Determine the probability of at least eight customers entering the store in a 15-minute interval.

9 Markov Chains

The outcomes of chance events are often influenced by prior chance outcomes; for examples, tomorrow's weather may be influenced by today's weather or perhaps yesterday's; gasoline usage this month may be affected by last month's usage; and one's blood pressure can be altered by earlier medication.

This contrasts with, say, coin tossing described by Bernoulli Trials in Chapter 8. With the assignment of probabilities for head and tail, we supposed that those probabilities not only remained the same for future tosses but also the outcome of any toss had no bearing on future tosses. It is as if the coin has no memory; mathematically, successive tosses are *independent* events.

Finite Mathematics: Models and Applications, First Edition. Carla C. Morris and Robert M. Stark.

Companion Website: http://www.wiley.com/go/morris/finitemathematics

This chapter studies chance phenomena, known as **Markov chains**, in which the current outcome or **state** affects chances for the next outcome.

We begin by clarifying further the distinction between Bernoulli Trials and **Markov processes**. Two models of weather forecasts for the next 2 days appear in two tree diagrams below, each beginning with rain (today).

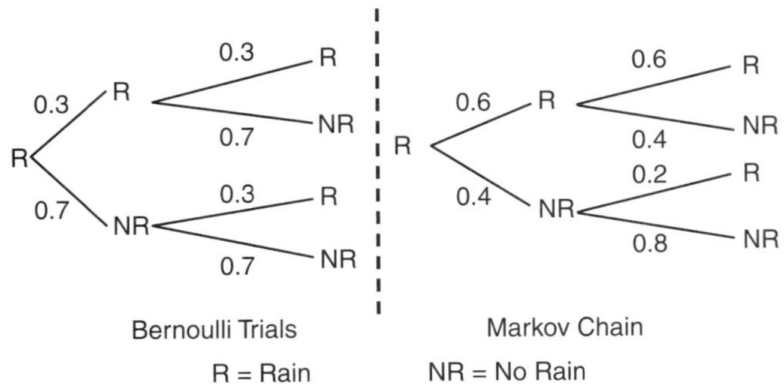

For Bernoulli Trials the chance of "No Rain" tomorrow is constant at 0.7 (and "Rain," 0.3), independent of yesterday's weather. For a Markov chain, the chance of "No Rain" tomorrow is influenced by today's weather. If there is "No Rain" today, then there is an 80% chance of "No Rain" tomorrow. However, if there is "Rain" today, the chance of "No Rain" tomorrow falls to 40%. These are decimals on the "trees."

For the day after tomorrow (the next stage), weather depends on tomorrow's weather, and so on. These are chance processes whose outcomes depend on a number of prior outcomes and not simply the single outcome that characterizes a Markov chain. Their study is beyond our scope. However, the Markov chain as a model is often realistic and useful and its mathematical description is well developed. Markov chain models have been applied to studies of learning, memory, social mobility, ecological systems, decision-making under uncertainty and other areas.

9.1 TRANSITION MATRICES AND DIAGRAMS

The probabilities of moving from one state to another in a Markov chain are called **transition probabilities**. The transition probability for "Rain" to "No Rain" in the weather example is 0.4, while the transition probability for "No Rain" to "No Rain" is 0.8, and so on. Note that transition probabilities are simply conditional probabilities.

Transition probabilities are usually depicted pictorially or using matrices. They are illustrated below. Choose either or both as you prefer!

The **transition diagram** below pictorially displays transition probabilities.

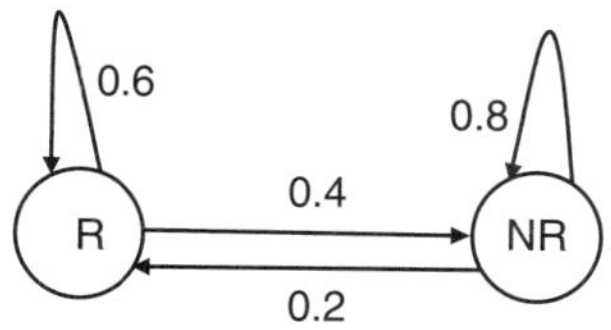

Arrows indicate the one step transition from one state to the next.

NR = "No Rain"

The **transition matrix** below also depicts transition probabilities.

$$\text{From}\quad \begin{array}{c} \\ R \\ NR \end{array} \overset{\text{To}}{\begin{array}{cc} R & NR \end{array}} \begin{bmatrix} 0.6 & 0.4 \\ 0.2 & 0.8 \end{bmatrix}$$

Transition matrices are square matrices whose rows sum to unity.

R = "Rain"

The one-step dependence of probabilities is called a **Markov property**.

Markov Property

The outcome of any trial depends only on the outcome of the directly preceding trial.

Example 9.1.1 Chess Matches

Two chess players, A and B, are equally matched at the start. However, winning or losing has a psychological effect on A. If A wins, the chance of winning the next game rises to 3/5. If A loses, the probability of winning the next game falls to 1/3.

Use a tree diagram to depict A's probabilities of winning the second game. Also, express the transition probabilities in a transition diagram and matrix.

Solution:

Tree diagram:

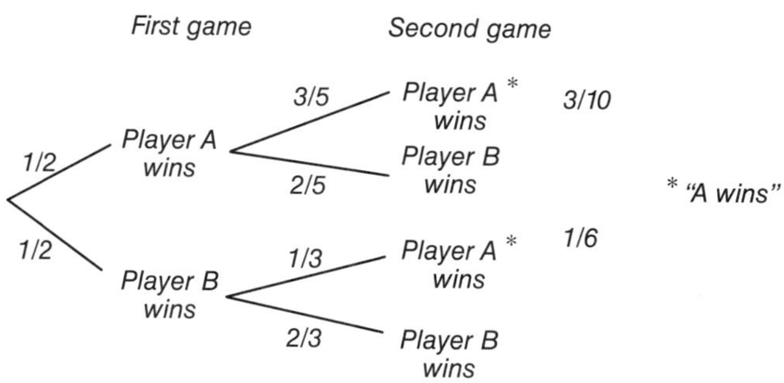

The probability that A wins the second game (denoted by asterisks) is

$$3/10 + 1/6 = 7/15$$

→

Transition diagram:

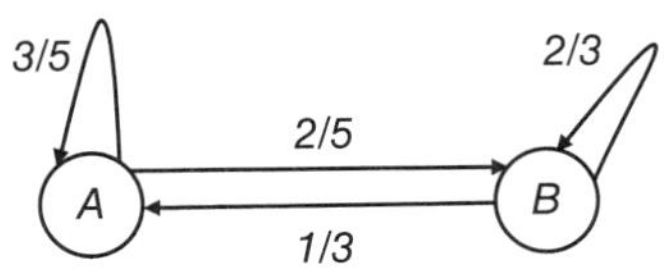

Transition matrix:

$$
\begin{array}{ccc} & & \text{To} \\ & & \begin{array}{cc} A & B \end{array} \\ \text{From} & \begin{array}{c} A \\ B \end{array} & \begin{bmatrix} 3/5 & 2/5 \\ 1/3 & 2/3 \end{bmatrix} = T \end{array}
$$

Tree diagrams are not practical for a larger number of events. Such trees develop many branches and depicting them is unwieldy, increasing by a power of 2 at each stage. Transition diagrams and matrices are usually more practical.

Later, we discuss how transition probabilities aid in predicting outcomes.

Example 9.1.2 A Transition Matrix

Express this transition diagram as a transition matrix.

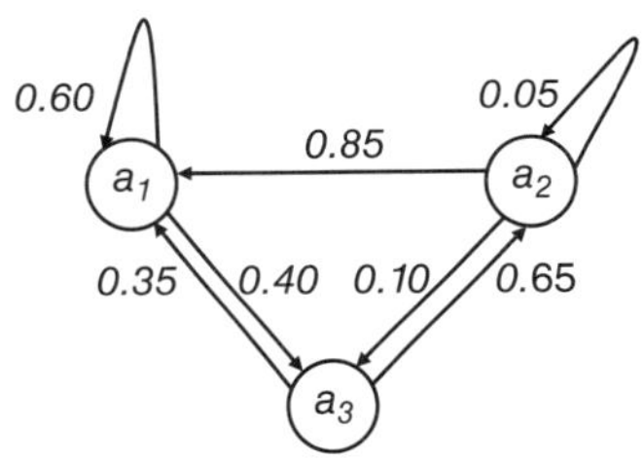

Solution:

Aside the arrows are probabilities of moving among states. States appear along the rows and columns as a guide for the matrix. For example, the probability of leaving State a_1 and to enter State a_3 is 0.40. The (1, 3) entry of the transition matrix is 0.40. As we proceed, the transition matrix becomes

$$
\begin{array}{ccc} & & \textbf{To} \\ & & \begin{array}{ccc} a_1 & a_2 & a_3 \end{array} \\ \textbf{From} & \begin{array}{c} a_1 \\ a_2 \\ a_3 \end{array} & \begin{bmatrix} 0.60 & 0.00 & 0.40 \\ 0.85 & 0.05 & 0.10 \\ 0.35 & 0.65 & 0.00 \end{bmatrix} \end{array}
$$

Example 9.1.3 A Transition Diagram

Draw the transition diagram corresponding to the transition matrix:

$$\textbf{From}\ \overset{\textbf{To}}{\begin{bmatrix} 0.30 & 0.35 & 0.25 & 0.10 \\ 0.60 & 0.25 & 0.00 & 0.15 \\ 0.20 & 0.00 & 0.80 & 0.00 \\ 0.00 & 0.95 & 0.00 & 0.05 \end{bmatrix}}$$

Hint: Label rows and columns as States $a_1, \ldots, a_4$.

Solution:

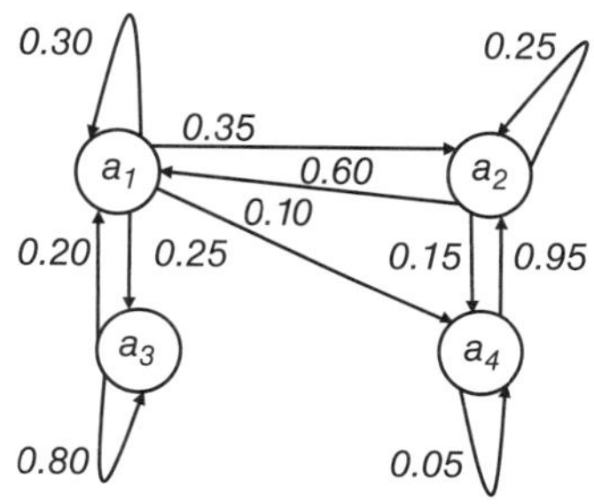

Of the 16 elements in the transition matrix, only 11 arrows appear on the transition diagram as the five zero-transition probabilities require no "State to State" arrows.

EXERCISES 9.1

1. What values of x and y complete these Markov chain transition probability matrices?

a) $\begin{bmatrix} 0.20 & x \\ y & 0.30 \end{bmatrix}$ b) $\begin{bmatrix} 0.40 & 0.50 & x \\ x & y & 0.30 \\ y & 0.30 & x \end{bmatrix}$ c) $\begin{bmatrix} 0.70 & 2y & y \\ x & x & 2y \\ 0.60 & y & x-y \end{bmatrix}$

2. Which transition matrices apply to Markov chains?

a) $\begin{bmatrix} 0.3 & 0.7 \\ 0.6 & 0.4 \end{bmatrix}$ b) $\begin{bmatrix} 1.00 & 0.00 & 0.00 \\ 0.10 & 0.60 & 0.30 \\ 1/3 & 1/3 & 1/3 \end{bmatrix}$ c) $\begin{bmatrix} 0.7 & 0.2 & 0.1 \\ 0.4 & 0.4 & 0.3 \\ 0.9 & 0.1 & 0.0 \end{bmatrix}$

3. Draw the transition diagram corresponding to this Markov chain matrix.

$$\begin{bmatrix} 0.4 & 0.5 & 0.1 \\ 0.1 & 0.6 & 0.3 \\ 0.8 & 0.0 & 0.2 \end{bmatrix}$$

4. Draw the transition diagram corresponding to this Markov chain matrix.

$$\begin{bmatrix} 0.4 & 0.3 & 0.1 & 0.2 \\ 0.1 & 0.5 & 0.0 & 0.4 \\ 0.2 & 0.0 & 0.8 & 0.0 \\ 0.5 & 0.2 & 0.0 & 0.3 \end{bmatrix}$$

5. Write the transition matrix corresponding to these transition diagrams.

a)

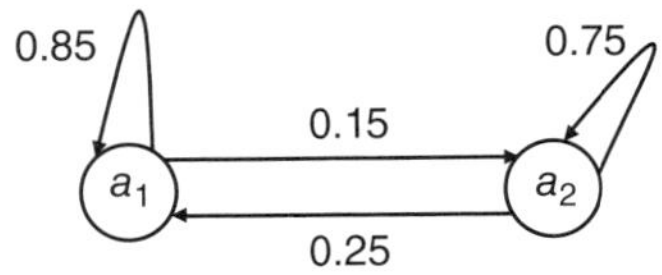

b)

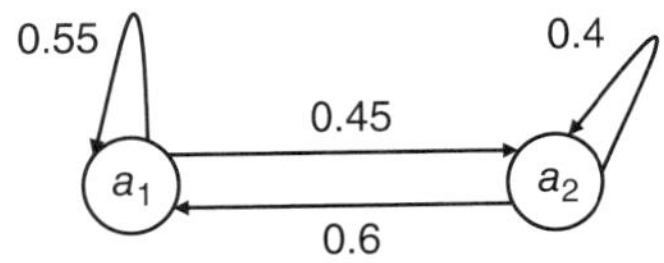

6. Find the transition matrix for the diagram.

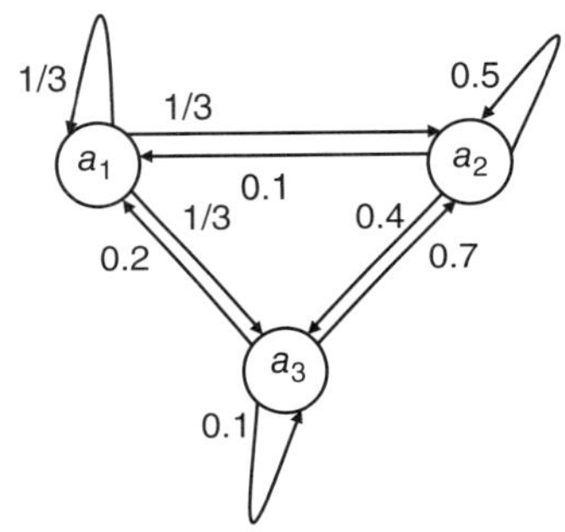

7. Find the transition matrix for the diagram.

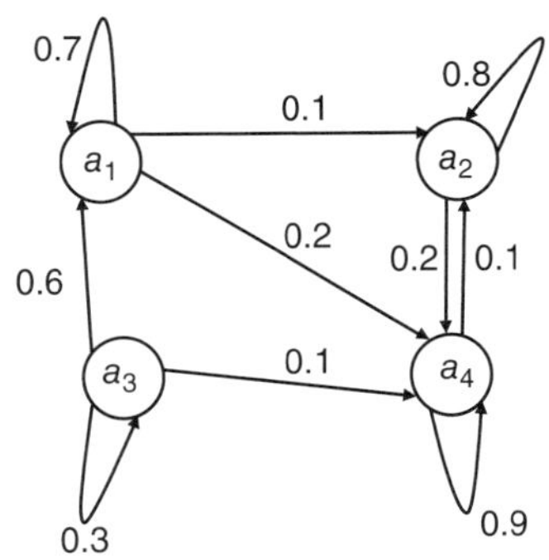

8. What is the probability that this Markov chain, currently in State 2, will next be in State 1?

$$\begin{bmatrix} 0.4 & 0.6 \\ 0.2 & 0.8 \end{bmatrix}$$

9. What is the probability that this Markov chain, currently in State 3, will next be in State 1?

$$\begin{bmatrix} 0.1 & 0.5 & 0.4 \\ 0.2 & 0.5 & 0.3 \\ 0.7 & 0.1 & 0.2 \end{bmatrix}$$

10. A Markov chain with this transition matrix is currently in State 2. What is the probability that it is in State 1 on the third observation? On the fifth observation?

$$\begin{bmatrix} 1.0 & 0.0 & 0.0 \\ 1.0 & 0.0 & 0.0 \\ 0.4 & 0.3 & 0.3 \end{bmatrix}$$

9.2 TRANSITIONS

Probabilities in a transition matrix are **one-step** probabilities. That is, they represent the probabilities of moving from one state to another (or the same state) at the next step.

Suppose the probability of moving from State i to State j in two steps, three steps, or k steps is sought in an n-state Markov chain, $i, j, k = 1, \ldots, n$. The conditional probability of moving from a State i to a State j in k steps is denoted by $t_{ij}(k)$. The matrix of the $t_{ij}(k)$, $\mathbf{T}(k)$, is called the ***k*-step transition probability matrix** for a Markov chain.

Example 9.2.1 A Two-Step Transition Matrix (Two States)

Compute T(2) for the Markov chain whose transition matrix is

$$\boldsymbol{T} = \begin{bmatrix} 0.9 & 0.1 \\ 0.2 & 0.8 \end{bmatrix}$$

Solution:

$\boldsymbol{T}(2)$ is square of $\boldsymbol{T}$, that is, $\boldsymbol{T}^2$.

$$\boldsymbol{T}(2) = \boldsymbol{T} \cdot \boldsymbol{T} = \begin{bmatrix} 0.9 & 0.1 \\ 0.2 & 0.8 \end{bmatrix} \begin{bmatrix} 0.9 & 0.1 \\ 0.2 & 0.8 \end{bmatrix} = \begin{bmatrix} 0.83 & 0.17 \\ 0.34 & 0.66 \end{bmatrix}$$

These values can also be found using a transition tree: for instance, to move from State 1 to State 1 in two steps can happen in two ways. First, move from State 1 to State 1. Then, from State 1 return to State 1. The second step moves from State 1 to State 2 and from State 2 back to State 1.

Note that a transition tree is needed for each possible starting state.

⟶

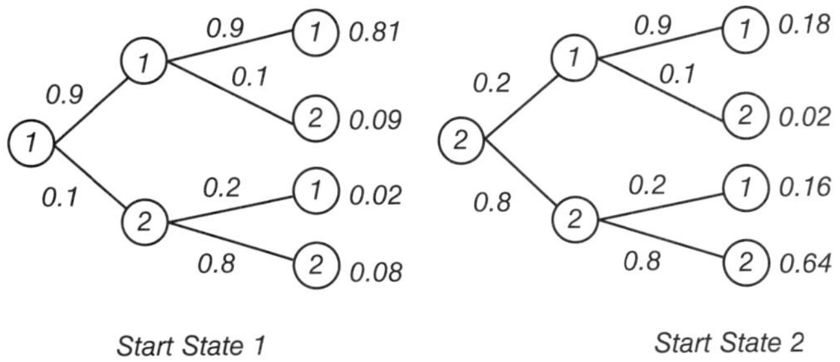

This can be represented algebraically as

$$p_{11}p_{11} + p_{12}p_{21} = 0.81 + 0.02 = 0.83 \qquad p_{21}p_{11} + p_{22}p_{21} = 0.18 + 0.16 = 0.34$$
$$p_{11}p_{12} + p_{12}p_{22} = 0.09 + 0.08 = 0.17 \qquad p_{21}p_{12} + p_{22}p_{22} = 0.02 + 0.64 = 0.66$$

Example 9.2.2 A Two-Step Transition Matrix (Three States)

For the Markov chain,

$$T = \begin{bmatrix} 0.1 & 0.7 & 0.2 \\ 0.3 & 0.4 & 0.3 \\ 0.5 & 0.1 & 0.4 \end{bmatrix}$$

a) *What is the two-step transition matrix T(2)?*
b) *From State 3 what is the probability of reaching State 2 in two steps?*

Solution:

a) $$T(2) = \begin{bmatrix} 0.1 & 0.7 & 0.2 \\ 0.3 & 0.4 & 0.3 \\ 0.5 & 0.1 & 0.4 \end{bmatrix}\begin{bmatrix} 0.1 & 0.7 & 0.2 \\ 0.3 & 0.4 & 0.3 \\ 0.5 & 0.1 & 0.4 \end{bmatrix} = \begin{bmatrix} 0.32 & 0.37 & 0.31 \\ 0.30 & 0.40 & 0.30 \\ 0.28 & 0.43 & 0.29 \end{bmatrix}$$

b) *Here,* $t_{32}(2)$ *is 0.43.*

The entries in each transition matrix row form a probability distribution and thus add to unity. This is a useful check when multiplying transition matrices.

Example 9.2.3 A Three-Step Transition Matrix

Compute ***T****(3), the third-step transition probability matrix, when the one-step transition matrix is*

$$T = \begin{bmatrix} 0.6 & 0.4 \\ 0.3 & 0.7 \end{bmatrix}$$

→

Solution:
After two steps, the transition matrix is

$$\boldsymbol{T}(2) = \boldsymbol{T} \cdot \boldsymbol{T} \begin{bmatrix} 0.6 & 0.4 \\ 0.3 & 0.7 \end{bmatrix} \begin{bmatrix} 0.6 & 0.4 \\ 0.3 & 0.7 \end{bmatrix} = \begin{bmatrix} 0.48 & 0.52 \\ 0.39 & 0.61 \end{bmatrix}$$

Similarly, after two steps, these values can also be found using a tree diagram.

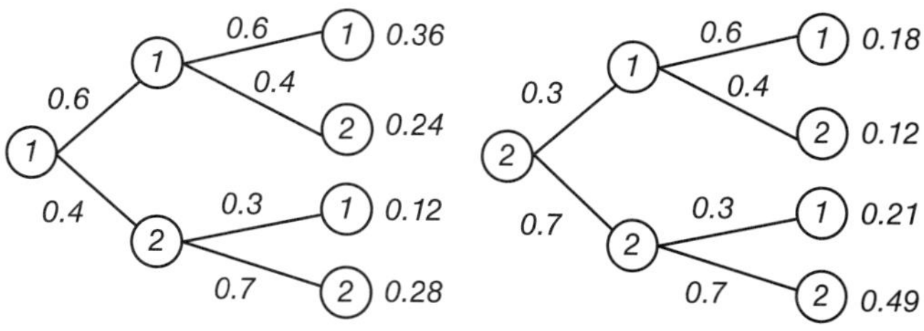

Start State 1

Start State 2

For instance, to move from State 1 to State 1 in two steps can happen in two ways. One is to move from State 1 to State 1 and then return to State 1. The second is to move from State 1 to State 2 and then to State 1.

Algebraically,

$$p_{11}p_{11} + p_{12}p_{21} = 0.36 + 0.12 = 0.48$$

The entries for the transition matrix are

$$\boldsymbol{p_{11}p_{11} + p_{12}p_{21} = 0.36 + 0.12 = 0.48} \qquad \boldsymbol{p_{21}p_{11} + p_{22}p_{21} = 0.18 + 0.21 = 0.39}$$

$$\boldsymbol{p_{11}p_{12} + p_{12}p_{22} = 0.24 + 0.28 = 0.52} \qquad \boldsymbol{p_{21}p_{12} + p_{22}p_{22} = 0.12 + 0.49 = 0.61}$$

For the third step,

$$\boldsymbol{T}(3) = \boldsymbol{T}(2) \cdot \boldsymbol{T} = \begin{bmatrix} 0.48 & 0.52 \\ 0.39 & 0.61 \end{bmatrix} \begin{bmatrix} 0.6 & 0.4 \\ 0.3 & 0.7 \end{bmatrix} = \begin{bmatrix} 0.444 & 0.556 \\ 0.417 & 0.583 \end{bmatrix}$$

Again, a tree diagram or a matrix method can be used. However, a transition tree is somewhat unwieldy. Algebraically, we have

$$p_{11}(3) = p_{11}(2)p_{11} + p_{12}(2)p_{21} = 0.48(0.6) + 0.52(0.3) = 0.444$$

$$p_{12}(3) = p_{11}(2)p_{12} + p_{12}(2)p_{22} = 0.48(0.4) + 0.52(0.7) = 0.556$$

$$p_{21}(3) = p_{21}(2)p_{11} + p_{22}(2)p_{21} = 0.39(0.6) + 0.61(0.3) = 0.417$$

$$p_{22}(3) = p_{21}(2)p_{12} + p_{22}(2)p_{22} = 0.39(0.4) + 0.61(0.7) = 0.583$$

Note the repeating pattern of subscripts—simply replace p_{11} by $p_{11}(2)$ and so on.

More generally, for an arbitrary step transition:

k-Step Transition Probability Matrix

The k-step transition matrix, T(k), is defined as the matrix product

$$\mathbf{T}(k) = \mathbf{T} \cdot \mathbf{T} \cdots \mathbf{T} = \mathbf{T}^k, \qquad k = 1, 2, \ldots$$

where T is the one-step transition probability matrix.

♦ Researchers in artificial intelligence and information science have addressed the issue of learning a musical style automatically. Music files, or MIDI files, compute transition probabilities of Markov chains and use them to generate musical notes.

EXERCISES 9.2

1. Compute $\mathbf{T}(2)$ for

$$\mathbf{T} = \begin{bmatrix} 0.2 & 0.8 \\ 0.9 & 0.1 \end{bmatrix}.$$

2. Compute $\mathbf{T}(3)$ for

$$\mathbf{T} = \begin{bmatrix} 1/4 & 3/4 \\ 2/3 & 1/3 \end{bmatrix}.$$

3. A Markov chain has the transition matrix

$$\mathbf{T} = \begin{bmatrix} 0.5 & 0.3 & 0.2 \\ 0.2 & 0.4 & 0.4 \\ 0.6 & 0.1 & 0.3 \end{bmatrix}.$$

   a) What is the two-step transition matrix, $\mathbf{T}(2)$?
   b) From State 1 what is the probability of being in State 2 after two steps?
   c) From State 2 what is the probability of being in State 3 after two steps?

4. Write the transition matrix and the three-step transition probabilities for this transition probability diagram.

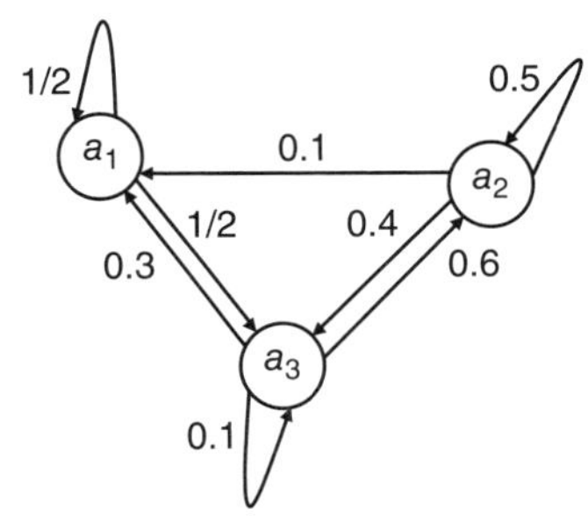

5. For the transition matrix

$$\mathbf{T} = \begin{bmatrix} 0.2 & 0.3 & 0.5 \\ 0.4 & 0.5 & 0.1 \\ 0.7 & 0.1 & 0.2 \end{bmatrix}$$

   a) determine **T(3)**.
   b) determine the probability of transition from State 1 to State 3 in three steps.
   c) determine the probability of transition from State 3 to State 2 in three steps.

6. The Barbecue Pit, a restaurant, does not open on rainy days. Past records indicate that on a rainy day the chance is 35% of rain the next day. On a day it does not rain, the chance of rain the next day is only 10%. If it rains on Tuesday, what is the chance the Barbecue Pit is closed on Thursday? On Friday?

7. An insurance company survey found, on average, that 25% of drivers in a city involved in an accident in any year also experienced one the next year. It also determined that 90% of drivers without an accident didn't have one the next year. What is the probability that a person in an accident this year will have one two years hence?

8. A grocery store seeks to estimate market share with a new advertising campaign. The store expects that 90% of its customers are loyal from week to week. It also anticipates a 40% chance that the campaign will bring customers of other grocery stores. What percentage of other store customers can the grocery store hope to acquire in three weeks with its new campaign?

9. Probabilities for changes in weekly unemployment rates are shown in the transition matrix below. The three states are from one week to the next week are "rate increases," "rate remains steady," and "rate decreases." If this week has an increase in unemployment from last week, what is the probability that unemployment rates will increase two weeks hence?

	I	S	D
Increase	0.1	0.7	0.2
Steady	0.2	0.6	0.2
Decrease	0.4	0.5	0.1

10. Is movement from State 3 to State 1 in four steps possible with this transition matrix? Explain.

$$\begin{bmatrix} 0.4 & 0.6 & 0.0 & 0.0 \\ 0.4 & 0.4 & 0.2 & 0.0 \\ 0.0 & 0.0 & 0.0 & 1.0 \\ 0.0 & 0.0 & 0.5 & 0.5 \end{bmatrix}$$

9.3 REGULAR MARKOV CHAINS

Once begun, a Markov chain's transition probability matrix, **T**, determines each subsequent state as we learned in the last section. There an initial state was specified. More likely, the initial state is not specified but rather the outcome of some other chance event.

For example, the initial state for a six-state Markov chain may be determined by a toss of a die, the number on the face of the die corresponding to the initial state.

For a fair die, each state has a 1/6 probability. This is expressed by a matrix called an **initial state probability vector** $\mathbf{X}_0 = [1/6 \;\; 1/6 \;\; 1/6 \;\; 1/6 \;\; 1/6 \;\; 1/6]$. When this (1×6) matrix is multiplied by the (6×6) transition probability matrix, the resulting (1×6) $(= (1 \times 6)(6 \times 6))$ vector is the first step in the Markov chain. Note that the elements of a state vector must sum to unity.

Example 9.3.1 Probability Vectors

Which of these are probability vectors?

a) [1/3 1/6 1/3 1/6]
b) [0.2 0.4 0.3]
c) [0 0.5 0.4 0.2]
d) [0.1 0.2 0.35 0.25 0.10]

Solution:
All components of probability vectors must be nonnegative and sum to unity.

a) It is a probability vector.
b) The components do not sum to unity. It is not a probability vector.
c) The components do not sum to unity. It is not a probability vector.
d) It is a probability vector.

Example 9.3.2 State Vectors

For the initial state vector $\boldsymbol{X_0} = [0.3 \quad 0.2 \quad 0.5]$ *and the transition matrix*

$$\boldsymbol{T} = \begin{bmatrix} 0.4 & 0.3 & 0.3 \\ 0.6 & 0.1 & 0.3 \\ 0.2 & 0.1 & 0.7 \end{bmatrix}$$

find the state vector after two transitions.

Solution:
To determine $\boldsymbol{X_1} = \boldsymbol{X_0 T}$*, multiply*

$$\boldsymbol{X_1} = \boldsymbol{X_0 T} = [0.3 \quad 0.2 \quad 0.5]\begin{bmatrix} 0.4 & 0.3 & 0.3 \\ 0.6 & 0.1 & 0.3 \\ 0.2 & 0.1 & 0.7 \end{bmatrix} = [0.34 \quad 0.16 \quad 0.50]$$

To determine $\boldsymbol{X_2} = \boldsymbol{X_1 T} = \boldsymbol{X_0 T^2}$*, first find* $\boldsymbol{T^2}$

$$\boldsymbol{T^2} = \begin{bmatrix} 0.4 & 0.3 & 0.3 \\ 0.6 & 0.1 & 0.3 \\ 0.2 & 0.1 & 0.7 \end{bmatrix}\begin{bmatrix} 0.4 & 0.3 & 0.3 \\ 0.6 & 0.1 & 0.3 \\ 0.2 & 0.1 & 0.7 \end{bmatrix} = \begin{bmatrix} 0.40 & 0.18 & 0.42 \\ 0.36 & 0.22 & 0.42 \\ 0.28 & 0.14 & 0.58 \end{bmatrix}$$

⟶

Next, multiply X_0 by T^2

$$X_0T^2 = [0.3 \quad 0.2 \quad 0.5]\begin{bmatrix} 0.40 & 0.18 & 0.42 \\ 0.36 & 0.22 & 0.42 \\ 0.28 & 0.14 & 0.58 \end{bmatrix} = [0.332 \quad 0.168 \quad 0.500]$$

Sociologists study the mobility of social, occupational, and demographic groups. The examples here are simplified illustrations of how such studies utilize Markov chains and matrices.

Example 9.3.3 Job Mobility

People's current employment status and that of a year ago had these probabilities:

Employed now, employed a year ago	*0.70*
Employed now, unemployed a year ago	*0.40*
Unemployed now, employed a year ago	*0.30*
Unemployed now, unemployed a year ago	*0.60*

Express these transition probabilities as a transition diagram and as a transition matrix.

Solution:
The transition diagram and matrix are

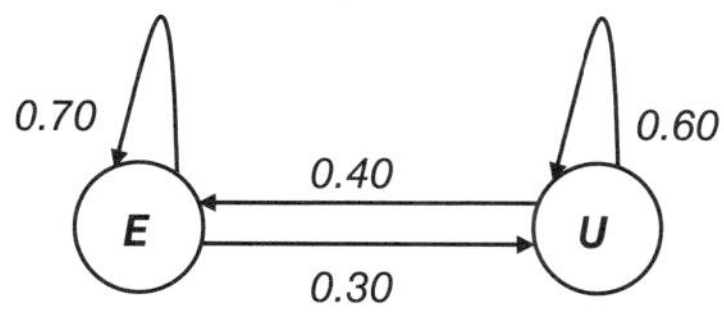

$$T = \begin{array}{c} \\ E \\ U \end{array}\begin{array}{c} \begin{array}{cc} E & U \end{array} \\ \begin{bmatrix} 0.70 & 0.30 \\ 0.40 & 0.60 \end{bmatrix}\end{array}$$

Example 9.3.4 Job Mobility (Continued)

*What is the probability that a person employed now is employed 3 years hence? Use the mobility matrix, **T**, in the above example.*

Solution:
Let $X_0 = (1 \quad 0)$ represent the initial state vector (currently employed). Then,

$$X_3 = X_0T^3 = (1 \quad 0)\begin{bmatrix} 0.583 & 0.417 \\ 0.556 & 0.444 \end{bmatrix} = (0.583 \quad 0.417)$$

There is a 0.583 probability that a person employed now is employed 3 years hence.

⟶

Compare column values of X_3 with X_1. Note that the row values are much closer to each other.

We have shown that $X_1 = X_0 T$, where T is the transition probability matrix. As we proceed, the state vector for the second step is $X_2 = X_1 T = X_0 T^2$. Thus, it leads to a theorem for state vectors.

Theorem on State Vectors

For a transition matrix T and state vectors X_k and X_{k+1} after k and $k + 1$ transitions, respectively, $X_{k+1} = X_k T$, where k = 1, 2, 3,.... So

$$X_1 = X_0 T$$

$$X_2 = X_1 T = X_0 T^2$$

$$\vdots$$

$$X_k = X_{k-1} T = X_0 T^k$$

That is, the state vector X_k that describes the system after k steps is the product of the initial state vector, X_0, and the kth power of the transition matrix T^k.

If a transition matrix **T**, or any power of **T**, has no zero elements, then the Markov chain is said to be **regular**.

Example 9.3.5 A Regular Matrix

---

*Show that transition matrix **T** is regular.*

$$T = \begin{bmatrix} 0 & 1 \\ 1/4 & 3/4 \end{bmatrix}$$

Solution:

*The transition matrix **T** has a zero element. We next examine T^2.*

$$T^2 = \begin{bmatrix} 0 & 1 \\ 1/4 & 3/4 \end{bmatrix} \begin{bmatrix} 0 & 1 \\ 1/4 & 3/4 \end{bmatrix} = \begin{bmatrix} 1/4 & 3/4 \\ 3/16 & 13/16 \end{bmatrix}$$

There are no zero elements here. The matrix is regular.

The importance of a regular Markov chain is that its long-term behavior is mathematically predictable. Probabilities are stable in the long term. One property of the stable probabilities is that they satisfy a steady-state condition in which additional transitions no longer alter the final state probabilities. This is summarized in the next theorem.

Theorem on Stable Probabilities

Let T be a transition matrix for a regular Markov chain. Then there is a unique probability vector w = $[w_1, w_2, \ldots, w_n]$ such that wT = w.

The probability vector, w, is the stable vector whose components are long-run stable probabilities of the Markov chain.

♦ Anthropologist Hans Hoffmann studied the stability of the age-grade system of Ethiopian Galla tribes. He modeled the system as a three-state finite Markov chain with transition probabilities. Sons entered at age-grades depending on the age-grade of the father. (Hoffmann H., Markov chains in Ethiopia. Explorations in Mathematical Anthropology. MIT Press; 1971.)

Example 9.3.6 Stable Probability Vector

Verify that **w** = [3/5 2/5] *is a stable (fixed) probability vector for the transition matrix*

$$\boldsymbol{T} = \begin{bmatrix} 1/3 & 2/3 \\ 1 & 0 \end{bmatrix}$$

Solution:

First, we verify that ***T*** *is a regular matrix.*

Here, $\boldsymbol{T}^2 = \begin{bmatrix} 7/9 & 2/9 \\ 1/3 & 2/3 \end{bmatrix}$ *has no zero elements, so* ***T*** *is a regular matrix.*

Next, we show that ***wT*** = ***w***.

$$\boldsymbol{wT} = [3/5 \quad 2/5]\begin{bmatrix} 1/3 & 2/3 \\ 1 & 0 \end{bmatrix} = [3/5 \quad 2/5] = \mathbf{w}.$$

Clearly, **w** as a steady-state vector is important and often sought to describe the eventual states of the chain.

Now, since $\mathbf{wT} = \mathbf{w}$, it follows that $\mathbf{wT} - \mathbf{w} = \mathbf{0}$. This can be written as $\mathbf{w}(\mathbf{T} - \mathbf{I}) = \mathbf{0}$ where **I** is an identity matrix. This is a $(1 \times n)$ $(= (1 \times n)(n \times n))$ system of equations in which all but one of the n equations are independent. Replacing any one of the equations in the $(1 \times n)$ system by the normality restriction $\mathbf{w}_1 + \mathbf{w}_2 + \cdots + \mathbf{w}_n = 1$ results in a system of independent equations to be solved for $\mathbf{w}_1, \mathbf{w}_2, \ldots, \mathbf{w}_n$.

The next example illustrates the procedure.

Example 9.3.7 Finding a Stable Probability Vector

Consider the transition probability matrix:

$$T = \begin{bmatrix} 1/2 & 2/5 & 1/10 \\ 1/10 & 3/5 & 3/10 \\ 3/10 & 3/5 & 1/10 \end{bmatrix}$$

Find the stable probability vector, $\mathbf{w}$.

Solution:

First, note that none of the elements of $\mathbf{T}$ *are zero so it is a regular Markov chain.*

Next, form $\mathbf{T} - \mathbf{I}$

$$\mathbf{T}-\mathbf{I} = \begin{bmatrix} 1/2 & 2/5 & 1/10 \\ 1/10 & 3/5 & 3/10 \\ 3/10 & 3/5 & 1/10 \end{bmatrix} - \begin{bmatrix} 1.0 & 0.0 & 0.0 \\ 0.0 & 1.0 & 0.0 \\ 0.0 & 0.0 & 1.0 \end{bmatrix}$$

$$= \begin{bmatrix} -1/2 & 2/5 & 1/10 \\ 1/10 & -2/5 & 3/10 \\ 3/10 & 3/5 & -9/10 \end{bmatrix}$$

Next, $\mathbf{w}(\mathbf{T} - \mathbf{I}) = \mathbf{0}$

$$(w_1 \quad w_2 \quad w_3) \begin{bmatrix} -1/2 & 2/5 & 1/10 \\ 1/10 & -2/5 & 3/10 \\ 3/10 & 3/5 & -9/10 \end{bmatrix} = (0 \ 0 \ 0)$$

This yields the system

$$\begin{aligned} (-1/2)w_1 + (1/10)w_2 + (3/10)w_3 &= 0 \\ (2/5)w_1 + (-2/5)w_2 + (3/5)w_3 &= 0 \\ (1/10)w_1 + (3/10)w_2 + (-9/10)w_3 &= 0 \end{aligned}$$

Now, we must include the restriction that $w_1 + w_2 + w_3 = 1$.

In the above system of equations, the first row can be eliminated (it is the negative sum of row 2 and row 3) or, indeed, any one of the three rows.

$$\begin{aligned} w_1 + \quad w_2 + \quad w_3 &= 1 \\ (2/5)w_1 + (-2/5)w_2 + (3/5)w_3 &= 0 \\ (1/10)w_1 + (3/10)w_2 + (-9/10)w_3 &= 0 \end{aligned}$$

⟶

The augmented matrix for the system to be solved becomes

$$\left[\begin{array}{ccc|c} 1 & 1 & 1 & 1 \\ 2/5 & -2/5 & 3/5 & 0 \\ 1/10 & 3/10 & -9/10 & 0 \end{array}\right]$$

Now, using rules for row reduction, to solve we find

$$\left[\begin{array}{ccc|c} 1 & 1 & 1 & 1 \\ 2/5 & -2/5 & 3/5 & 0 \\ 1/10 & 3/10 & -9/10 & 0 \end{array}\right] \to \left[\begin{array}{ccc|c} 1 & 1 & 1 & 1 \\ 0 & -4/5 & 1/5 & -2/5 \\ 0 & 1/5 & -1 & -1/10 \end{array}\right] \to$$

$$\left[\begin{array}{ccc|c} 1 & 1 & 1 & 1 \\ 0 & 1 & -1/4 & 1/2 \\ 0 & 1/5 & -1 & -1/10 \end{array}\right] \to \left[\begin{array}{ccc|c} 1 & 1 & 1 & 1 \\ 0 & 1 & -1/4 & 1/2 \\ 0 & 0 & -19/20 & -1/5 \end{array}\right] \to$$

$$\left[\begin{array}{ccc|c} 1 & 1 & 1 & 1 \\ 0 & 1 & -1/4 & 1/2 \\ 0 & 0 & 1 & 4/19 \end{array}\right] \to \left[\begin{array}{ccc|c} 1 & 1 & 0 & 15/19 \\ 0 & 1 & 0 & 21/38 \\ 0 & 0 & 1 & 4/19 \end{array}\right] \to \left[\begin{array}{ccc|c} 1 & 0 & 0 & 9/38 \\ 0 & 1 & 0 & 21/38 \\ 0 & 0 & 1 & 4/19 \end{array}\right]$$

and the vector of stable probabilities is $\mathbf{w} = [9/38 \quad 21/38 \quad 4/19]$.

It is possible to find the vector of stable probabilities for a 2×2 transition matrix using algebraic methods as in the next example.

Example 9.3.8 Finding a Stable Probability Vector (2 × 2)

Derive the stable probability vector of a 2×2 *transition matrix,* $(a,\ b \neq 0)$.

$$\boldsymbol{T} = \begin{bmatrix} a & 1-a \\ b & 1-b \end{bmatrix}$$

Solution:

Let $w = [w_1 \quad w_2]$. *Since* $\mathbf{w}\boldsymbol{T} = \mathbf{w}$ *must hold for* $\mathbf{w}$ *to be the steady-state probability vector,* $[w_1 \quad w_2]\begin{bmatrix} a & 1-a \\ b & 1-b \end{bmatrix} = [w_1 \quad w_2]$ *is equivalent to the system of equations*

$$\begin{aligned} aw_1 + \qquad bw_2 &= w_1 \\ (1-a)w_1 + (1-b)w_2 &= w_2 \end{aligned}$$

Replacing the second equation by $w_1 + w_2 = 1$, *we have the system*

$$\begin{aligned} aw_1 + bw_2 &= w_1 \\ w_1 + w_2 &= 1 \end{aligned}$$

⟶

and therefore,

$$bw_2 = (1 - a)w_1$$
$$b(1 - w_1) = (1 - a)w_1$$
$$b = (1 - a + b)w_1$$

Hence,

$$w_1 = \frac{b}{1 - a + b}$$

and

$$w_2 = \frac{1 - a}{1 - a + b}$$

♦ Utah is known for larger and stable families and genealogical records. It is an important source for genetic studies of human kinship, links to diseases (as diabetes and asthma), and characteristics (left-handedness and longevity) propagating through family trees. ("By Accident, Utah Is Providing an Ideal Genetic Laboratory," *The New York Times.*)

EXERCISES 9.3

1. If the initial state vector is [0.7 0.3] and $\mathbf{T} = \begin{bmatrix} 0.6 & 0.4 \\ 0.2 & 0.8 \end{bmatrix}$, find the state vector after three transitions.

2. If the initial state vector is [1/3 1/3 1/3] and $\mathbf{T} = \begin{bmatrix} 0.7 & 0.2 & 0.1 \\ 0.6 & 0.2 & 0.2 \\ 0.3 & 0.4 & 0.3 \end{bmatrix}$, find the state vector after two transitions.

3. If the supermarket of Exercise 9.2.8 initially has a 30% share of the market, what share might it have after 3 weeks into the campaign?

4. Which of these transition matrices are regular?

   a) $\begin{bmatrix} 0.7 & 0.3 \\ 0.2 & 0.8 \end{bmatrix}$ b) $\begin{bmatrix} 0.0 & 1.0 \\ 0.2 & 0.8 \end{bmatrix}$ c) $\begin{bmatrix} 0.25 & 0.75 \\ 0.0 & 1.0 \end{bmatrix}$

5. Determine whether these transition matrices are regular.

   a) $\begin{bmatrix} 0.7 & 0.3 & 0.0 \\ 0.0 & 0.8 & 0.2 \\ 0.5 & 0.3 & 0.2 \end{bmatrix}$ b) $\begin{bmatrix} 1.0 & 0.0 & 0.0 \\ 0.4 & 0.2 & 0.4 \\ 0.6 & 0.1 & 0.3 \end{bmatrix}$ c) $\begin{bmatrix} 0.8 & 0.1 & 0.1 \\ 0.2 & 0.1 & 0.7 \\ 0.4 & 0.5 & 0.1 \end{bmatrix}$

6. Signs at state park entrances alert visitors to the daily potential of fire danger using one of three levels: High, Medium, and Low. A record of the daily level changes is organized into a transition probability matrix for fire danger.

⟶

$$\begin{array}{c} \\ \text{High} \\ \text{Medium} \\ \text{Low} \end{array} \begin{array}{c} \begin{array}{ccc} \text{High} & \text{Medium} & \text{Low} \end{array} \\ \begin{bmatrix} 0.60 & 0.40 & 0.00 \\ 0.20 & 0.50 & 0.30 \\ 0.10 & 0.30 & 0.60 \end{bmatrix} \end{array}$$

If fire danger on Tuesday is posted as "Medium," what is the probability for a "High" on Friday?

7. In Exercise 9.2.7, assume that 5% of the drivers in the city had an accident this year. What is the probability that a driver chosen at random will have an accident next year? The following year?

8. Determine whether these are probability vectors.

   a) [0.7 0.25 0.05] b) [1/8 1/4 3/8 1/8 1/8] c) [0 0.4 0.5 0.1 0]

9. Determine the value of x for which each is a probability vector.

   a) $[0.3 \quad x \quad 0.2 \quad 0.4]$ b) $[x \quad 0 \quad x \quad 0 \quad x]$ c) $[x \quad 2x \quad 3x]$

10. In each case, verify that **w** is the stable probability vector for the regular Markov chain with matrix **T**.

    a)

$$\mathbf{w} = [7/9 \quad 2/9] \qquad \mathbf{T} = \begin{bmatrix} 0.8 & 0.2 \\ 0.7 & 0.3 \end{bmatrix}$$

    b)

$$\mathbf{w} = [0.36 \quad 0.24 \quad 0.40] \qquad \mathbf{T} = \begin{bmatrix} 0.90 & 0.10 & 0.00 \\ 0.15 & 0.60 & 0.25 \\ 0.00 & 0.15 & 0.85 \end{bmatrix}$$

11. Find the stable probability vector for the regular Markov chain with transition matrix

$$\mathbf{T} = \begin{bmatrix} 0.1 & 0.9 \\ 0.7 & 0.3 \end{bmatrix}$$

12. Find the stable probability vector for the regular Markov chain with transition matrix

$$\mathbf{T} = \begin{bmatrix} 0.7 & 0.3 \\ 0.4 & 0.6 \end{bmatrix}$$

    Calculate $\mathbf{T}^5$ and compare its row values to **w**.

13. Find the stable probability vector for the regular Markov chain with transition matrix

$$\mathbf{T} = \begin{bmatrix} 0.00 & 0.00 & 1.00 \\ 0.25 & 0.75 & 0.00 \\ 0.00 & 0.50 & 0.50 \end{bmatrix}.$$

14. Find the stable probability vector for the regular Markov chain with transition matrix

$$\mathbf{T} = \begin{bmatrix} 1/5 & 2/5 & 2/5 \\ 2/5 & 1/5 & 2/5 \\ 2/5 & 2/5 & 1/5 \end{bmatrix}.$$

15. Find the stable probability vector for the regular Markov chain with transition matrix

$$\mathbf{T} = \begin{bmatrix} 1/8 & 3/4 & 1/8 \\ 1/4 & 1/4 & 1/2 \\ 1/2 & 1/4 & 1/4 \end{bmatrix}.$$

16. Assigning Democrats to State 1 and Republicans to State 2, use the results of a century of presidential elections (1904–2004) to construct a transition probability matrix. Then:
    a) Find the probability that a Republican is in office two elections hence, if a Republican is in office now.
    b) Assuming that the transition probability matrix does not change, what is the long-run probability of a Democrat in office?

17. Repeat the previous exercise for your state's governors (if necessary, use State 3 for a third party).

9.4 ABSORBING MARKOV CHAINS

Although some Markov chains can continue indefinitely for events that are essentially endless (e.g., weather change), other Markov chains terminate naturally; for example, a gambler loses all or a mouse successfully exits a maze. Such Markov chains are said to be **absorbing**.

A state in a Markov chain is an **absorbing state** when no exit from the state is possible. If it is possible to leave a state, then it is **nonabsorbing**.

A Markov chain is an **absorbing chain** if the transition probability matrix has at least one absorbing state that can be reached from every nonabsorbing state.

Example 9.4.1 Absorbing Chains

Is this a transition matrix for an absorbing Markov chain?

$$T = \begin{bmatrix} 0.3 & 0.5 & 0.2 \\ 0.0 & 1.0 & 0.0 \\ 0.1 & 0.0 & 0.9 \end{bmatrix}$$

Solution:

An absorbing chain must have at least one absorbing state. The simplest way to detect an absorbing state in a transition probability matrix is to seek a "1" on the main diagonal. The above transition matrix has State 2 as a lone absorbing state.

→

The "1" on the main diagonal is a clue to State 2 being an absorbing state. From State 1, one can go directly to State 2 with probability 0.50. From State 3, the other nonabsorbing state, one cannot go directly to State 2. However, from State 3 it is possible to reach State 1, which then can lead to State 2.

Therefore, an absorbing state can be reached from each nonabsorbing state and the Markov chain is an absorbing state.

Note the clue to an absorbing state: a "1" on the main diagonal of the transition matrix.

Example 9.4.2 More on Absorbing Chains

Which diagrams and matrices describe absorbing Markov chains?

a)

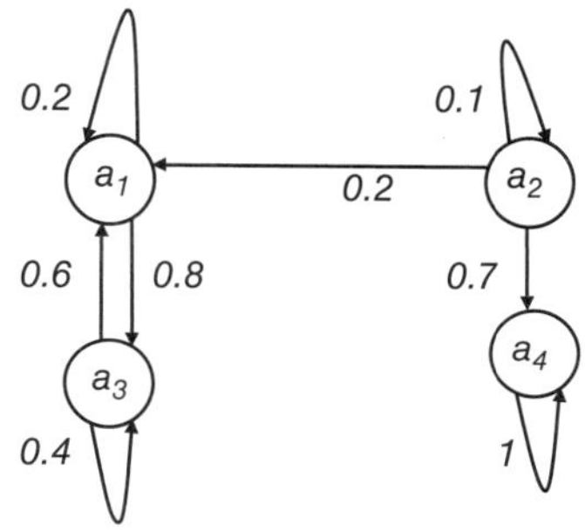

b) $$T = \begin{bmatrix} 0.1 & 0.2 & 0.3 & 0.4 \\ 0.0 & 1.0 & 0.0 & 0.0 \\ 0.4 & 0.1 & 0.3 & 0.2 \\ 0.7 & 0.0 & 0.1 & 0.2 \end{bmatrix}$$

c) $$T = \begin{bmatrix} 0.4 & 0.1 & 0.1 & 0.2 & 0.2 \\ 0.0 & 0.7 & 0.3 & 0.0 & 0.0 \\ 0.0 & 0.1 & 0.9 & 0.0 & 0.0 \\ 0.0 & 0.0 & 0.0 & 1.0 & 0.0 \\ 0.1 & 0.6 & 0.3 & 0.0 & 0.0 \end{bmatrix}$$

Solution:

a) There is an absorbing state. State 4 has a unit probability of return to that state. State 4 can be reached from State 2. However, it cannot be reached from State 1 or State 3 so the Markov chain is not absorbing.

b) From nonabsorbing States 1 and 3, absorbing State 2 is directly accessible. From nonabsorbing State 4, absorbing State 2 is accessible via States 1 or 3. The Markov chain is absorbing. Note the "1" on the diagonal as a clue.

c) From nonabsorbing State 1, absorbing State 4 is directly accessible. Also, from nonabsorbing State 5, absorbing State 4 can be reached via State 1. However, State 2 and State 3 loop without exit. Since State 4 cannot be reached from either of these nonabsorbing states, the Markov chain cannot be absorbing. Note again, the absorbing State 4 has a "1" on the diagonal.

The absorbing states of a transition probability matrix of an absorbing Markov chain have a special role. Absorbing states are relabeled and reordered to appear in rows preceding the nonabsorbing states. This new format for the transition probability matrix is called the **canonical form**.

A canonical form has four submatrices labeled, counterclockwise, **I**, **R**, **S**, and **0**.

$$\mathbf{T} = \left[\begin{array}{c|c} \mathbf{I} & \mathbf{0} \\ \hline \mathbf{R} & \mathbf{S} \end{array}\right]$$

Here, **I** is a identity matrix with dimension equal to the number of absorbing states. Matrix **R** has transition probabilities from nonabsorbing states to absorbing states. The square matrix **S** has transition probabilities among nonabsorbing states and **0** is a zero matrix.

Example 9.4.3 Canonical Form

Rewrite this transition probability matrix of an absorbing chain in canonical form.

$$T = \begin{array}{c} \\ 1 \\ 2 \\ 3 \\ 4 \end{array} \begin{array}{c} \begin{array}{cccc} 1 & 2 & 3 & 4 \end{array} \\ \begin{bmatrix} 0.0 & 0.3 & 0.2 & 0.5 \\ 0.1 & 0.4 & 0.2 & 0.3 \\ 0.0 & 0.0 & 1.0 & 0.0 \\ 0.6 & 0.0 & 0.3 & 0.1 \end{bmatrix} \end{array}$$

Solution:

State 3 is an absorbing state (indicated by 1 on the main diagonal). State 3 is repositioned to the top. It helps to actually write the "former" state labels along rows and columns as a guide.

$$\begin{array}{c} \\ 3 \\ 1 \\ 2 \\ 4 \end{array} \begin{array}{c} \begin{array}{c|ccc} 3 & 1 & 2 & 4 \end{array} \\ \left[\begin{array}{c|ccc} 1.0 & 0.0 & 0.0 & 0.0 \\ \hline 0.2 & 0.0 & 0.3 & 0.5 \\ 0.2 & 0.1 & 0.4 & 0.3 \\ 0.3 & 0.6 & 0.0 & 0.1 \end{array}\right] \end{array}$$

The identity matrix, ***I****, is (1 × 1) for the only absorbing state.*

Here, the transition matrix, ***R****, to absorbing states is* $\begin{bmatrix} 0.2 \\ 0.2 \\ 0.3 \end{bmatrix}$ *and the transition matrix to nonabsorbing states,* ***S****, is* $\begin{bmatrix} 0.0 & 0.3 & 0.5 \\ 0.1 & 0.4 & 0.3 \\ 0.6 & 0.0 & 0.1 \end{bmatrix}$.

Example 9.4.4 Canonical Form (Revisited)

Rewrite this transition probability matrix of an absorbing chain in canonical form.

$$T = \begin{array}{c} \\ 1 \\ 2 \\ 3 \\ 4 \\ 5 \end{array} \begin{array}{c} \begin{array}{ccccc} 1 & 2 & 3 & 4 & 5 \end{array} \\ \begin{bmatrix} 0.0 & 0.6 & 0.3 & 0.1 & 0.0 \\ 0.0 & 1.0 & 0.0 & 0.0 & 0.0 \\ 0.1 & 0.1 & 0.2 & 0.3 & 0.3 \\ 0.0 & 0.0 & 0.0 & 1.0 & 0.0 \\ 0.7 & 0.0 & 0.1 & 0.2 & 0.0 \end{bmatrix} \end{array}$$

Solution:

First, note the two absorbing States 2 and 4. They are repositioned in the first two rows for a canonical form. Again, "former" labels are placed along the rows and columns as a guide.

$$\begin{array}{c} \\ 2 \\ 4 \\ 1 \\ 3 \\ 5 \end{array} \begin{array}{c} \begin{array}{cc:ccc} 2 & 4 & 1 & 3 & 5 \end{array} \\ \left[\begin{array}{cc:ccc} 1.0 & 0.0 & 0.0 & 0.0 & 0.0 \\ 0.0 & 1.0 & 0.0 & 0.0 & 0.0 \\ \hdashline 0.6 & 0.1 & 0.0 & 0.3 & 0.0 \\ 0.1 & 0.3 & 0.1 & 0.2 & 0.3 \\ 0.0 & 0.2 & 0.7 & 0.1 & 0.0 \end{array}\right] \end{array}$$

*The identity matrix here is a (2×2) matrix because there are two absorbing states. Here, the **R** matrix is* $\begin{bmatrix} 0.6 & 0.1 \\ 0.1 & 0.3 \\ 0.0 & 0.2 \end{bmatrix}$ *and the **S** matrix is* $\begin{bmatrix} 0.0 & 0.3 & 0.0 \\ 0.1 & 0.2 & 0.3 \\ 0.7 & 0.1 & 0.0 \end{bmatrix}$.

*For example, the probability of reaching absorbing State 2 from nonabsorbing State 1 is the element 0.6 in the **R** matrix. Similarly, the chance of reaching nonabsorbing State 1 from nonabsorbing State 5 is 0.7 in the **S** matrix.*

A **fundamental matrix N** defined as $\mathbf{N} = (\mathbf{I} - \mathbf{S})^{-1}$ is useful in the study of absorbing chains. Its elements are the *expected (average) number* of times a nonabsorbing state is visited from another nonabsorbing state before being absorbed. That is, starting in a nonabsorbing State j, the expected number of times another nonabsorbing State k is visited before absorption is the jk^{th} element of **N**, $\mathbf{N}_{j,k}$: $j, k = 1, \ldots, n$, in an n-state Markov chain.

The expected (average) number of steps before reaching an absorbing state, when starting in a nonabsorbing state, can be found by summing the entries in the row of **N** corresponding to the starting nonabsorbing state.

A matrix **C** where **C** = **NR** determines the probability that the chain will enter a particular absorbing state having begun in a given nonabsorbing state. The row corresponding to the nonabsorbing state and column corresponding to the absorbing state gives the entry of **C** that is the associated probability.

These ideas are illustrated in the next example.

Example 9.4.5 Rat in a Maze

A rat is in Compartment #2 of the maze below and is equally likely to use any exit from a compartment. Traps set in Compartments #1 and #5 are sure to catch any rat.

a) Calculate the expected number of visits the rat enters Compartment #3 before being trapped.
b) Calculate the probability the rat will eventually be trapped in Compartment #5.

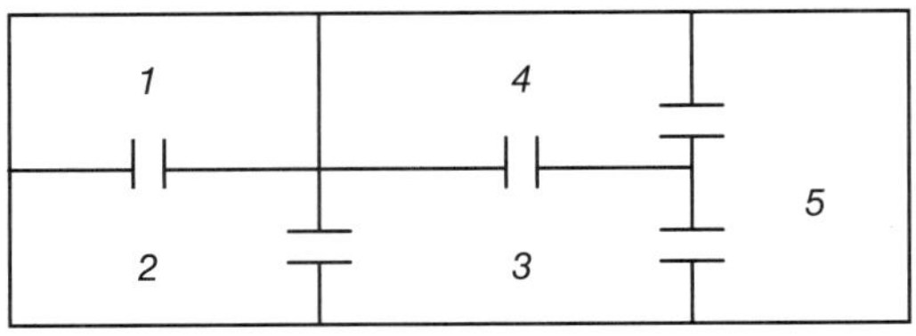

Solution:

Since Compartments #1 and #5 are absorbing states, relabel states in the order #1, #5, #2, #4, #3 to better track the rat's possible movements and have the transition matrix in canonical form.

(Note the states can be ordered several ways having the absorbing states first followed by the nonabsorbing states.)

$$\begin{array}{c} \\ 1 \\ 5 \\ 2 \\ 4 \\ 3 \end{array}\begin{array}{c} \begin{array}{ccccc} 1 & 5 & 2 & 4 & 3 \end{array} \\ \left[\begin{array}{cc:ccc} 1 & 0 & 0 & 0 & 0 \\ 0 & 1 & 0 & 0 & 0 \\ \hdashline 1/2 & 0 & 0 & 0 & 1/2 \\ 0 & 1/2 & 0 & 0 & 1/2 \\ 0 & 1/3 & 1/3 & 1/3 & 0 \end{array}\right] \end{array}$$

Next, determine $\boldsymbol{I} - \boldsymbol{S}$.

$$\boldsymbol{I} - \boldsymbol{S} = \begin{bmatrix} 1 & 0 & 0 \\ 0 & 1 & 0 \\ 0 & 0 & 1 \end{bmatrix} - \begin{bmatrix} 0 & 0 & 1/2 \\ 0 & 0 & 1/2 \\ 1/3 & 1/3 & 0 \end{bmatrix} = \begin{bmatrix} 1 & 0 & -1/2 \\ 0 & 1 & -1/2 \\ -1/3 & -1/3 & 1 \end{bmatrix}$$

Next, determine $\boldsymbol{N} = (\boldsymbol{I} - \boldsymbol{S})^{-1}$. *We use the procedure of Chapter 3 for finding inverses.*

$$\left[\begin{array}{ccc|ccc} 1 & 0 & -1/2 & 1 & 0 & 0 \\ 0 & 1 & -1/2 & 0 & 1 & 0 \\ -1/3 & -1/3 & 1 & 0 & 0 & 1 \end{array}\right] \rightarrow \left[\begin{array}{ccc|ccc} 1 & 0 & -1/2 & 1 & 0 & 0 \\ 0 & 1 & -1/2 & 0 & 1 & 0 \\ 0 & -1/3 & 5/6 & 1/3 & 0 & 1 \end{array}\right]$$

$$\rightarrow \left[\begin{array}{ccc|ccc} 1 & 0 & -1/2 & 1 & 0 & 0 \\ 0 & 1 & -1/2 & 0 & 1 & 0 \\ 0 & 0 & 2/3 & 1/3 & 1/3 & 1 \end{array}\right] \rightarrow \left[\begin{array}{ccc|ccc} 1 & 0 & -1/2 & 1 & 0 & 0 \\ 0 & 1 & -1/2 & 0 & 1 & 0 \\ 0 & 0 & 1 & 1/2 & 1/2 & 3/2 \end{array}\right]$$

$\longrightarrow$

$$\rightarrow \left[\begin{array}{ccc|ccc} 1 & 0 & -1/2 & 1 & 0 & 0 \\ 0 & 1 & 0 & 1/4 & 5/4 & 3/4 \\ 0 & 0 & 1 & 1/2 & 1/2 & 3/2 \end{array}\right] \rightarrow \left[\begin{array}{ccc|ccc} 1 & 0 & 0 & 5/4 & 1/4 & 3/4 \\ 0 & 1 & 0 & 1/4 & 5/4 & 3/4 \\ 0 & 0 & 1 & 1/2 & 1/2 & 3/2 \end{array}\right]$$

Therefore,

$$\mathbf{N} = \begin{array}{c} \\ 2 \\ 4 \\ 3 \end{array} \begin{array}{c} \begin{array}{ccc} 2 & 4 & 3 \end{array} \\ \begin{bmatrix} 5/4 & 1/4 & 3/4 \\ 1/4 & 5/4 & 3/4 \\ 1/2 & 1/2 & 3/2 \end{bmatrix} \end{array}$$

*For a), the expected number of times Compartment #3 is visited is determined from this matrix, **N**. Since the rat started in Compartment #2, the answer is 3/4 as determined by the first row and third column. This means that if a rat starts in Compartment #2, it can expect on average to have 3/4 visits to Compartment #3 before being trapped.*
*For b), **C** = **NR** is required. Therefore,*

$$\mathbf{C} = \mathbf{NR} = \begin{bmatrix} 5/4 & 1/4 & 3/4 \\ 1/4 & 5/4 & 3/4 \\ 1/2 & 1/2 & 3/2 \end{bmatrix} \begin{bmatrix} 1/2 & 0 \\ 0 & 1/2 \\ 0 & 1/3 \end{bmatrix} = \begin{array}{c} \\ \mathbf{2} \\ \mathbf{4} \\ \mathbf{3} \end{array} \begin{array}{c} \begin{array}{cc} \mathbf{1} & \mathbf{5} \end{array} \\ \begin{bmatrix} 5/8 & \boxed{3/8} \\ 1/8 & 7/8 \\ 1/4 & 3/4 \end{bmatrix} \end{array}$$

*The first column of **C** has the probabilities for being trapped in Compartment #1 while those in the second column are for being caught in Compartment #5. If the rat started in Compartment #2, the probability is 3/8 that it is trapped in Compartment #5. (Recall that the rat started in Compartment #2, which is represented by the top row and Compartment #5 is the second column – nonabsorbing states are labeled alongside the rows and absorbing states by the columns.)*

EXERCISES 9.4

1. Determine whether these are absorbing Markov chains:

a) $\begin{bmatrix} 1 & 0 \\ 1/3 & 2/3 \end{bmatrix}$ b) $\begin{bmatrix} 1/2 & 1/4 & 1/4 \\ 1 & 0 & 0 \\ 1/3 & 1/3 & 1/3 \end{bmatrix}$ c) $\begin{bmatrix} 0.1 & 0 & 0.7 & 0.2 \\ 1 & 0 & 0 & 0 \\ 0 & 0 & 1 & 0 \\ 0 & 0.4 & 0.4 & 0.2 \end{bmatrix}$

d) $\begin{bmatrix} 0.1 & 0 & 0.6 & 0.3 & 0 \\ 0 & 1 & 0 & 0 & 0 \\ 0 & 0 & 1 & 0 & 0 \\ 0 & 0.4 & 0.4 & 0.2 & 0 \\ 0.5 & 0 & 0 & 0 & 0.5 \end{bmatrix}$

2. Place each absorbing Markov chain from Exercise 9.4.1 into canonical form.
3. Identify the absorbing states for the transition diagram and determine whether it represents an absorbing Markov chain.

⟶

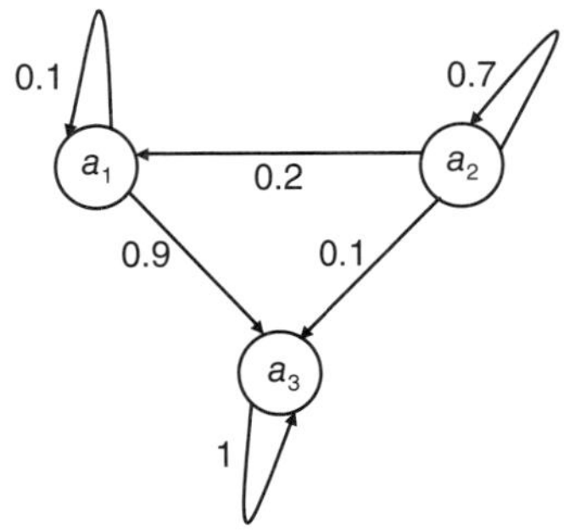

4. A person rolls a die until a "six" ends the game. What is the average number of rolls to end the game?

5. An absorbing Markov chain has the transition matrix

$$\begin{bmatrix} 0.2 & 0.4 & 0.2 & 0.2 \\ 0 & 1 & 0 & 0 \\ 0.5 & 0 & 0.3 & 0.2 \\ 0 & 0.6 & 0 & 0.4 \end{bmatrix}$$

a) Place in canonical form. b) Find the fundamental matrix.

6. An absorbing Markov chain has the transition matrix

$$\begin{bmatrix} 1/6 & 2/3 & 1/6 & 0 \\ 1/9 & 2/3 & 1/9 & 1/9 \\ 0 & 0 & 1 & 0 \\ 0 & 1/3 & 0 & 2/3 \end{bmatrix}$$

a) Place in canonical form. b) Find the fundamental matrix.

7. A rat is placed in Compartment #1 of the maze below. Traps are set in Compartments #2 and #5 and are sure to trap any rat there. The rat is equally likely to use any exit.
   a) Find the expected number of times the rat enters Compartment #3 before being trapped.
   b) Find the probability the rat will eventually be trapped in Compartment #5.

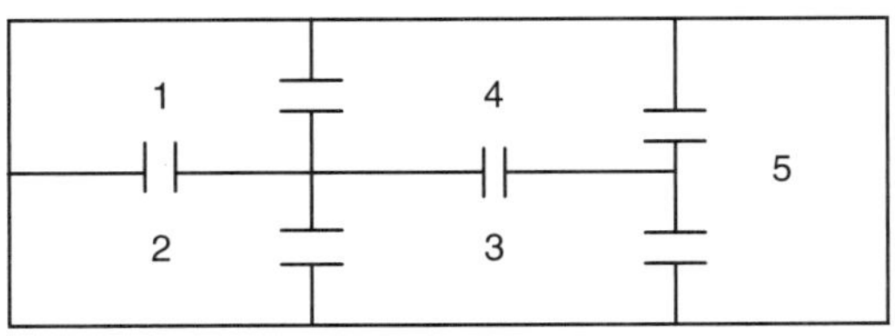

8. Suppose two people each have \$2 to play a simple die tossing game. A fair die is tossed and if even, Player A pays Player B \$1; if odd, Player B pays Player A \$1. The game continues until a player is broke. How many expected plays for a winner? *This is the famous Gambler's Ruin problem.*

HISTORICAL NOTES

◆ The development of new prosthetics is both an exciting and an uplifting topic of research. Newer models adapt to the movement of users and are able to "learn" idiosyncratic behavior.

One imagines electronic gear that nearly continuously "update" Markov chain transition matrices in the mechanical analyses of movements. Search the Internet for recent medical progress.

Andrei Andreyevich Markov (1856–1922)—was born in Russia. Studying at the University of St. Petersburg, he focused on mathematical topics including the evaluation of limits for functions, integrals, and derivatives. He studied probability theory in his later years.

Markov's most notable contribution was the development of Markov chains and the theory of stochastic processes. Although he found few practical applications for his work, they are now numerous in the biological sciences, physical sciences, and technology.

◆ Theories of stock market behavior abound. You might try your hand by forming a daily or weekly transition matrix for the percentage changes in the closing prices of a few stocks. Then apply your transition matrix to predict price movements (to keep calculations manageable you might designate states by percentage ranges: −2 to 0%, 0 to 2%.)

This, being a very simple model, it is not likely to be very effective. However, you can experiment with modifications as, for example, taking changes in market averages into account. If these ideas interest you, "A Random Walk on Wall Street" by Bernard Malakiel is a classic.

CHAPTER 9 SUPPLEMENTARY EXERCISES

1. A Markov chain has the transition matrix **T** shown below. Find the transition diagram for this Markov chain.

$$\mathbf{T} = \begin{bmatrix} 0.1 & 0.2 & 0.0 & 0.7 \\ 0.4 & 0.3 & 0.2 & 0.1 \\ 0.8 & 0.0 & 0.0 & 0.2 \\ 0.0 & 0.4 & 0.1 & 0.5 \end{bmatrix}$$

2. Determine the transition matrix for the transition diagram.

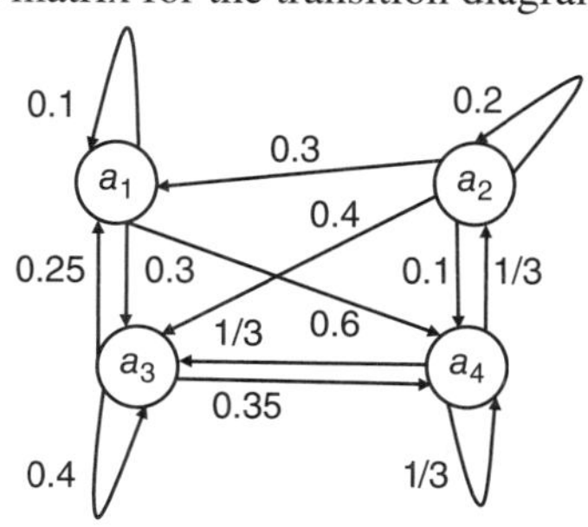

3. A Markov chain has transition matrix:

$$\mathbf{T} = \begin{bmatrix} 0.1 & 0.2 & 0.7 \\ 0.4 & 0.5 & 0.1 \\ 0.8 & 0.0 & 0.2 \end{bmatrix}$$

   a) Find the three-step transition matrix $\mathbf{T}(3)$.
   b) If the system is initially in State 2, what is the probability that it is in State 1 three observations hence?

4. A basketball player practices three-point shots. She finds that when a shot is made, the next shot is made 3/5 of the time. If she misses a shot, the next shot is made 1/4 of the time.
   Identify the appropriate states and a transition matrix for this Markov chain.

5. Determine whether a Markov chain with the given transition matrix is regular.

$$\text{a) } \mathbf{T} = \begin{bmatrix} 0.4 & 0.0 & 0.6 \\ 0.2 & 0.7 & 0.1 \\ 0.8 & 0.2 & 0.0 \end{bmatrix} \quad \text{b) } \mathbf{T} = \begin{bmatrix} 1.0 & 0.0 \\ 0.5 & 0.5 \end{bmatrix} \quad \text{c) } \mathbf{T} = \begin{bmatrix} 1/4 & 3/4 & 0 \\ 0 & 0 & 1 \\ 1/2 & 0 & 1/2 \end{bmatrix}$$

6. Determine the stable probability vector for the regular Markov chain:

$$\mathbf{T} = \begin{bmatrix} 0.25 & 0.75 \\ 0.55 & 0.45 \end{bmatrix}$$

7. Determine the stable probability vector for the regular Markov chain:

$$\mathbf{T} = \begin{bmatrix} 1/2 & 0 & 1/2 \\ 0.0 & 1/2 & 1/2 \\ 3/4 & 0 & 1/4 \end{bmatrix}$$

8. Determine the stable probability vector for the regular Markov chain:

$$\mathbf{T} = \begin{bmatrix} 0.4 & 0.6 & 0.0 & 0.0 \\ 0.2 & 0.3 & 0.5 & 0.0 \\ 0.0 & 0.2 & 0.4 & 0.4 \\ 0.0 & 0.0 & 0.7 & 0.3 \end{bmatrix}$$

9. Determine the stable probability vector for the regular Markov chain when $2a + b = 1$.

$$\mathbf{T} = \begin{bmatrix} b & a & a \\ a & b & a \\ a & a & b \end{bmatrix}$$

10. Show that the transition matrix is that of an absorbing chain. Place the matrix in canonical form.

$$\mathbf{T} = \begin{bmatrix} 1/3 & 0 & 1/3 & 1/3 \\ 1/4 & 1/2 & 1/8 & 1/8 \\ 0 & 0 & 1 & 0 \\ 0.2 & 0.1 & 0.3 & 0.4 \end{bmatrix}$$

11. Show that the transition matrix is that of an absorbing chain. Place the matrix in canonical form.

$$\mathbf{T} = \begin{bmatrix} 1/3 & 0.0 & 1/3 & 0.0 & 1/3 \\ 0.0 & 1.0 & 0.0 & 0.0 & 0.0 \\ 1/4 & 1/4 & 1/2 & 0.0 & 0.0 \\ 0.1 & 0.3 & 0.4 & 0.2 & 0.0 \\ 0.0 & 0.0 & 0.0 & 1.0 & 0.0 \end{bmatrix}$$

12. Find the vector of stable probabilities for the Markov chain with transition matrix:

$$\mathbf{T} = \begin{bmatrix} 1-a & a \\ 1 & 0 \end{bmatrix} \qquad 0 < a < 1$$

13. A rat is placed in Compartment #2 of the maze below. Traps set in Compartments #3 and #5 are sure to trap any rat there. The rat is equally likely to use any exit from a compartment.

a) Find the expected number of times the rat enters Compartment #4 before being trapped.

b) Find the probability the rat will eventually be trapped in Compartment #3.

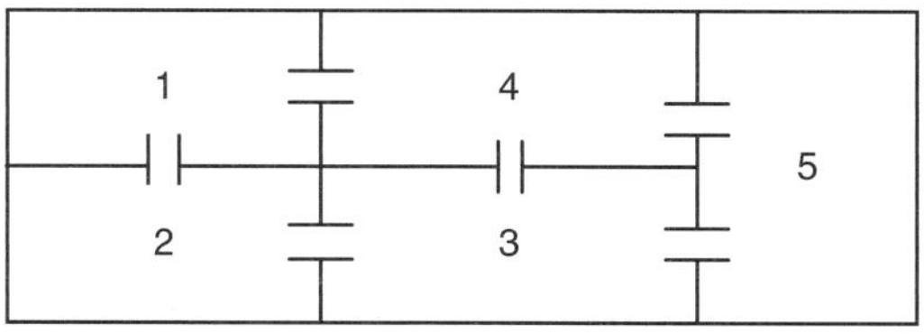

14. Two people each have \$3 to play a simple coin-tossing game. A fair coin is tossed. If "heads," Player A pays Player B \$1. If "tails," Player B pays Player A \$1. The game continues until a player is broke. What is the average duration of the game?

10 Mathematical Statistics

Finite Mathematics: Models and Applications, First Edition. Carla C. Morris and Robert M. Stark.

Companion Website: http://www.wiley.com/go/morris/finitemathematics

A **population** is the totality of possible outcomes. Data, composed of observations and measurements, are a **sample** from the population. **Mathematical statistics** is the science of inferring information about a population from analyzing sample data. For example, periodic traffic counts of highway traffic (a sample) are used for inferences about annual traffic (a population); testing several items (a sample) from a day's production (a population) to control quality; and sampling oranges yields crop information (the population).

10.1 GRAPHICAL DESCRIPTIONS OF DATA

The branch of statistics that deals with the presentation, organization, and summarization of sample data is called **descriptive statistics**. The descriptive statistics in this section is primarily graphical representations of data.

> ***Elements of Descriptive Statistics***
>
> - **A population and a sample.**
> - **A measured (or observed) variable.**
> - **Representation in tables, graphs, and numbers.**
> - **Patterns revealed and conclusions.**

Data may be **qualitative** or **quantitative.** Qualitative data are categorical and nonnumerate. Student classification (freshman, sophomore, etc.) or marital status (married, single, etc.) are examples.

Qualitative data can be classified as **nominal** or **ordinal**. Nominal data relates to names such as brands of shoes or marital status. Ordinal data can be ranked such as stars in rating television programs, restaurants, or hotels.

Quantitative data, on the other hand, is numerate. Quantitative data are measurements on **interval** or **ratio** scales. Interval data do not have an absolute zero. Celsius and Fahrenheit temperatures are from interval scales. Ratio scales have a true or absolute zero such as height, weight, and Kelvin absolute temperatures.

Example 10.1.1 Classifying Data

Classify the following data as qualitative or quantitative. Also, determine whether the data can be described on a nominal, ordinal, interval, or ratio scale.

a) The brand of a computer.
b) An outdoor Celsius temperature.
c) Your height.
d) The place finish in a car race.

Solution:

a) Brands are categorical and therefore qualitative. It is nominal data.
b) Temperature is quantitative. Celsius temperatures are measured on an interval scale. For example, 10 degrees Celsius temperature is not "twice as warm" as 5 degrees Celsius. (Kelvin temperatures relate directly to molecular movement, so 10 Kelvin is "twice as warm" as 5 Kelvin and is, therefore, on a ratio scale).
c) Height is quantitative: measured on a ratio scale.
d) Positioning is qualitative and ordinal scaled. Finishing first may be 1 second or 10 seconds, etc., before the runner up.

Statistical work begins with data. Often, graphs are used to pictorially illustrate relevant information. Graphs also help detect relationships among data. A disadvantage is that graphs can require precision.

Pie charts or **circle charts** are widely used graphical methods to display qualitative data. Pie charts should be titled and the total number of items or value clear so percentages can be converted to observational quantities. The 360° of the "pie" is divided into "slices," each representing a category of data. Slice sizes are proportionate and are arranged in decreasing size either clockwise or counterclockwise on the circle. Pie charts are obviously not useful for very many "slices."

Example 10.1.2 XYZ Monthly Expenditures

The XYZ Company plans to spend $80,000 this month. Payroll is $30,000; $20,000 for supplies; $10,000 for rent; $10,000 for advertising; and the remaining $10,000 for utilities. Depict this information on a pie chart.

⟶

Solution:

First, good practice calls for a clear explanatory title including the total dollar of expenditures. Now, "slices" can be translated into dollar amounts.

Payroll represents \$30,000/\$80,000 or 37.5% of monthly expense. Therefore, payroll on the pie chart uses 0.375(360°) or 135° for the "slice." "Slices" for other expenditures are calculated similarly for the pie or circle chart shown.

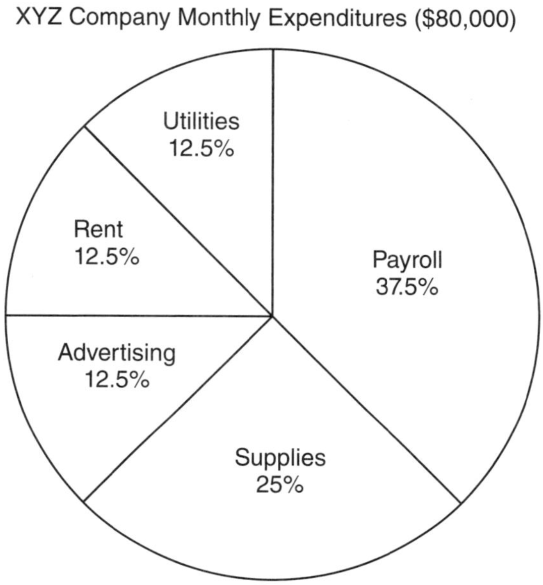

Bar charts are another widely used graphical means to describe qualitative data. The "bars" (rectangles) can be arranged vertically or horizontally. Rectangles are of equal width and separated by category. The bar chart title should clearly indicate the information depicted.

The number of times a particular event occurs is its **frequency**. Here, vertical bars are used with the vertical axis representing frequencies.

Relative frequencies, f, are frequencies divided by the total sample size, n, to normalize them to a unit scale. Therefore, relative frequencies are fractions on the unit interval. The vertical (relative frequency) axis can be scaled so the rectangular height matches graph paper gridlines.

A **relative frequency percentage** is the relative frequency multiplied by 100% ($f/n \times 100\%$). As relative frequencies sum to unity, so relative frequency percentages sum to 100%. A bar chart is the same for both relative frequencies and percentages.

Example 10.1.3 Teacher Classification

A smaller university has 150 teachers of which 30 are professors; 60 are associate professors; 45 are assistant professors; and 15 are instructors. Depict the teacher classifications with a frequency bar chart.

⟶

Solution:

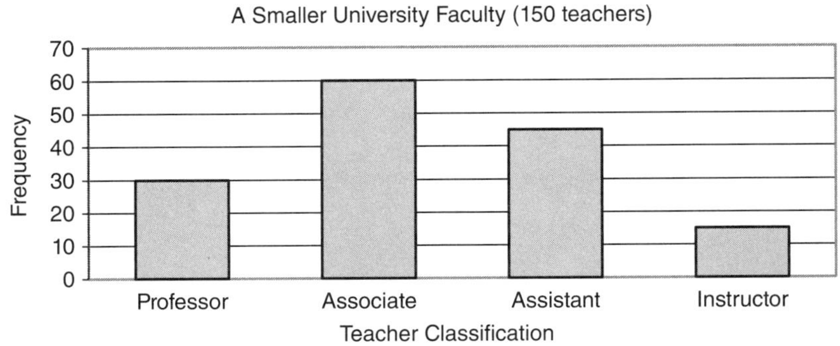

Titling includes faculty cohort size and labeled axes. While the proportion in each category is known, the actual number of teachers is not known without the population size. For instance, 1/5 being professors does not reveal their actual number unless the total (150) is indicated.

Histograms are commonly used for quantitative data. They are a special case of vertical bar graphs whose horizontal axis is a continuous variable. The number of measurements within a **class boundary** (the bar height) is the **class frequency**. Typically, the class boundaries are placed on the horizontal axis. The distance between the class boundaries (the bar width) is the **class interval**. Frequency tables are convenient to keep track of data depicted in a histogram.

Histogram Construction

1. **Form a frequency table.**
2. **Between 5 and 20 classes are used depending on the data.**
3. **A measurement must belong to exactly one class.**
4. **The vertical axis should be about 3/4 of the horizontal axis.**
5. **Plot frequencies on the vertical axis and class boundaries on the horizontal axis.**
6. **Use rectangles of equal width.**
7. **Use an informative title and label axes.**

Example 10.1.4 Annual Income at XYZ

The annual bonuses (in thousands of dollars) of 40 randomly selected employees of the XYZ Corporation appear below. Construct a frequency histogram using 27.25 as the lowest class

⟶

boundary (0.05 less than the smallest bonus, 27.3) and a class interval of 1.2 thousands of dollars.

27.3	*27.7*	*27.9*	*28.4*	*28.5*	*28.6*	*28.7*	*28.8*	*29.1*	*29.3*
29.5	*29.7*	*30.0*	*30.2*	*30.4*	*30.7*	*31.1*	*31.3*	*31.5*	*31.8*
32.1	*32.2*	*32.4*	*32.5*	*32.7*	*32.8*	*33.0*	*33.3*	*33.8*	*33.9*
34.1	*34.5*	*34.6*	*35.2*	*36.2*	*36.4*	*36.6*	*36.8*	*37.3*	*37.5*

Solution:

First, form a frequency table (check that no items have been missed or counted twice).

Class Boundaries	*Tally*	*Frequency*
27.25–28.45	||||	*4*
28.45–29.65	~~||||~~ ||	*7*
29.65–30.85	~~||||~~	*5*
30.85–32.05	||||	*4*
32.05–33.25	~~||||~~ ||	*7*
33.25–34.45	||||	*4*
34.45–35.65	|||	*3*
35.65–36.85	||||	*4*
36.85–38.05	||	*2*
		40

As the data appear to be rounded to the first (tenths) decimal place, using 0.05 less than the smallest value and increments in tenths assures that no datum is in an overlapping category.

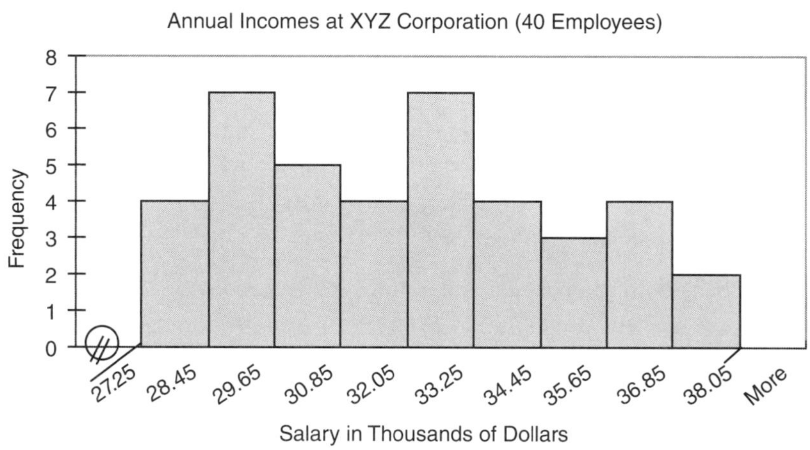

Note the gap (circled) on the salary axis.

⟶

Excel graphs, as above, tend to have values on the independent axis shifted to the left. The number 27.25 corresponds to the left-hand side of the first rectangle and 38.05 to the right-hand side of the last rectangle.

Choosing a graphical method to display information is often optional. There are few absolute rules. Regardless of choice, data should be displayed in a clear and meaningful way.

♦ An excellent data resource is the National Vital Statistics Reports (NVSS) from the Centers for Disease Control National Center for Health Statistics.

♦ A research study by Howard Markman, a Denver University psychologist, showed that the average divorce rate in cities that have a major league baseball team is 28% lower than in cities that do not. Markman says this finding is just a coincidence but perhaps there is some logic behind the results.

Another statistic about the divorce rate is that it can be predicted by knowing the number of days out of the year that women in the area can wear swimsuits (http://www.divorceform.org/mel/rgeog.html).

EXERCISES 10.1

1. Public opinion is polled for an indoor smoking ban in public places. Two thousand adults in the community were polled.
   a) What is the population of interest?
   b) What is the sample?
   c) What is the variable of interest?
2. A statistician seeks to estimate average household income. Five thousand households are polled at random and asked their annual income.
   a) What is the variable of interest?
   b) What is the population of interest?
3. For each baby born at a hospital, the following were recorded:
   a) Length b) Weight c) Gender d) Ethnicity

   Classify these variables as qualitative or quantitative.
4. Classify these data as qualitative or quantitative.
   a) The circumference of an apple.
   b) Brand of jeans you prefer.
   c) Today's outdoor temperature.
   d) Your job classification.
   e) Your highest educational level.

5. Classify and type these data as (qualitative, quantitative) and as (nominal, ordinal, interval, ratio).
   a) The height of your desk.
   b) The Kelvin temperature in your building.
   c) The brand of your calculator.
   d) The finished position of three racers at the Miller 500.
   e) The Fahrenheit temperature outdoors at noon today.

6. Complete the following table for 200 people surveyed for their highest educational attainment. Next, depict in a frequency bar chart.

Grades on an Exam	Frequency	Relative Frequency
H.S. Diploma		0.10
Associate	54	
Bachelors		0.32
Masters	46	
Doctorate		
Total	200	1.00

7. A company has 160 workers: 32 secretarial, 24 managerial, 40 custodial, and 64 others. Depict these on a frequency bar chart.

8. Forty business managers were surveyed for the highest university degree attained. Five had an associate's degree; 20 a bachelor's degree; 10 a master's degree; and the others, a doctorate. Depict in a pie chart.

9. A company spends \$240,000 each month. Payroll is \$120,000; utilities \$60,000; rent \$30,000; and the rest on advertising. Display in a pie chart.

10. Construct a frequency histogram of the 7 AM blood sugar levels for these 50 patients.

Blood Sugar Level	Frequency
59.5 – 79.5	2
79.5 – 99.5	5
99.5 – 119.5	20
119.5 – 139.5	8
139.5 – 159.5	2
159.5 – 179.5	5
179.5 – 199.5	0
199.5 – 219.5	3
219.5 – 239.5	4
239.5 – 259.5	1

11. Construct a frequency histogram for patient reaction time (in minutes) to a drug.

Reaction Time	Frequency
3.45 – 3.95	14
3.95 – 4.45	16
4.45 – 4.95	13
4.95 – 5.45	17
5.45 – 5.95	20
5.95 – 6.45	10
6.45 – 6.95	8
6.95 – 7.45	2
Total	100

12. Construct a frequency histogram for the GPA data of graduating seniors below.

GPA	Frequency
1.95 – 2.15	50
2.15 – 2.35	120
2.35 – 2.55	80
2.55 – 2.75	110
2.75 – 2.95	170
2.95 – 3.15	150
3.15 – 3.35	100
3.35 – 3.55	90
3.55 – 3.75	80
3.75 – 3.95	50
Total	1000

13. As part of a study of property values in a small town, the ages (years) of 68 buildings appears below.

10	71	64	69	85	21	26	14	26	31	81	27	86	78	73	64	72
12	73	24	56	86	18	78	62	83	54	21	64	55	65	72	58	52
84	75	85	77	22	33	26	75	74	77	58	83	67	83	13	79	31
86	65	29	73	08	25	21	07	80	81	16	19	59	64	26	77	63

a) Use statistical software to construct a frequency histogram for this data using 6.5 as the lower class boundary and a boundary width of 8.0

b) Use your graph to determine the percentage of buildings less than 71 years.

10.2 MEASURES OF CENTRAL TENDENCY AND DISPERSION

Summation notation is used as a convenient shorthand in this section. The Greek letter $\sum$ (sigma) is used to indicate the sum of variable values to the right of the symbol, beginning with the first and ending with the nth measurements. In symbols,

$$x_1 + x_2 + \cdots + x_n = \sum_{i=1}^{n} x_i \text{ or simply } \sum x_i \text{ when the limits are understood.}$$

Example 10.2.1 Journey Mileage

The mileages of five segments of a journey are 10, 24, 17, 31, and 8 miles. Calculate the summation $\sum_{i=1}^{5} x_i$, representing the total mileage.

Solution:
Here, $x_1 = 10$, $x_2 = 24$, $x_3 = 17$, $x_4 = 31$, and $x_5 = 8$. The notation indicates addition of the five segments: $10 + 24 + 17 + 31 + 8 = 90$ miles.

Central Tendency

Given data, with a group of measurements, one instinctively seeks ways to summarize them. The most common and probably the most useful is the **mean**. A simple average! People often refer to average temperature, average miles per gallon, average travel time, average test score, and so on.

Calculating a Mean

If $x_1, x_2, \ldots, x_n$ represent a group of n measurements, their average or mean is

$$\frac{x_1 + x_2 + \cdots + x_n}{n} = \frac{\sum_{i=1}^{n} x_i}{n}$$

Example 10.2.2 SAT Math Scores

The scores of six students on the math portion of the SAT are 550, 480, 630, 700, 680, and 590. What is their average SAT math score?

⟶

Solution:
The six scores are added and then divided by six (by definition) for the mean. The mean is

$$\frac{\sum_{i=1}^{6} x_i}{6} = \frac{x_1 + x_2 + x_3 + x_4 + x_5 + x_6}{6} = \frac{550 + 480 + 630 + 700 + 680 + 590}{6}$$
$$= 605$$

Example 10.2.3 Calculating Means

Determine the mean for these samples:

a) 5, 11, 16, 24, 44
b) 10, 30, 50, 70, 90, 110, 130
c) 4, 6, 2, 2, 0, 0, 0, 0, 5, 7

Solution:
To determine the mean, the values are added and divided by their number.

a) $(5 + 11 + 16 + 24 + 44)/5 = 100/5 = 20$
b) $(10 + 30 + 50 + 70 + 90 + 110 + 130)/7 = 490/7 = 70$
c) $(4 + 6 + 2 + 2 + 0 + 0 + 0 + 0 + 5 + 7)/10 = 26/10 = 2.6$

The mean may not actually be a sample value. Also, note that zeros are taken into account.

Means that arise from including all members of a population are called **population means** and are universally denoted by the Greek letter μ. It is defined by $\sum_{\text{all } x} xp(x)$, where $p(x)$ is a probability for the particular value of x. Note that when all n elements of the population are equally likely, $p(x) = 1/n$ so $\sum xp(x)$ is identical to $\sum x/n$.

Means (averages) arising from random sampling in a population are called **sample means** and denoted by $\bar{x}$. In Example 10.2.1, the mean segment length for the trip $\mu = 90/5 = 18$ miles is a population mean as the five segments constitute a population. In Example 10.2.2, the six students are a sample and therefore $\bar{x} = 605$ is a sample mean; as are the means calculated in Example 10.2.3 as the data indicated they were samples.

Among the many other ways to describe quantitative (numerical) data are the measures of **central tendency**; a clustering or centering of data about certain values.

The mean or average value is one popular measure of central tendency; a single number describes the data. The mean value need not be among the data. It may not even be a possible outcome. Two other measures of central tendency are the median and the mode.

The **median** is the middle value of a ranked data (ascending or descending). If the sample size, n, is odd, then the median is unique and is at the $(n + 1)/2$ ranked position, the middle value in the ordered data. If n is even, there are two middle values. A common practice averages the two numbers in the $(n/2)$ and $(n/2) + 1$ ranked positions. Therefore, when n is an even number, the median may not belong to the data set.

Calculating the Median

1. **Rank quantitative data (ascending or descending).**
2. **Note their number, n.**
3. **When n is odd, the median is the unique middle value in the $(n + 1)/2$ position.**
4. **When n is even, the two middle values in the $(n/2)$ and the $(n/2) + 1$ positions are averaged.**

Note: complete ranking is not necessary; only as needed for the middle values.

Example 10.2.4 Calculating Medians

Determine the median of these samples:

a) 5, 3, 9, 7, 5, 8, 1
b) 20, 50, 40, 20, 50, 70, 0, 100

Solution:
First, rank the sample. Then, identify the "middle" number.

a) Rearrange as 1, 3, 5, 5, 7, 8, 9. Since $n = 7$, an odd number the middle value is $(7 + 1)/2$ or fourth ranked position. It is 5.
b) Rearrangement yields 0, 20, 20, 40, 50, 50, 70, 100. Here, $n = 8$, so the $(8/2)$ and $(8/2) + 1$ ranked position apply. They are 40 and 50, respectively, so the median is their average, or 45.

Note: Only four numbers are needed to be ranked in a) and five in b) for the median.

The **mode** is the most frequent value in the data. The mode, if it exists, is always is in the data, unlike the mean and median, and need not be unique.

Example 10.2.5 Identifying Modes

Identify the modes.

a) 10, 20, 20, 30, 30, 40, 50, 50, 50, 50
b) 1, 1, 2, 2, 3, 3, 4, 4, 5, 5
c) 100, 100, 100, 200, 300, 400, 400, 400, 500, 600, 600

Solution:

For a), 50 occurs most often, so it is the mode.
For b), each number has the same frequency. Therefore, each is a mode. Some might say that there is no mode in this situation.
For c), 100 and 400 occur most frequently so both are modes. The data are ***bimodal****.*

Variability

Variability is another important descriptor of data; a measure of its **dispersion** or **spread**. The simplest measure of variability is the **range**; the difference between its largest and smallest values.

Example 10.2.6 Calculating Ranges

Calculate ranges:

a) 1, 9, 7, 11, 3, 10, 3, 5, 9
b) 100, 800, 600, 400, 200
c) 23, 23, 23, 25, 35, 45, 62, 19, 74

Solution:

For a), the largest is 11 and the smallest is 1. The range is $11 - 1 = 10$.
For b), the largest is 800 and the smallest is 100. The range is $800 - 100 = 700$.
For c), the largest is 74 and the smallest is 19. The range is $74 - 19 = 55$.

Although the range is easy to calculate, it is not particularly useful for statistical theory. It is easy to see why! The range depends on only two observations, and hence is independent of the sample size. An unusual value for either the largest or the smallest observation dramatically alters the range.

Another, more useful measure for variability is the **variance**. Again, as with the mean, there is a population value and a sample value. A **population variance** is denoted by the square of the lower case Greek letter sigma, σ^2, and is defined by

$$\sigma^2 = \sum_{\text{all } x} (x - \mu)^2 p(x)$$

where $p(x)$ is the probabilities for x. When all n elements of a population are equally likely, $p(x)$ is simply $1/n$. Then, the population variance is $\sigma^2 = \dfrac{\sum_{i=1}^{n} (x_i - \mu)^2}{n}$.

The **sample variance**, s^2, is defined as

$$s^2 = \frac{\sum_{i=1}^{n} (x_i - \bar{x})^2}{n - 1}$$

where $\bar{x}$, the sample mean, replaces the unknown population mean μ. A sample variance is derived entirely from the sample data. An equivalent form for the sample variance is:

$$s^2 = \frac{\sum_{i=1}^{n} x_i^2 - 2n\bar{x}^2 + n\bar{x}^2}{n - 1} = \frac{\sum_{i=1}^{n} x_i^2 - n\bar{x}^2}{n - 1} = \frac{\sum_{i=1}^{n} x_i^2 - \dfrac{\left(\sum_{i=1}^{n} x_i\right)^2}{n}}{n - 1}$$

Calculating Sample Variance

The sample variance s^2 is defined as

$$s^2 = \frac{\sum_{i=1}^{n} (x_i - \bar{x})^2}{n - 1}$$

It is often easier to use the equivalent form:

$$s^2 = \frac{\sum_{i=1}^{n} x_i^2 - \dfrac{\left(\sum_{i=1}^{n} x_i\right)^2}{n}}{n - 1} \quad \textbf{(shortcut)}$$

Example 10.2.7 Calculating Sample Variances

Calculate the variance using both the definition and the shortcut for the sample 5, 11, 16, 24, and 44.

Solution:
The data sum is 100, and since the sample size is 5 the sample mean $\bar{x} = 20$.

$$\sum_{i=1}^{5} x_i = 5 + 11 + 16 + 24 + 44 = 100, \left(\sum_{i=1}^{5} x_i\right)^2 = 100^2 = 10{,}000$$

$$\sum_{i=1}^{5} x_i^2 = 5^2 + 11^2 + 16^2 + 24^2 + 44^2 = 2914$$

Using the definition, $s^2 = \dfrac{(5-20)^2 + (11-20)^2 + \cdots + (44-20)^2}{5-1} = \dfrac{914}{4} = 228.5.$

Using the shortcut expression,

$$s^2 = \frac{2914 - \dfrac{(100)^2}{5}}{4} = \frac{914}{4} = 228.5$$

Note that $\sum_{i=1}^{5} x_i^2$ *is not the same as* $\left(\sum_{i=1}^{5} x_i\right)^2$. *The first expression is a sum of the squares and the second is the square of a sum. The shortcut is sometimes handier for calculation when the mean or the data have decimal values.*

It is useful to think of the variance as an average of square fluctuations about the mean. Note that all data are used in the variance while the range has only the two extreme values. For this reason, the variances are preferred as measures of variability.

The square root of the variance, the **standard deviation** or **standard error**, is used in applications. There is a **population standard deviation**, σ, and a **sample standard deviation**, s.

Mean, Variance, and Standard Deviation

SAMPLE		POPULATION	
Mean	$\bar{x}$	**Mean**	μ
Variance	s^2	**Variance**	σ^2
Standard deviation	s	**Standard deviation**	σ

Note: Greek letters are customary for population values.

Example 10.2.8 Variance and Standard Deviation

Calculate the variance and standard deviation for data in Examples 10.2.1 and 10.2.2.

Solution:

For Example 10.2.1, the data are 10, 24, 17, 31, and 8 with $\mu = 18$.

The population variance, using its definition, is

$$\sigma^2 = \frac{(10-18)^2 + (24-18)^2 + (17-18)^2 + (31-18)^2 + (8-18^2)}{5} = 74.$$

The population standard deviation is $\sigma = \sqrt{74} = 8.60$.

In Example 10.2.2, a sample of six measurements of 550, 480, 630, 700, 680, and 590 has a sample mean $\bar{x} = 605$. *The sample variance, using its definition, is*

$$s^2 = \frac{(550-605)^2 + (480-605)^2 + \cdots + (590-605)^2}{6-1} = \frac{34{,}150}{5} = 6830$$

and the sample standard deviation, s, is $\sqrt{6830} = 82.64$.

♦ Policing statistically? Many police departments in larger cities deploy police statistically. One management program, called **Compstat,** assigns police based on crime statistics.

Empirical Rule

A useful **empirical rule** for larger data sets is that about 68% of the data lie within one standard deviation about the sample mean; 95% of measurements lie within two standard deviations; and virtually all lie within three standard deviations.

Although these percentages derive from a **Normal Distribution**, they apply remarkably to non-normal, "mound-shaped", distributions. The empirical rule is a rough indication of the normality of data.

EXERCISES 10.2

1. A sample consists of: 5, 10, 25, 30, and 30.
   Determine

   a) $\sum_{i=1}^{5} x_i$ b) $\sum_{i=1}^{5} x_i - 5$ c) $\sum_{i=1}^{5} (x_i - 4)$ d) $\sum_{i=1}^{5} x_i^2$ e) $\sum_{i=1}^{5} (x_i - 10)^2$

2. A sample consists of: 5, 8, 12, 15, 19, and 25.
Determine

a) $\sum_{i=1}^{6} x_i$ b) $\sum_{i=1}^{6} x_i + 3$ c) $\sum_{i=1}^{6} (x_i - 2)$ d) $\sum_{i=1}^{6} x_i^2 - 20$ e) $\sum_{i=1}^{6} x_i^3$

3. Calculate the mean, median, and mode:

20, 30, 40, 40, 30, 10, 40, 60, 70, and 50

4. Calculate the mean, median, and mode:

8, 5, 11, 9, 1, 4, 0, 7, 7, 3, and 11

5. Calculate the median:

378, 408, 382, 406, 400, 404, 361, 376, 364, 371, 382, 362, 371, 388, and 381

6. Calculate the median:

78, 58, 82, 16, 40, 44, 36, 6, 64, 71, 12, 62, 71, 88, 81, and 10

7. Calculate the means for these summary statistics:

a) $n = 10$ and $\sum x = 200$
b) $n = 4$ and $\sum x = 4040$
c) $n = 50$ and $\sum x = 20$
d) $n = 25$ and $\sum x = 700$

8. Calculate the variance and standard deviation for a sample with
a) $n = 10$, $\sum x^2 = 95$, and $\sum x = 20$.
b) $n = 20$ and $\sum (x_i - \bar{x})^2 = 950$.
c) $n = 50$, $\bar{x} = 30$, and $\sum x^2 = 46{,}290$.

9. Calculate the range, variance, and standard deviation for data in

a) Exercise 10.2.3
b) Exercise 10.2.4

10. Determine six measures of central tendency and dispersion for the sample:

16, 21, 33, 42, 18, 33, 33, 24, and 59

11. Determine six measures of central tendency and dispersion for the sample:

15, 13, 30, 25, 17, 13, 21, 19, 30, 67, 23, 19, 43, 47, and 23

12. Determine six measures of central tendency and dispersion for the sample:

655	568	595	615	599	628	692	688	706	707
599	599	662	576	679	600	564	595	583	593
668	627	680	702	666	695	684	688	701	706

13. For the sample of 40 measurements:

13.6	8.7	9.4	5.4	7.6	6.8	5.9	9.6	8.3	5.8
6.7	6.8	6.9	5.8	5.9	8.4	6.8	7.3	7.2	7.9
10.2	10.0	8.9	9.1	6.9	8.9	7.5	7.6	8.0	8.4
9.8	8.1	7.7	5.7	6.0	6.1	7.5	7.4	8.4	8.5

a) Calculate the mean and standard deviation.
b) How many measurements lie within the intervals $\bar{x} \pm s$, $\bar{x} \pm 2s$, and $\bar{x} \pm 3s$? Does the data seem to follow the empirical rule?

10.3 THE UNIFORM DISTRIBUTION

When all values of a random variable have a common frequency or probability they are said to follow the **uniform distribution**. The graph of the uniform frequency function is a rectangle.

When probabilities are represented as areas on a probability distribution curve, the total area is unity. Thus, the area of the uniform distribution rectangle is unity. If the random variables range from $x = c$ to $x = d$, then the uniform distribution $f(x) = 1/(d - c)$, $c \leq x \leq d$, is the height of the rectangle. The mean of a uniform random variable, $(c + d)/2$, is the midpoint of c and d.

Uniform Distribution

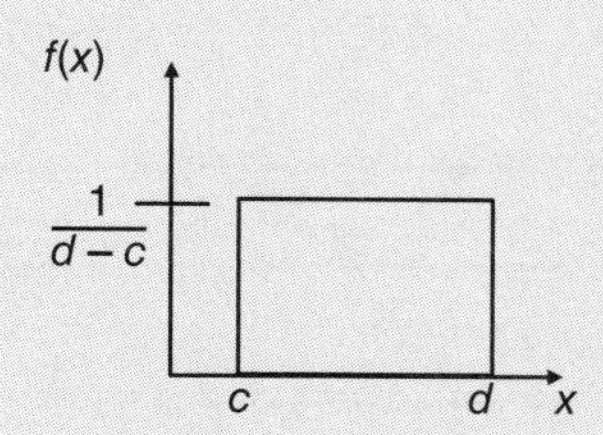

$$f(x) = \frac{1}{d - c} \qquad c \leq x \leq d$$

$$\mu = \frac{c + d}{2}$$

$$\sigma^2 = \frac{(d - c)^2}{12}$$

The probability of a single value is zero, as with any continuous distribution; as for a rectangle of zero width.

Example 10.3.1 Uniform Distribution

A random variable has a uniform distribution with $c = 20$ and $d = 70$:

a) Calculate μ and σ^2.
b) Calculate these probabilities:

$P(x \geq 60)$ $\quad P(25.5 \leq x \leq 55.5)$ $\quad P(15 \leq x \leq 60)$ $\quad P(x = 35)$

Solution:

a) $\mu = (20 + 70)/2 = 45$ and $\sigma^2 = (70 - 20)^2/12 = 2500/12$
b) To calculate probabilities, use a sketch of the distribution.

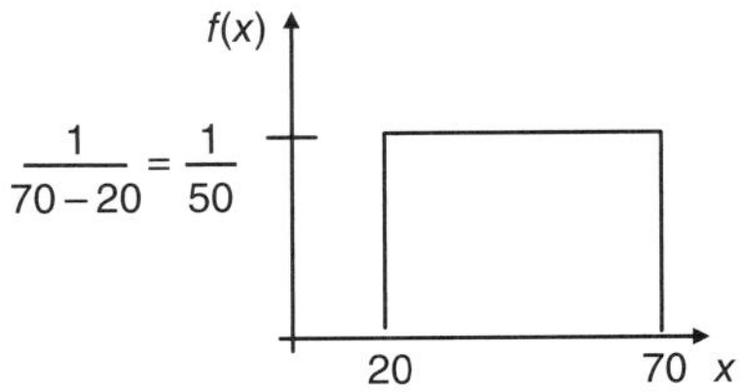

The $P(x \geq 60)$ is the rectangular area bounded by $60 \leq x \leq 70$. The probability is $\frac{70 - 60}{50} = \frac{10}{50}$ *or 0.20.*

The $P(25.5 \leq x \leq 55.5)$ is the rectangular area between 25.5 and 55.5. The probability is $\frac{55.5 - 25.5}{50} = \frac{30}{50}$ *or 0.60.*

For $P(15 \leq x \leq 60)$, first note that the probability associated with values of x between 15 and 20 is zero. We seek the area between $x = 20$ and $x = 60$. The probability is $\frac{60 - 20}{50} = \frac{40}{50}$ *or 0.80.*

The probability that a continuous variable takes on a specific value is zero.
So, $P(x = 30) = 0$.

Example 10.3.2 Uniform Distribution Probabilities

Consider a uniform distribution with $c = 15$ and $d = 35$. Find a value a such that

a) $P(x < \mathbf{a}) = 0.4$
b) $P(x > \mathbf{a}) = 0.95$
c) $P(17 \leq x \leq \mathbf{a}) = 0.60$
d) $P(\mathbf{a} \leq x \leq 20) = 0.75$

→

Solution:
To calculate the probabilities, use the sketch below.

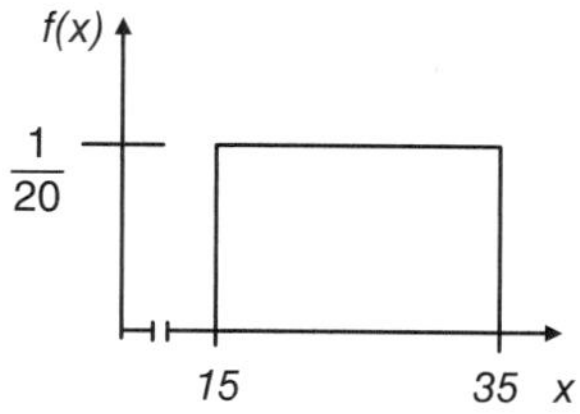

a) $P(x < \mathbf{a}) = 0.40$ *is the fraction of the unit rectangle between 15 and* $\mathbf{a}$. *The resulting equation is*

$$\frac{\mathbf{a} - 15}{35 - 15} = \frac{\mathbf{a} - 15}{20} = 0.40 \quad \text{so} \quad \mathbf{a} - 15 = 8 \quad \text{and} \quad \mathbf{a} = 23$$

b) $P(x > \mathbf{a}) = 0.95$ *is the fraction of the unit rectangle between* $\mathbf{a}$ *and 35. The resulting equation is*

$$\frac{35 - \mathbf{a}}{20} = 0.95$$

$$35 - \mathbf{a} = 19 \quad \text{so} \quad \mathbf{a} = 16$$

c) $P(17 \le x \le \mathbf{a}) = 0.60$ *is the fraction of the unit rectangle between 17 and* $\mathbf{a}$. *The resulting equation is*

$$\frac{\mathbf{a} - 17}{20} = 0.60$$

$$\mathbf{a} - 17 = 12 \quad \text{so} \quad \mathbf{a} = 29$$

d) $P(\mathbf{a} \le x \le 20) = 0.75$ *is the fraction of the unit rectangle between* $\mathbf{a}$ *and 20. The resulting equation is*

$$\frac{20 - \mathbf{a}}{20} = 0.75$$

$$20 - \mathbf{a} = 15$$

$$\mathbf{a} = 5$$

This is not possible as $15 < x < 35$ *so values less than 15 cannot increase the probability.*

EXERCISES 10.3

1. For a uniform random variable with $c = 10 \le x \le d = 25$,
   a) find $f(x)$.
   b) calculate the mean, variance, and standard deviation of x.

2. For the uniform distribution in Exercise 10.3.1,
   a) calculate $P(\mu - \sigma \leq x \leq \mu + \sigma)$.
   b) calculate $P(\mu - 1.5\sigma \leq x \leq \mu + 1.5\sigma)$.

3. Assume a uniform random variable with $c = 15 \leq x \leq d = 30$.
   Find the probabilities.

   a) $P(x > 17)$
   b) $P(x = 19)$
   c) $P(x \geq 12)$
   d) $P(x < 27.5)$
   e) $P(20 \leq x \leq 25)$
   f) $P(23 \leq x \leq 33)$

4. Assume a uniform random variable with $c = 100 \leq x \leq d = 200$.
   Find the probabilities.

   a) $P(x > 117)$
   b) $P(x = 139)$
   c) $P(x \geq 158)$
   d) $P(x < 162)$
   e) $P(140 \leq x \leq 195)$
   f) $P(123 \leq x \leq 203)$

5. Assume a uniform random variable with $c = 100 \leq x \leq d = 150$.
   Calculate the value of $\boldsymbol{a}$ (if possible) such that

   a) $P(x > \boldsymbol{a}) = 0.30$
   b) $P(x > \boldsymbol{a}) = 0.10$
   c) $P(\boldsymbol{a} < x < 140) = 0.20$
   d) $P(x < \boldsymbol{a}) = 0.80$
   e) $P(150 \leq x \leq \boldsymbol{a}) = 0.40$
   f) $P(128 \leq x \leq \boldsymbol{a}) = 0.70$

6. Assume a uniform random variable with $c = 20 \leq x \leq d = 70$.
   Calculate the value of $\boldsymbol{a}$ (if possible) such that

   a) $P(x > \boldsymbol{a}) = 0.40$
   b) $P(x < \boldsymbol{a}) = 0.20$
   c) $P(\boldsymbol{a} < x < 60) = 0.60$
   d) $P(x < \boldsymbol{a}) = 0.80$
   e) $P(30 \leq x \leq \boldsymbol{a}) = 0.50$
   f) $P(32 \leq x \leq \boldsymbol{a}) = 0.90$

7. A bakery seeks a thickness of about 0.30 inch for its rolled pie crust dough. Its pie crust roller rolls a random thickness between 0.25 and 0.35 inches. The random thickness of rolled dough, x, has a uniform distribution.

   a) Find the mean and standard deviation for the thickness of rolled dough.
   b) Find $P(0.28 < x < 0.32)$.
   c) Find $\boldsymbol{a}$ so that $P(0.27 < x < \mathbf{a}) = 0.40$.

10.4 THE NORMAL DISTRIBUTION

Imagine a marathon with thousands of runners; their run times recorded as each crosses the finish line. Forming a frequency diagram of run times, it is virtually a mathematical certainty that it resembles a **bell-shaped curve**!

♦ Indeed, such a frequency diagram for the New York City Marathon often appears in the *New York Times* after the event.

Similarly, in a large randomly chosen group of people, one can expect that a frequency diagram of their heights, shoe sizes, arm lengths, and so on form a bell-shaped curve. It is known as the **Normal Distribution** or **Gaussian distribution.**

The normal distribution is the most important among probability distributions. It faithfully describes an immense variety of phenomena – natural, technologic, and societal.

To better illustrate many interesting applications and statistical uses of the normal distribution, it is necessary to know formal and numerical workings of this amazing and ubiquitous probability distribution.

The figure illustrates several bell curves for varied means and standard deviations.

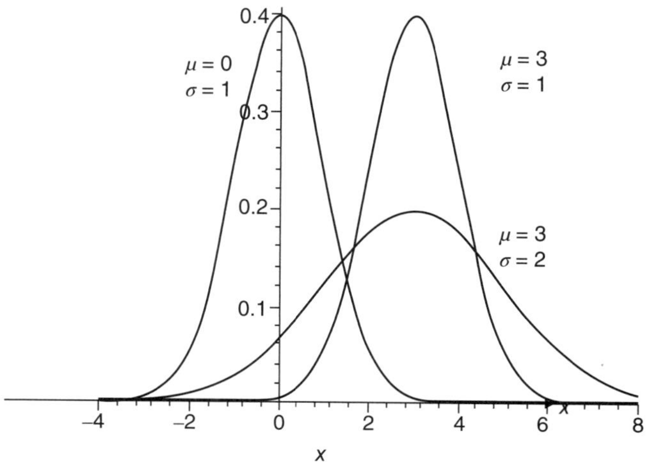

While appearances differ, each is a normal distribution. Note that each is symmetric about its mean μ at the maximum height (the mode).

Every normal distribution is uniquely defined by its mean μ and variance σ^2. The mean and mode coincide for the normal distribution. The variance is a measure of the "spread" or dispersion of the bell curve, as the figure illustrates.

A mathematical representation of the normal distribution, somewhat complicated, appears next. It is not necessary to memorize!

Probability Distribution for a Normal Random Variable X

$$f(x) = \frac{1}{\sigma\sqrt{2\pi}} e^{-(1/2)[(x-\mu)/\sigma]^2}$$

where x = Random variable ($-\infty < x < \infty$)

μ = Mean

σ = Standard deviation

π and e are transcendental numbers (Section 1.6).

Areas under the normal curve correspond to probabilities. For example, the probability that the random variable x takes a value between a and b, $P(a \leq x \leq b)$, is the shaded area in the figure.

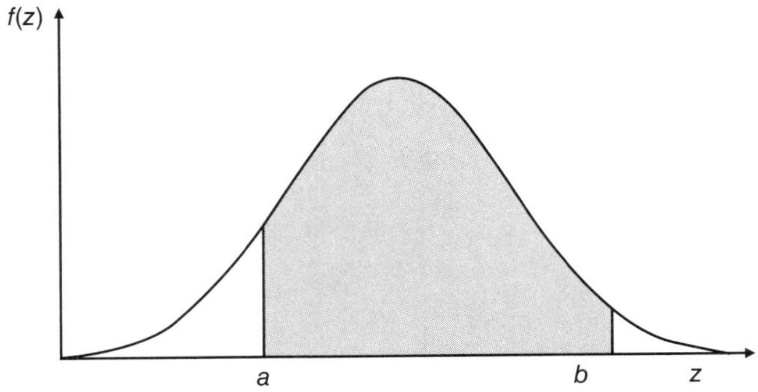

The Standard Normal Distribution

The normal distribution for $\mu = 0$ and $\sigma = 1$ is called a **Standard Normal Distribution**. It is a special normal distribution for the random variable Z.

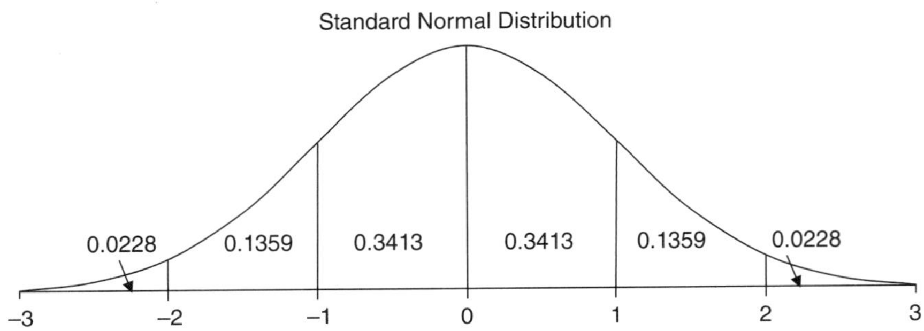

Standard Normal Random Variable Z

$$f(z) = \frac{1}{\sqrt{2\pi}} e^{-(1/2)z^2} \qquad (-\infty < z < \infty)$$

where *Z* is a Standard Normal random variable with $\mu = 0$, $\sigma = 1$

Values of the standard normal distribution, corresponding to areas under the normal curve, have been tabulated. Fortunately, it is not necessary to calculate areas under the curve to obtain probabilities; they comprise the **Standard Normal Table**. Note the arrangement of table values – second decimal place values of z are set horizontally. The probability that z lies between 0 and 1.23, for example, is 0.3907.

A few important features of the table require explanation. First, since the area under every probability distribution is unity, the area for $0 \leq z \leq \infty$ is 1/2. Likewise, the area for $-\infty < z \leq 0$ is also 1/2.

Second, for negative values of z use the symmetry of the normal distribution. The probability that z is between 0.00 and −1.23, for example, is 0.3907; exactly the same as for z between 0.00 and 1.23.

Third, while z can, in theory, take very large values, the table shows that the area bounded by $0 \leq z \leq 3.09$ is 0.4990 or nearly (1/2)! Therefore, in fact, larger values of z, while theoretically possible, are most unlikely. These features are illustrated in the examples that follow. The box summarizes these properties.

To Obtain Probabilities for a Standard Normal Random Variable

1. **Sketch a normal curve noting the origin (mean).**
2. **Shade the area of interest under the curve.**
3. **Obtain areas (probabilities) from the standard normal table. Use the symmetry of the normal distribution for negative values of z.**
4. **Remember these properties of the standard normal random variable:**
   a) $P(z \geq 0) = (z \leq 0) = 0.50$
   b) $P(-a \leq z \leq 0) = P(0 \leq z \leq a)$
   c) $P(z = a) = 0$ where $a > 0$, $\mu = 0$, and $\sigma = 1$.

Example 10.4.1 Standard Normal Probabilities

What probabilities correspond to one, two, and three standard deviations from the mean for a standard normal variable?

Solution:

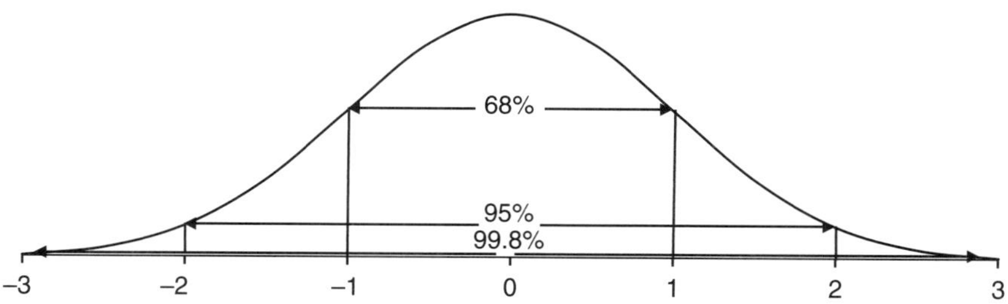

The table indicates 0.3413 as the area of the standard normal curve for $0 \leq z \leq 1$. The area between −1 and +1 standard deviations is twice that value or 0.6826, about 68% of the area under the curve. Similarly, between $z = 0$ and $z = 2$, the area is 0.4772.

So $P(-2 < z < 2) = 0.4772 + 0.4772 = 0.9544$ or about 95% of the area.

Finally, $P(-3 < z < 3) = 0.4990 + 0.4990 = 0.9980$ or nearly the entire area.

⟶

Recall these percentages from the empirical rule cited earlier. Note that the strictness of inequalities, or lack of it, is irrelevant for continuous distributions as the probability for any single value is zero.

Standard Normal Probabilities

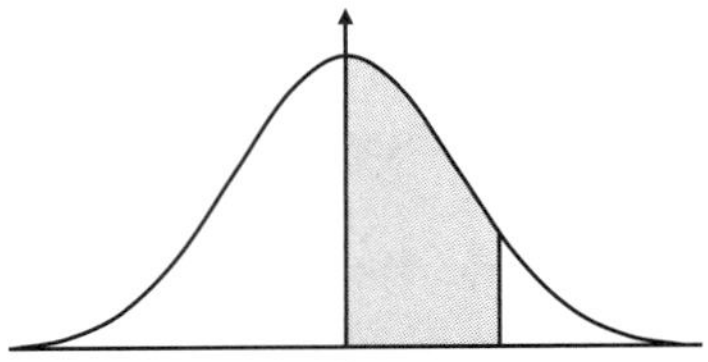

Z	0.00	0.01	0.02	0.03	0.04	0.05	0.06	0.07	0.08	0.09
0.0	0.0000	0.0040	0.0080	0.0120	0.0160	0.0199	0.0239	0.0279	0.0319	0.0359
0.1	0.0398	0.0438	0.0478	0.0517	0.0557	0.0596	0.0636	0.0675	0.0714	0.0753
0.2	0.0793	0.0832	0.0871	0.0910	0.0948	0.0987	0.1026	0.1064	0.1103	0.1141
0.3	0.1179	0.1217	0.1255	0.1293	0.1331	0.1368	0.1406	0.1443	0.1480	0.1517
0.4	0.1554	0.1591	0.1628	0.1664	0.1700	0.1736	0.1772	0.1808	0.1884	0.1879
0.5	0.1915	0.1980	0.1985	0.2019	0.2054	0.2088	0.2123	0.2157	0.2190	0.2224
0.6	0.2257	0.2291	0.2324	0.2357	0.2389	0.2422	0.2454	0.2486	0.2517	0.2549
0.7	0.2580	0.2611	0.2642	0.2673	0.2704	0.2734	0.2764	0.2794	0.2823	0.2852
0.8	0.2881	0.2910	0.2939	0.2967	0.2995	0.3023	0.3051	0.3078	0.3106	0.3133
0.9	0.3159	0.3186	0.3212	0.3238	0.3264	0.3289	0.3315	0.3340	0.3365	0.3389
1.0	0.3413	0.3438	0.3461	0.3485	0.3508	0.3531	0.3554	0.3577	0.3599	0.3621
1.1	0.3643	0.3665	0.3686	0.3708	0.3729	0.3749	0.3770	0.3790	0.3810	0.3830
1.2	0.3849	0.3869	0.3888	0.3907	0.3925	0.3944	0.3962	0.3980	0.3997	0.4015
1.3	0.4032	0.4049	0.4066	0.4082	0.4099	0.4115	0.4131	0.4147	0.4162	0.4177
1.4	0.4192	0.4207	0.4222	0.4236	0.4251	0.4265	0.4279	0.4292	0.4306	0.4319
1.5	0.4332	0.4345	0.4357	0.4370	0.4382	0.4394	0.4406	0.4418	0.4429	0.4441
1.6	0.4452	0.4463	0.4474	0.4484	0.4495	0.4505	0.4515	0.4525	0.4535	0.4545
1.7	0.4554	0.4564	0.4573	0.4582	0.4591	0.4599	0.4608	0.4616	0.4625	0.4633
1.8	0.4641	0.4649	0.4656	0.4664	0.4671	0.4678	0.4686	0.4693	0.4699	0.4706
1.9	0.4713	0.4719	0.4726	0.4732	0.4738	0.4744	0.4750	0.4756	0.4761	0.4767
2.0	0.4772	0.4778	0.4783	0.4788	0.4793	0.4798	0.4803	0.4808	0.4812	0.4817
2.1	0.4821	0.4826	0.4830	0.4834	0.4838	0.4842	0.4846	0.4850	0.4854	0.4857
2.2	0.4861	0.4864	0.4868	0.4871	0.4875	0.4878	0.4881	0.4884	0.4887	0.4890
2.3	0.4893	0.4896	0.4898	0.4901	0.4904	0.4906	0.4909	0.4911	0.4913	0.4916
2.4	0.4918	0.4920	0.4922	0.4925	0.4927	0.4929	0.4931	0.4932	0.4934	0.4936
2.5	0.4938	0.4940	0.4941	0.4943	0.4945	0.4946	0.4948	0.4949	0.4951	0.4952
2.6	0.4953	0.4955	0.4956	0.4957	0.4959	0.4960	0.4961	0.4962	0.4963	0.4964
2.7	0.4965	0.4966	0.4967	0.4968	0.4969	0.4970	0.4971	0.4972	0.4973	0.4974
2.8	0.4974	0.4975	0.4976	0.4977	0.4977	0.4978	0.4979	0.4979	0.4980	0.4981
2.9	0.4981	0.4982	0.4982	0.4983	0.4984	0.4984	0.4985	0.4985	0.4986	0.4986
3.0	0.4987	0.4987	0.4987	0.4988	0.4988	0.4989	0.4989	0.4989	0.4990	0.4990

Example 10.4.2 Standard Normal Probabilities (Continued)

Calculate the indicated probabilities for the standard normal random variable z. Illustrate in a sketch.

a) $P(0 < z < 1.56)$ b) $P(z > -1.45)$ c) $P(-1.31 < z < 2.93)$

Solution:

a) *The area of interest is shaded on the sketch of the standard normal curve. The tabulated probability for $z = 1.56$ is 0.4406.*

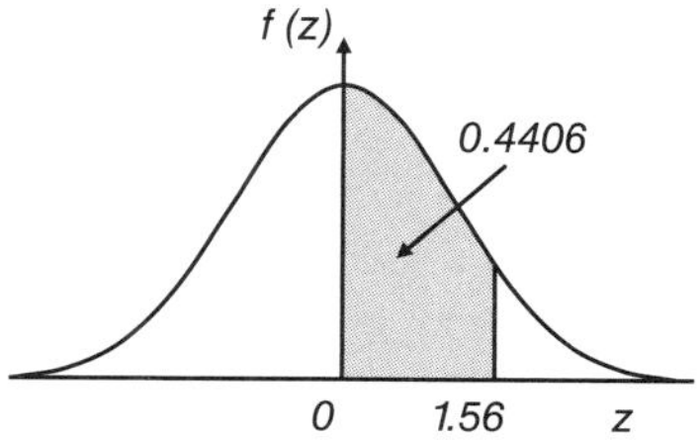

b) *For $z > -1.45$, the probability is the area between $-1.45 < z < 0$ and is added to the entire right-side area of 0.5000. The desired probability is $0.4265 + 0.5000 = 0.9265$, as shown.*

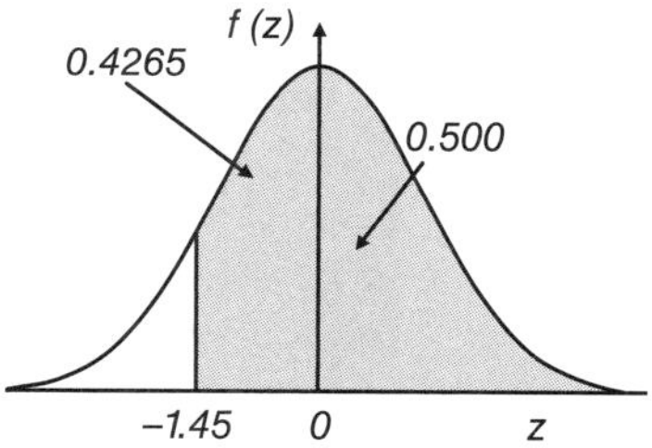

c) *Since z values are opposite in sign (on opposite sides of zero), the probabilities associated with $-1.31 \leq z \leq 0$ and $0 \leq z \leq 2.93$ must be added. Here, we have $0.4049 + 0.4983 = 0.9032$, as shown in the sketch.*

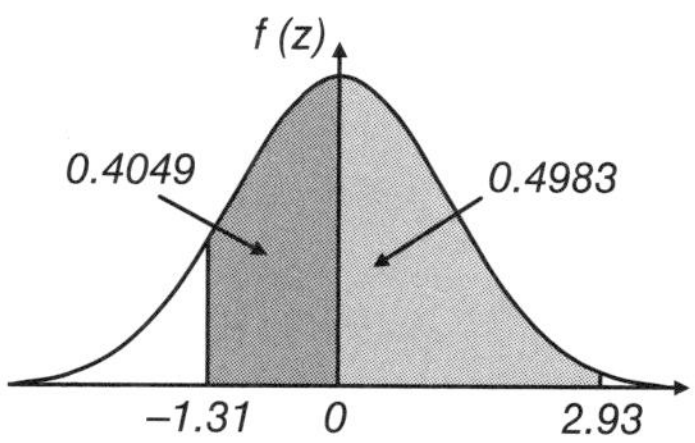

Example 10.4.3 More on Standard Normal Probabilities

Find the following probabilities for the standard normal random variable z and illustrate in a sketch.

a) $P(-1.25 < z < 1.25)$ *b)* $P(z \geq 2.14)$ *c)* $P(z \leq -3.86)$

Solution:

a) The area of interest is shaded in the sketch. The probability from $0 \leq z \leq 1.25$ *is doubled in view of the symmetry. The desired probability is* $2(0.3944) = 0.7888$.

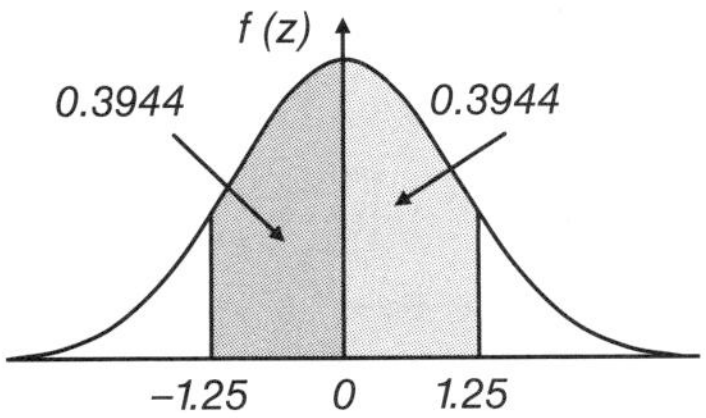

b) For the probability that $z > 2.14$*, which is in the "upper tail" of the normal curve, the probability of* $0 \leq z \leq 2.14$ *must be subtracted from 0.5000 (since each side has an area of* $1/2$*). Here, the probability is* $0.5000 - 0.4838 = 0.0162$*, as shown in the sketch.*

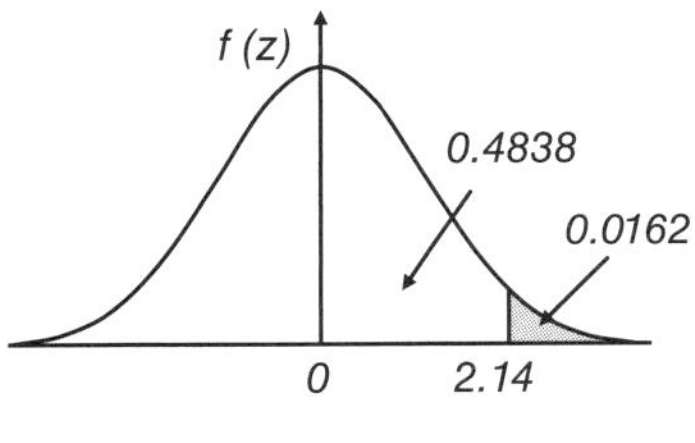

c) For $z \geq 0$*, the normal table has 0.4990 – nearly the entire area for* $z \geq 0$*. The same for* $-3.09 \leq z \leq 0$*. For* $z \leq -3.86$*, the lower tail area is beyond table accuracy. The probability is zero for practical purposes.*

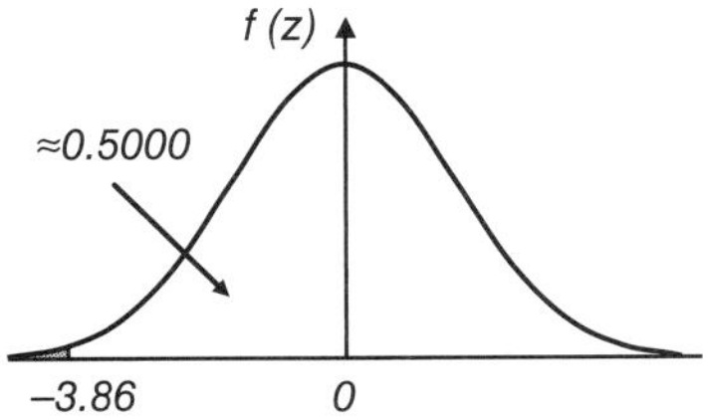

Transforming to a Standard Normal Distribution

Not all normal distributions are standard having $\mu = 0$ and $\sigma = 1$. Fortunately, and remarkably, every normal distribution, regardless of the values of μ and σ, is easily transformed into a standard normal distribution. Therefore, one can determine areas and hence probabilities, for any normal distribution using the standard normal table, with which you are already familiar.

The linear relation

$$z = \frac{x - \mu}{\sigma}$$

transforms any normal random variable with mean μ and standard deviation σ into a standard normal variable, z, with $\mu = 0$ and $\sigma = 1$. The value of z is also known as a **z-score**.

To apply the normal probability table to a normal random variable x, we transform the value of x to its corresponding z-score. The value of z is the number of standard deviations between a value of x and the mean μ. A negative z-score indicates that x is less than the mean while a positive z-score indicates that x is more than the mean.

Example 10.4.4 $\mu = 2$ *and* $\sigma = 1/2$

Find the probability that $1 \le x \le 2.5$ *for the normal random variable, x, with* $\mu = 2$ *and* $\sigma = 1/2$.

Solution:
The interval, $1 \le x \le 2.5$, *must be transformed into an equivalent interval for z. First, subtract the mean from each term in the inequality to yield*

$$1 - 2 \le (x - \mu) \le 2.5 - 2 \quad or \quad -1 \le x \le 0.5$$

Next, divide each term by the standard deviation to yield

$$\frac{-1}{1/2} \le \frac{x - \mu}{\sigma} \le \frac{0.5}{1/2} \quad or \quad -2 \le z \le 1$$

Therefore,

$P(1 \le x \le 2.5) = P(-2 \le z \le 1)$. *The two z values are on opposite sides of zero so the desired probability is found by adding*
$P(-2 \le z \le 0) + P(0 \le z \le 1) = 0.4772 + 0.3413 = 0.8185$. *(A sketch is helpful)*

Example 10.4.5 $\mu = 1$ *and* $\sigma = 2$

Find the probability that $0 \le x \le 2$ *for the normal variable, x, with* $\mu = 1$ *and* $\sigma = 2$.

Solution:
Again, the corresponding z variable is obtained in two steps:

⟶

First, subtract the mean

$$0 - 1 \leq (x - \mu) \leq 2 - 1, \text{ or } -1 \leq (x - \mu) \leq 1$$

and, next, divide by σ to yield

$$\frac{-1}{2} \leq \frac{x-\mu}{\sigma} \leq \frac{1}{2} \text{ or } \frac{-1}{2} \leq z \leq \frac{1}{2}$$

Using symmetry, $P(\frac{-1}{2} \leq z \leq \frac{1}{2}) = 2P(0 \leq z \leq \frac{1}{2}) = 2(0.1915) = 0.3830.$

Example 10.4.6 $\mu = 8$ **and** $\sigma = 2$

The random variable x is normal with $\mu = 8$ *and* $\sigma = 2$. *Find the probability that* $x > 10$.

Solution:
Transforming to a z value, $x > 10$ *is equivalent to* $z > (10 - 8)/2 = 1$. *This is the upper tail of the normal curve. Using* $P(z > 1) = 0.50 - P(0 \leq z \leq 1) = 0.500 - 0.3413 = 0.1587$ *is the required probability.*

Example 10.4.7 $\mu = 1$ **and** $\sigma = 2$

Find the probability that $2 \leq x \leq 5$ *for the normal random variable x with* $\mu = 1$ *and* $\sigma = 2$.

Solution:
Again, the corresponding z variable is obtained in two steps:
First, subtract the mean

$$2 - 1 \leq (x - \mu) \leq 5 - 1, \text{ or } 1 \leq (x - \mu) \leq 4$$

and, next, divide by σ to yield

$$\frac{1}{2} \leq \frac{x-\mu}{\sigma} \leq \frac{4}{2} \text{ or } 1/2 \leq z \leq 2$$

Both z values are positive so

$$P(1/2 \leq z \leq 2) = P(0 \leq z \leq 2) - P(0 \leq z \leq 1/2) = 0.4772 - 0.1915 = 0.2857$$

Our occasional alternatives in notations or calculations have sought to encourage you to select a comfortable format. We recommend sketching of the required area under the normal curve. This will assist you in checking your work in more complicated problems.

The normal distribution is the most important of all the theoretical probability distributions. It is at the base of most statistical theory and practical applications. The normal curve provides a good approximation for the binomial distribution for large values of n. It is a good approximation for many other probability distributions. For example, the normal distribution is often a good approximation for relative frequency distributions of people's heights or weights, intelligence quotients (IQ) and other test scores, measurement errors in manufacturing parts and in laboratory experiments, and so on.

Earlier, the run time in the New York City Marathon was cited. Actually, deviations from normality were found a few minutes before 4 hours and just before 5 hours. These were interpreted as the extra efforts of runners to achieve goals such as "I'll try to break four hours, etc." Sometimes, the deviations from normality can be used to detect abnormalities, as in a manufacturing run.

Example 10.4.8 IQ Testing

Scores from a Wechsler test, used by psychologists to indicate intelligence, follow a normal distribution with mean 100 and variance 225. Determine probabilities that an individual

a) scores less than 121. *b) scores between 82 and 97.*

Solution:

a) Here, $\sigma = \sqrt{225} = 15$ and $\mu = 100$. The corresponding z score is $z = (121 - 100)/15 = 1.40$ so we seek $P(z < 1.40)$. Again, use a rough sketch and shade the appropriate region. The probability is $0.5000 + 0.4192 = 0.9192$. There is about a 92% chance that an individual score is less than 121.

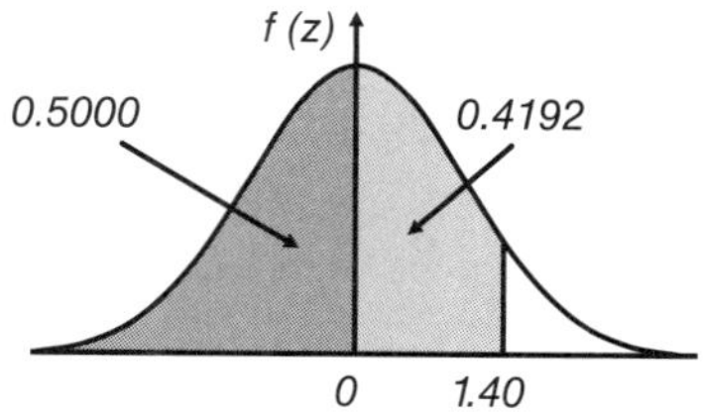

b) Both x values must be converted to z-scores. Both z values exceeded zero so their probabilities are subtracted for the solution.
The two z values are $(82 - 100)/15 = -1.2$ and $(97 - 100)/15 = -0.20$.
The corresponding difference in probabilities is $0.3849 - 0.0793 = 0.3056$, as in the sketch.

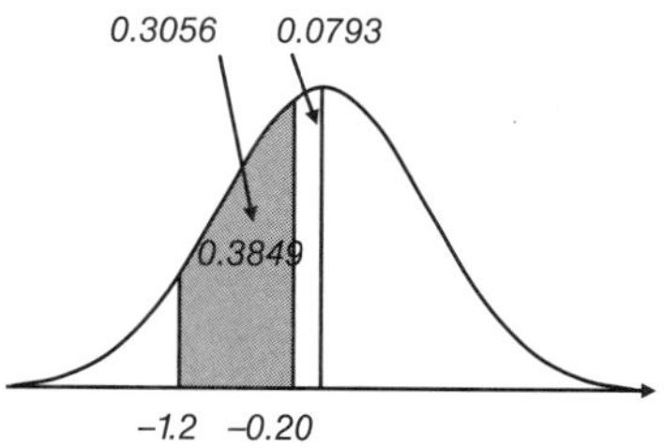

Here, normal probabilities have been determined by using a Standard Normal Table, converting nonstandard variables to z variables when the mean and standard deviation are known. Sometimes, the probabilities are given, as in a quality standard of, say, 95%, and other parameters are to be calculated. Actually, there are four relevant quantities – probability, variable value, mean, and standard deviation. The above examples have used the last three to obtain probabilities. Actually, any three can be used to determine the fourth, as illustrated in the next examples.

Example 10.4.9 The Standard Normal Random Variable

Find a value of the standard normal random variable z, call it z_0, such that

a) $P(z > z_0) = 0.7764$

b) $P(-z_0 < z < z_0) = 0.9500$

c) $P(-1.64 < z < z_0) = 0.1915$

Solution:

a) To determine z_0 such that $P(z > z_0) = 0.7764$, so the probability exceeds 1/2 (figure below). Therefore, z_0 must be negative. There is a 0.2764 probability (0.7764 − 0.5000) between z_0 and 0. Looking for 0.2764 or its closest in the table interior, find the corresponding z score as 0.76. Here, $z_0 = -0.76$.

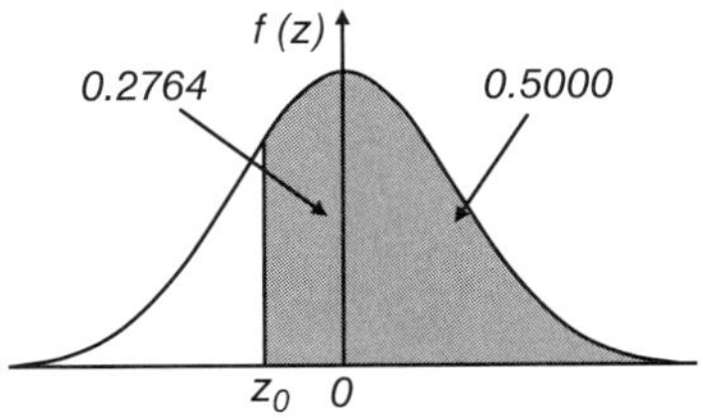

b) To determine z_0 such that $P(-z_0 < z < z_0) = 0.9500$, first note the symmetry. Each side of the zero has the same probability, 0.9500 / 2 = 0.4750. Reading from 0.4750 in the table interior (and working to the margin), we find $z_0 = 1.96$. There is a 95% probability that z lies between −1.96 and +1.96.

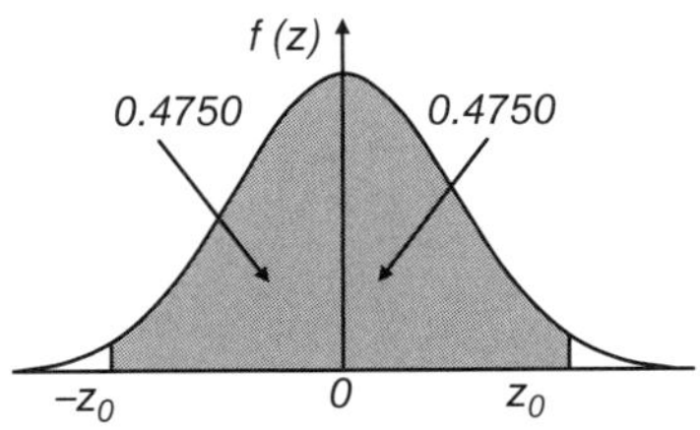

c) To determine z_0 such that $P(-1.64 < z < z_0) = 0.1915$, first determine the probability for the known z-score. The probability that z_0 lies between −1.64 and 0 is 0.4495, as z_0 is negative. Regions a and b combined (figure) represent a total

⟶

*probability of 0.4495. However, region **a** represents a probability of 0.1915, leaving region **b** to a probability of 0.2580. This probability, in the interior of the table, has z_0 as -0.70.*
As a check, $P(-1.64 < z < -0.70) = 0.1915$!

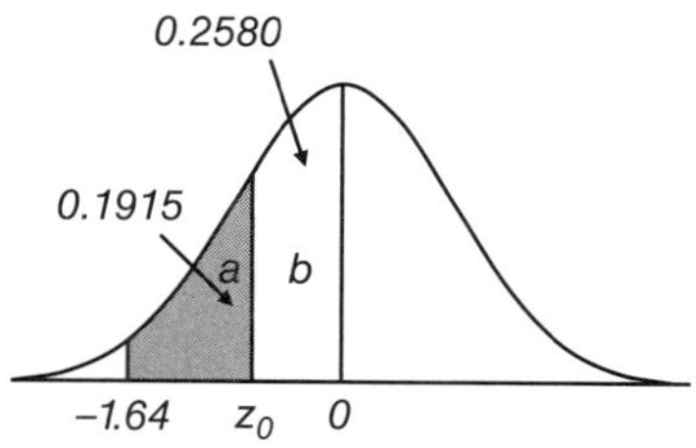

Example 10.4.10 Regulating Machine Settings

For a normal distribution, the probability that x is less than 118 is to be 4/5.

a) *If $\sigma = 12$, find μ.*
b) *If $\mu = 100$, find σ.*

Solution:

a) *Here, $P(x < 118) = 0.80$. Transforming to a z variable*

$$P\left(\frac{x-\mu}{\sigma} < \frac{118-\mu}{12}\right) = 0.80$$

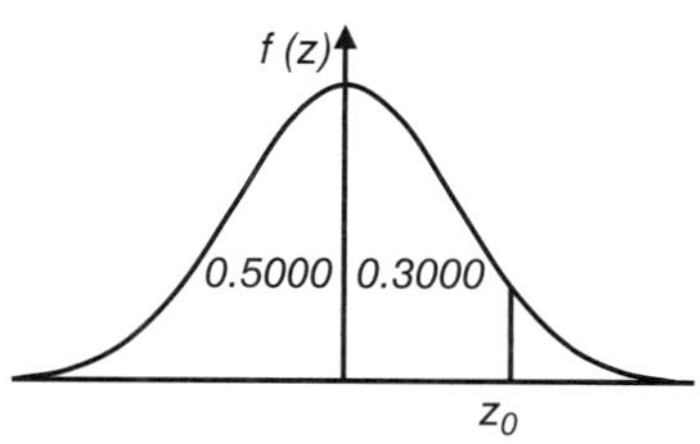

The normal table yields a z_0 value of about 0.84. Therefore, set

$$\frac{118-\mu}{12} = 0.84 \text{ to yield } \mu = 118 - 12(0.84) = 107.92.$$

($z = 0.84$ corresponds to a probability of 0.2995, the closest table value to 0.30).

⟶

b) Now, we have $P(x < 118) =$

$$P\left(\frac{x-\mu}{\sigma} < \frac{118-100}{\sigma}\right) = 0.80$$

Here,

$$\frac{118-100}{\sigma} = 0.84 \quad so \quad \sigma = 18/0.84 = 21.43$$

EXERCISES 10.4

1. Find these probabilities for the standard normal random variable Z:

   a) $P(0 \leq z \leq 1.39)$
   b) $P(0 \leq z \leq 0.99)$
   c) $P(0 < z < 2.16)$
   d) $P(-1.14 < z < 0)$
   e) $P(-2.38 \leq z \leq 0)$
   f) $P(-0.06 \leq z \leq 0)$

2. Find these probabilities for the standard normal random variable Z:

   a) $P(-1.26 \leq z \leq 0)$
   b) $P(0 \leq z \leq 1.83)$
   c) $P(-2.36 < z < 0)$
   d) $P(0 < z < 1.55)$
   e) $P(-0.78 \leq z \leq 0)$
   f) $P(-0.19 \leq z \leq 0)$

3. Find these probabilities for the standard normal random variable Z:

   a) $P(-1.23 \leq z \leq 2.45)$
   b) $P(-2.59 \leq z \leq 1.04)$
   c) $P(-0.16 \leq z \leq 2.75)$
   d) $P(-2.67 \leq z \leq 0.83)$
   e) $P(-1.02 \leq z \leq 1.02)$
   f) $P(-2.03 \leq z \leq 4.07)$

4. Find these probabilities for the standard normal random variable Z:

   a) $P(-1.08 \leq z \leq 1.08)$
   b) $P(-1.40 \leq z \leq 0.58)$
   c) $P(-0.36 \leq z \leq 1.99)$
   d) $P(-4.60 \leq z \leq 0.80)$
   e) $P(-2.37 \leq z \leq 1.69)$
   f) $P(-2.73 \leq z \leq 2.07)$

5. Find these probabilities for the standard normal random variable Z:

   a) $P(z > -1.45)$ b) $P(z \geq 2)$ c) $P(z < -1.36)$ d) $P(z < 1.5)$

6. Find these probabilities for the standard normal random variable Z:

   a) $P(z > -1.25)$
   b) $P(z < 2.68)$
   c) $P(z > 0)$
   d) $P(z = -1.64)$
   e) $P(z \leq -2.33)$
   f) $P(z \geq 1.40)$

7. Find these probabilities for the standard normal random variable Z:
   a) $P(1.23 < z < 2.45)$
   b) $P(-1.99 \leq z \leq -1.11)$
   c) $P(1.03 \leq z \leq 2.84)$

8. The random variable x has a normal distribution with mean of 500 and standard deviation of 50.
   Determine the z-scores that correspond to each of these x values.
   a) 350 b) 470 c) 500 d) 625 e) 740

9. The random variable x has a normal distribution with mean of 80 and standard deviation of 5.
   Determine the z-score that corresponds to each of these x values.
   a) 85 b) 77 c) 69 d) 71 e) 86 f) 80

10. If SAT math scores are normally distributed with a mean of 500 and standard deviation of 100, determine the probability that a score is
    a) between 460 and 640.
    b) below 720.
    c) at least 390.

11. Grades on a test are normally distributed with mean of 80 and variance of 16, find the probability that a student's grades are
    a) more than 78 on the test.
    b) between 74 and 84 on the test.
    c) below 85 on the test.

12. Grades are normally distributed with a mean of 75 and standard deviation of 5, determine the grade for which
    a) 80% of the class did worse.
    b) 60% of the class did better.

13. Suppose x is a normally distributed random variable with mean of 70 and variance of 9. Determine each of these probabilities:
    a) $P(x < 76)$ b) $P(61 < x < 73)$ c) $P(67 \leq x \leq 77)$

14. Find a value of the standard normal random variable z, call it z_0, such that
    a) $P(z \geq z_0) = 0.9656$
    b) $P(-z_0 \leq z \leq z_0) = 0.8740$
    c) $P(z \geq z_0) = 0.3015$
    d) $P(-1.1 \leq z \leq z_0) = 0.8000$

15. Find a value of the standard normal random variable z, call it z_0, such that
    a) $P(z \geq z_0) = 0.1314$
    b) $P(-z_0 \leq z \leq z_0) = 0.8530$
    c) $P(z \leq z_0) = 0.9000$
    d) $P(-1 \leq z \leq z_0) = 0.7977$

16. A machine that regulates paint for a gallon container is set to dispense 128.5 ounces on average. The paint dispensed is normally distributed. If less than 128 ounces is dispensed, the product is unacceptable as it is a false claim. Determine the standard deviation so that no more than 2.5% of the paint cans are underfilled.

17. A machine used to regulate sugar dispensed into a soda bottle is set to μ grams per liter of soda, on average. The amount of sugar dispensed is normally distributed with a standard deviation of 0.36 g. If more than 6.5 grams of sugar is dispensed, the liter of soda is unacceptable. What setting of μ will allow no more than 5% of the soda output to be unacceptable?

10.5 NORMAL DISTRIBUTION APPLICATIONS

That the normal distribution has a role in virtually every facet of life – natural and otherwise – has already been emphasized. Even the variation and adaptation accompanying natural selection track the bell curve. The examples in this section illustrate a variety of practical uses for the technical skills acquired in the previous section.

◆ The colorings in variations among some species follow a bell curve. The coloring of gypsy moths is an example. On a color scale, a majority of gypsy moths' colorings form the central orange and green sections of a bell curve color scale. Darker colored moths, forming the lower end of the bell curve color scale, are further from the average. In earlier times, darker colored gypsy moths were an exception. However, emergence of a predator resulted in a majority of the lighter colored moths (near the mean of the bell curve) being killed while the less discernible darker moths escaped. Over time, the mean of the bell curve shifted toward the darker colored moths. A similar evolutionary phenomenon has been noted among male peacocks as females seek partners with larger colored tails and greater physical prowess.

Example 10.5.1 Airline Boarding Denied

Customers arrive in random numbers for a noon shuttle flight according to a normal distribution with mean of 85 passengers and a standard deviation of 10 passengers. An aircraft is fitted to seat 100 passengers. What is the chance an awaiting passenger is denied boarding?

Solution:

Here, x is the number of passengers who seek to board

$$P(x < 100) = P\left(\frac{x-\mu}{\sigma} < \frac{100-85}{10} = 1.50\right) = 0.8332$$

There is almost a 17% chance, on average, that boarding is denied a passenger.

Example 10.5.2 More Seating

Airline management seeks to reduce the chance of denied boarding in the previous example to no more than 5%. How many seats are needed to reach this goal?

Solution:

Now the probability is specified, not the aircraft's capacity, C.

Therefore, $P(x < C) = P\left(\frac{x-\mu}{\sigma} < \frac{C-85}{10}\right) \geq 0.95$. *Using the normal table, a z-value between 1.64 and 1.65, say, 1.645 includes 45% of the area above the mean at zero and, hence, the desired 95% of the area under the bell curve.*

Setting

$$\frac{C-85}{10} = 1.645$$

and solving for C,

$C = 10(1.645) + 85 = 101.45$ *or 102 seats are necessary for the specified goal.*

Example 10.5.3 Dam Height Design

The height of a dam is to be such that there is only a 4% chance of overtopping in any year. The distribution of the yearly maximum river height, x, is approximately normal with a mean of 12 feet and a standard deviation of 3 feet. What should be the design height?

Solution:

Let H be the unknown design height.

Then $P(x > H) = p\left(\frac{x-\mu}{\sigma} > \frac{H-12}{3}\right) = 0.04$ *and the complement,* $P(12 < x < H) = 0.46$. *The normal table has a value of nearly 0.46 for* $z = 1.75$. *Therefore, setting* $\frac{H-12}{3} = 1.75$ *and solving yields* $H = 17.25$ *feet as the required design height.*

Example 10.5.4 Choosing a Vending Machine

A vending machine is needed to dispense six ounces of liquid. Various models are available. An important factor in their price is the standard deviation, related to the variation, of the normal random amount of dispensed liquid. The lower the standard deviation, the higher the price of the machine, generally, since more precise mechanisms are required to ensure that the liquid dispensed is closer to the mean. What values of σ *are required to ensure that a 7 ounce cup will have no less than 5.5 ounces and no more than 6.5 ounces, 95% of the time on average? 99%?*

⟶

Solution:
Here, x is the random amount of dispensed liquid and the mean is given as 6 ounces. We seek to have a 95% probability of dispensing between 5.5 and 6.5 ounces.

$$\text{Therefore, } P(5.5 < x < 6.5) = P\left(\frac{5.5-6}{\sigma} < \frac{x-\mu}{\sigma} < \frac{6.5-6}{\sigma}\right) = 0.95$$

The figure illustrates the required 95% probability and corresponding values of z. Note that the 100% − 95% = 5% has been divided equally between the two "tails" or "ends" of the bell curve.

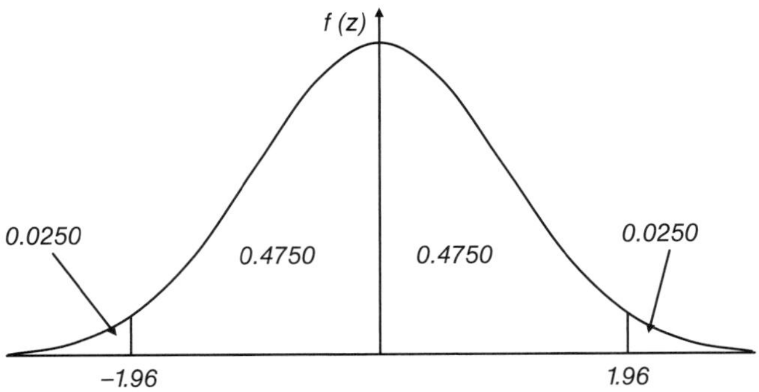

One may set either $\frac{5.5-6}{\sigma}$ *or* $\frac{6.5-6}{\sigma}$ *to* $z = \mp 1.96$*, respectively, to calculate σ. That is,* $\frac{6.5-6}{\sigma} = 1.96$ *and, therefore,* $\sigma = \frac{0.5}{1.96} = 0.255$ *ounces.*

For a 99% probability, 1% of the area is divided equally between the two tails as shown in the next figure. Here, the z-score becomes 2.575.

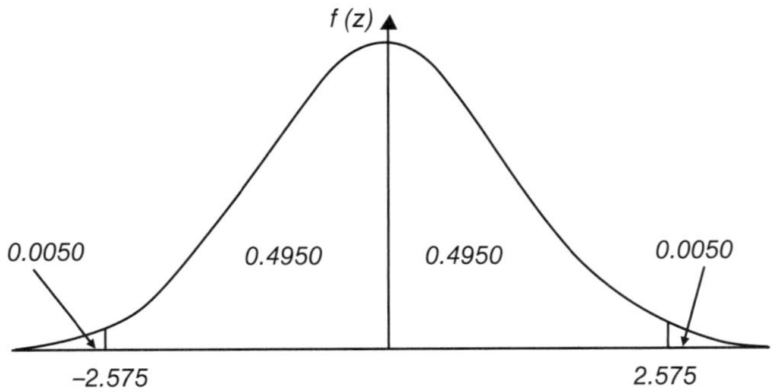

So, $\frac{6.5-6}{\sigma} = 2.575$ *and* $\sigma = 0.194$*.*

As a rule, other things equal the price of higher quality machines and equipment rises rapidly with decreasing σ.

The above examples clarify, again, that knowledge of any three of the four parameters of the normal distribution – the threshold constant (such as the airline capacity or dam height), mean, standard deviation, and probability – permits determination of the fourth parameter.

Another interesting situation can arise when comparing alternative strategies. Suppose that completion times follow a normal distribution on two alternative routes of travel or ways to execute a construction project. Let the respective means and standard deviations of the two alternatives be μ_1 and μ_2 and σ_1 and σ_2 and assume that an earlier completion time is desirable.

Clearly if $\mu_1 < \mu_2$ and $\sigma_1 < \sigma_2$, the first alternative should be chosen since not only it is shorter on average but also there is less variability because $\sigma_1 < \sigma_2$. If $\mu_1 > \mu_2$ and $\sigma_1 > \sigma_2$, the second alternative is preferred.

The interesting situation is when, say, $\mu_1 < \mu_2$ and $\sigma_1 > \sigma_2$. Now, on average, the first alternative is quicker. However, the higher standard deviation for the first alternative indicates greater variability, and, hence, greater risk of a large upward deviation from the mean. How can a choice be rationally made? The next example shows a way!

Example 10.5.5 Which Construction Plan?

A construction superintendent is considering two alternative plans to undertake a project. The time of completion is approximately normally distributed for each alternative and with the following parameters:

Plan I: $\mu_I = 33.9$ *months and* $\sigma_I = 11.8$ *months*

Plan II: $\mu_2 = 29.3$ *months and* $\sigma_2 = 17.5$ *months*

The target completion time is 42 months. Which plan is the better choice?

Solution:

Let x_I *and* x_2 *be the respective random completion times. The text indicated the need to choose a means to measure a "better choice" of plans. The means and variances, by themselves, are conflicting signals. However, since a target completion date is specified, it is possible to calculate the probabilities of meeting it. We seek the probabilities:*

$$P(x_I < 42) = P\left(\frac{x_I - \mu_I}{\sigma_I} < \frac{42 - 33.9}{11.8} = 0.69\right) = 0.7549$$

$$P(x_2 < 42) = P\left(\frac{x_2 - \mu_2}{\sigma_2} < \frac{42 - 29.3}{17.5} = 0.73\right) = 0.7673*$$

The two probabilities are close so that other factors, not considered to here, may influence the superintendent's choice. However, other things equal, Plan II has a better chance of meeting the 42-month target.

Example 10.5.6 New Target Date

The superintendent in the prior example has just been granted an extension to a 44-month completion target date. Should the construction plan be reconsidered?

⟶

Solution:
Now, we have

$$P(x_1 < 44 = P\left(\frac{x_1 - \mu_1}{\sigma_1} < \frac{44 - 33.9}{11.8} = 0.86\right) = 0.8051^*$$

$$P(x_2 < 44) = P\left(\frac{x_2 - \mu_2}{\sigma_2} < \frac{44 - 29.3}{17.5} = 0.84\right) = 0.7995$$

While the probabilities are even closer than earlier, surprisingly, a change of plan is suggested. Now, Plan I is the preferred choice!

♦ The last two examples warrant careful study. Their counterintuitive results can be a guide to decisions in everyday life.

Statistical Samples

Statisticians deal with **samples,** and the **sample mean** is the foremost in the information derived from them. Recall that

$$\bar{x} = \frac{x_1 + \cdots + x_n}{n} \quad \text{sample mean}$$

Now, it is a mathematical fact that if $x_1, x_2, \ldots, x_n$ are independent observations on a normal random variable, then $\bar{x}$, also a random variable, has a normal distribution. Actually, the **central limit theorem** ensures that even if $x_1, x_2, \ldots, x_n$ are observations from a non-normal distribution, under rather general conditions, $\bar{x}$ remains approximately normally distributed. The approximation tends to improve with increasing sample size, n. Many statisticians consider the approximation to be "exact" for practical purposes for sample sizes of 30 or more. It is this theorem, one of the most remarkable in mathematics, which makes the normal distribution so important and is the basis for much of mathematical statistics.

Central Limit Theorem

The distribution of the sample mean, $\bar{x} = \frac{x_1 + \cdots + x_n}{n}$, tends to a normal distribution regardless of the distribution of the independent observations $x_1, x_2, \ldots, x_n$; the approximation to normality tending to improve as n increases.

Since $\bar{x}$ is a normal random variable, it has a mean and variance. If x, the normal variable on which observations are made, has a mean μ and standard deviation σ, then $\bar{x}$ has the same mean μ and a standard deviation of $\sigma/\sqrt{n}$, where n is the sample size. The proofs of these, while not difficult, are omitted here.

Mean and Variance of $\overline{x}$

If $x_1, x_2, \ldots, x_n$ is a random sample of size n from a distribution with mean μ and standard deviation σ, its sample mean is

$$\overline{x} = \frac{x_1 + \cdots + x_n}{n}$$

having a mean $\mu_{\overline{x}} = \mu$ and standard deviation $\sigma_{\overline{x}} = \sigma / \sqrt{n}$.

As sample size increases one intuitively expects $\overline{x}$, the sampling average, to approach μ. In fact, it does according to a law of large numbers!

For the standard deviation of $\overline{x}$, a measure of its variability about the mean of $\overline{x}$, one expects its variability to decrease as the sample size increases. After all, the larger the sample size, the nearer one expects $\overline{x}$ to estimate μ. This is clear from the standard deviation of $\overline{x}$, $\sigma_{\overline{x}} = \sigma / \sqrt{n}$, varying inversely as the square root of the sample size n.

Significance Tests

These properties of $\overline{x}$ enable the development of important statistical concepts and tests. Statisticians evaluate claims such as the mean lives of products or the efficacy of drugs. In the next examples a claim for the mean life of tires is evaluated. To evaluate the claim for a mean life of 50,000 miles, a group of nominally identical tires is wear tested and $\overline{x}$ calculated. If the sample mean were, say, 55,000 or 42,000 miles, the claim might seem questionable. However, if $\overline{x}$ were 49,500 or 50,500 miles, intuitively, one might reasonably believe the claim of a 50,000 mile mean since some random variation about the mean is expected. For one thing, the sample mean of itself is an incomplete guide. The standard deviation also has a role; it can indicate limits of tolerable variation.

Where, then, is the boundary between the "believable" and the "unbelievable?" When can a claim be "accepted" and when should it be "not accepted?" The next examples develop **significance tests**, also called **tests of hypothesis**, to help answer such questions.

Example 10.5.7 Tire Mileage

A tire manufacturer claims that a tire has an average life of 50,000 miles. Previous experience indicates that true tire life approximates a normal distribution with a standard deviation of 1000 miles. A sample of 10 tires is tested and $\overline{x} = 49{,}760$ miles. Is the manufacturer's claim credible?

Solution:

While there cannot be a definitive proof of the claim, we can assess its likelihood. The sample mean, $\overline{x}$, at 49,760 miles is 240 miles less than the manufacturer's claim of a 50,000-mile mean.

⟶

It seems reasonable to ask, "What is the chance that a sample mean is less than the claimed mean by 240 miles?"

That is,

$$P(\overline{x} \leq 49{,}760)$$
$$= P\left(\frac{\overline{x}-\mu}{\sigma/\sqrt{n}} \leq \frac{49{,}760-50{,}000}{1000/\sqrt{10}}\right) = P(Z_{\overline{x}} \leq -0.76)$$
$$= P(z_{\overline{x}} \leq -0.76) = 0.2236$$

Note the use of $\sigma/\sqrt{n}$ for the single standard deviation, where σ is the standard deviation for the life of a random single tire.

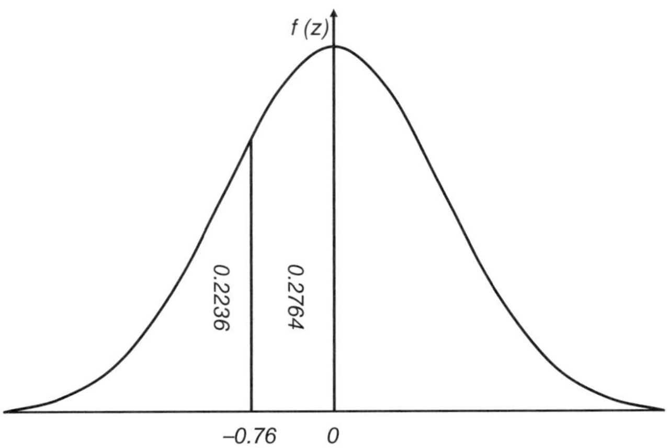

To interpret the results in the context of what is known as a **hypothesis test**, interest lies in the probability in the "tails", or "lower tail" in this case. The smaller a tail probability, the more reason to doubt the claim. The reason is that the z value becomes larger (indicating a larger deviation from the mean). Here, there is about a 22% probability in the lower tail. The result suggests that it is not unreasonable to believe the claim of 50,000 miles although the sample mean is 49,760 miles.

Example 10.5.8 A Larger Sample

To continue the last example, an additional 15 tires are tested. Coincidentally, imagine that the sample mean remains unchanged at $\overline{x} = 49{,}760$ miles. How does this influence belief in a 50,000 mile mean?

⟶

Solution:
We still seek $P(49,760 < \bar{x} < 50,000)$. However, the standard deviation of $\bar{x}$ is $1000/\sqrt{25}$ since the sample size is now 25. Consequently,

$$P(\bar{x} < 49{,}760) = P\left(\frac{\bar{x}-\mu}{\sigma/\sqrt{n}} < \frac{49{,}760 - 50{,}000}{1000/\sqrt{25}}\right) = P(z_{\bar{x}} < -1.20)$$
$$= P(z_{\bar{x}} < -1.20) = 0.1151$$

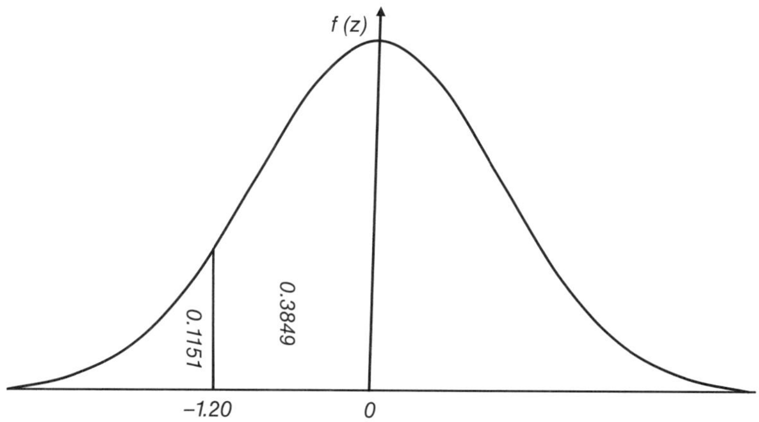

Now, there is about an 11.5% chance, not 22%, that a tire life would have a sample mean of 49,760 or less when in actuality the mean is 50,000. This indicates less likelihood that the 50,000-mile claim is correct.

Example 10.5.9 A Still Larger Sample

Continuing the above example, an additional 200 tires are tested. Amazingly, the sample mean remains unchanged at $\bar{x} = 49,760$ miles. How does this new information influence belief?

Solution:
We still seek $P(\bar{x} < 49,760)$. However, now the standard deviation of $\bar{x}$ is $1000/\sqrt{225}$ since the sample size is now 225. Consequently,

$$P(49,760 < \bar{x} < 50,000) = P\left(\frac{\bar{x}-\mu}{\sigma/\sqrt{n}} < \frac{49,760 - 50,000}{1000/\sqrt{225}}\right) = P(z_{\bar{x}} < -3.60)$$
$$P(z_{\bar{x}} < -3.60) \approx 0$$

⟶

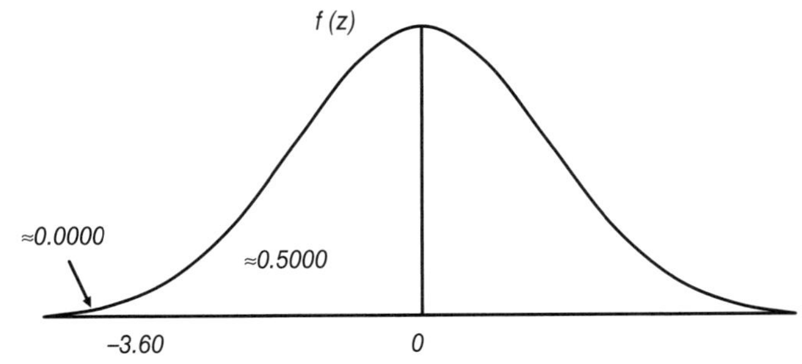

A $z_{\bar{x}}$ is, literally, "off the chart". This normal table lacks a value of z of -3.60. It has a "tail" probability less than 0.1%. There should be serious doubts about the claim.

Keeping the sample mean fixed at 49,760 miles and increasing the sample size enable us to interpret its effect in the last examples.

As the sample size increased and so the standard deviation of $\bar{x}$ decreased, the tail probabilities decreased indicating more doubt about the claimed mean as z differed significantly from zero. Theoretically, there is no reason to doubt the claim if $z \approx 0$. Doubts begin as z differs from zero and, thus, tail probabilities become smaller.

The foregoing examples, in which $\bar{x}$ was less than the presumed mean, should not suggest that a value of $\bar{x}$ above $\mu = 50{,}000$ miles is a guarantee of sorts. Certainly, it strengthens the belief that the mean, μ, is not less than 50,000 miles – it does not ensure it.

In the previous examples, the concern was that mean tire life not be less than the claimed 50,000 miles. There was no concern expressed that the mean could be larger. A larger mean would indicate the tire was of a better quality than claimed. Sometimes, there are limits, upper and lower, that require consideration as in the next example.

Example 10.5.10 Medication Dosage

A pharmaceutical firm seeks to establish a recommended mean dosage for a new drug. Since people respond differently, dosage is varied until the desired therapeutic effect is observed in each of 100 subjects. The average dose per subject was $\bar{x} = 504$ milligrams and a sample standard deviation of $s = 25$ milligrams. Is it reasonable to specify 500 milligrams as a mean dosage? A "mound-shaped" distribution is reasonable.

Solution:

First, note that there is no mention of the distribution of $\bar{x}$, and that the value of σ is not given. However, with the relatively large sample size of 100 (recall, a suggested, $n \geq 30$ rule), it may be reasonable to use s as an approximation for σ. Second, and importantly, the central limit theorem supports an assumption of normality for the distribution of $\bar{x}$ for this sample size and the "mound-shaped" distribution.

We might regard as reasonable an interval about the hypothesized mean of 500 that has, say, a 95% chance of including a sample mean.

⟶

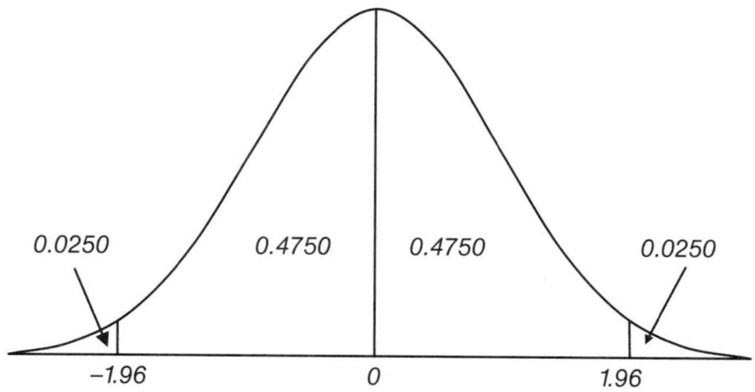

Since dosages either too large or too small are of concern, the 95% is equally divided about the mean, a normal table indicates $z = \pm 1.96$.

Now, the sample mean of 504 milligrams, translated into a z value, is

$$z = \frac{504 - 500}{25/\sqrt{100}} = \frac{4}{2.5} = 1.6$$

A z-score of 1.60 clearly falls within the interval −1.96 and 1.96. There is insufficient evidence here not to believe that the true mean is 500 milligrams. Note the care in wording! There is no claim that 500 milligrams is the correct mean – another sample may have a different outcome – only that there is no reason at the 95% level to reject $\mu = 500$ *milligrams.*

The 95% figure is called the **test level** or **level of significance** and is often, as here, rather arbitrary. The next example illustrates a common situation in which a test level is not specified in advance. The **observed significance level** or ***p*-value**, as it is called, measures the extent to which the test statistic disagrees with the hypothesized value of the mean. It is the probability in the tail(s) that we have been using to show the doubts about claims in the last few examples. Often this value is reported on statistical data analyzed with a statistical computer package. The idea being that it is for the user to judge the significance or its lack for the purpose at hand.

Example 10.5.11 A p-value

Determine the observed significance level, or ***p-value****, for the previous example.*

Solution:

Here, instead of specifying 95% in advance as the test level, we seek the limiting test level at which the value $\mu = 500$ *milligrams can be supported. For the sample mean,* $\bar{x} = 504$ *milligrams,* $z = \pm 1.60$ *are the corresponding z values for an "acceptance" interval.*

⟶

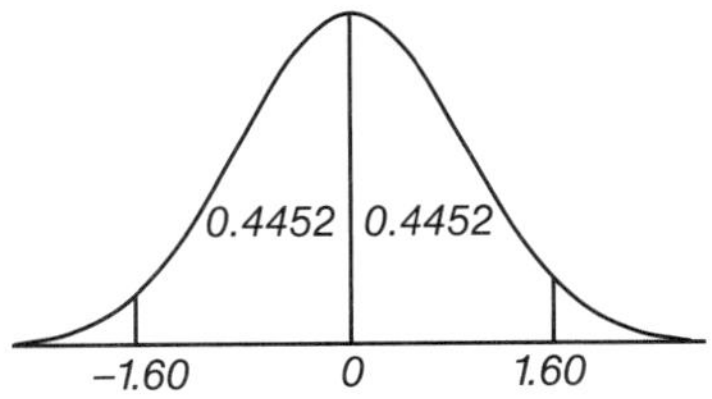

Using a normal probability table, this corresponds to a probability of $0.4452 + 0.4452 = 0.8904$ in the "acceptance" region. The p-value here is the probability in the two tail regions, 0.1096 in this case. This is the complement to 0.8904 or $1 - 0.8904$.

The preceding examples are suggestive of **tests of hypothesis** to students of statistics. Our discussion has omitted some technical notations to better and informally suggest the underlying concepts in the brief space here.

We have mentioned several times that statisticians regard $\bar{x}$ as an estimate of μ and, to be sure, an estimate that generally improves with sample size. Then, a **maximum-likelihood estimate** was briefly considered in an example of the hypergeometric distribution (Chapter 8). In both instances, the mean is specified as a single value – the so-called **point estimate**.

For many practical purposes, the exact value of the mean is not especially important. Rather, some measure of confidence for an interval in which the mean might lie is more useful.

The so-called **confidence intervals** have four components: the sample mean, the standard deviation, the sample size, and a **confidence level**. The confidence level is a probability. It expresses the likelihood that a confidence interval actually includes the (unknown) mean μ. After all, the only interval that certainly includes the mean is the real line, $-\infty$ to ∞. Of course, this is not useful. However, it does suggest that a narrowing of an interval entails risk that the chosen interval will not include the mean μ.

Forming a confidence interval begins with $\bar{x}$. The interval about $\bar{x}$ is proportional to the confidence level, $1 - \alpha$, expressed by the corresponding z value, $z_{\alpha/2}$, the interval is also proportional to the standard deviation for $\bar{x}$. In symbols, a confidence interval is expressed as

$$\bar{x} \pm z_{\alpha/2}\frac{\sigma}{\sqrt{n}}$$

This can be visualized using the next figure. We seek a z value such that the confidence level (shaded area) is $1 - \alpha$. That is,

$$P(-z_{\alpha/2} < z < z_{\alpha/2}) = 1 - \alpha$$

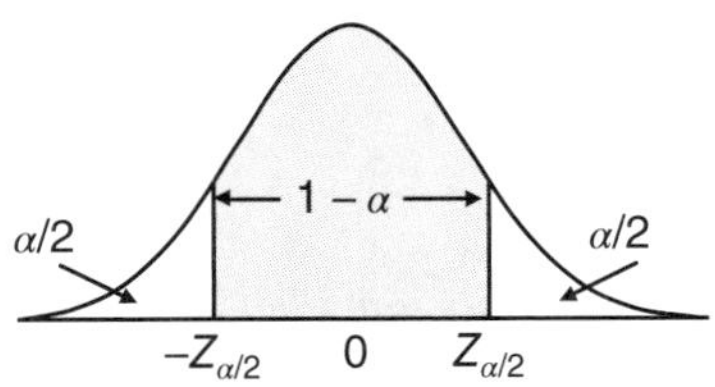

Replacing z by $\dfrac{\bar{x}-\mu}{\sigma/\sqrt{n}}$ and simplifying:

$$\begin{aligned}
P(-z_{\alpha/2} < z < z_{\alpha/2}) &= P\left(-z_{\alpha/2} < \frac{\bar{x}-\mu}{\sigma/\sqrt{n}} < z_{\alpha/2}\right) \\
&= P(-z_{\alpha/2}(\sigma/\sqrt{n}) < \bar{x}-\mu < z_{\alpha/2}(\sigma/\sqrt{n})) \\
&= P(-\bar{x}-z_{\alpha/2}(\sigma/\sqrt{n}) < -\mu < -\bar{x}+z_{\alpha/2}(\sigma/\sqrt{n})) \\
&= P(\bar{x}-z_{\alpha/2}(\sigma/\sqrt{n}) < \mu < \bar{x}+z_{\alpha/2}(\sigma/\sqrt{n})) \\
&= 1-\alpha
\end{aligned}$$

The above derivation is instructive although there is some awkwardness in the last probability expression. Readers might note that loosely worded expressions as "The probability that the mean, μ, … " have no meaning and are avoided here. There is no "probability" about μ. The mean μ is a fixed quantity. We simply may not know its value. One is reminded of the "number of jellybeans in a jar" contests. Probabilities of the number of jellybeans are meaningless as their number is fixed. However, it is the estimates of their numbers that are subject to chance.

Our use of z implies an assumption of normality for $\bar{x}$. Other situations require more advanced statistical theory than is presented here.

100(1 − α) % Confidence Interval for μ

$$\bar{x} \pm z_{\alpha/2}\frac{\sigma}{\sqrt{n}}$$

♦ "A statistical method for estimating animal populations is proving increasingly useful in documenting human-right abuses" according to the report listed below The method that had been used to count wild animals and adjusted to census estimates of hard-to-count groups was used to reconcile the conflicting figures given regarding the death toll of the 1980–2000 civil war in Peru. A report concluded that 69,280 people (a confidence interval from 61,007 to 77,552) were killed or disappeared during the Peruvian conflict.

(Knight, J., "Statistical model leaves Peru counting the cost of civil war." Nature, September 4, 2003.)

Example 10.5.12 Confidence Intervals

A scientific field expedition requires batteries with a mean life of 35 hours. A sample of 36 batteries is chosen at random from a large batch and life tested. The result yields an

⟶

$\bar{x} = 35$ hours. Since the population standard deviation, σ, is not known, an estimate is obtained by calculating the sample standard deviation, s, as 1 hour. Construct confidence intervals for the mean battery life at 90% and 80% confidence levels.

Solution:
A 90% level means that α is of 0.10 and we seek $z_{0.10/2} = z_{0.05}$, a z value with a 0.05 probability in each tail. The $z_{0.05}$ value is ± 1.645 for an assumed normal distribution (as shown).

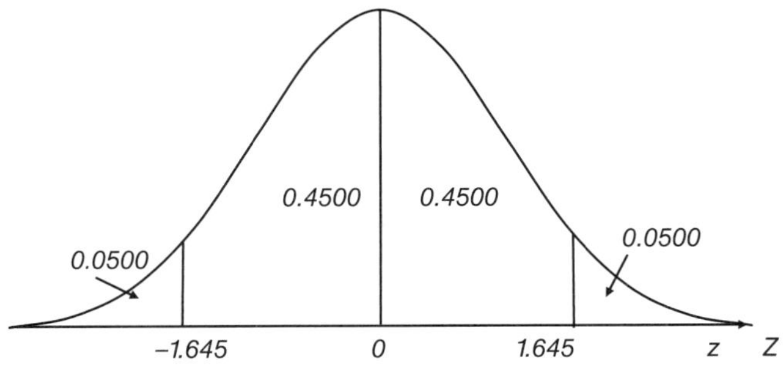

Using the expression for a confidence interval yields

$$35 \pm 1.645\left(\frac{1}{\sqrt{36}}\right) = 35 \pm 0.27 = (34.73,\ 35.27)$$

For the 80% level, the $z_{\alpha/2} = z_{0.10}$ value is $\pm$ 1.28.
The interval is $35 \pm 1.28(1/\sqrt{36}) = 35 \pm 0.21 = (34.79,\ 35.21)$.
In this example, both the 90% and 80% levels include 35 hours.

Note that the higher the confidence level, the wider the confidence interval. In a sense, the higher the chosen confidence level, the wider the interval needed to cover the risk that it includes the mean μ.

Also, note that the width of the confidence interval is $2z_{\alpha/2}(\sigma/\sqrt{n})$. Narrower confidence intervals are usually preferable to wider intervals for a given confidence level. There are three ways by which an interval can be narrowed: reduce the confidence level, $z_{\alpha/2}$; and/or increase the sample size, n; and/or reduce the population standard deviation.

Example 10.5.13 Confidence Interval Width

In the previous example, the two confidence intervals (34.73, 35.27) and (34.79, 35.21) have respective widths of 0.54 and 0.42 hours. Assuming that the sample standard deviation, s, remains unchanged, how many additional samples are required to reduce the interval width to 20 minutes?

⟶

Solution:
For the 90% interval, its width of 1/3 hour is

$$W = 1/3 = 2(1.645)(1)/\sqrt{n}$$

Solving for n yields 97.4 (rounded to 98). Thus, 62 additional samples in addition to the earlier 36 are required.

For the 80% interval, its width is $W = 1/3 = 2(1.28)(1)/\sqrt{n}$.

Solving for n and rounding up yields 59, so 23 additional samples are required.

◆ Baseball "stats" have been a perennial attraction for amateurs and professionals over many decades. From "gb" to "rbi" and "era," clever means have been devised to measure the performance of players, teams, and even managers.

A recent measure identifies a manager's ability to squeeze a few extra wins from each season. Start with the "Pythagorean theory" devised by "stat guru" Bill Jones, which estimates the number of wins expected from a team – called "P-wins."

P-wins are subtracted from actual wins to equal a "P-Diff," which purports to measure a manager's ability to win close games over a long career.

Years ago, a *Wall Street Journal* article ("Einstein on the Bench") noted Bobby Cox of the Atlanta Braves among other long-term managers with a 1.90 P-Diff per season over his 21-year career.

Here is an untried idea! Recall that $\frac{x-\mu}{\sigma}$ is a basic statistical measure. P-Diff is roughly comparable to the numerator. By tracking the manager's record by seasons, σ can be estimated. This could be a clue to a manager's consistency and improvement over the years.

EXERCISES 10.5

Assume a normal probability distribution unless otherwise specified.

1. In many applications, there is an interest in the probability that a normal random variable is within "plus or minus" one, two, and three standard deviations from the mean. Calculate the associated probabilities:
   a) $P(-\sigma < (x - \mu) < \sigma)$
   b) $P(-2\sigma < (x - \mu) < 2\sigma)$
   c) $P(-3\sigma < (x - \mu) < 3\sigma)$

2. Manufacturers know that the machines used to fill containers with product actually dispense amounts slightly different than their setting. The variation, usually slight, might reasonably follow a normal law with mean μ and standard deviation σ. To help assure that the amount dispensed will not overrun the container, a slightly larger size is used. While one cannot absolutely guarantee no overrun, a container size, C, can be determined that reduces the chances of overrun to an acceptable level.

   Determine C for the case $\mu = 10$, $\sigma = 2$, and $\alpha = 1\%$.

3. In the previous exercise, containers of size $C = 14$ are available for a machine with a standard deviation of unity. If $\alpha = 2\%$, how should the machine's mean content be set?

4. The amount released by a dispensing machine is a normal random variable with a mean of two units. The product is released into containers of size 2.1 units. Unfortunately, the machine loses precision with usage as its standard deviation increases. If $\alpha = 5\%$ is the maximum tolerable overrun policy, how large can σ become before the machine requires overhaul?

5. Two routes of travel, A and B, require random times described by normal probabilities.
   Suppose

$$\mu_{\mathrm{A}} = 39 \text{ minutes} \qquad \sigma_{\mathrm{A}} = 4 \text{ minutes}$$

$$\mu_{\mathrm{B}} = 40 \text{ minutes} \qquad \sigma_{\mathrm{B}} = 3 \text{ minutes}$$

   If 45 minutes can be allocated for the journey, which route should be chosen? Suppose that only 42 minutes can be allocated. Does this change the choice of route?

6. A preventive health screening to test for a certain malady is being planned for a large number of people. A choice is to be made between two types of tests that are available. Each test requires that it be administered until a certain indicator appears. The time for this is a random variable that varies from one person to the next person according to a normal distribution. For the first test, the normal distribution has a mean of 12 minutes with a standard deviation of 2 minutes. In the second test, the mean is 10 minutes and the standard deviation is 3 minutes. To manage the screening, planners have allowed up to 15 minutes per person. Which test should be chosen for the screening?

7. The height of a dam should be such that there is only a 3% chance of overtopping in any year. The distribution of the yearly maximum river height is normal with a mean of 5 feet and a standard deviation of 2 feet. How high should the dam be built?

8. Modern road-paving equipment can be programmed to produce a selected pavement thickness. The actual thickness ("on the ground") is a normal random variable. A 2 inch thickness is specified for a certain road. Standard deviations tend to be about 1 / 10 of the (intended) mean thickness. Inspectors can assess penalties for lapses in the 2 inch requirement. What thickness should be programmed so that the chance is at least 90% that a pavement thickness measured at random will meet the 2 inch requirement?

9. Peak hour airline seat demand is approximately normal with a mean of 1000 passengers per hour and a standard deviation of 100 passengers per hour. What seat capacity is needed so that no more than 1% of demand is unfilled?

10. Find $z_{\alpha/2}$ for these values of α:

    a) 0.15 b) 0.10 c) 0.08 d) 0.005

11. What is the confidence level for each of these intervals?

a) $\bar{x} \pm 1.96\left(\frac{\sigma}{\sqrt{n}}\right)$

b) $\bar{x} \pm 2.575\left(\frac{\sigma}{\sqrt{n}}\right)$

c) $\bar{x} \pm 1.28\left(\frac{\sigma}{\sqrt{n}}\right)$

d) $\bar{x} \pm 1.645\left(\frac{\sigma}{\sqrt{n}}\right)$

12. Given $n = 45$, $\bar{x} = 100$, and $s = 20$, find a 99% confidence interval for μ.

10.6 DEVELOPING AND CONDUCTING A SURVEY

Marketing professionals, public officials, health and social planning agencies, and pollsters seek reliable information about populations of interest. Usually, time and cost preclude the practicality of a **census**, the polling of an entire population. Sample surveys fill the need! In a **simple random survey**, each member of the population has an equal chance of selection. While there are many types of surveys, simple random surveys are the most common and the subject of this section.

A sample survey is a subset (sample) of a larger set, the **population**. Characteristics of interest are observed or otherwise measured on each member of the sample from which summary statistics such as the mean and variance are calculated. Extrapolations to the population are made from sample data. The validity and reliability of these extrapolations depend on the care with which samples are chosen and measurements taken. Achieving competent samples is usually not easy!

Some polls were such flops that they are famous. Can you determine why they failed?

- *The Literary Digest*, a famous magazine of its day, predicted that in the 1936 presidential election Republican Alfred Landon would win a landslide victory over Democrat Franklin Roosevelt. Actually, Landon won in just two states and lost the election by a huge electoral margin.
- The Gallup Poll, along with almost all other leading national polls of the time, predicted a huge victory for Republican Thomas Dewey over incumbent Harry Truman in the 1948 presidential election. In the end, Truman was the victor.
- In the 2000 presidential election, television network polling called Florida for Democrat Albert Gore in early evening. Later in the evening it "uncalled" Florida as "too close to call." In the end, the US Supreme Court decided the election in favor of Republican George W. Bush.
- In 1985, the Coca-Cola Company launched a huge marketing effort. It announced that it was reformulating its famous drink based on extensive marketing surveys. Soon after the change, public outcry caused the company to restore the original formula as "Classic Coke."

While designing competent surveys can be difficult, explaining why they sometimes fail can be a greater challenge.

In conducting its poll, *The Literary Digest* sent postcards to people having telephones, magazine subscriptions, car owners, and some registered voters. This sample was biased toward the more affluent who could afford such conveniences. In addition, the low 25% response rate may have been a factor. The poll was clearly not a representative of the voting population and the magazine's loss of credibility so great that it ceased publication soon afterward.

Although pollsters had learned much from their 1936 *Literary Digest* experience, the 1948 election is another example. Again, pollsters and political writers confidently, and incorrectly, predicted that Thomas E. Dewey would handily defeat incumbent Harry S Truman. Still, there seems no clear consensus on what went wrong with the polling! The strong Dewey lead for almost the entire campaign, and the complacency it engendered may have led to diminished sampling in the 2 weeks before the election. Meanwhile, Truman traveled relentlessly in his "Give 'em Hell" speeches to growing crowds.

The pollsters' quite visible mistakes have led to improved sampling techniques. Greater care is taken for samples to be representative of the population. Exit polls assure that opinions are taken into account right through Election Day.

Questions to consider along with poll results include the following:

- Who conducted the poll?
- Is the sample representative of the population?
- How were samples chosen?
- How was interviewing done?
- What questions are asked and in what sequence?
- When was the poll conducted?
- What factors could skew results?
- What is the margin of error?

Professional pollsters use surveys to inform manufacturers who test market new products or try to identify potential markets of the feasibility of an endeavor. Various organizations conduct political polls to predict the outcomes of elections or to help in planning campaign strategies. The Nielsen Survey seeks information on TV viewing choices. Surveys not only can provide useful information in a timely and efficient manner, they may be the only practical alternative.

Surveys can be conducted by mail, telephone, in person, Internet, and so on. In person, interviews are more expensive and may be necessary for complex or sensitive information. Telephone surveys may be less expensive and most useful when attempting to measure a respondent's depth of feeling. While mail surveys tend to cost less, they usually have a poor response rate as compared to other surveys. They have the advantage of allowing respondents time to think about their answers to complex questions. The use of computers, the Internet, and other forms of telecommunications are changing the cost, speed, and other factor for conducting surveys.

In any case, a representative sample of the pertinent population is essential to properly use inferential statistics for a reliable survey. Some statisticians believe that a low survey response may raise more questions than a small sample. The number of nonrespondents hinders inferences about the population.

Some have noted an inverse relationship between the length of a survey and its response rate. The merits of each survey question should be justified. Questions should be concise, easily interpreted, and carefully worded to be free of bias and ambiguities.

Sensitive questions may be inappropriate! For instance, some may be reluctant to respond to personal questions as to age, health, or income. Sometimes, ranges such as under 30 and 30–49 are more successful.

When constructing a survey, choose broad topics that reflect the theme of the survey. Decide on the mode of response before development and consider how the collected data is analyzed. Gathering quantities of data unnecessary to the analysis phase or gathering insufficient data for reliable conclusions are drawbacks.

Developing a Survey

Most surveys include *demographic* questions appropriate to the population of interest. They could include questions about the respondent's income, age, race/ethnicity, marital status, gender, or level of education. In addition to simple "yes/no" and multiple choice questions, a researcher might include "open-ended" or "short response" questions as part of a survey. Although such questions are more difficult to tally and less likely to elicit responses, they might suggest items for the researcher to consider.

It may be important to preface a survey with some information about its purpose; how replies will be used; and estimates of the time to complete the survey. Assurances of confidentiality in an individual's response help to improve a response rate.

Consider a survey on smoking bans, for example, a timely topic in many places. Relevant information might be a person's age, whether or not they smoke in addition to how much, whether family members smoke, and opinions on a smoking ban. There may be additional useful questions. The next few examples help develop some survey questions.

Example 10.6.1 Age Category

Rephrase the survey question "How old are you?" as less intrusive?

Solution:
While some people are willing to answer so direct a question, it may elicit more responses in the form:

"What is your age category?"

a) 25 or under *b) 26–40* *c) 41–60* *d) Over 60 years*

Here, categories alleviate some discomfort of directness. Also, note the potential for imprecision in aggregating results when one person who is, say, 40 years and 9 months responds. The person may possibly answer b) or c) here. Can you improve on the question?

Example 10.6.2 Smoker Status

Decide on a rephrasing of the question and cite your reasoning.

"Are you a heavy or a light smoker?"

Solution:
The first potential difficulty can be a nonsmoker offended by the presumption of the questions. Clearly, a prior question asking whether the respondent smokes seems helpful. If the answer is "yes," then a follow-up question can be asked.

⟶

A second difficulty is the terms "heavy smoker" and "light smoker." What one regards as "heavy" another may not. Some may even misinterpret the question. Here is an improvement:

"On average how many packs of cigarettes do you smoke in a day?"

a) 0 (nonsmoker)   b) Less than 1/2   c) 1/2 to 1   d) More than 1

Can you improve the question?

Example 10.6.3 Smoking Ban Opinion

What are biases in these questions?

"Studies have shown that second-hand smoke can have ill effects on people nearby. Do you favor a ban on smoking in public places?"

(Circle one)   Yes   No   Maybe

"The government wants to violate your personal freedom by enacting a ban on smoking. Are you in favor of such a ban?"

(Circle one)   Yes   No   Maybe

Solution:

The first question may have a respondent feel possible harm to others as an influence.

The second question would have the respondent feel their rights are violated.

Sometimes, survey questions involve scaling. There are two types of scaling questions: rating and ranking. In rating, there is a precoded set of ordered responses. In ranking, the respondent is asked to rank a set of options. A relatively small list is needed for rankings. The larger the list, the more difficult to rank reliably.

Example 10.6.4 Selecting a Doctor

A physician, new to town, seeks an office site. Form a survey question to rank factors that influence people's choice of physician.

Solution:

One possibility is

In choosing a physician's rank the following considerations: use 1, for the most important; 2, for the next most, etc.

a) Office location
b) Availability of appointment

⟶

c) Short waiting time
d) Cost of services
e) Quality of car

Sometimes, a pilot survey is useful. A subset of subjects is used for a trial run and conducted as intended for the entire group. For example, for a telephone survey, the test should also be by telephone. The pilot phase is valuable to judge appropriate length, identify ambiguities, clarify questions, and note potential pitfalls. Based on the pilot survey, revisions can be made and if another significant pilot test may be prudent. Mistakes in data collection and recording are common. Finally, people who are to implement the survey should have a trial run and anticipate possible difficulties.

EXERCISES 10.6

1. Improve these survey questions:
   a) How often do you shop on eBay?
   b) What is your annual income?
   c) How many texts have you purchased online?
2. Improve the question:
   Ex-President Bush once proposed changes in Social Security benefits. Do you
   a) agree   b) have no opinion   c) disagree
   with these changes?
3. Write a short survey (10–20 questions) on smoking bans. Include demographic questions and at least one question for the two types of scaling discussed in the text.
4. Pilot test a short survey on a small sample and then reconstruct the survey to reflect the information gained.
5. Create a survey on a topic of interest in your class, pilot test it, and collect the results. Individually or in groups, briefly explain what was learned about creating surveys.

HISTORICAL NOTES

The word "statistics" has several meanings in different contexts. An American Heritage dictionary describes statistics as "the mathematics of the collection, organization, and interpretation of numerical data, especially the analysis of population characteristics by inference from sampling."

While an interest in numbers and data can be traced to many centuries past, the recognition of the ubiquitousness of the normal distribution in many diverse phenomena and data marks a beginning for the modern development of mathematical statistics.

By the late nineteenth century, the maturation of statistical studies is evident in such efforts as those of William Gosset. Gosset, a sometime student of chemistry and mathematics, employed by the Guinness Brewing Company, noticed small deviations from the normal distribution in the quality control sampling of brews. He correctly reasoned that even if the sampling was from a normal distribution, the small samples and unknown population variance accounted for the small discrepancies he observed. His "near normal" distribution is known today as the *t*-distribution and, sometimes, as the "student's distribution" in recognition of the pseudonym Gosset used in reporting his work.

In the early twentieth century, the vigorous mathematical development of the science continued in Britain. Foremost among a number of workers was Sir Ronald Fisher. He is credited with many important advances such as the principle of maximum likelihood, the design of experiments, and the analysis of variance. In addition, his applications of statistics to agriculture were of such importance and value that to this day some of the outstanding schools of statistics are housed in universities' agricultural colleges. "*The Lady Tasting Tea*" by David Salsburg is a book about the development of mathematical statistics.

Advances in statistical science and application continue. Professor George E. P. Box of the University of Wisconsin, a son-in-law of Sir Ronald Fisher, was among the foremost researchers and practitioners of the subject. His "*Statistics for Experimenters,*" coauthored with Stuart and Hunter, is one of the mainstays of statistics, which along with advances in automated computation with an accuracy and speed unknown before the end of the twentieth century.

In a real sense, mathematical statistics has attained an importance in the social sciences, commerce, and economics, in addition to genetics, medicine, environmental science, and other life sciences, which appear to be leading sciences of the twenty-first century.

♦ The science of Biostatistics is one that applies statistical theory and methods to determine the solutions to problems in the biological and health sciences. Biostatisticians can answer questions such as what are the risk factors associated with certain diseases and what preventive steps can people take to reduce the risk of these diseases.

Mathematical modeling and new experimental methods in nutrition are gaining recognition as powerful research tools in the life sciences. A group of mathematical modelers has been meeting to share and exchange ideas about mathematical modeling methods in nutrition and health sciences since 1980.

George E. P. Box (1919–2013) was born and raised in England. His interest in statistics began with biochemical experiments on the effect of poison gases on small animals for the British Army during World War II (WWII). After the war, he went on to receive a degree in mathematics and statistics from University College, London.

From 1948 to 1956, Box was a statistician at Imperial Chemical Industries (ICI), Britain's chemical giant. In 1960, Box joined the University of Wisconsin in Madison to begin their Department of Statistics, now a leading center.

Sir Ronald Aylmer Fisher (1890–1962) — geneticist and statistician, born in London. He studied mathematics and astronomy at Cambridge with an interest in biology. He taught mathematics and physics from 1915 to 1919. He became chief statistician at Britain's Galton laboratories as well as a statistician at the Rothamsted Agricultural Experiment Station, the oldest agricultural research institute in the United Kingdom.

At Rothamsted, Fisher contributed to both genetics and the statistical design and analysis of experiments. He introduced concepts of randomization, the analysis of variance, and the method of maximum likelihood, and is the "father" of modern statistics.

William Sealy Gosset (1876–1937) — English statistician was educated at New College, Oxford, where he studied chemistry and mathematics obtaining his degrees in 1897 and 1899. Gosset, employed as a chemist by the Guinness brewery in Ireland, acquired his interest in statistics in the sampling of brews. He noted unexplained small deviations from the normal distribution for smaller samples and devised the t-distributions to explain them. Gosset published under the pseudonym "Student" and the t-distribution is often referred to by that name.

CHAPTER 10 SUPPLEMENTARY EXERCISES

1. Classify these data by class (qualitative, quantitative) and by type (nominal, ordinal, interval, ratio).
   a) Your political party affiliation.
   b) A taste testers ranking (best, worst, etc.) of five restaurants.
   c) Your preferred brand of soft drink.
   d) The weight of your textbooks.
2. Collect gasoline prices at several stations and calculate measures of central tendency and dispersion.
3. Have classmates share data (above) to decide on a lower class boundary and class interval for a relative frequency histogram of gas prices.
4. Find the measures of central tendency for the data given below:

   75, 45, 51, 87, 32, 95, 90, 44, 56, 56, 33, 63, 32, 39

5. A sample of the time (in minutes) for 20 people to assemble a rocking chair kit follows. Determine the measures of central tendency and dispersion.

50	70	70	60	30	85	80	80	100	40
70	80	90	100	45	75	50	100	75	35

6. Assume a uniform random variable, x, $40 \leq x \leq 90$. Find these probabilities.

   a) $P(x > 37)$ b) $P(x \geq 68)$ c) $P(x < 82)$ d) $P(45 \leq x \leq 78)$

7. Assume a uniform distribution with $50 \leq x \leq 80$. Find $\boldsymbol{a}$ such that
   a) $P(x < \boldsymbol{a}) = 0.20$
   b) $P(x \geq \boldsymbol{a}) = 1$ (there is more than one correct answer)
   c) $P(\boldsymbol{a} < x < 70) = 0.35$
   d) $P(x > \boldsymbol{a}) = 0$ (there is more than one correct answer)

8. Find the standard normal probabilities:

a) $P(-1.43 \leq z \leq 2.15)$
b) $P(1.09 \leq z \leq 2.74)$
c) $P(z \leq 2.75)$
d) $P(z \geq 0.83)$
e) $P(-1.92 \leq z \leq 1.92)$
f) $P(0 \leq z \leq 1.87)$

9. Find a value of the standard normal random variable z, call it z_0, such that
a) $P(z \leq z_0) = 0.7357$
b) $P(-z_0 \leq z \leq z_0) = 0.3400$
c) $P(z \geq z_0) = 0.9099$
d) $P(z_0 \leq z \leq 1.10) = 0.8000$

10. Find a value of the standard normal random variable z, call it z_0, such that
a) $P(z \geq z_0) = 0.9875$
b) $P(-z_0 \leq z \leq z_0) = 0.8740$
c) $P(z \geq z_0) = 0.1611$
d) $P(z_0 \leq z \leq 2.10) = 0.7494$

11. The random variable x has a normal distribution with standard deviation 25. It is known that the probability that x exceeds 145 is 0.8413. What is the mean?

12. If $n = 50$, $\sum x = 5000$, and $\sum (x - \bar{x})^2 = 980$, find a
a) 80% confidence interval for μ.
b) 90% confidence interval for μ.
c) 95% confidence interval for μ.

11 Enrichment in Finite Mathematics

Finite Mathematics: Models and Applications, First Edition. Carla C. Morris and Robert M. Stark.

Companion Website: http://www.wiley.com/go/morris/finitemathematics

11.1 GAME THEORY

Life abounds with choices! Different choices lead to different outcomes! Usually, there is an "opponent," imaginable, natural, or human, whose action may be met as a pilot's response to a weather change, a general's tactical change, a politician's response to a poll, a company's advertising campaign, an athlete's reaction, and so on.

Game theory is the mathematics of competitive conflict. It aids choices of strategies among opponents.

The subject has attracted a wide following among military strategists and economists. From beginnings in the 1920s, a great advance was the publication of *Games and Economic Behavior* by John Von Neumann and Oscar Morganstern in 1944. A half-century later, Nobel Prizes in Economics were won by such thinkers as John Nash, John Harsanyi, and Reinhard Selten. Their 1994 prize was awarded for "their pioneering analysis of equilibria in the theory of non-cooperative games."

♦ The unusually popular movie and book of the late 1990s, *A Brilliant Mind* told of John Nash's game theory discoveries.

♦ The 2005 Nobel Prize in Economics was awarded to Robert I. Aumann and Thomas C. Schelling. Their game theory work "helped defense analysts use models to map out options available to an adversary and thus predict what the opponent must do in a confrontation," according to the Royal Swedish Academy of Science.

Game theorists study **two person zero sum games**. That is, a "*player*" and an "*opponent*" engage strategies or "*plays*" such that gain (loss) of one party is exactly the loss (gain) of the opposing party: a "zero sum!"

Payoff Matrix

In a simple coin match, each player simultaneously displays a coin face. If the faces match, both "heads" or both "tails," say, *Player A* wins the opponent's coin. If the faces do not match, Player A surrenders the coin to *Player B*. Each play of the game can be represented by a "**payoff matrix**." The payoff matrix for Player A is at the left. The payoff matrix for Player B, on the right, is its opposite.

Payoff Matrix for Player A

$$\text{Player A}\;\begin{matrix} & \text{Player B} \\ & \begin{matrix}\textbf{H} & \textbf{T}\end{matrix} \\ \begin{matrix}\textbf{H}\\ \textbf{T}\end{matrix} & \begin{bmatrix} +1 & -1 \\ -1 & +1 \end{bmatrix}\end{matrix}$$

Payoff Matrix for Player B

$$\text{Player A}\;\begin{matrix} & \text{Player B} \\ & \begin{matrix}\textbf{H} & \textbf{T}\end{matrix} \\ \begin{matrix}\textbf{H}\\ \textbf{T}\end{matrix} & \begin{bmatrix} -1 & +1 \\ +1 & -1 \end{bmatrix}\end{matrix}$$

Clearly, if the payoff matrix for one player is known in zero sum games, the payoff matrix for the opponent is also known. Therefore, it is only necessary to display a single matrix, and the payoff matrices here will henceforth be for Player A.

To interpret the payoff matrix for the simple-coin match, suppose Player A consistently plays its first strategy (Heads) hoping that Player B also plays Heads. Once Player B realizes that Player A consistently plays Heads, a switch is made to the second strategy, Tails. In this manner, Players switch strategies seeking an advantage.

Example 11.1.1 Payoff Matrix

Determine Player B's payoff matrix if the payoff matrix for Player A is

$$\begin{bmatrix} 2 & 3 & 1 \\ -1 & 6 & -5 \\ 3 & -2 & 1 \end{bmatrix}$$

Solution:
Player B's payoff matrix is the negative of Player A's matrix.
The payoff matrix for Player B is

$$\begin{bmatrix} -2 & -3 & -1 \\ 1 & -6 & 5 \\ -3 & 2 & -1 \end{bmatrix}$$

In a mixed strategy, players choose from a multiple of strategies, usually at random, with frequencies (probabilities) for each strategy.

Example 11.1.2 "Rock, Paper, Scissors"

Recall the childhood game in which "rock crushes scissors, scissors cut paper, and paper smothers rock." Express the game in a payoff matrix.

Solution:
The payoff matrix for Player A can be written as

$$\begin{array}{cc} & \textbf{Player B} \\ & \begin{array}{ccc} \textbf{R} & \textbf{P} & \textbf{S} \end{array} \\ \textbf{Player A} \begin{array}{c} \textbf{R} \\ \textbf{P} \\ \textbf{S} \end{array} & \begin{bmatrix} 0 & -1 & 1 \\ 1 & 0 & -1 \\ -1 & 1 & 0 \end{bmatrix} \end{array}$$

where, of course, **R** *represents "rock;"* ***P****, "paper;"* ***S****, "scissors."*

♦ RPS – "An ancient game for two players flinging a hand at each other on a 1-2-3 count. A fist is rock, a flat hand is paper, and two extended fingers signify scissors. Rock smashes scissors, paper covers rock, scissors slices paper. Typically, the winner takes two of three throws" is the way RPS was described in the *Wall Street Journal.*

To avoid giving an opponent an obvious advantage when only a single play is used, the symmetry of the game suggests that each Player randomly choose among R, P, and S equally on average. This game is **fair**, that is, its value is zero to each Player as shown later.

♦ Think Rock–Paper–Scissors is only an ancient children's game? A recent USA Rock Paper Scissors League, formed by Hollywood executives, competes with an old time World R.P.S. Society based in Toronto, Canada. Avid players develop strategies of feigning and finger selection to prepare for a championship tournaments.

Example 11.1.3 A Pure Strategy

What strategies are evident for this payoff matrix?

$$\begin{bmatrix} 1 & 2 \\ -2 & -1 \end{bmatrix}$$

Solution:
Player A will never choose the second strategy since it can only lose (either two or one unit, respectively, depending on B's play). Player B, on the other hand, will never choose the second strategy since the first strategy has uniformly better consequences: losing, at

⟶

most, one unit if A plays the first strategy and gaining two units if A foolishly plays the second strategy.

As a result, each Player will choose their first strategy: Player B is resigned to lose one unit at each play and Player A is resigned to never gaining more than one unit at each play. This results in a single choice is called a ***pure strategy****.*

Fundamental Principles

1. Each player, acting rationally, chooses the best possible strategy.
2. Each player assumes that the opponent also chooses the best possible strategy.
3. A best possible strategy maximizes a player's objective.

The size of a payoff matrix may be reduced by eliminating rows or columns by inspection. That is, if each element of a row is smaller than the corresponding element in another row, it is said to be **dominated** and can be deleted from the matrix as a rational player would not make that choice. A column is dominated if its elements are larger than the corresponding elements in another column.

Guidelines for Reduction by Dominance

1. Are elements of a row of the payoff matrix dominated by another row? If affirmative, delete the dominated row.
2. Is there a column of the payoff matrix whose elements are dominated by another column? If affirmative, delete the dominated column.
3. Repeat these steps in any order until no row or column dominates another.

Although pure strategies occasionally occur, it is more likely that a mixed strategy is needed. The choice of a mixed strategy usually is not as obvious as in the last two examples. The **value** of a game is the average return to each Player per play of the game. In a zero-sum game, if the value is positive for one player it is negative for the opponent.

An example may help understand concepts.

Example 11.1.4 Reduction by Dominance

Reduce this payoff matrix using dominance

$$\begin{bmatrix} 5 & 3 & 5 & 4 \\ 3 & 2 & 3 & 3 \\ 4 & 4 & 1 & 6 \end{bmatrix}$$

Solution:

Row 1 dominates Row 2 since each element of Row 1 is larger than the corresponding element of Row 2. Therefore, Row 2 can be eliminated as a rational row player would never use it.

$$\begin{bmatrix} 5 & 3 & 5 & 4 \\ 4 & 4 & 1 & 6 \end{bmatrix}$$

⟶

$$\text{from} \begin{bmatrix} 5 & 3 & 5 & 4 \\ \cancel{3} & \cancel{2} & \cancel{3} & \cancel{3} \\ 4 & 4 & 1 & 6 \end{bmatrix}$$

Column 2 dominates both Columns 1 and 4 (since its values are less than or equal to each of these column values). Therefore, Columns 1 and 4 can be eliminated as Player B would not choose either of these strategies so beneficial to Player A.

The result is

$$\begin{bmatrix} 3 & 5 \\ 4 & 1 \end{bmatrix} \qquad \text{from} \begin{bmatrix} \cancel{5} & 3 & 5 & \cancel{4} \\ \cancel{4} & 4 & 1 & \cancel{6} \end{bmatrix}$$

Neither remaining row nor column dominates so there are no further reductions.

Minimax Criterion

A **best possible strategy** depends on what players decide is "best." In a **minimax** strategy, Players opt for strategies to minimize their worst possible outcomes. Therefore, Player A, the row player, seeks to maximize the minimum payoff. Player B, the column player, on the other hand, seeks to minimize the maximum payoff.

Example 11.1.5 The Minimax Criterion

Determine each player's minimax strategy for this payoff matrix.

$$\begin{array}{cc} & \textbf{\textit{Player B}} \\ & \begin{array}{ccc} p & q & r \end{array} \\ \textbf{\textit{Player A}} \begin{array}{c} s \\ t \\ u \end{array} & \begin{bmatrix} -3 & -1 & 3 \\ 4 & 0 & -1 \\ 3 & 2 & 4 \end{bmatrix} \end{array}$$

Solution:

The game cannot be reduced since there is no row or column dominance. To determine the minimax strategy for Player A, place a circle about the minimum gain for each row strategy. In case of a tie choose either.

$$\begin{array}{cc} & \textbf{\textit{Player B}} \\ & \begin{array}{ccc} p & q & r \end{array} \\ \textbf{\textit{Player A}} \begin{array}{c} s \\ t \\ u \end{array} & \begin{bmatrix} \enclose{circle}{-3} & -1 & 3 \\ 4 & 0 & \enclose{circle}{-1} \\ 3 & \enclose{circle}{2} & 4 \end{bmatrix} \end{array}$$

The row minima are −3, −1, and 2 (circled). The largest is 2, so Player A chooses strategy u.

$\longrightarrow$

Next, Player B notes the column maximum values. These are the worst outcomes for him (boxed).

$$\textbf{Player A}\quad \begin{array}{c} \\ s \\ t \\ u \end{array}\overset{\textbf{Player B}}{\begin{array}{c} \begin{array}{ccc} p & q & r \end{array} \\ \begin{bmatrix} -3 & -1 & 3 \\ \boxed{4} & 0 & -1 \\ 3 & \boxed{2} & \boxed{4} \end{bmatrix} \end{array}}$$

The smallest boxed value for the columns is 2, indicating that Player B should use strategy q.

The best that Player A can do is to choose strategy u while for Player B strategy q minimizes his loss.

In the previous example, the payoff to Player A of 2 with the (u, q) strategy was simultaneously a row minimum and a column maximum. It is a **saddle point**. Any game that has one or more saddle points is said to be **strictly determined**. That is, no other strategy can improve either player's return. Saddle points have the same payoff value, called the **value of the game**. They result from **pure strategies** for Players A and B. If a game is *fair,* its value is zero. If the value is other than zero, the game is *biased* or *unfair*.

Often, there is no pure strategy that is optimal and the game is **nonstrictly determined** as the next example illustrates.

Example 11.1.6 Nonstrictly Determined Game

Verify that the payoff matrix is for a nonstrictly determined game.

$$\textbf{Player A}\quad \begin{array}{c} \\ s \\ t \\ u \end{array}\overset{\textbf{Player B}}{\begin{array}{c} \begin{array}{ccc} p & q & r \end{array} \\ \begin{bmatrix} 4 & 3 & -5 \\ 3 & -4 & 2 \\ -6 & 5 & 3 \end{bmatrix} \end{array}}$$

Solution:

Placing circles and boxes about "best" choices in the rows and columns, as earlier, yields

$$\textbf{Player A}\quad \begin{array}{c} \\ s \\ t \\ u \end{array}\overset{\textbf{Player B}}{\begin{array}{c} \begin{array}{ccc} p & q & r \end{array} \\ \begin{bmatrix} \boxed{4} & 3 & \textcircled{-5} \\ 3 & \textcircled{-4} & 2 \\ \textcircled{-6} & \boxed{5} & \boxed{3} \end{bmatrix} \end{array}}$$

There being no overlap among circles and boxes, the game has no saddle point and no pure strategy (nonstrictly determined).

The **value** of a game is the (average) amount won (or lost) per play. It follows that for a mixed strategy the value of the game is a chance outcome. In randomly chosen mixed strategies, the value of the game is an **average** or **expected value**. Determining the probabilities with which mixed strategies should be executed can be challenging. Remarkably, such games can be solved using linear programming.

♦ Oncologists apply game theory to predict conflict and cooperation in cell-to-cell interactions with tumors as they seek energy to mutate and grow. Game theory calculations accounted for mutation rates when a tumor suddenly switches between energy metabolic strategies. The tumor is thought to be especially vulnerable within the window of strategy switching to having its cell relationships disrupted. (Use a Google search of game theory–oncology).

The Game as a Linear Program

It is a surprising fact that linear programming can solve every two person zero sum game. A solution yields each player's best strategy and the expected value of the game.

Consider a general $m \times n$ payoff matrix, $\boldsymbol{P}$.

$$\boldsymbol{P} = \begin{bmatrix} a_{11} & a_{12} & \cdots & a_{1n} \\ a_{21} & a_{22} & \cdots & a_{2n} \\ & & \cdots & \\ a_{m1} & a_{m2} & \cdots & a_{mn} \end{bmatrix}$$

Again, $\boldsymbol{P}$ is the payoff matrix for Player A (row Player) and the same matrix with each element's sign reversed is Player B's (column Player) payoff matrix.

Let $x_1, x_2, \ldots, x_m$ and $y_1, y_2, \ldots, y_n$ be the (to be determined) frequencies or probabilities, comprising Player A's and Player B's respective mixed strategies. Of course, as all probability sums, $x_1 + x_2 + \cdots + x_m = 1$ and $y_1 + y_2 + \cdots + y_n = 1$.

If Player A plays his first strategy (with probability x_1) the expected payoff, being the payoff multiplied by the associated probability, is

$$a_{11}y_1 + a_{12}y_2 + \cdots + a_{1n}y_n$$

as Player B plays her mixed strategy $y_1, y_2, \ldots, y_n$. However, Player A only plays his first strategy with probability x_1. The contribution to the expected payoff to Player A from the first strategy is obtained simply by multiplying by x_1 as

$$a_{11}y_1x_1 + a_{12}y_2x_2 + \cdots + a_{1n}y_nx_n$$

Continuing in this manner for each of Player A's strategies, and adding them, yields the total expected payoff, Z, for Player A. That is, the double sum (over rows and columns) is

Total expected payoff: $Z = \sum_{j=1}^{n} \sum_{i=1}^{m} a_{ij}y_jx_i.$

Example 11.1.7 Total Expected Payoff

*For the 2 × 2 payoff matrix, **P**, write the expected payoff for each player's strategy and the total expected payoff*

$$\boldsymbol{P} = \begin{bmatrix} 2 & 3 \\ 1 & 4 \end{bmatrix}$$

Solution:

For the row player, Player A, using the mixed strategy x_1 and x_2 $(x_1 + x_2 = 1)$ and y_1 and y_2 $(y_1 + y_2 = 1)$, and for the column player, Player B, the expected payoffs are

Row player	$x_1(2y_1 + 3y_2)$
	$x_2(y_1 + 4y_2)$
Column player	$y_1(2x_1 + x_2)$
	$y_2(3x_1 + 4x_2)$

The total expected payoff is

$$Z = (2y_1x_1 + 3y_2x_1) + (y_1x_2 + 4y_2x_2) + (2x_1y_1 + x_2y_1) + (3x_1y_2 + 4x_2y_2)$$

To formulate a game as a linear program, consider Player A's situation. If Player B plays its first strategy while Player A plays its mixed strategy, the expected payoff, using the general payoff matrix, **P**, is

$a_{11}x_1 + a_{21}x_2 + \cdots + a_{m1}x_m$ Player B plays strategy 1 and Player A plays mixed strategy

Similarly,

$a_{12}x_1 + a_{22}x_2 + \cdots + a_{m2}x_m$ Player B plays strategy 2 and Player A plays mixed strategy

Let v be the (unknown) expected value of the game to Player A. Then the expected payoffs to Player A can be written as constraints that represent the expected return for a choice of strategy (assuming $v \geq 0$)

$$a_{11}x_1 + a_{21}x_2 + \cdots + a_{m1}x_m \geq v$$
$$a_{12}x_1 + a_{22}x_2 + \cdots + a_{m2}x_m \geq v$$
$$\cdots$$
$$a_{1n}x_1 + a_{2n}x_2 + \cdots + a_{mn}x_m \geq v$$

Since constraints are the expected return when A selects successive strategies, each is at least the expected value of the game, v. Hence, the use of the "≥" symbol.

Player A's objective is to maximize v subject to the (above) constraints.

That is, dividing the constraints by v, one has

$$a_{11}X_1 + a_{21}X_2 + \cdots + a_{m1}X_m \geq 1$$
$$a_{12}X_1 + a_{22}X_2 + \cdots + a_{m2}X_m \geq 1$$
$$\cdots$$
$$a_{1n}X_1 + a_{2n}X_2 + \cdots + a_{mn}X_m \geq 1$$

where $X_1 = \frac{x_1}{v}$, $X_2 = \frac{x_2}{v}$, … , $X_m = \frac{x_m}{v}$ $(v > 0)$.

For the objective, maximizing v is equivalent to minimizing $\frac{1}{v}$

$$\text{Minimize}: Z = \frac{1}{v} = \frac{x_1}{v} + \frac{x_2}{v} + \cdots + \frac{x_m}{v} = X_1 + X_2 + \cdots + X_m$$

Also, $x_1, x_2, \ldots, x_n \geq 0$. This completes the linear program formulation for Player A.

Example 11.1.8 Player A's Linear Program

Express the game in Example 11.1.7 as a linear program for Player A.

Solution:
Let x_1 and x_2 be Player A's probabilities and v the expected payoff. Replacing x_1 and x_2 by $X_1 = \frac{x_1}{v}$ and $X_2 = \frac{x_2}{v}$, respectively, yields

$$\begin{aligned} \textit{Minimize:} \quad & Z = \frac{1}{v} = X_1 + X_2 \\ \textit{Subject to:} \quad & 2X_1 + X_2 \geq 1 \\ & 3X_1 + 4X_2 \geq 1 \\ & X_1, X_2 \geq 0 \end{aligned}$$

The solution appears in Example 11.1.10.

Now, a formulation for Player B seeks to keep Player A's maximum expected payoff, v, as small as possible; that is, to maximize Z (smaller v, larger $Z = 1/v$). Letting $Y_1 = \frac{y_1}{v}$, … , $Y_n = \frac{y_n}{v}$, we write

$$\text{Maximize:} \quad Z = \frac{1}{v} = Y_1 + \cdots + Y_n$$

(Again, $y_1 + y_2 + \cdots + y_n = 1$ so $Y_1 + \cdots + Y_n = \frac{1}{v}$ after substitution.)

Subject to:

$$a_{11}Y_1 + a_{12}Y_2 + \cdots + a_{1n}Y_n \leq 1$$

$$\cdots$$

$$a_{m1}Y_1 + a_{m2}Y_2 + \cdots + a_{mn}Y_n \leq 1$$

$$Y_1, \ldots, Y_n \geq 0$$

The constraints were formed by noting that if Player A plays its first strategy while Player B plays a mixed strategy, $y_1, \ldots, y_n$, the expected payoff is

$$a_{11}y_1 + a_{12}y_2 + \cdots + a_{1n}y_n \leq v$$

Since Player B wants to lose no more than v, a "$\leq$" symbol applies. Dividing by v yields the desired format.

Example 11.1.9 Player B's Linear Program

Express the game in Example 11.1.7 as a linear program for Player B.

Solution:
Let y_1 and y_2 be B's respective probabilities and v the expected payoff to be minimized. Then

$$\text{Maximize: } Z = Y_1 + Y_2$$

$$\text{Subject to: } 2Y_1 + 3Y_2 \leq 1$$

$$Y_1 + 4Y_2 \leq 1$$

$$Y_1,\ Y_2 \geq 0$$

expresses the linear program for Player B.

It is instructive to compare the two linear programs.

Player A	Player B
Minimize: $Z = X_1 + \cdots + X_m$	Maximize: $Z = Y_1 + \cdots + Y_n$
Subject to:	Subject to:
$a_{11}X_1 + a_{21}X_2 + \cdots + a_{m1}X_m \geq 1$	$a_{11}Y_1 + a_{12}Y_2 + \cdots + a_{1n}Y_n \leq 1$
$a_{12}X_1 + a_{22}X_2 + \cdots + a_{m2}X_m \geq 1$	$a_{21}Y_1 + a_{22}Y_2 + \cdots + a_{2n}Y_n \leq 1$
$\cdots$	$\cdots$
$a_{1n}X_1 + a_{2n}X_2 + \cdots + a_{mn}X_m \geq 1$	$a_{m1}Y_1 + a_{m2}Y_2 + \cdots + a_{mn}Y_n \leq 1$
$X_1, \ldots, X_m \geq 0$	$Y_1, \ldots, Y_n \geq 0$

These are dual linear programs! Either can be the primal program and the other its dual. These are primal and dual, having the solution to one yields the solution to the other.

In the sense of game theory, having the optimal strategy for Player A, the optimal strategy for Player B is at hand and vice versa. It follows, as well, that if the optimal value of the game is v for Player A, then $-v$ is optimal for Player B.

Example 11.1.10 Player B's Optimal Strategy

Use the Simplex Method to solve Example 11.1.9.

Solution:
Forming a tableau from the example

Row	Y_1	Y_2	s_1	s_2	RHS
1	2	3	1	0	1
2	1	4	0	1	1
Objective	−1	−1	0	0	0

Row	Y_1	Y_2	s_1	s_2	RHS
1	1	3/2	1/2	0	1/2
2	0	5/2	−1/2	1	1/2
Objective	0	1/2	1/2	0	1/2

An optimal solution! The tableau indicates $Y_1 = 1/2$ and $Y_2 = 0$. Therefore,

$$v = \frac{1}{Y_1 + Y_2} = \frac{1}{1/2 + 0} = 2$$

The optimal strategy for Player B is $y_1 = Y_1 v = (1/2)(2) = 1$ and $y_2 = 0$. The optimal value, 1/2, is the value of $1/v$. Therefore, $v = 2$. Also, $Y_1 = y_1/v$, etc.

Note that the slack-related variables have coefficients $s_1 = 1/2$ and $s_2 = 0$. These correspond to $x_1 = s_1 v = (1/2)(2) = 1$ and $x_2 = s_2 v = 0$. This result follows from the nature of the dual program. Optimal strategies are row 1 for Player A and column 1 for Player B. Alternatively, you may have noted that the game in 11.1.7 had a saddle point "2" as it was both a row minimum and column maximum. This would also indicate Player A chooses roe 1 and Player B chooses column 1.

We have assumed that the expected payoffs are positive valued and, in particular, that the optimal expected value of the game is positive for Player A. This is an easily removed restriction as shown in the next example.

Example 11.1.11 Negative Payoffs

Use the Simplex Method to solve the game

$$\boldsymbol{P} = \begin{bmatrix} -3 & +2 \\ +1 & -2 \end{bmatrix}$$

⟶

Solution:
Since negative values appear in the payoff matrix, it is possible that the value of the game to the row player is negative. Since negative-valued variables can be an inconvenience, simply add a quantity to each element so that all of them become positive valued. In this instance, adding +4 to each element yields

$$\boldsymbol{P}^{(+4)} = \begin{bmatrix} 1 & 6 \\ 5 & 2 \end{bmatrix}$$

Remember that +4 must be subtracted later from the optimal value of the linear program solution.

One can form a linear program for either player. We use the formulation for Player B as it yields a familiar SMP (Chapter 5). Let v represent expected value to Player A and Player B's mixed strategy, y_1 and y_2, then

$$\begin{aligned} \text{Maximize:} \quad & Z = Y_1 + Y_2 \\ \text{Subject to:} \quad & Y_1 + 6Y_2 \le 1 \\ & 5Y_1 + 2Y_2 \le 1 \\ & Y_1,\ Y_2 \ge 0 \end{aligned}$$

where $Y_1 = y_1/v \ge 0$ and $Y_2 = y_2/v \ge 0$.

Forming a tableau and using slack variables s_1 and s_2

Row	Y_1	Y_2	s_1	s_2	RHS
1	1	6	1	0	1
2	(5)	2	0	1	1
Objective	−1	−1	0	0	0

Row	Y_1	Y_2	s_1	s_2	RHS
1	0	(28/5)	1	−1/5	4/5
2	1	2/5	0	1/5	1/5
Objective	0	−3/5	0	1/5	1/5

Row	Y_1	Y_2	s_1	s_2	RHS
1	0	1	5/28	−1/28	1/7
2	1	0	−1/14	3/14	1/7
Objective	0	0	3/28	5/28	2/7

The last tableau indicates $Y_1 = 1/7$ and $Y_2 = 1/7$. Therefore,

$$v = \frac{1}{Y_1 + Y_2} = \frac{1}{1/7 + 1/7} = \frac{7}{2}$$

However, we must subtract the +4 that was added to the payoff matrix at the start so $v = 7/2 - 4 = -1/2$.

⟶

Recall that v is the optimal expected value of the game to Player A. Therefore, in this instance, the value to Player B is $-v = +1/2$ when the mixed strategy of "half and half" is played. For Player A, the best strategy is to play $x_1 = s_1 = 3/8$ and $x_2 = s_2 = 5/8$ although it results in a loss of 1/2 unit on average at each play.

♦ Game theory may have a role in recent research on human altruism – believed to be unique among mammals. Scientists' theories have had difficulty explaining pure altruism. In experiments, players are given a sum of money and individually decide how much to contribute to a "kitty." The amount in the kitty is doubled and distributed equally among the players. Clearly, freeloaders have an advantage. Researchers have found that some players try to exact some punishment from freeloaders. Various game situations and payoff matrices are devised to study the phenomenon whereby players punish freeloaders, even at some expense to themselves, resulting in a continuing civilization. (Google: Game theory-altruism).

EXERCISES 11.1

In Exercises 1–3, write the opponents payoff matrix

1. $\begin{bmatrix} 6 & 3 & -1 \\ -5 & 3 & -4 \\ 3 & 2 & 1 \end{bmatrix}$ 2. $\begin{bmatrix} -1 & 3 & -2 \\ 1 & -4 & 5 \\ -3 & 2 & -1 \end{bmatrix}$ 3. $\begin{bmatrix} 1 & 3 & 6 \\ -2 & -4 & 5 \\ -1 & 3 & 1 \end{bmatrix}$

In Exercises 4–7, use dominance to reduce the payoff matrices.

4. $\begin{bmatrix} 4 & 1 & 3 \\ 2 & 0 & 1 \\ 5 & 5 & 3 \end{bmatrix}$

5. $\begin{bmatrix} 5 & 3 & 5 & 4 \\ 2 & 1 & -1 & 3 \\ 6 & 1 & 4 & 2 \end{bmatrix}$

6. $\begin{bmatrix} 2 & 1 & -1 & 5 \\ 3 & 2 & 2 & 5 \\ 3 & 2 & 2 & 3 \\ 2 & 0 & 1 & 3 \end{bmatrix}$

7. $\begin{bmatrix} 8 & 2 & 4 & 8 & 6 \\ 1 & 4 & 5 & 9 & 4 \\ 6 & 4 & 1 & 5 & 5 \\ 9 & 5 & 4 & 7 & 5 \end{bmatrix}$

In Exercises 8 and 9, determine a minimax strategy.

8. $\begin{bmatrix} 3 & 2 & 5 \\ 6 & 1 & 3 \\ 2 & 1 & 0 \end{bmatrix}$

9. $\begin{bmatrix} 2 & 3 & 3 & 4 \\ 1 & 4 & -2 & 2 \\ 2 & 1 & 3 & 4 \\ 0 & 3 & -3 & 1 \end{bmatrix}$

10. Verify that this game is nonstrictly determined (no saddle point).

$$\begin{bmatrix} -3 & 2 & 5 \\ 6 & 4 & 3 \\ 2 & -1 & 4 \end{bmatrix}$$

11. Verify that the game for this payoff matrix is nonstrictly determined (no saddle point).

$$\begin{bmatrix} 5 & 2 & 3 & 4 \\ 3 & 1 & 2 & 3 \\ 4 & 2 & 3 & 1 \\ 5 & 3 & 3 & 2 \end{bmatrix}$$

12. Use linear programming to solve the game for this payoff matrix.

$$\begin{bmatrix} 5 & 3 \\ 4 & 6 \end{bmatrix}$$

13. Use linear programming to solve the game for this payoff matrix.

$$\begin{bmatrix} 3 & 2 \\ -2 & 3 \end{bmatrix}$$

HISTORICAL NOTES

Émile Borel (1871–1956) – French mathematician was educated at the École Normale Supérieure in Paris. At age 22 Borel was appointed to the chair of mathematics at Lille University. He is credited with creating the first effective theory of the measure of sets of points. His work helped begin the modern theory of functions of a real variable. From 1921 to 1927, Borel published a series of papers on game theory and became the first to define games of strategy. Borel authored almost 300 papers and over 30 books.

John Von Neumann (1903–1957) – a foremost mathematician of the twentieth century. Some credit him with setting much of the course of twentieth-century mathematics. His fundamental work in non-Aristotelian logic and the mathematics of game theory had a role in the development of quantum physics.

11.2 APPLICATIONS IN FINANCE AND ECONOMICS

Finances: "Up Close and Personal"

Here are a few financial principles of everyday personal life.

Start Saving Early Starting in 1963, "Early Saver" invested $2000 each year in an S&P 500 Index fund for 10 years and then let the money ride for a total investment of $20,000. Unfortunately, "Early" had unbelievably terrible timing. Each year, the $2000 investment was made on the day the S&P 500 Index was at its highest for the whole year. (The object is to "buy low" and "sell high").

"Later Saver" also invested $2000 each year in the same S&P 500 Index fund. However, "Later" began in 1973 and continued to invest for 20 years (twice as long as "Early Saver") and, also, let the money ride – a total investment of $40,000. Fortunately, "Later" had unbelievable timing luck. Each year the investment was made on the day that the S&P 500 was at its the lowest for the year! How did they do?

As of May, 1999 "Later's" account was worth \$848,917. A supergrowth of money! Amazingly, "Early's" account is worth more at \$876,004 -with half the investment!

Such is the power of saving regularly and "early" rather than "later"! It is geometric versus arithmetic growth.

(adapted from a 1999 Louis Rukeyser Publication)

How to Calculate an Average Annual Rate of Return Newspapers and magazines carry advertising from mutual funds touting their average annual return. Often the choice of "average" isn't cited. There are arithmetic averages and geometric averages – and they differ. Which is a more useful measure of money over time?

◆ Given a sum of money at some instant, its value is clear. However, when monies received are spent at varying times, there is not a single way to assess value. This is quite apart from changing currency values. Rather, it relates to the ability of money to earn money.

It can be a challenge to compare sums invested for longer periods when returns fluctuate. Consider the implications of simple percentages. Imagine a gain of 50% one year and a 50% loss the next. Your average percentage return is 0% (= 50% + (−50%))! Right? Wrong!

Let us check it out!

Start with \$1000. Add 50% to have \$1500. Lose 50% to have \$750 – a loss of \$250 over two periods; an average of 25% over two periods or, roughly, 12.5% per period. Clearly, 0%, and the **arithmetic mean** it represents are not a useful measure.

Many people arithmetically average annual rates of return. For example, if the annual percentage returns for four consecutive years are 7%, 3%, 5%, and 5%, the average arithmetic annual return is 5% (= (7% + 3% + 5% + 5%)/4). Clearly, this is not a useful measure of total return.

Another and more useful method is a **geometric average**, usually known as the *compound annual return* in the financial literature. Generally, an initial principal, P_0, invested for n years at annual rates of return $r_1\%$, $r_2\%$, … , $r_n\%$ in the respective years becomes a final principal, P_n, given by

$$P_n = P_0\left(1+\frac{r_1}{100}\right)\left(1+\frac{r_2}{100}\right)\cdots\left(1+\frac{r_n}{100}\right)$$

Let us try another mean – the **geometric mean**, representing the percentage as a fraction, for example, 50% by 1/2. The geometric mean for a 50% gain and a 50% loss over two periods yields the annual average rate of

$$\sqrt{\left(1+\frac{1}{2}\right)\left(1-\frac{1}{2}\right)}-1=\sqrt{\frac{3}{2}\left(\frac{1}{2}\right)}-1=-0.134$$

or −13.4% per period.

As a check – reduce \$1000 by 13.4% to yield \$866 after 1 year. Then reduce \$866 by another 13.4% for the second year to yield \$750. This agrees with our earlier calculation!

The lesson? Be cautious in accepting average annual returns uncritically. It is well to ask, "*What* ***mean*** *do you mean*?"

The geometric average, or compound annual return, is:

Geometric Mean – (Compound Annual Return)

$$\left(\frac{P_n}{P_0}\right)^{1/n} = \sqrt[n]{\left(1+\frac{r_1}{100}\right)\left(1+\frac{r_2}{100}\right)\cdots\left(1+\frac{r_n}{100}\right)} - 1$$

♦ John Allen Paulos, in his popular "A Mathematician Plays the Stock Market," cites an example using a $10,000 stock investment that rises 80% one week and loses 60% the next, repeatedly over a year. The average weekly return might be stated as an "average gain of 10% per week"; 1/2 [(80% + (−60%)], the arithmetic mean. At that rate, the original investment is worth about $1.4 million after a year.

That is, $10,000(1.10)^{52} = \$1,420,429.30$! However, a more likely outcome is only $1.95 – Yes! Less than $2 for the entire year!

Using the geometric mean,

$\sqrt{(1+.8)(1-.6)} - 1 = -0.1515$ or an average loss of about 15% each week.

Therefore, $\$10,000(1 - 0.1515)^{52} = \$1.95$.

The importance of the compound annual return, the geometric mean, is that it considers both compounding and volatility. The arithmetic average return is particularly vulnerable to distortion by volatile returns as the next example illustrates.

Example 11.2.1 A Volatile Speculation

A $1000 speculation in a volatile security returns 100% the first year and a 50% decline in the second year. Calculate the arithmetical average annual return and compare with the compound annual return.

Solution:

After the first year, the original principal becomes $2000 (100% return).

After the second year, the previous year's principal becomes $1000 (−50%) return. The arithmetic average annual return is

$$[(100\% + (-50\%)]/2 = 25\%$$

Of course, the actual return is zero!

Using the compound annual return,

$$\left[\sqrt{\left(1+\frac{100}{100}\right)\left(1+\frac{-50}{100}\right)} - 1\right] 100\% = \left(\sqrt{(1+1)(1-0.50)} - 1\right) 100\% = 0\%$$

A proper result!

♦ An instructive article in the *Wall Street Journal* some years ago noted that the S&P 500 Index from 1927 to 2003 returned about 12% using arithmetic averages and 10% using geometric averages. That 2% difference is a substantial sum when compounded over a working life.

Financial Planning A widespread financial concern is how much can safely be withdrawn **periodically** from savings, retirement plans, insurance settlements, inheritances, lottery winnings, and such and still meet some long-term need or objective. It is the flip side of saving for some goal: a new car, home, or university education.

If only matters were as simple as dividing the total available sum by the number of required years! In reality interest rates can fluctuate, so earnings vary substantially over longer periods and life spans are uncertain, to say nothing of inflationary effects, tax law changes, etc.

To begin, we explore the effect of interest rates to gain a "ball park" estimate of annual income.

Example 11.2.2 Effect of Interest Rates

A newly retired 65-year-old has a $1 million lifetime retirement fund. With good health and habits, a 20-year life span is quite likely.

A reasonable long-term average for interest rates may be about 5% with a 2% variation that provides likely upper and lower boundaries. What annual incomes will these rates provide?

Solution:

This is equivalent to an amortization. In Section 2.3, a loan L at interest rate i was shown to be retired by n annual payments of

$$\frac{iL}{\left(1 - \frac{1}{(1+i)^n}\right)}$$

Here, $L = \$1,000,000$, $n = 20$ years. We calculate for $i = 3\%$, 5%, and 7%.

$$Y_{3\%} = \frac{(0.03)(1,000,000)}{\left(1 - \frac{1}{(1+0.03)^{20}}\right)} = \$67,215.71$$

$$Y_{5\%} = \frac{(0.05)(1,000,000)}{\left(1 - \frac{1}{(1+0.05)^{20}}\right)} = \$80,242.59$$

$$Y_{7\%} = \frac{(0.07)(1000000)}{\left(1 - \frac{1}{(1+0.07)^{20}}\right)} = \$94,392.93$$

Note the significant dependence of annual income, Y, on the interest rate over a relatively long period.

Unknown life expectancy, of course, is a serious impediment to planning. This raises the query as to the variation for different life spans.

Example 11.2.3 Effect of Life Span

Assuming a 5% annual rate, how will annual income vary for remaining life spans of 15, 20, and 25 years?

Solution:
Again, consider a $1 million retirement fund. As in the last example, the annual income is found for n = 15, 20, and 25 years.

$$Y_{5\%} = \frac{(0.05)(1,000,000)}{\left(1 - \frac{1}{(1+0.05)^{15}}\right)} = \$96,342.29 \quad \textit{15 years}$$

$$Y_{5\%} = \frac{(0.05)(1,000,000)}{\left(1 - \frac{1}{(1+0.05)^{20}}\right)} = \$80,242.59 \quad \textit{(from last example) 20 years}$$

$$Y_{5\%} = \frac{(0.05)(1,000,000)}{\left(1 - \frac{1}{(1+0.05)^{25}}\right)} = \$70,952.46 \quad \textit{25 years}$$

♦ Many mutual fund companies provide web sites to aid in calculating retirement benefits using different strategies and assuming different estimates for returns. Vanguard Funds, Fidelity Investments, and T. Rowe Price are examples of such major mutual funds.

Actually, both interest rates and life expectancy being uncertain, Monte Carlo Method using probability distributions permits a variety of experimentation as suggested in the exercises. Here, we illustrate Monte Carlo Method using uniform probability distributions. You may defer the next example until you study Section 11.4.

Example 11.2.4 Monte Carlo

Assume that an annual interest rate, i, is described by a probability $P(i) = 1/5$, $i = 3\%, 4\%, 5\%, 6\%$, and 7%, and the probability of a life expectancy, l, in years, by $P(l) = 1/3$, $l = 15, 20$, and 25 years. Organize a Monte Carlo estimate of average annual income. (Section 11.4 for the Monte Carlo Method.)

Solution:
One way to proceed is to choose a random number (RN) to fix a life span for a single trial. Next, choose an RN to fix the interest rate. The year and income assume that the selected

⟶

interest rate applies for the remainder of life. The next year begins with $1 million plus the year's interest and minus the first-year income. Next, a new interest rate is randomly selected and the process repeated for the principal resulting from the first year. You may use the table on page 407.

Maximizing Cash Flow Much commerce is carried on by contracts that specify when the work is to be completed and how (periodic) payments are to be made.

It is usually to the *contractor's* advantage to receive payments as early as possible since monies are required to continue the work.

Example 11.2.5 Cash Flow

Consider the simplified and hypothetical excavation project description below provided by the ***owner*** *(sponsor).*

Item #	*Description Contractor's*	*Estimate*
1	Clearing	25,000 square yards
2	Earth excavation	60,000 cubic yards
3	Rock excavation	40,000 cubic yards
4	Clean up	25,000 square yards

The owner forms estimates of time and cost as

Item #	*Estimated Unit Price, $*	*Amount, $*	*Estimated Completion Time, Months*
1	2.00	50,000	3
2	1.00	60,000	12
3	3.50	140,000	12
4	2.00	50,000	15
	Total	300,000	

Assuming that payment is made to the contractor by the owner at the completion of each phase, and the contractor borrows money at a 1% monthly rate to perform the work, the present worth of the job to the contractor is

$(50,000)(1.01)^{-3} + (60,000)(1.01)^{-12} + (140,000)(1.01)^{-12} + (50,000)(1.01)^{-15}$
$= \$269,086.83.$

How might the contractor increase cash flow?

Now the contractor considers ways to increase return from the project. Unit prices are estimates and the actual prices can depend on labor, weather, and equipment availability, and such factors are not known in advance. One can experiment on the effect of varying unit prices.

⟶

Solution:
Let X_1 be the unit price (per square yard) of clearing, X_2 per cubic yard of earth excavation, etc. First, form the present worth to be maximized as

$$Z = (25,000)(1.01)^{-3}\ X_1 + (60,000)(1.01)^{-12}\ X_2 + (40,000)(1.01)^{-12}\ X_3 + (25,000)(1.01)^{-15} X_4$$

where, again, X_1 is the unit price of item #1, etc., and quantity estimates are assumed to be correct.

The maximization is subject to constraints. For example, a unit price of earth excavation should not exceed that for rock excavation. To ensure this, use the constraint

$$X_2 - X_3 \leq 0$$

Also, since the job is "priced" at \$300,000, write

$$25,000X_1 + 60,000X_2 + 40,000X_3 + 25,000X_4 = 300,000$$

Solving the associated linear program, in this case by inspection, yields

$$X_1^* = 12,\ X_2^* = X_3^* = X_4^* = 0,\ \text{and}\ Z^* = \$291,177.04$$

a sizable increase over the \$269,086.83 calculated earlier. However, this solution essentially asks for full payment upon completion of item #1. Clearly, this is not a practical solution. Indeed, it is a risky one since quantity estimates may not be reliable. For example, if clearing item #1 should prove to be 20,000 square yards rather than 25,000 square yards, the contractor would be in the difficult position of having to complete the job for \$240,000 (20,000 × \$12). One can avoid such extreme results with additional constraints. For example, a reasonable estimate for the cost of clearing might be between \$1.50 and \$2.50 per square yard. This suggests the constraint $1.50 \leq X_1 \leq 2.50$ and so on. The discussion of maximizing cash flow is carried on in the exercises.

♦ This simplified example is an instance of *unbalanced bidding*, prevalent in bidding for contracts in engineered and highway construction, lumbering in national forests, and so on. Contracts are usually awarded on the basis of competitive bids for the entire project. However, unit prices are also specified since in virtually all instances the quantities in the bid proposal are, of necessity, only estimates. After the contract is awarded, periodic payments are made on the basis of the unit prices. Clearly, unbalancing bids can be risky and, often, resisted. A Google search lists over 400,000 references to this financial topic.

Bank Portfolio Management A commercial bank's income and profitability depend on decisions to deploy its assets or *portfolio*.

♦ An article "Linear Programming: A New Approach to Bank Portfolio Management" in the Monthly Review of the Federal Reserve Bank of Richmond (November 1972) is an illustration. The following simple example is adapted from the article.

Example 11.2.6 A Bank Portfolio

A bank has \$100 million available of which \$45 million are from demand deposit accounts, \$45 million from time deposit accounts, and \$10 million of capital and surplus. For simplicity of discussion, suppose that securities yield average 5% while loans (since they are less "liquid" and riskier) have average yield of 10%. However, the bank has constraints upon its actions. While there are many in reality, consider these three:

1. *Total Funds Constraint:* $L + S \leq 100$ *million*
   where L is the dollar amount of loans and S is of securities.
   The bank recognizes that some quantity of its securities should be "liquid," that is easily turned into cash to meet unanticipated withdrawals. The bank makes it a rule to maintain some minimum ratio of securities to total assets, say 25% here. That is,
2. *Liquidity constraint:* $S \geq 0.25(L + S)$
   Or, equivalently,
   $$S \geq (1/3)L$$
   Lending being a primary activity, the bank attempts to satisfy all of the requests for loans by its principal customers. Suppose that loan needs total \$30 million. That is,
3. *Loan balance constraint:* $L \geq 30$ *million*
   Illustrate the effect of each constraint graphically. Then illustrate the feasible region. Finally, select an asset portfolio that maximizes the bank's total returns.

Solution:
The graphs of the three constraints are placed on a single graph (as illustrated) with the feasible region shaded and the corner points noted.

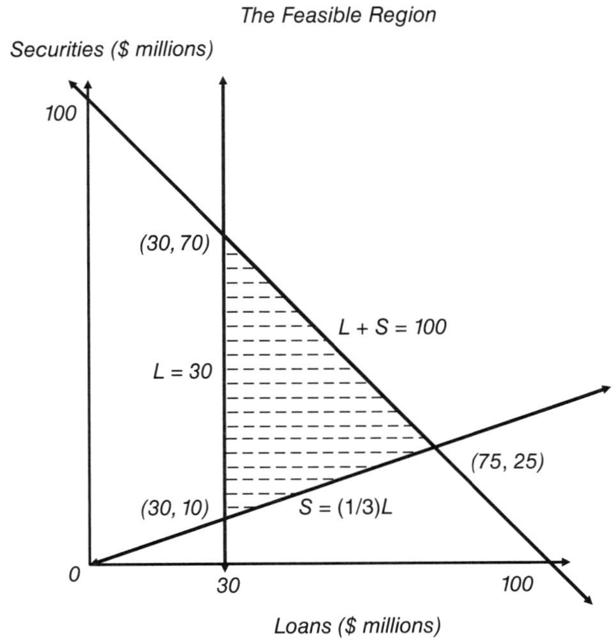

→

The feasible region(shaded), in which the three constraints just outlined are all relevant, shows how as a group they restrict the bank's range of choice. Any asset portfolio represented by a point outside the shaded region violates one or more of the constraints. Conversely, any portfolio represented by a point within or on one of the boundaries of this shaded region satisfies all of the constraints. The portfolio selected must be a point within the shaded region including its boundaries.

Since loans yield 10% and securities 5%, the bank's total income, I, is

$$I = 0.10\,L + 0.05\,S$$

The desired linear program is

$$\begin{aligned} \text{Maximize:}\quad & I = 0.10\,L + 0.05\,S \\ \text{Subject to:}\quad & L + S \leq 100 \text{ million} \\ & S \geq (1/3)L \\ & L \geq 30 \end{aligned}$$

The optimum portfolio has $L^* = \$75$ *million,* $S^* = \$25$ *million, and* $I^* = \$8.75$ *million.*

◆ The Federal Reserve Bank of Richmond article cited earlier discusses such matters as the central importance of bank portfolio management to balance (i) income, (ii) liquidity to meet unanticipated loan demands and withdrawals, and (iii) the risk of default. It reviews commercial bank practices, describing their limitations in dealing with such banking realities as account stability and shifts in asset yields, as well as the advantages of centralizing operations to recognize important interactions. It claims that the linear programming approach avoids these difficulties.

The article also describes actual banking applications of its time. These are interesting discussions of actual situations such as ones in which there are more than two categories of assets and liabilities, with detailed and realistic sets of liquidity constraints reflecting cash flow patterns, coping with seasonal fluctuations in loan demand, and maintaining legal reserve requirements. A distinct advantage of the linear programming framework, according to the author, is that it can be easily applied to rapidly changing market conditions.

EXERCISES 11.2

1. Imagine a stock that doubles every odd-numbered week of the year and loses one-half its value each even-numbered week. What is the arithmetic average weekly return and how much will a \$1000 investment be worth in a year? What is the geometric average weekly return and how much will a \$1000 investment be worth in a year?
2. Imagine a volatile investment that gains 75% a year and loses 75% the next year. Is the average gain (loss) actually zero?

3. A car is financed for \$25,000 at 6% interest for 6 years. What is the monthly payment?
4. Credit card issuers require a minimum monthly payment on unpaid balances. Using a \$1000 unpaid balance, ascertain the minimum payment for your favorite credit card and determine how long it will take to retire the balance.
5. Compare total loan costs on the purchase of a new car with loans of 3, 4, 5, or 6 years.
6. Car leasing is a common alternative to purchase. Compare the economics of lease versus purchase for a choice of vehicle.
7. For the cash flow problem in Example 11.2.5, suppose that other constraints are added to reduce risks of unfavorable unit prices. Solve the resulting linear program with the additional constraints:

$$1 \leq X_1 \leq 3, \qquad 1 \leq X_4 \leq 3, \qquad 1 \leq X_2$$

8. Unbalancing of bids can help to time income. Suppose that payments on items 2 and 3 in the text cash flow, Example 11.2.5, are on the basis of units completed (table below) even if the entire excavation is incomplete. Also suppose the contractor seeks no more than \$125,000 income this tax year. How are the optimal unit prices altered and what is the optimal contract yield?

Item #	Required Time Months	Begin	End
1	3	July 1, 2005	September 30, 2005
2	9	October 1, 2005	June 30, 2006
3	9	October 1, 2005	June 30, 2006
4	3	July 1, 2006	September 30, 2006

9. Suppose that the contractor of the cash flow, Example 11.2.5, has a better estimate of quantities and finds that items #2 and #3 are each 50,000 cubic yards; other quantities being accurate. How might the unit pricing be changed?
10. The bank in Example 11.2.6 gains access to an additional \$10 million of time deposits raising the total to \$55 million. The reduced risk from the larger amount of time deposits permits a reduction to 20% in the liquidity ratio. How shall assets be deployed for an optimum return? If the bank were to decline the increase in its income, by how much can it afford to increase the yield to its depositors?

11.3 APPLICATIONS IN SOCIAL AND LIFE SCIENCES

Social Sciences

Kinship Structure Social scientists have a keen interest in how social groups form and evolve. The following three examples touch on aspects of the subject and illustrate the importance of matrices to social scientists.

Example 11.3.1 Kinship

A hypothetical three clan system has these marriage rules:

Men from Clan A only marry women from Clan C
Men from Clan B only marry women from Clan A
Men from Clan C only marry women from Clan B

For children the rules are

Children in Clan B have fathers from Clan A
Children in Clan C have fathers from Clan B
Children in Clan A have fathers from Clan C

Express these rules in a matrix format.

Solution:
Use a "1" to indicate a "positive" response and a "0" otherwise.

$$\begin{array}{cc} & \text{Wife's Clan} \\ & \begin{array}{ccc} A & B & C \end{array} \\ \text{Husband's Clan} & \begin{array}{c} A \\ B \\ C \end{array}\begin{bmatrix} 0 & 0 & 1 \\ 1 & 0 & 0 \\ 0 & 1 & 0 \end{bmatrix} = \mathbf{M} \quad \text{Marriage rules} \end{array}$$

and

$$\begin{array}{cc} & \text{Child's Clan} \\ & \begin{array}{ccc} A & B & C \end{array} \\ \text{Father's Clan} & \begin{array}{c} A \\ B \\ C \end{array}\begin{bmatrix} 0 & 1 & 0 \\ 0 & 0 & 1 \\ 1 & 0 & 0 \end{bmatrix} = \mathbf{P} \quad \text{Progeny rules} \end{array}$$

Next, we use these matrices to study this society.

Example 11.3.2 Next of Kin

*Use the **M** and **P** matrices in the last example to describe this society.*

Solution:
*From the **M** matrix,*

- *The three clans are exogamous – no marriages within the clan.*
- *Men (row) can only choose wives (column) from a single clan.*

→

- *Women can only choose husbands from a single clan.*
- *Men from different clans must marry women from different clans.*

Consequently, ***M*** *cannot be an identity matrix; otherwise people could marry within their clan.*
From the ***P*** *matrix,*

- *All children of a couple belong to the same single clan.*
- *Children do not share a clan with either parent.*

The ***P*** *matrix cannot be an identity matrix as a child cannot share a clan with a parent.*

Whether two people, related by marriage or descent, are in the same clan depends only on the kind of kinship and not the clan to which either belongs. For example, the wife of a grandson in B shares A with his grandfather. This follows since the son of a father in A belongs to B. As a man in B, his wife will also belong to A. Thus, the sharing of a clan by a grandfather and a grandson's wife depend on kinship, not clan. Mathematically, this means that if any diagonal element of P is unity, then all its diagonal elements are unity. (If this were not true, some men would have children in their own clan and other men might not).

Next, we trace progeny affiliation!

Example 11.3.3 In What Clan?

The sister of a man in Clan A has a child. In what clan does the child belong?

Solution:
Since all the children of a couple are in the same clan, the sister of the man in Clan A is also in Clan A. Referring to column A in the ***M*** *matrix, the sister's husband's clan corresponding to the row containing a 1 is B.*

Now, knowing that the father belongs to B, the ***P*** *matrix indicates the child's clan as C. The child of a sister of a man in clan A belongs to clan C.*

Example 11.3.4 In What Clan? Revisited

Solve the last example using matrix operations.

Solution:
The row vector $\mathbf{S} = [1 \quad 0 \quad 0]$ *expresses the sister's and brother's clan.*

The "backward" or "reverse" path into ***M*** *to obtain the husband's clan is equivalent to a "forward" look through its transpose,* $\mathbf{M}^t$. *That is,* $\mathbf{SM}^t$

$$\mathbf{SM}^t = [1 \quad 0 \quad 0]\begin{bmatrix} 0 & 1 & 0 \\ 0 & 0 & 1 \\ 1 & 0 & 0 \end{bmatrix} = [0 \quad 1 \quad 0]$$

⟶

indicating that her husband belongs to B. Since the husband is the child's father, apply $\boldsymbol{SM^t}$ *to* $\boldsymbol{P}$ *as*

$$\boldsymbol{SM^t\,P} = [0 \quad 1 \quad 0]\begin{bmatrix} 0 & 1 & 0 \\ 0 & 0 & 1 \\ 1 & 0 & 0 \end{bmatrix} = [0 \quad 0 \quad 1]$$

The result, as before, indicates that the child belongs to C.

It is interesting that transposes of matrices **M** and **P**, $\mathbf{M^t}$ and $\mathbf{P^t}$, are also equal to their respective inverses. That is,

$$\mathbf{M^t} = \mathbf{M^{-1}} \text{ and } \mathbf{P^t} = \mathbf{P^{-1}}$$

This follows, as the elements are 0's and 1's with exactly a single-unit element in each row and column. Such matrices are called **permutation matrices**.

In a similar manner, multiplications with the four matrices, **M**, $\mathbf{M^{-1}}$, **P**, and $\mathbf{P^{-1}}$, in appropriate sequences, identify clans of kin by blood or by marriage.

Another interesting consequence is that in a social system of n clans, there is a value of k so that $\mathbf{P}^k = \mathbf{I}$ for $1 \leq k \leq n$. If this were not so, no descendent (or ancestor) could be in the same clan. The same is true for powers of **M**. One inference is that the rules that divide some societies into mutually exclusive groups eventually tend to integrate them.

♦ Various rules for societal formations can be studied using the methods illustrated here. A primary source is *An Anatomy of Kinship* by Harrison C. White (1963). Also, *Mathematical Sociology* by Robert L. Leik and B. F. Meeker (1975), from which examples have been adapted here. *An Introduction to Finite Mathematics* by John Kemeny, L. Snell, and G. Thompson (1957) is useful.

Occupational Mobility The movement of workers among jobs is also of considerable interest to sociologists. The job mobility examples in Chapter 9 illustrated the use of Markov chains in their study. While such models reasonably predict occupational transitions, they may overestimate the numbers that are willing to change occupations. A more sophisticated model assumes that workers are one of two types: "stayers" or "movers."

Imagine four categories of occupations denoted by A, B, C, and D for convenience. Consider the proportions of "stayers" as elements of a matrix **S**. Note that the elements form a diagonal matrix.

For example, in the matrix **S**,

$$\mathbf{S} = \begin{array}{c} \\ \mathbf{A} \\ \mathbf{B} \\ \mathbf{C} \\ \mathbf{D} \end{array} \begin{array}{c} \begin{array}{cccc} \mathbf{A} & \mathbf{B} & \mathbf{C} & \mathbf{D} \end{array} \\ \begin{bmatrix} 0.50 & 0.00 & 0.00 & 0.00 \\ 0.00 & 0.80 & 0.00 & 0.00 \\ 0.00 & 0.00 & 0.60 & 0.00 \\ 0.00 & 0.00 & 0.00 & 0.70 \end{bmatrix} \end{array} \quad \textbf{Stayer's matrix}$$

For instance, 80% of workers in occupation B remain there indefinitely.

It follows that $\mathbf{M} = \mathbf{I} - \mathbf{S} = \begin{bmatrix} 1.00 & 0.00 & 0.00 & 0.00 \\ 0.00 & 1.00 & 0.00 & 0.00 \\ 0.00 & 0.00 & 1.00 & 0.00 \\ 0.00 & 0.00 & 0.00 & 1.00 \end{bmatrix} - \begin{bmatrix} 0.50 & 0.00 & 0.00 & 0.00 \\ 0.00 & 0.80 & 0.00 & 0.00 \\ 0.00 & 0.00 & 0.60 & 0.00 \\ 0.00 & 0.00 & 0.00 & 0.70 \end{bmatrix}$

Therefore,

$$\mathbf{M} = \begin{bmatrix} 0.50 & 0.00 & 0.00 & 0.00 \\ 0.00 & 0.20 & 0.00 & 0.00 \\ 0.00 & 0.00 & 0.40 & 0.00 \\ 0.00 & 0.00 & 0.00 & 0.30 \end{bmatrix} \quad \textbf{Mover's matrix}$$

Since "movers" move, we need to know to where they are likely to move. These probabilities form the transition matrix **J**. For example,

$$\mathbf{J} = \begin{array}{c} \\ \mathbf{A} \\ \mathbf{B} \\ \mathbf{C} \\ \mathbf{D} \end{array} \begin{array}{c} \begin{array}{cccc} \mathbf{A} & \mathbf{B} & \mathbf{C} & \mathbf{D} \end{array} \\ \begin{bmatrix} 0.75 & 0.10 & 0.02 & 0.13 \\ 0.09 & 0.76 & 0.12 & 0.03 \\ 0.01 & 0.03 & 0.69 & 0.27 \\ 0.02 & 0.04 & 0.08 & 0.86 \end{bmatrix} \end{array} \quad \textbf{Job change matrix}$$

Example 11.3.5 Workforce Composition

For the data above, what is the workforce composition proportions one period hence?

Solution:
The proportions of "movers" are the elements of $\boldsymbol{M} = \boldsymbol{I} - \boldsymbol{S}$. *Multiplying by the transition matrix* ***J*** *yields the workforce composition of "movers" after one period.*

$$\boldsymbol{MJ} = \begin{bmatrix} 0.50 & 0.00 & 0.00 & 0.00 \\ 0.00 & 0.20 & 0.00 & 0.00 \\ 0.00 & 0.00 & 0.40 & 0.00 \\ 0.00 & 0.00 & 0.00 & 0.30 \end{bmatrix} \begin{bmatrix} 0.75 & 0.10 & 0.02 & 0.13 \\ 0.09 & 0.76 & 0.12 & 0.03 \\ 0.01 & 0.03 & 0.69 & 0.27 \\ 0.02 & 0.04 & 0.08 & 0.86 \end{bmatrix}$$

$$= \begin{bmatrix} 0.375 & 0.050 & 0.010 & 0.065 \\ 0.018 & 0.152 & 0.024 & 0.006 \\ 0.004 & 0.012 & 0.276 & 0.108 \\ 0.006 & 0.012 & 0.024 & 0.258 \end{bmatrix}$$

As expected, the transition reduces the proportion of "movers" who do not move in the period. To this, add the "stayer" matrix, ***S****, to obtain the entire workforce composition.*

⟶

$$\boldsymbol{S}+\boldsymbol{MJ}=\begin{bmatrix}0.50 & 0.00 & 0.00 & 0.00\\ 0.00 & 0.80 & 0.00 & 0.00\\ 0.00 & 0.00 & 0.60 & 0.00\\ 0.00 & 0.00 & 0.00 & 0.70\end{bmatrix}+\begin{bmatrix}0.375 & 0.050 & 0.010 & 0.065\\ 0.018 & 0.152 & 0.024 & 0.006\\ 0.004 & 0.012 & 0.276 & 0.108\\ 0.006 & 0.012 & 0.024 & 0.258\end{bmatrix}$$

$$=\begin{bmatrix}0.875 & 0.050 & 0.010 & 0.065\\ 0.018 & 0.952 & 0.024 & 0.006\\ 0.004 & 0.012 & 0.876 & 0.108\\ 0.006 & 0.012 & 0.024 & 0.958\end{bmatrix}$$

In a study (cited below), occupations were divided into 11 categories (agriculture and forestry, food and kindred products, transportation and communications, etc.) and an 11×11 job change matrix, **J**, was formed. The proportions of "stayers" and "movers" formed the matrices **S** and $\mathbf{M} = \mathbf{I} - \mathbf{S}$, respectively.

♦ An early reference is *The Industrial Mobility of Labor as a Probability Process* by I. Blumen, et al. (Cornell University Press, 1955). The examples have been adapted from R. K. Leik and B. F. Meeker, *Mathematical Sociology* (Prentice Hall, 1975).

Marriage Problem Marital alliances have varied, and still do, with eras and cultures. The advent of linear programming provides sociologists with mathematical answers to what form of marriage is "best."

To maximize "happiness," let m_{ij} be the ith man's "desire" to spend a unit of time (e.g., a day) with the jth woman. The women's opinion must also be considered. Let w_{ij} be the jth woman's desire to spend a unit of time with the ith man; $i, j = 1, \ldots, n$. Their desires count equally, we suppose, so we use their average $C_{ij} = \dfrac{m_{ij} + w_{ij}}{2}$ as the couple's "happiness" in spending a unit of time together.

Next, let X_{ij} be the time units that the couple (i^{th} man and the ith woman) choose to spend together. Then $C_{ij}X_{ij}$ is the "happiness" of that couple. Total "happiness" of all $2n$ people is the double sum

$$Z = \sum_{i=1}^{n}\sum_{j=1}^{n} C_{ij}X_{ij}$$

Clearly, we seek the X_{ij} that maximizes "total happiness," Z.

Now, that maximization is subject to constraints. A person cannot allocate more than his or her total time, which we take as one unit (e.g., 1 day). That is, the total time that the ith man spends with all of $j = 1, \ldots, n$ women cannot exceed unity. In symbols,

$$\sum_{j=1}^{n} X_{ij} \le 1, \quad \text{all } i = 1, \ldots, n$$

Similarly, for women, the total time the jth female spends with all men cannot exceed unity. In symbols,

$$\sum_{i=1}^{n} X_{ij} \leq 1, \text{ all } j = 1, \ldots, n$$

The linear program to maximize total "total happiness" is

$$\text{Maximize:} \quad Z = \sum_{i=1}^{n} \sum_{j=1}^{n} C_{ij} X_{ij}$$

$$\text{Subject to:} \quad \sum_{j=1}^{n} X_{ij} \leq 1, \quad \text{all } i = 1, \ldots, n$$

$$\sum_{i=1}^{n} X_{ij} \leq 1, \quad \text{all } j = 1, \ldots, n$$

$$\text{and} \qquad 0 \leq X_{ij} \leq 1, \quad \text{all } i, j$$

Mathematically, the optimal values of X_{ij}, all i, j, must be either 0 or 1. This can be interpreted as "Monogamy is best!" Of course, to "minimize" happiness, maximize $-Z$ to conclude "Monogamy is worst." That the X_{ij} are either zero or unity follows from a theorem in the theory of linear programming that states, roughly, that when the coefficients and the right-hand sides of each constraint are unity, the variables X_{ij} must take values of zero or one. It is mathematically impossible for the X_{ij} to have fractional values.

Life Sciences

Setting Medical Dosages Ever wonder how drug dosages are set? Clearly, dosages are intended to be safe and produce therapeutic concentrations in the body.

A base assumption is that the rate of decrease in drug concentration in the body is directly proportional to the instantaneous concentration.

Let $C(t)$ be the drug concentration in the patient at a time t (days), $t = 0, 1, 2, \ldots$ The declining concentration with time can be expressed by a difference equation with k as a proportionality constant known as the *elimination constant.*

$$\Delta C(t) = C(t+1) - C(t) = -kC(t), \qquad t = 0, 1, 2, \ldots$$

where the proportionality factor k is called the *elimination constant.*

The solution to this difference equation can be verified as

$$C(t) = C_0(1-k)^t, \quad t = 0, 1, 2, \ldots$$

where $C_0 = C(0)$ is the initial dosage at $t = 0$.

While the initial dosage decays over the course of time, there is usually a measurable (less than therapeutic) residual at the time for the next dosage.

Example 11.3.6 Residual Concentration

A patient is prescribed a daily medication with an elimination constant 0.80. What percentage of the dose remains the next day when the dose is renewed? The day after?

Solution:
Here, $R_1 = C_0(1 - 0.80)$ corresponds to the drug remaining. Therefore, 20% of the first-day dosage is present on the second day when the next dosage is received.

The day after that, $C(2) = C_1(1 - 0.80) = [C_0 + C_0(1 - 0.8)](1 - 0.8) = 0.24C_0$ or about a quarter of a dose remains before the second day's dosage is received.

While some of the drug remains in the blood stream when it is time for the next dose, the amount may not have much therapeutic value, as noted earlier. However, over long periods undesirably large concentrations might develop.

The drug concentration remaining, R_1, when the next dose is due, is

$$R_1 = C_0(1 - k)$$

Therefore, after the next dose, the concentration, C_1, to begin the second day is

$$C_1 = C_0 + R_1 = C_0 + C_0(1 - k)$$

The remainder after the second dose, R_2, at the end of the second day and just before the third dose is due is

$$R_2 = C_1(1 - k)$$

$$C_2 = C_0 + R_2 = C_0 + (C_0 + C_0(1 - k))(1 - k)$$

Just after the third dose, the drug concentration is $C_2 = C_0 + R_2$. When the fourth dose is due, the remainder is

$$R_3 = C_2(1 - k)$$

$$\text{where } C_2 = C_0 + C_1(1 - k)$$

Generalizing, the cumulative remainder after n doses is

$$R_n = C_0(1 - k)(1 + r + r^2 + \cdots + r^{n-1}) \quad \text{where } r = (1 - k)$$

This is a geometric series with $r < 1$ whose sum is

$$R_n = \frac{C_0(1 - k)(1 - (1 - k)^n)}{1 - (1 - k)}$$

The long-term limit is also of interest as $n \to \infty$ is

$$\lim_{n\to\infty} R_n = \frac{C_0(1 - k)}{k}$$

Therefore, in the long term, when the next dose is due, the concentration is

$$\frac{C_0(1-k)}{k} + C_0 = \frac{1}{k}C_0$$

This is the point where the amount of drug that is due will exactly replace the amount of drug eliminated the day before. It is a "steady state." Here, C_0 is the "maintenance dosage" and $\frac{1}{k}C_0$ is the "therapeutic dosage."

Example 11.3.7 Two Drug Residuals

Two drugs a and b are prescribed to be taken simultaneously every day. If the respective elimination constants are k_a and k_b, what is the long run ratio of their residual concentrations?

Solution:
With k_a and k_b as

$$\lim_{n\to\infty} \frac{R_a}{R_b} = \frac{\dfrac{C_a(1-k_a)}{k_a}}{\dfrac{C_b(1-k_b)}{k_b}} = \frac{C_a\left(\dfrac{1}{k_a}-1\right)}{C_b\left(\dfrac{1}{k_b}-1\right)}$$

For instance, if the initial dosages are equal and $k_a = 0.80$ and $k_b = 0.90$, the long-term ratio of the residuals would be

$$\frac{C_a\left(\dfrac{1}{0.80}-1\right)}{C_a\left(\dfrac{1}{0.90}-1\right)} = \frac{9}{4}$$

Adapted from B. Horelick and S. Koomb, "UMAP 72" and Giordano and Weir, "First Course in Mathematical Modeling."

Example 11.3.8 Setting a Dosage

Devise a strategy for a daily dosage, C_0, so that the concentration at any time is always therapeutic ($C(t) > L$) and is never excessive ($C(t) < H$) where L and H are respective low and high limits.

Solution:
Here, L is the smallest therapeutic concentration and H is the highest safe concentration. If R is the limiting residual drug in the bloodstream, we might set $R = L$ and $H = C_0 + R$. That is, $C_0 = H - L$.

Since $\lim_{n\to\infty} R_n = \dfrac{C_0(1-k)}{k} = R,$

⟶

and substituting

$$L = \frac{(H - L)(1 - k)}{k}$$

Solving for k

$$k = \frac{H - L}{H}$$

This is an elimination constant that keeps the drug therapeutic and the dosage safe.

EXERCISES 11.3

Exercises 1–3 relate to kinship structure

1. a) In what clans are the grandson and great grandson of the sister in Example 11.3.4?
   b) Outline matrix operations for the answer to a)?
2. a) In some societies it is preferred that a man marries his mother's brother's daughter. What matrix operations determine the daughter's clan?
   b) Show that the matrix from (a) is equal to **M**.
3. Verify that a value of k, $1 \leq k \leq n$, results in $\mathbf{P}^k$ to becoming an identity matrix. Use $n = 3$ for convenience.

Exercises 4–6 relate to occupational mobility

4. Calculate the workforce composition at the end of the next period for Example 11.3.5.
5. Develop a "stayer"–"mover" model for voter trends in selected communities.
6. Analyze experimental behavioral data using a "stayer"–"mover" model.

Exercises 7–9 relate to the marriage problem

7. Choose arbitrary values for the C_{ij} and $i = j = n$ and actually solve the linear program using a computer program.
8. The text assumed that the numbers of men and women are equal. Verify that the same conclusion is reached if there are m men and n women $(m \neq n)$. Of course, some X_{ij} are zero. Again, assign arbitrary values for the C_{ij}, and for $i = 1, \ldots, m$ and $j = 1, \ldots, n$.
9. One observer noted that this "marriage model" can be used to argue that monogamy is the "worst" situation. What prompted the observer's conclusion?

11.4 MONTE CARLO METHOD

Monte Carlo Method is arguably the most widely used mathematics in industry and commerce! Better known as **simulation** (perhaps, incorrectly), Monte Carlo is a newer twist on **random sampling** schemes.

Monte Carlo methods are widely used in simulations to evaluate alternatives in the operation of systems too complex for mathematical analyses. Among examples are weather forecasting, managing production lines, traffic on busy highways, multilane supermarket and highway queues, routing delivery trucks, scheduling railroad and airline flights, evaluating stock market portfolio strategies, and operating telephone networks and much more. While these situations are mathematically complex, their Monte Carlo analyses are fairly easy to execute.

Random Numbers

A **random number table** on an accompanying page consists of 1600 random digits arranged in groups of four. Among truly RNs, there are no statistically meaningful patterns among digits.

Example 11.4.1 Family Planning

Imagine a young couple planning for a family of three children. What is the chance that they are girls?

Solution:

Clearly, the chance is one in eight, assuming single births and equally probable male and female offspring.

This can be verified experimentally! Toss three fair coins repeatedly. Assigning, say, head for the outcome "a girl" and tail for "a boy," the number of times three heads appear in each trial of three tosses divided by the number of trials is an estimate of the desired probability.

This is one way to simulate the chance that the three children are girls. Tossing coins repeatedly can be tedious. Random sampling is more convenient!

To begin the random sampling, choose any four digit (for convenience) RN from a random number table, considering only the first three digits. Arbitrarily, assign the event "a girl" to an odd digit, and a zero or even digit to "a boy." For example, the random number 943 represents the family "two girls and one boy." Continuing in this way, tally the number of families with three girls to the total number. In general, as the sample size increases the ratio tends to 1/8 or 0.125. Ten trials are shown in the following table.

Trial Number	*Random Number (RN)*	*Truncated RN*	*Corresponding Event*	*Number of Girls*	*Probability Estimate for Three Girls*
1	9436	943	GBG	2	0/1
2	9083	908	GBB	1	0/2
3	1467	146	GBB	1	0/3
4	5147	514	GGB	2	0/4
5	4058	405	BBG	1	0/5
6	7338	733	GGG	3	1/6
7	0440	044	BBB	0	1/7
8	3394	339	GGG	3	2/8
9	9955	995	GGG	3	3/9
10	2003	200	BBB	0	3/10
⋮	⋮	⋮	⋮	⋮	⋮

⟶

Sampling of Families of Three Children

The last column, a probability average, fluctuates widely. That is an indication that 10 trials are not nearly adequate for a useful probability estimate. You may wish to continue the example to verify the probability and gain an appreciation of the numbers of trials to achieve two-place decimal accuracy. A simple computer program is useful to automate the tally.

Example 11.4.2 *A Perambulating Drunk*

Imagine a drunk standing directly below a streetlight in a town square. The drunk begins to walk, taking steps at random. How far will the drunk travel from the starting point after n steps?

Solution:

Suppose that the drunk begins at the origin of a Cartesian coordinate system. We use Monte Carlo Method to estimate the drunk's distance from the origin after, say, six ($n = 6$) steps of equal unit length. To initiate the random walk, flip two coins. If the first coin lands heads, it is interpreted as a positive unit (increase) in X, and a negative unit (decrease) if tails. Similarly, use the second coin for Y. The table organizes the results:

A Perambulating Drunk

Trial	*Step N*	*First Coin*	*Second Coin*	*Drunk's Location* (X_n, Y_n)	*Distance*^{2} *(From Origin)*	*Distance From Origin*	*Trial Average*	*Moving Average*
1	1	H	H	(1, 1)	$(1)^2 + (1)^2 = 2$	$\sqrt{2}$		
	2	H	T	(2, 0)	$(2)^2 + (0)^2 = 4$	2		
	3	T	T	(1, −1)	$(1)^2 + (-1)^2 = 2$	$\sqrt{2}$		
	4	T	H	(0, 0)	$(0)^2 + (0)^2 = 0$	0		
	5	H	T	(1, −1)	$(1)^2 + (-1)^2 = 2$	$\sqrt{2}$		
	6	H	T	(2, −2)	$(2)^2 + (-2)^2 = 8$	$2\sqrt{2}$	$2\sqrt{2}$	$2\sqrt{2}$
2	1	T	H	(−1, 1)	$(-1)^2 + (1)^2 = 2$	$\sqrt{2}$		
	2	H	T	(0, 0)	$(0)^2 + (0)^2 = 0$	0		
	3	T	H	(−1, 1)	$(-1)^2 + (1)^2 = 2$	$\sqrt{2}$		
	4	T	H	(−2, 2)	$(-2)^2 + (2)^2 = 8$	$2\sqrt{2}$		
	5	H	T	(−1, 1)	$(-1)^2 + (1)^2 = 2$	$\sqrt{2}$		
	6	T	T	(−2, 0)	$(-2)^2 + (0)^2 = 4$	2	2	$1 + \sqrt{2}$

⟶

The drunk's position, (2, −2), represents a distance (from the origin) of $2\sqrt{2}$ *in the first trial consisting of six steps. Here, we use the well-known distance formula between two points in a plane,* (x_1, y_1) *and* (x_2, y_2)*, as* $d = \sqrt{(x_2 - x_1)^2 + (y_2 - y_1)^2}$.

The procedure is repeated, flipping coins many times in groups of six tosses and calculating the distance d. The average of the values of d obtained by repeated trials, called a trial average, is an estimate of the average distance of the drunk from the starting point. The first trial average was just the distance of the drunk after six steps, $2\sqrt{2}$*. The second trial of six steps yielded a distance from the starting point (origin) of 2. Averaged with* $2\sqrt{2}$ *from the first trial, the updated trial average after the two trials (moving average column) is* $1 + \sqrt{2}$ *and so on.*

To use the random number table, select an arbitrary position and read the numbers horizontally or vertically (or in any other pattern that does not compromise randomness). One can truncate the four-digit random numbers to three-digit, two-digit, or single-digit numbers for ease of calculation. The random numbers are truncated to the first two digits for this example. We use the scheme: numbers 00–49 are considered "a head" and numbers 50–99 are "a tail." For example, reading 12 numbers (to represent the six pairs of tosses) organize the result as

Trial	Step N	First RN	Second RN	Drunk's Location (X_n, Y_n)	Distance² (From Origin)	Distance From Origin	Trial Average	Moving Average
	1	94 (T)	90 (T)	(−1, −1)	$(-1)^2 + (-1)^2 = 2$	$\sqrt{2}$		
	2	14 (H)	51 (T)	(0, −2)	$(0)^2 + (-2)^2 = 4$	2		
3	3	40 (H)	73 (T)	(1, −3)	$(1)^2 + (-3)^2 = 10$	$\sqrt{10}$		
	4	04 (H)	33 (H)	(2, −2)	$(-2)^2 + (2)^2 = 8$	$2\sqrt{2}$		
	5	99 (T)	20 (H)	(1, −1)	$(1)^2 + (-1)^2 = 2$	$\sqrt{2}$		
	6	79 (T)	95 (T)	(0, −2)	$(0)^2 + (-2)^2 = 4$	2	2	$\frac{2\sqrt{2}+4}{3}$

Note the moving average after this third trial is calculated by determining the average of the three trial averages $2\sqrt{2}$*, 2, and 2 found thus far.*

♦ A random number table is a useful tool in imaginative hands.

Freight destined to go across the country would be carried by different railroads. The charge was paid at the originating station, and periodically railroads would settle among themselves by examining each "way bill." Clearly, a sizable clerical expense was required. Eventually, management was persuaded that random selection of a fraction of the way bills was adequate to divide the revenue within a tiny fraction of the actual amount and at much less expense.

Table of random numbers (four blocks of 100 four-digit numbers)

9436	7981	9771	0288	0428	1170	7010	6742	2860	4638
9083	9519	2783	8378	1139	2404	8822	7980	8169	9252
1467	2210	2752	6181	6965	8531	3709	5522	9817	6244
5147	3695	1218	2051	9854	3736	1176	3377	4920	3335
4058	0078	0526	9847	6326	5973	4506	2187	4042	4547
7338	9351	7224	7245	3123	6063	9891	6217	2291	8049
0440	1080	0126	3017	0857	0772	6977	8811	6279	8676
3394	0063	4211	2493	9972	0170	5441	1243	6174	2571
9955	5490	3076	9462	1981	0124	6378	4500	8068	7170
2003	8556	9722	9252	3984	6960	9295	4666	7734	4681

4772	0111	8197	3885	7297	5218	3688	0710	9431	4036
2129	9206	9343	3672	1321	3266	5069	9721	3887	5405
0945	8359	4919	2330	3772	7000	4866	3167	3856	1383
3614	2553	0947	1869	5671	9628	9173	5014	6237	1536
6347	7257	4020	2353	5834	9246	3984	4050	2246	7931
1375	2465	6455	7841	3381	6486	1747	2323	5564	4270
6665	1008	3516	4515	2228	6977	4316	1058	2610	5937
0125	7309	8146	3312	2795	0152	9881	9607	7352	4087
5978	2216	7294	7176	8974	1018	2371	2226	3512	1551
2771	5939	0203	9421	1903	7916	4870	1566	7292	3852

2840	5297	8031	7174	8243	8124	2517	3567	0014	8940
1107	2740	1495	9285	9803	0913	2330	0719	3249	8867
1177	8259	0056	6071	0833	8938	6754	8661	6631	5925
1220	2516	9948	1431	4115	0201	1905	7198	8944	5222
5388	8566	1081	9962	8959	8918	4285	2760	6545	7381
2396	7111	0739	5630	5088	9619	9963	4108	0591	3439
0891	0159	5158	6557	3502	1713	1666	7780	3819	6509
4581	8453	0499	7749	2859	0357	2228	2150	4517	5301
2401	3572	8235	7415	9222	6325	6885	0348	2593	0928
9366	2262	9478	9728	5187	4276	0489	5673	1563	9817

5570	8310	6945	6857	4196	6805	9039	4672	5719	6800
3742	5209	9140	6261	4879	2738	2485	8180	6997	3729
9000	0974	6421	3833	5663	6625	1940	7995	1960	3950
2716	7925	1530	3260	2715	6810	4348	3701	8094	5299
4248	7929	9010	3759	3955	2816	7712	0133	3131	5297
9071	3879	4618	2460	4736	3757	0922	3958	4951	5484
6833	8407	1000	5860	8694	3625	2353	2772	4070	8146
4606	3653	4348	5467	8958	8596	0798	3024	7369	5329
1447	8510	1032	2720	2820	1283	3697	6615	6239	6691
9469	8518	4185	6966	4635	6119	8481	6365	6263	8504

♦ One of the world's larger engineering construction firms was awarded a contract to build a long coastal highway. Soon after construction began, it became apparent that something had gone badly wrong. Equipment brought to the site based on contract specifications proved undersized and underpowered causing project delays. Project delays resulted in specialized crews, contracted earlier, idled, waiting to work. Eventually, many millions of dollars in claims were paid. An investigation revealed that the soil borings for subsurface rock were grossly under representative. The crews deployed to take soil borings along the proposed route, which became part of the contract, apparently chose sites that were easily accessible. A more knowledgeable manager would have advantageously used a random number table to select soil-boring sites in advance of deploying the crews

Example 11.4.3 Estimates of π

Use Monte Carlo to estimate the value of π.

Solution:

The transcendental number $\pi \approx 3.14158\ \ldots$ is well known and widely recognized (Section 1.4). Probably, most, if not nearly all, take the value of π on faith. Here, you can actually estimate its value!

First, imagine a "perfectly awful" dart thrower who repeatedly throws darts at a square in which a quarter circle is inscribed. Some darts will lie within the quarter circle and some in the area outside (figure). Since the dart thrower is "perfectly awful," we assume that any one point of the square is as likely to be hit as any other point.

It seems reasonable that if a sufficiently large number of darts are thrown, the numbers landing within the quarter circle area, n, to the total number within the square, $N\,(N \geq n)$, is an estimate of the fraction of the square's area occupied by the quarter circle. Since the square area is known, the quarter circular area can be estimated by the proportion:

$$\text{Circular area} = \frac{n}{N} \cdot \text{Square area}$$

Perfectly awful dart throwers, in the sense here, do not exist. In addition, a random number table is safer!

Consider a quarter circle in the first quadrant and inscribed in a unit square with a vertex at the origin of a Cartesian system as shown.

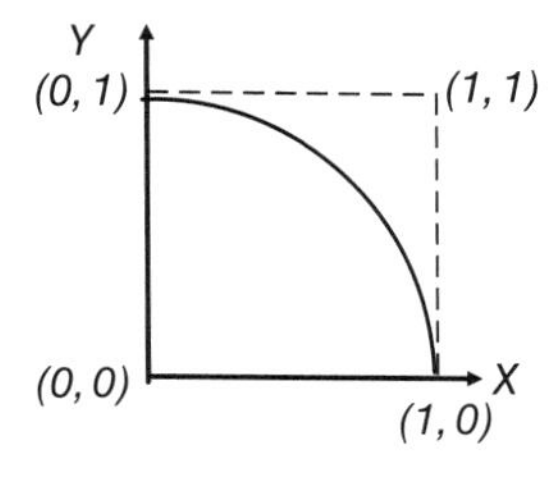

→

Select two RNs from the random number table to represent the x and y coordinates of a point within the unit square. For instance, arbitrarily, the first pair of numbers in the fifth column of the table is the decimalized ordered pair (0.04, 0.28). By inspection, it lies within the quarter circle. Generally, a random point lies within the quarter circle if it satisfies

$$X^2 + Y^2 \leq 1$$

Continuing in this manner, the trial results of selecting random-ordered pairs are tabulated and a trial average calculated as the "current" estimate of $\pi/4$, the area of a quarter circle with unit radius. The process continues until the trial average stabilizes. That may require thousands of trials for even modest accuracy. Clearly, a computer or programmable calculator is a necessity for even a rough estimate.

Scaling Random Numbers

So far, the examples have needed little alteration of the RNs in the table. Suppose, however, that RNs are needed for some interval of interest.

To illustrate, suppose that RNs are needed on the interval $-5 \leq X \leq 15$; each point being equally likely. To properly scale the RNs from the table, note that the interval width is 20. Let R be a decimalized RN. The smallest number on the interval is −5, so the scaled value, X, is

$$X = -5 + 20R$$

For example, if the two-digit RN is 38, decimalize it as $R = 0.38$. $X = -5 + 20(0.38) = 2.6$. In this manner, the random number table can be adapted to provide RNs on any desired interval. The next example illustrates the use of scaling.

Example 11.4.4 Estimating Areas

Use Monte Carlo to estimate the area of a circle with radius 10 inches. Compare your result with the actual value. Explain how Monte Carlo can be used to estimate irregular areas.

Solution:

The area of a circle of radius r is, of course, πr^2. Therefore, a circle of radius 10 inches centered at the origin of a Cartesian coordinate system (represented by $X^2 + Y^2 = 100$) has an area of 100π inches^2.

In this instance, the area enclosed by the circle is known from geometry. However, one can imagine an irregular figure whose area is sought; a lake, forest region, or map area, for example. There usually is no formula to calculate such an irregular area.

⟶

One way to estimate such irregular areas is to surround them with a figure of known area. For convenience of illustration, imagine the 20 inch diameter circle inscribed in a 20 inch square, as shown. (Simple areas are used here for clarity. A more realistic and irregular area is left for the exercises).

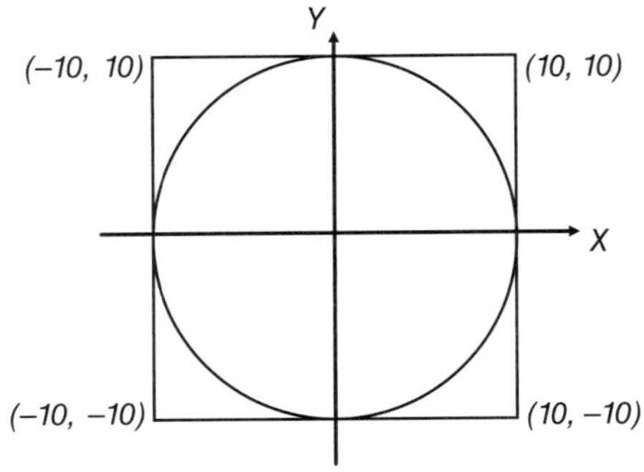

For a Monte Carlo, choose 2 two-digit RNs, say, R_x *and* R_y*. Note that* $-10 \leq X \leq 10$ *and* $-10 \leq Y \leq 10$*. Since the smallest value of X is* -10 *and the largest is* $+10$*, the range of X is 20.*

Therefore, since any value in the interval is equally likely, the random X coordinate is

$$X_r = \text{Smallest value} + \text{Range (decimalized random number)}$$
$$= -10 + 20\ R_x.$$

A similar reasoning for Y yields $Y_r = -10 + 20\ R_y$*. For example, if* $R_x = 94$ *and* $R_y = 90$*, then* $X_r = -10 + 20(0.94) = 8.8$ *and* $Y_r = -10 + 20(0.90) = 8.0$*.*

Once the ordered random pair (X_r, Y_r) *is generated, we determine whether or not it lies within the circle. If the point lies on or within the circle, then it must satisfy* $X_r^2 + Y_r^2 \leq 100$*. If the relation is not satisfied, then the point lies between the circular boundary and the edges of the square. For example,* $X_r = 8.8$ *and* $Y_r = 8.0$ *result in* $X_r^2 + Y_r^2 = 141.4$*, indicating a point outside the circle.*

Now, of course, many repetitions are required to obtain a reliable estimate. It is important at the outset to organize a table for the sampling and calculations. The table shows the results for several trials. Once again, the RNs have been truncated to two digits for ease of calculation. A point lying on or within the circle is designated a "hit."

Estimating an area

Trial	R_x	R_y	X_r	Y_r	X_r^2	Y_r^2	$X_r^2 + Y_r^2$	*Hit?*	*Trial Average*
1	0.94	0.90	8.8	8.0	77.4	64.0	141.4	No	0/1
2	0.14	0.51	−7.2	0.2	51.84	0.04	51.88	Yes	1/2
3	0.40	0.73	−2.0	4.6	4.00	21.16	25.16	Yes	2/3
4	0.04	0.33	−9.2	−3.4	84.64	11.56	96.2	Yes	3/4
5	0.99	0.20	9.8	−6.0	96.04	36	132.04	No	3/5
⋮	⋮	⋮	⋮	⋮	⋮	⋮	⋮	⋮	⋮

⟶

The actual fraction of the 20 × 20 square area occupied by the circle is $\pi/4$ (= $100\pi/400$) or the area of the circle divided by the area of the rectangle ≈ 0.785. In practical usage, if the actual values are known, there is no need for a Monte Carlo. Usually, a very large number of trials are required for the trial average to settle. Even for the example here, several thousand trials may be needed for an adequate estimate.

Example 11.4.5 More on Estimating Areas

Use Monte Carlo to estimate the area of a circle whose equation is

$$(x-4)^2 + (y+1)^2 = 36$$

Solution:

The area of a circle of radius 6 is 36 π.

The center of this circle is (4, −1). The values of x must lie between 4 ± 6 while y values lie between −1 ± 6.

For a Monte Carlo, choose 2 two-digit (decimalized) random numbers, say, R_x and R_y. Note that $-2 \leq x \leq 10$ and $-7 \leq y \leq 5$.

Therefore, since any value in the interval is equally likely, the random x coordinate is

$$X_r = \text{Smallest value} + \text{Range}(\text{decimalized random number})$$
$$= -2 + 12\,R_x$$

A similar reasoning for Y yields $Y_r = -7 + 12\,R_y$.

Once the ordered random pair (X_r, Y_r) is generated, we determine whether or not it lies within the circle. If the point lies on or within the circle, then it must satisfy $(x-4)^2 + (y+1)^2 \leq 36$.

The table below has results for several trials. Once again, the RNs have been truncated to two digits for ease of calculation. A point lying on or within the circle is designated a "hit."

Estimating an area

Trial	R_x	R_y	X_r	Y_r	$(X_r-4)^2$	$(Y_r+1)^2$	$(X_r-4)^2 + (Y_r+1)^2$	Hit?
1	0.25	0.23	1.0	−4.24	9.00	10.5	19.5	Yes
2	0.67	0.19	6.04	−4.72	4.16	13.84	18.9	Yes
3	0.42	0.99	3.04	4.88	0.92	34.57	35.49	Yes
4	0.16	0.22	−0.08	−4.36	16.65	11.29	27.94	Yes
5	0.68	0.04	6.16	−6.52	4.67	30.47	35.14	Yes
⋮	⋮	⋮	⋮	⋮	⋮	⋮	⋮	⋮

The area then can be determined by the proportion of hits multiplied by the area of the rectangle (12 × 12 = 144).

Stopping Rules and Probabilities

The examples in this chapter can be misleading in that only a few trials are illustrated. In actual problems, it is not uncommon for the numbers of Monte Carlo trials to reach tens and hundreds of thousands and more for useful estimates. Clearly, when to stop the sampling is an important issue and a subject of considerable research interest. Unfortunately, there are few simple rules having a strong mathematical foundation. Usually, one continues until the estimate appears to stabilize, and continued sampling results in no meaningful change. Sometimes, tolerances are set in advance and trials are continued until they are reached. Others set the number of trials in advance from experience.

Whatever discipline is used, there usually is no guarantee that if the trials had continued, a different result might be reached. Generally, one relies on experience, particularly when repeated situations of comparable similarity are the case.

True randomness is an idealization in much the sense of frictionless surfaces. The quality of tables of RNs is determined by statistical measures. Random number tables of high quality are available. Computers and programmable calculators store complicated formulas, which, primed by RNs as "seeds," produce numbers that are quite acceptable surrogates for RNs. Such formulae circumvent the need to use large sections of memory to store random number tables. The expansions of transcendental and irrational numbers to many decimal places have been used as RNs. Expansions of rational fractions cannot be used since there are periodicities to the digits.

You may have noted in the previous examples that the direct use of the random number table implies that all outcomes are equally likely; the RNs may even be the sample value itself. In many applications, this is not the case as some outcomes are more likely than others.

Mapping Random Numbers

To cope with outcomes that are not equally likely, instead of using RN directly from tables as sample values, they are mapped against the cumulative distribution of the sample values.

The figure illustrates an arbitrary cumulative distribution against which an RN is mapped to yield a sample value (3, in the figure)

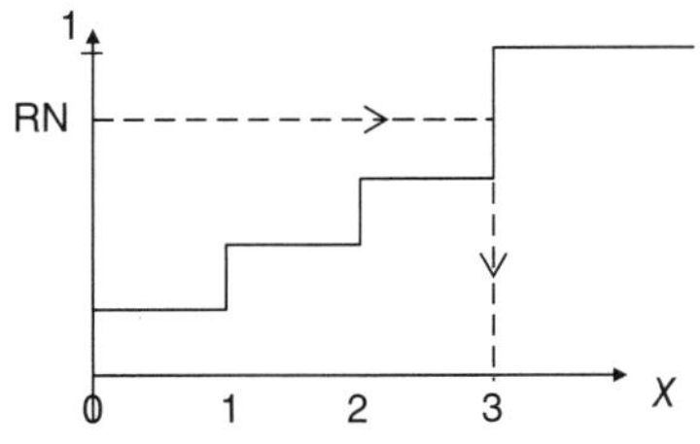

While, of course, the mapping can be accomplished "on paper," it is more common to program for automated computation.

Two important general observations can be made. First, a Monte Carlo depends on a cumulative probability distribution. In the previous examples, cumulative distributions did not appear explicitly. The reason is that sample values were obtained directly from the random number table. In those cases, the cumulative distribution is a 45° line passing through the origin and bounded by the ordinates 0 and 1 (figure). That is, the sample value equaled or was directly proportional to the random number so that the cumulative distribution's role was "hidden" and, so, required no mention.

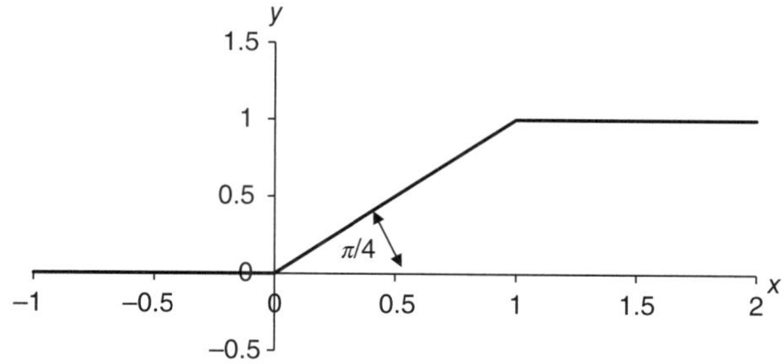

A second observation is that every cumulative distribution takes values inclusively from 0 to 1 by definition. By choosing and decimalizing random numbers, so that they have values on the unit interval, they "map" into sample values. To the extent that the cumulative distribution faithfully represents the random phenomena, the use of random numbers results in correspondingly faithful mapping of sample values. On a plot of a cumulative distribution, a random decimal locates a point on the vertical (probability) axis. A horizontal line from that axis intersects the cumulative distribution curve. At that point, a perpendicular to the horizontal (sample value) x-axis determines the random sample value. Again, the process repeats throughout the sampling. The next example illustrates the use of the cumulative distribution.

Example 11.4.6 Fast Paced Fast Food

A number of customers arriving at a multistation fast food counter are recorded in 2-minute intervals for 40 minutes. The 20 recorded values are

$$2,\ 3,\ 6,\ 4,\ 7,\ 3,\ 8,\ 2,\ 0,\ 1,\ 4,\ 8,\ 7,\ 6,\ 6,\ 5,\ 4,\ 3,\ 7,\ 4$$

Management seeks economic staffing levels (number of stations manned) while avoiding unreasonable customer delays. In addition, assume that, on average, a server can service three customers per 2-minute period (two servers can service six customers, etc.). Assume that the 40-minute sample is representative of the period.

Solution:

First, arrange the data in a frequency histogram (Section 10.1). Using the histogram, form a cumulative distribution diagram (graphs below). Next, choose random numbers to simulate customer traffic by mapping the random numbers against the cumulative distribution.

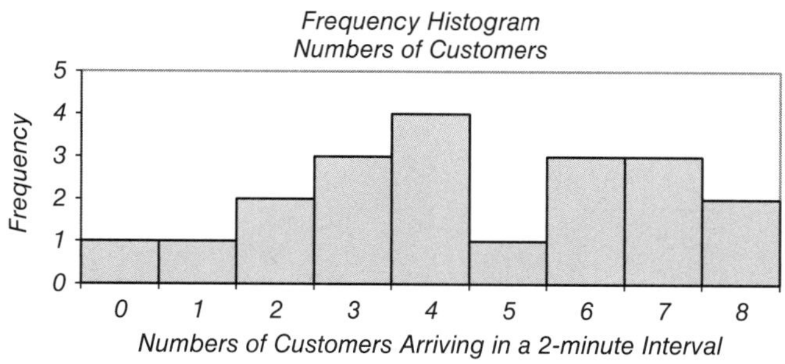

→

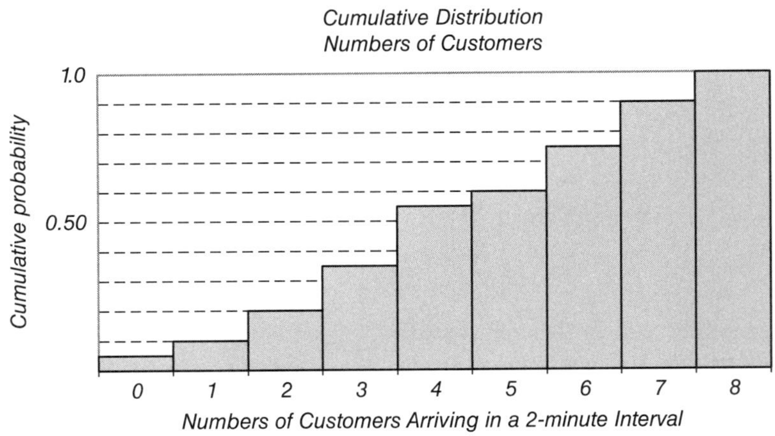

The following table illustrates the results of choosing 20 random numbers to simulate 20 two-minute intervals.

Multistationed fast food servers

Time, min	*Random Numbers*	*Number of Customers*	*One-Manned Station*		*Cumulative # Delayed/ Period*	*Two-Manned Station*		*Cumulative # Delayed/ Period*
			Number Served	*Number Delayed*		*Number Served*	*Number Delayed*	
1	0.47	4	3	1	1/1	4	0	0/1
2	0.21	3	3	1	2/2	3	0	0/2
3	0.09	1	2	0	2/3	1	0	0/3
4	0.36	4	3	1	3/4	4	0	0/4
5	0.63	6	3	4	7/5	6	0	0/5
6	0.13	2	3	3	10/6	2	0	0/6
7	0.66	6	3	6	16/7	6	0	0/7
8	0.01	0	3	3	19/8	0	0	0/8
9	0.59	5	3	5	24/9	5	0	0/9
10	0.27	3	3	5	29/10	3	0	0/10
11	0.01	0	3	2	31/11	0	0	0/11
12	0.92	8	3	7	38/12	6	2	2/12
13	0.83	7	3	11	49/13	6	3	5/13
14	0.25	3	3	11	60/14	6	0	5/14
15	0.72	6	3	14	74/15	6	0	5/15
16	0.24	3	3	14	88/16	3	0	5/16
17	0.10	2	3	13	101/17	2	0	5/17
18	0.73	6	3	16	117/18	6	0	5/18
19	0.22	3	3	16	133/19	3	0	5/19
20	0.59	5	3	18	151/20	5	0	5/20

(The random numbers are from the first two columns of second block in the random number table of page 407)

Our first decimalized RN is 0.47. Using the cumulative probability distribution table locate 0.47 on the vertical axis and then read horizontally to the cumulative distribution. Reading on the x -axis, we find the sample value of four arriving customers. Because only three customers can be served if there is only a single server, one customer is delayed, as indicated in the table. With two servers, there is no delay.

With two servers, there is little or no delay because up to six people can be served per 2-minute period. Only a few customers have to wait more than a 2-minute period for service(first come, first serve policy). The "one server" policy causes many customers to be delayed. When there are three servers, there is never any delay because up to nine people can be served, and the largest number arriving in any time interval is only eight.

Note that when a theoretical choice for the cumulative probability distribution is not available, one can use an actual sampling or history as an estimate. In the last example, the actual numbers of arrivals are used to form the cumulative distribution from their histogram.

Example 11.4.7 Portfolio Performance

Investors often refer to the ownership of a group of stocks as a "portfolio." Its selections seek a selected investment objective.

Use a Monte Carlo to show how an imaginary investor could "experiment" or "try out" the result of having a portfolio of three stocks, A, B, and C as shown. (Note that stock market variation is assumed to be random within the limits cited).

Corporation	*Number of Shares*	*Current Price per Share, $*	*Current Value, $*	*Annual Variation (AV), %*
A	*100*	*40*	*4,000*	*± 10*
B	*200*	*30*	*6,000*	*± 8*
C	*300*	*20*	*6,000*	*± 12*
		Total	*16,000*	

Solution:

While short-term stock price movements are generally unpredictable, our investor seeks an estimate of the portfolio value after 1 year. At the end of any year, a stock's price may have increased or decreased or remained unchanged. Assume that a study of each stock's annual long-term price variation indicated respective values of ±10%, ±8%, and ±12%.

The first three digits of the RNs for A, B, and C were chosen from the fourth block of 100 numbers in the random number table of this section. Using the scaling as in an earlier example, as [minimum value + (range)(decimalized RN)], we calculate the annual variations for A, B, and C as follows:

$$-10 + (20)(\textit{decimalized RN A}) = \%\textit{variation for A}$$

$$-8 + (16)(\textit{decimalized RN B}) = \%\textit{variation for B}$$

$$-12 + (24)(\textit{decimalized RN C}) = \%\textit{variation for C}$$

⟶

Therefore, for trial #1, the (random) annual percentage variation for Corporation A is $-10 + (20)(0.557) = +1.14$, or $+1.14\%$. The -10 arises since the smallest value of the $\pm 10\%$ variation is the -10%: the 20 is the range from -10% to $+10\%$; and, here, the 0.557 is decimalized random number. Now, 1.14% of the \$40 per share gain is \$0.456 or \$45.60 for 100 shares. This is rounded to \$46 for convenience. The value of Corporation A's stock in the portfolio at the end of the year is \$4046.

In a similar way, for Corporation B: $-8 + 16(0.374) = -2.02$, or -2.02%, while for Corporation C: $-12 + 24(0.900) = +9.60$, or $+9.60\%$. For Corporation B, the per-share loss is (\$0.0202)(30) = \$0.606. For 200 shares the loss is \$121.20, rounded to \$121. At the end of the year, the value of the portfolio Corporation B stock is \$6000 – \$121 = \$5879. For Corporation C, the 9.60% translates into a value of \$6576 for the value of the 300 shares.

Stock Portfolio Returns

Trial #	*RN A*	*RN B*	*RN C*	*Random Annual Variation, %*			*Value, \$, Corporation*			*Portfolio Value, \$*	*Annual Average, \$*
				A	*B*	*C*	*A*	*B*	*C*		
1	*557*	*374*	*900*	*1.14*	*−2.01*	*9.60*	*4,046*	*5,879*	*6,576*	*16,501*	*16,501*
2	*271*	*424*	*907*	*−4.58*	*−1.22*	*9.77*	*3,817*	*5,927*	*6,586*	*16,330*	*16,415*
3	*683*	*460*	*144*	*3.66*	*−0.64*	*−8.54*	*4,146*	*5,962*	*5,488*	*15,596*	*16,142*
⋮	⋮	⋮	⋮	⋮	⋮	⋮	⋮	⋮	⋮	⋮	⋮

The portfolio total for a single-year simulation is \$16,501. The procedure is repeated, shown as trials #2 and #3, in the table, and the annual average recorded in the right-hand column. If a more stable value of the actual average is sought, many more trials are needed.

♦ eBay can be a real-world laboratory for students of auctions for Monte Carlo experiments.

Estimating Combinatorics

Likely, you can recall the challenges of combinatoric probability problems in Chapter 3. One reason they can be difficult is that a useful intuitive notion for the correct value is elusive. Usually, combinatoric numbers are very large and outside the range of our experiences. For example, using combinatorics it is easy to calculate that there are ${}_{52}C_5 = 2,598,960$ five-card poker hands. Not many people can estimate this number from experience. To make matters worse, reliable estimates or approximations are not usually apparent.

Monte Carlo Method may help to estimate combinatoric values. A simple example illustrates an idea:

Example 11.4.8 An Elevator Query

An elevator with three passengers stops at three floors. Use Monte Carlo to estimate the number of ways a passengers can leave such that exactly one passenger leaves at each floor.
(The answer is 3! = 6, assuming all floors are equally likely.)

Solution:
For a Monte Carlo, choose sets of three random numbers from a random number table. If less than 1/3, it can represent a passenger who exits on the first floor; if between 1/3 and 2/3, the passenger exits on the second floor; and if the random number exceeds 2/3, the passenger exits on the third floor. If all three floors result in a leaving passenger, it is considered a "success" and assigned a unit score (hit), otherwise a zero is assigned (miss).

Elevator Passenger Leavings

Trial #	*RN1*	*RN2*	*RN3*	*Hit?*	*Trial Average*
1	*0.94*	*0.90*	*0.14*	*No*	*0/1*
2	*0.51*	*0.40*	*0.79*	*No*	*0/2*
3	*0.95*	*0.22*	*0.36*	*Yes*	*1/3*

First, note that there are $3^3 = 27$ possible ways for the passengers to exit. One seeks to estimate the fraction of these in which each passenger leaves at a different floor. The table organizes a few trials. Many more trials are needed for a more reliable trial average, which is multiplied by 27 to attain the desired estimate.

♦ Recent research highlights causes of tie-ups in stop-and-go traffic. Studies show that erratic lane changing and unnecessary braking considerably reduce the effective capacity of a crowded highway. A new technology, adaptive cruise control (ACC), already standard on some luxury vehicles, can pilot cars to avoid excessive braking and the resulting compression wave that induces traffic jams. The ACC, by not stopping a car, unless the vehicle ahead is stopped, permits safe tailgating. The effect is roughly similar to pouring from a narrow-necked bottle. Pour at a shallow angle and the flow is smooth. At a larger angle, flow becomes erratic and uneven.

The surprising research result is that if only 20% of cars on a highway have ACC, traffic jams may be reduced by as much as 75%.

An exploratory research project relies on conducting a Monte Carlo simulation (*Science Journal* in the July 30, 2004, *Wall Street Journal*).

EXERCISES 11.4

1. In a two-player game, Jack and Jill take turns tossing a fair coin. If the outcome is heads, Jack pays Jill \$1 and for tails, Jill pays Jack \$1. Simulate the game using random numbers. How close to zero is your result after 10 tosses? After 20 tosses?

2. Suppose that the normal rate of inflammation from a vaccination is 1/4. A family of five is vaccinated. Use a Monte Carlo to estimate the probability that two or more family members develop an inflammation. Compare your estimate with the calculated value using a binomial distribution.

3. A currency trader specializes in three world currencies. Assume that each daily trading period results in a unit increase or decrease or remains unchanged in value for each currency independently with these probabilities:

		Advance	**Decline**	**Unchanged**
	A	0.45	0.35	0.20
Currency	**B**	0.40	0.45	0.15
	C	0.40	0.35	0.25

What is the trader's average weekly gain (loss) from equal shares of each currency? Assume five trading days per week.

4. Use Monte Carlo to estimate the probability of exactly three face cards in a 13-card hand. Compare with the actual value.

5. Does the sample point resulting from (R_x, R_y) given below lie within the circle $(x + 1)^2 + (y - 4)^2 = 9$?

a) (0.61, 0.20) b) (0.24, 0.71) c) (0.68, 0.84)

6. Does the sample point resulting from (R_x, R_y) given below lie within the circle $(x - 3)^2 + (y + 4)^2 = 25$?

a) (0.11, 0.24) b) (0.49, 0.33) c) (0.92, 0.50)

7. For the circle $(x - 3)^2 + (y - 4)^2 = 100$, estimate the area after $n = 5$ trials and after an additional $n = 10$ trials. Does the estimate appear to stabilize or are further trials desirable?

8. The area within the ellipse $\frac{x^2}{a^2} + \frac{y^2}{b^2} = 1$ is πab. Use Monte Carlo with $X_r = -a + 2aR_x$ and $Y_r = -b + 2bR_y$ to estimate the area of an ellipse with $a = 4$ and $b = 3$.

9. Use a Monte Carlo to estimate the shaded area below.

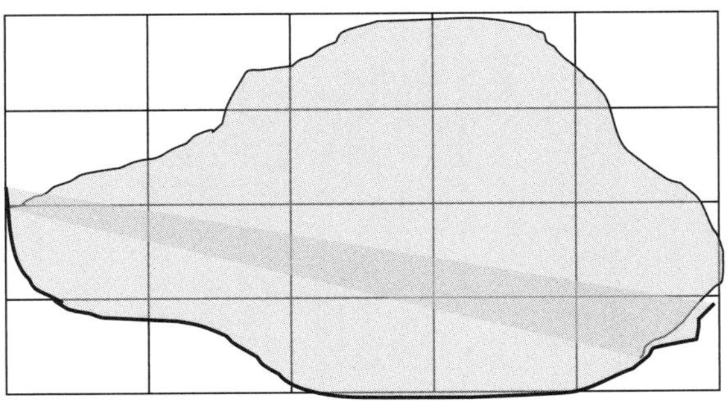

⟶

Hint: View the map as a coordinate system with $0 \le x \le 5$ and $0 \le y \le 4$. Assume that x and y values in the range are equally likely. Select two random numbers, R_x and R_y. Use $0 + 5R_x$ and $0 + 4R_y$ for coordinates to obtain a Monte Carlo estimate.

10. A square landing pad (or target) is to be designed for parachute group training (or missile landings, etc.). Actual landings deviate randomly from the target center.

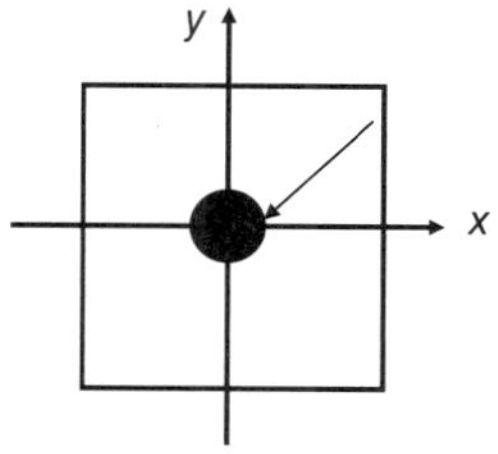

Data from previous attempts indicate that the x and y coordinates of the landing points follow a normal distribution about the origin with a standard deviation of 100 feet. How large should the landing pad be so that 95% of drops land on the pad?

a) Organize a Monte Carlo to estimate landing pad size. You can obtain normal random numbers on the web site below or from "A Million Random Digits and 100,000 Normal Deviates" by the Rand Corporation, MacMillan, New York 1955. http://www.rand.org/publications/classics/randomdigits/samples.txt.

b) Calculate the exact value to compare with your Monte Carlo estimate.

11. Execute a Monte Carlo to estimate the probability that five cards chosen at random from a standard deck have different denominations.

Hint: The exact answer is $4^5 \dfrac{\binom{13}{5}}{\binom{52}{5}}$.

12. Customers arrive at a rate between four and seven per hour according to the probability distribution below.

Customers per hour, x	4	5	6	7
$p(x)$	0.50	0.25	0.15	0.10

Assume that customers arrive at the beginning of the hour and remain until served at a facility that requires 12 minutes per service. Estimate the average number of customers not served within the hour of their arrival.

13. A production line is set to turn out 20 items per day. However, deviations occur for various reasons. Sampling has resulted in these probability estimates for daily production.

Daily production, x	17	18	19	20	21	22	23
$p(x)$	0.20	0.15	0.15	0.25	0.15	0.05	0.05

A single truck that can hold 21 items moves production daily. Items that cannot be taken on the day of production are held over to the next day. Estimate the average number of items waiting shipment and the average number of empty spaces on the truck.

14. A car rental agency has formed the following table for the probabilities of economic life for its fleet of three cars. Estimate the average replacements every 2 years.

Economic life, x	1/2 year	1 year	1 1/2 years	2 years
$p(x)$	0.15	0.40	0.30	0.15

15. Arrange a Monte Carlo to estimate the probability that in a group of five people at least two share the same birth month (assuming all months are equally likely). (The actual probability is 0.6189.)

    Hint: You can use the random number table with $0 \leq R_1 < 1/12$ representing January, the next 1/12 representing February, and so on. Each trial utilizes five random numbers. An alternative is to use 12 cards numbered 1–12. Five cards are drawn with replacement to represent a trial.

16. In the previous exercise, suppose that there are seven people. How does your estimate of the probability change? (The actual probability is 0.888.)

17. *Experimental Psychology* – Psychologists, and other scientists and engineers, have a need of larger scale experimentation that is often expensive and time consuming. Monte Carlo is a widely used supplement. To illustrate, in the maze (below), a mouse enters seeking food (the arrows leading to food occur at the eight intersections in the maze).

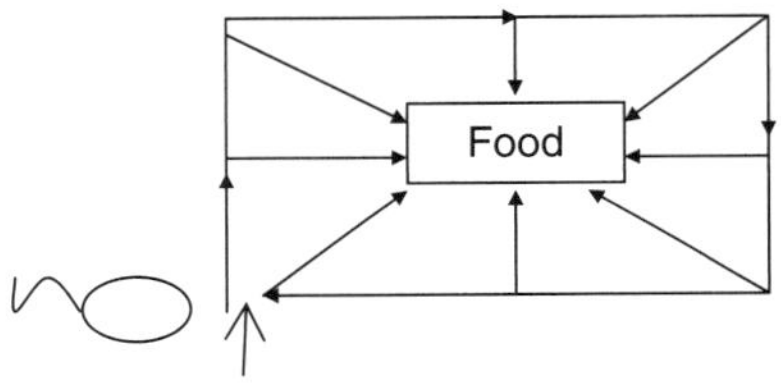

    A right turn at any intersection leads to food. The number of missed right turns is a measure of the mouse's lack of luck in reaching food more quickly. For a simple Monte Carlo, assume that the probability of a right turn at any intersection is 1/2 (i.e., pure chance).

    Organize and execute a Monte Carlo for the experiment.

    Note the potentialities here. An experimenter may conduct live trials to estimate a probability distribution for "right" turns. Then, a number of trials can be simulated with Monte Carlo and perhaps with different mazes and as such to analyze mouse "intelligence."

18. Three points are chosen at random on the circumference of a circle. Estimate the probability that all three lie on a semicircle. (Hint: actual value is 3/4.)

19. *Buffon's needle problem* is a classic example of the Monte Carlo Method to estimate π. In the experiment, on a surface with equally spaced parallel lines, a needle is dropped whose length is equal to the width of the spacings. If it crosses (or touches) one of the parallel lines, it is recorded as a "hit."

⟶

Let N = number of times the needle is dropped and
C = number of times a dropped needle crossed the parallel lines
Buffon concluded that $\pi \approx 2N/C$.

a) Repeat Buffon's needle experiment for at least 50 trials to determine a value of π.
b) Share data with your classmates to determine a 95% confidence interval for the average value of π. (Recall confidence intervals in Chapter 10.)

HISTORICAL NOTES

As noted earlier, random sampling has been known for centuries; Monte Carlo being a newer twist on an old idea. It differs from traditional simple random sampling where every item in the population has the same probability of being chosen. In Monte Carlo, items in the population are chosen by chances that mirror the actual chance situation being simulated.

Associated with games of chance about the time of World War II (WWII), it acquired the colorful name of the gambling center of the Principality of Monaco. The names of John von Neumann and Stanislaw Ulam are often mentioned in connection with the early years of Monte Carlo Method.

In the years since WWII, Monte Carlo Method has acquired an important and diverse following in many fields. Military defense systems; bridge and other structural designs; vehicular traffic patterns, signals, and speeds; surgical procedures; telephone network analyses; train and flight schedules; and engineered construction are some of the many roles in which Monte Carlo Method is indispensable.

By now you realize how easily Monte Carlo skills can be acquired and how much use you are likely to make of them. Using a computationally based Monte Carlo, one can simulate months and years of data in a few minutes. It permits the manipulation of factors that can be controlled without the time and risk of experimentation on an actual system.

♦ Earthwork is often a major expense in the construction of highways, airfields, dams, and so on. The quantities of earth to be "cut" or "filled" are required to estimate costs in advance of construction. An imaginative Monte Carlo scheme has been devised to estimate earthwork quantities inexpensively and with precision. The method is described in an interesting paper "Earthwork Quantities by Random Sampling" by M. Gates and A. Scarpa in the *Journal of Construction Division, American Society of Civil Engineering*, July 1969.

John Von Neumann (1903–1957) — see Section 11.1.

Stanislaw M. Ulam (1909–1984) — A Polish-American mathematician who made notable contributions to the WWII "Manhattan Project." He was among the earliest to seek approximate solutions in atomic theory using random sampling (Monte Carlo Method).

♦ The text "Markov Chain Monte Carlo in Practice" by W. R. Gilks has several examples of how Markov chain Monte Carlo (MCMC) methods are used by researchers in various actual studies. These include California epidemiologists studying breast cancer, a vaccination program in Gambia to reduce hepatitis B, Austrian archeologists to place a Bronze Age site in its proper temporal location, and French researchers in mapping a disease.

11.5 DYNAMIC PROGRAMMING

Better known as **dynamic programming**, optimization by recursion replaces a single optimization in, say, n variables by n maxima (minima) each in one variable. Optimization problems in a single variable are usually much easier to solve than those with many variables. Expertly developed by Richard Bellman from the 1950s, it has been applied to a variety of sequential decision problems in a variety of contexts and, sometimes, with astonishing results.

Dynamic programming solutions depend on casting problems into a stagewise or sequential format in which a solution at one stage becomes an input for a solution at the next stage. The process continues until the entire solution is obtained.

Dynamic programming depends on a **principle of optimality**. Roughly said, whatever the state of a system, an optimal policy depends only on the current state, independent of how it was reached.

Principle of Optimality

An optimal policy has the property that regardless of prior decisions, remaining decisions constitute an optimal policy with respect to the current state.

Mathematical work is often driven by theorems that relate to particular mathematical forms. With dynamic programming, there is no particular mathematical format other than a stagewise structure. The optimality principle has the effect of "uncoupling" the stages of the problem. In many respects, solving optimization and decision situations using dynamic programming is an art learned by example.

Example 11.5.1 *Shortest Distance between Two Locations*

A shortest path is sought through a network. Use dynamic programming to determine the shortest path between nodes labeled 1 and 10. Distances between nodes are related to the numbers on the links joining them.

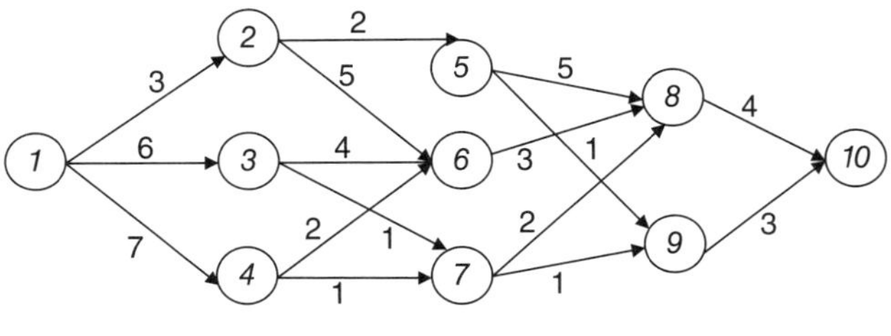

⟶

Solution:

The network is decomposed into these four subproblems or stages.

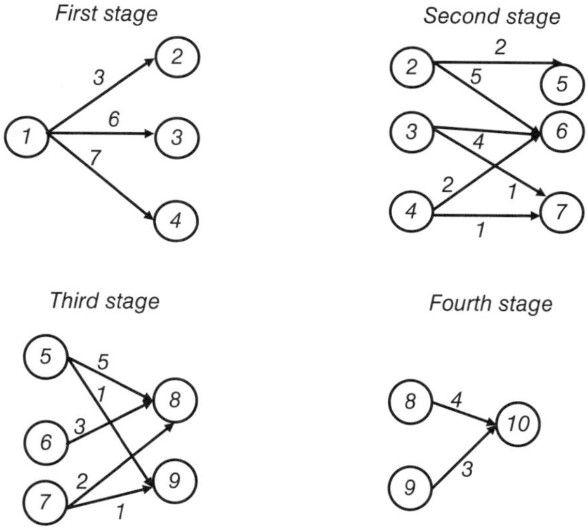

The first stage shortest paths and distances are

Node 2 (Path 1—2), distance 3
Node 3 (Path 1—3), distance 6
Node 4 (Path 1—4), distance 7

In the second stage, Node 5 can only be reached from Node 2. The path from Node 1 is 5 (= 3 + 2) units. Node 6 can be reached from Nodes 2, 3, or 4 for respective path lengths of 3 + 5 = 8, 6 + 4 = 10, and 7 + 2 = 9. The shortest path to Node 6 is 8 units.

Node 7 can be reached from either Node 3 or 4 for respective path lengths of 6 + 1 = 7 and 7 + 1 = 8. The shortest distance is 7 units.

Therefore, at the end of the second stage, the shortest paths to Nodes 5–7 are:

Node 5 (Path 1—2—5), distance 5
Node 6 (Path 1—2—6), distance 8
Node 7 (Path 1—3—7), distance 7

Note that as the recursion proceeds, some links are removed from further consideration. For example, to reach Node 7 from Node 1, no solution will use Node 4.

At the third stage, Node 8 can be reached from Nodes 5, 6, or 7and the shortest distances are (5 + 5 = 10, 8 + 3 = 11, and 7 + 2 = 9). Node 9 can be reached from Nodes 5 or 7, and the shortest distances are (5 + 1 = 6 and 7 + 1 = 8). Therefore, at the end of the third stage, we determine the shortest paths to Nodes 8 and 9 are

Node 8 (Path 1—3—7—8), distance 9
Node 9 (Path 1—2—5—9), distance 6

At the final stage, Node 10 can be reached from Nodes 8 and 9, yielding distances of $9 + 4 = 13$ and $6 + 3 = 9$. The shortest distance to Node 10 from Node 1 is 9 units.

Node 10 (Path 1—2—5—9—10), distance 9.

Recursion is a method of defining functions in which the subject function is applied in its own definition. Both forward and backward recursions yield the same result. Forward recursion may seem more logical, but backward recursion is usually applied in dynamic programming problems as it seems to be more efficient computationally in many cases. The next example uses backward recursion for the previous example.

Example 11.5.2 Shortest Distance between Two Locations (Revisited)

Solve Example 11.5.1 using backward recursion.

Solution:

For backward recursion, start at the destination, Node 10, and seek the shortest path to Node 1, the origination node.

Here, the stages are

First stage

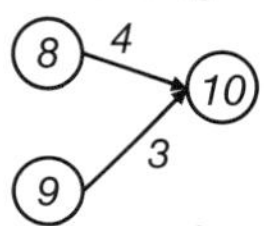

Second stage

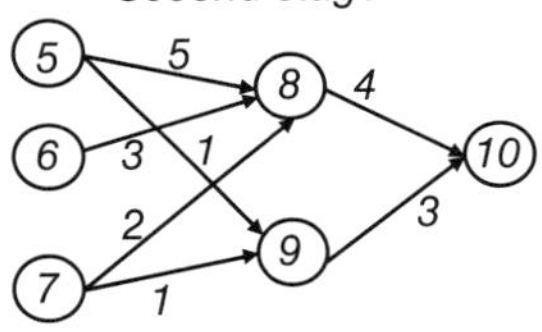

Third stage

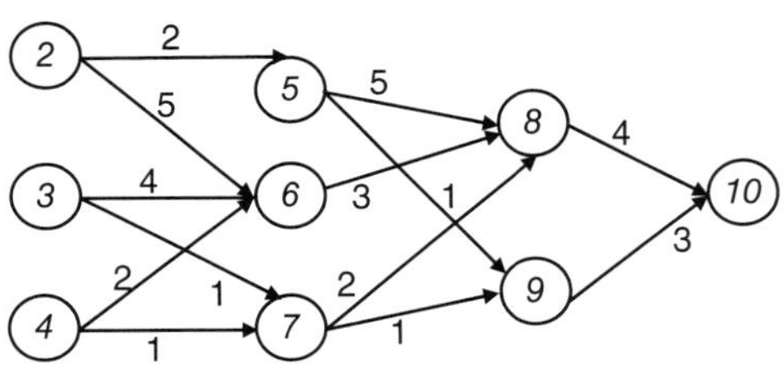

Fourth stage

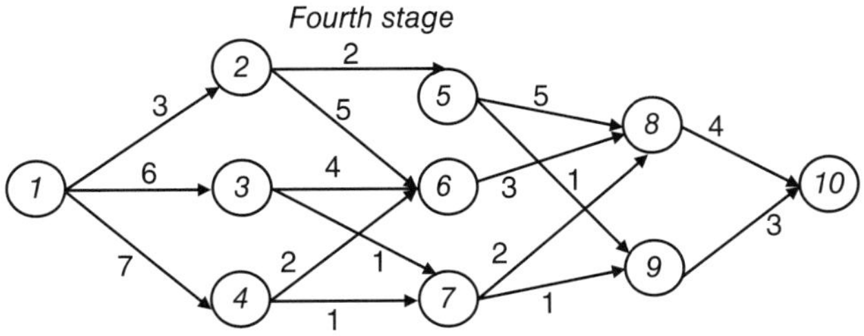

⟶

Number of Steps to Node 10	*Distance to Node 10*		*Minimum Distance to Node 10*	
	From Node	*Distance*	*Distance*	*Path*
	8	*4*		
1	*9*	*3*	*3*	*9–10*
2	*5*	$5+4=9$ $1+3=4$	*4*	*5–9–10*
	6	$3+4=7$	*7*	*6–8–10*
	7	$2+4=6$ $1+3=4$	*4*	*7–9–10*
3	*2*	$2+4=6$ $5+7=12$	***6***	*2–5–9–10*
	3	$4+7=11$ $1+4=5$	***5***	*3–7–9–10*
	4	$2+7=9$ $1+4=5$	***5***	*4–7–9–10*
4	*1*	$3+6=9$ $6+5=11$ $7+5=12$	***9***	*1–2–5–9–10*

The backward recursion gives the same shortest path as forward recursion. The shortest path from Node 1 to Node 10 is 9 units using the path Node 1 to Node 2 to Node 5 to Node 9 to Node 10.

Dynamic programming has been applied to areas of equipment replacement, inventory modeling, investment planning, production scheduling, protein sequence alignments, and the "knapsack" and "traveling salesman" problems. The increase in the number of state variables in the subproblem computations often causes a computational difficulty known as "*the curse of dimensionality.*"

The problems introduced in this chapter are **deterministic** in nature. By deterministic, we mean that the state at the next stage of the method is completely determined by the current state and policy decision and not on any chance events requiring probability distributions.

Example 11.5.3 Shortest Path in a Network

*Find a path from Node **M** to Node **D** in the network below so that the total cost of travel is a minimum. Assume that only eastbound and northbound steps are allowed. Numbers on the links are the associated travel costs.*

→

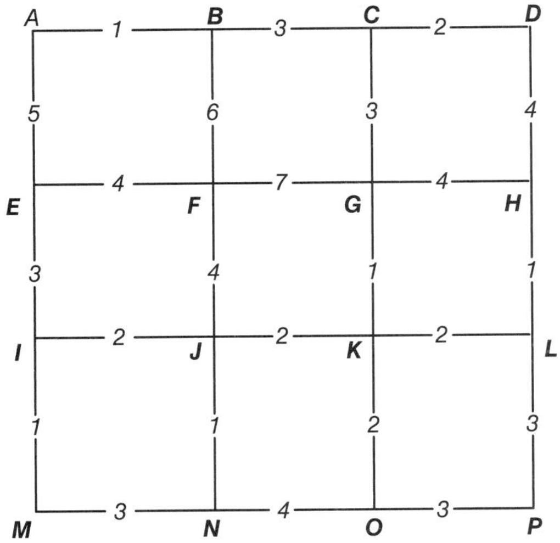

Solution:

It is often a characteristic of solutions by dynamic programming to begin at the "end" of the problem. Starting with the final node, ***D,*** *calculate the least cost path from each node in a "backward expansion." The node previous to* ***D*** *must have been either Node* ***C*** *or Node* ***H****. The cost on path* ***CD*** *is 2 units; while on* ***HD*** *it is 4 units. The accompanying table records this first (backward) step from Node* ***D****.*

Continuing in this way, the "two step" position must be one of the Nodes ***B****,* ***G****, or* ***L****. From Node* ***B****, the only available path is eastbound to* ***C*** *(at cost 3). Using the principle of optimality, coupled with the (one-step) least cost from Node* ***C*** *to Node* ***D****, it yields a (two-step) least cost of 5.*

A similar argument applies at Node ***L****.*

From the node at ***G****, however, two paths are possible. The eastbound path (at cost 4) coupled with the least cost path from* ***H*** *yields a cost of 8 units. This is to be compared with a northbound movement from Node* ***G*** *at a cost 3 added to a least cost of 2 from Node* ***C*** *to yield a total of 5. Clearly, the path* ***GCD*** *is preferred to* ***GHD****, as indicated in the final column of the table*

In a similar manner, the expansion continues backwards to starting Node ***M****, as indicated in the table. The solution to the problem is the path* ***MIJKGCD*** *at a cost of 11 units.*

Number of Steps from Node ***D***	*Cost to Node* ***D***		*Least Cost to Node* ***D***	
	From Node	*Cost*	*Cost*	*Path*
1	***C***	2	2	***CD***
	H	4	4	***HD***
2	***B***	$3 + 2 = 5$	5	***BCD***
	G	$3 + 2 = 5$ $4 + 4 = 8$	5	***GCD***
	L	$1 + 4 = 5$	5	***LHD***

→

3	***A***	*1 + 5 = 6*	***6***	***ABCD***
	F	*6 + 5 = 11* *7 + 5 = 12*	***11***	***FBCD***
	K	*1 + 5 = 6* *2 + 5 = 7*	***6***	***KGCD***
	P	*3 + 5 = 8*	***8***	***PLHD***
4	***E***	*5 + 6 = 11* *4 + 11 = 15*	***11***	***EABCD***
	J	*4 + 11 = 15* *2 + 6 = 8*	***8***	***JKGCD***
	O	*2 + 6 = 8* *3 + 8 = 11*	***8***	***OKGCD***
5	***I***	*3 + 11 = 14* *2 + 8 = 10*	***10***	***IJKGCD***
	N	*1 + 8 = 9* *4 + 8 = 12*	***9***	***NJKGCD***
6	***M***	*1 + 10 = 11* *3 + 9 = 12*	***11***	***MIJKGCD***

It is instructive to analyze the above solution. First, note that any path from Node M to Node D requires exactly six steps: three northbound and three eastbound. This means that the number of possible paths can be calculated from the combination

$$\binom{6}{3} = \frac{6!}{3!(6-3)!} = 20$$

In the "3 × 3" network of the example, the total cost along any path is calculated with five additions of the six costs. Therefore, solution by complete enumeration requires 100 additions for the 20 possible paths. The above dynamic programming solutions required only 24 additions (including evaluations at Nodes **C** and **H**).

For a "10 × 10" network, there are $\binom{20}{10} = 184{,}756$ paths with 19 (= 10 + 10 − 1) addition on each. That is, about 3.5 million additions required for complete enumeration.

For a "30 × 30" network, about 7×10^{18} additions are required for complete enumeration. A computer capable of 10^{12} additions per second requires almost 80 days for a complete enumeration. Even with computers, the need for alternatives to complete enumeration for larger problems is apparent.

The dramatic computational gains in dynamic programming arise because not every path is enumerated. The enumeration along a path continues only so long as that path holds promise of being part of an optimal solution. When such promise is lost, the path is abandoned and any further computational effort is spared. In general, the computational effort increases exponentially with the number of variables in complete enumeration and linearly in the dynamic programming formulation.

Example 11.5.4 Poll Taking Assignments

A Survey Research firm has been employed to take polls in three communities. The firm has only six skilled pollsters and seeks to deploy them optimally. The number of homes polled in each community depends on the number of pollsters assigned there.

The manager estimates these returns in the communities according to the number of pollsters assigned there; numbers in the table are numbers of homes polled.

Number of Polltakers

	0	1	2	3	4	5	6
Community A	0	3	7	10	18	25	35
Community B	0	2	8	12	17	25	30
Community C	0	4	7	15	19	22	29

How can pollsters best be deployed?

Solution:

For a stagewise format, let $S_i(y)$ = max return when y polltakers are assigned to i communities (in the sequence A, B, C), i = 1, 2, 3. Let $n_A(j)$, $n_B(j)$, and $n_C(j)$ represent the number of people polled in respective communities A, B, and C when j pollsters and j = 1, … ,6.

Starting with only Community A left to poll

$$S_1(y) = n_A(y)$$

However, since we are at the "end" of the problem with only one community left to poll, we do not know how many pollsters are available for assignment after the other communities have had allocations. Therefore, we must enumerate $S_1(y)$ for every possible value of y.

$$y = 1 \quad S_1(1) = n_A(1) = 3$$
$$y = 2 \quad S_1(2) = n_A(2) = 7$$
$$y = 3 \quad S_1(3) = n_A(3) = 10$$
$$y = 4 \quad S_1(4) = n_A(4) = 18$$
$$y = 5 \quad S_1(5) = n_A(5) = 25$$
$$y = 6 \quad S_1(6) = n_A(6) = 35$$

So, if we had only one community, A, to consider and if we knew how many pollsters are available for assignment, the values of $S_1(y)$ above are the optimal solutions.

Now, consider two communities, B as well as A.

⟶

Let $S_2(y)$ represent the maximum return when there are y pollsters available and x of them are assigned to Community B.

Then

$$S_2(y) = \max_{x \leq y}\{n_B(x) + S_1(y - x)\}$$

$$y = 0 \quad S_2(0) = \max_{x \leq 0}\{n_B(0) + S_1(0) = 0 + 0 = 0^*\} = 0$$

$$y = 1 \quad S_2(1) = \max_{x \leq 1}\left\{\begin{matrix} n_B(0) + S_1(1) = 0 + 3 = 3^* \\ n_B(1) + S_1(0) = 2 + 0 = 2 \end{matrix}\right\} = 3$$

$$y = 2 \quad S_2(2) = \max_{x \leq 2}\left\{\begin{matrix} n_2(0) + S_1(2) = 0 + 7 = 7 \\ n_2(1) + S_1(1) = 2 + 3 = 5 \\ n_2(2) + S_1(0) = 8 + 0 = 8^* \end{matrix}\right\} = 8$$

$$y = 3 \quad S_2(3) = \max_{x \leq 3}\left\{\begin{matrix} n_B(0) + S_1(3) = 0 + 10 = 10 \\ n_B(1) + S_1(2) = 2 + 7 = 9 \\ n_B(2) + S_1(1) = 8 + 3 = 11 \\ n_B(3) + S_1(0) = 12 + 0 = 12^* \end{matrix}\right\} = 12$$

$$y = 4 \quad S_2(4) = \max_{x \leq 4}\left\{\begin{matrix} n_B(0) + S_1(4) = 0 + 18 = 18^* \\ n_B(1) + S_1(3) = 2 + 10 = 12 \\ n_B(2) + S_1(2) = 8 + 7 = 15 \\ n_B(3) + S_1(1) = 12 + 3 = 15 \\ n_B(4) + S_1(0) = 17 + 0 = 17 \end{matrix}\right\} = 18$$

$$y = 5 \quad S_2(5) = \max_{x \leq 5}\left\{\begin{matrix} n_B(0) + S_1(5) = 0 + 25 = 25^* \\ n_B(1) + S_1(4) = 2 + 18 = 20 \\ n_B(2) + S_1(3) = 8 + 10 = 18 \\ n_B(3) + S_1(2) = 12 + 7 = 19 \\ n_B(4) + S_1(1) = 17 + 3 = 20 \\ n_B(5) + S_1(0) = 25 + 0 = 25^* \end{matrix}\right\} = 25$$

$$y = 6 \quad S_2(6) = \max_{x \leq 6}\left\{\begin{matrix} n_B(0) + S_1(6) = 0 + 35 = 35^* \\ n_B(1) + S_1(5) = 2 + 25 = 27 \\ n_B(2) + S_1(4) = 8 + 18 = 26 \\ n_B(3) + S_1(3) = 12 + 10 = 22 \\ n_B(4) + S_1(2) = 17 + 7 = 24 \\ n_B(5) + S_1(1) = 25 + 3 = 28 \\ n_B(6) + S_1(0) = 30 + 0 = 30 \end{matrix}\right\} = 35$$

Again, for example, if we knew that there are four pollsters available for allocation to the two Communities A and B, $S_2(4)$ tells us to assign none to Community B and four to Community A for a maximum return of 18.

Finally, let $S_3(y)$ represent the maximum return when all three communities require allocation and there are y pollsters available and x pollsters are assigned to Community C.

⟶

$$S_3(y) = \max_{x \le y}\{n_C(x) + S_2(y-x)\}$$

$$y = 0 \quad S_3(0) = \max_{x \le 0}\{n_C(0) + S_2(0) = 0 + 0 = 0^*\} = 0$$

$$y = 1 \quad S_3(1) = \max_{x \le 1}\left\{\begin{matrix} n_C(0) + S_2(1) = 0 + 3 = 3 \\ n_C(1) + S_2(0) = 4 + 0 = 4^* \end{matrix}\right\} = 4$$

$$y = 2 \quad S_3(2) = \max_{x \le 2}\left\{\begin{matrix} n_C(0) + S_2(2) = 0 + 8 = 8^* \\ n_C(1) + S_2(1) = 4 + 3 = 7 \\ n_C(2) + S_2(0) = 7 + 0 = 7 \end{matrix}\right\} = 8$$

$$y = 3 \quad S_3(3) = \max_{x \le 3}\left\{\begin{matrix} n_C(0) + S_2(3) = 0 + 12 = 12 \\ n_C(1) + S_2(2) = 4 + 8 = 12 \\ n_C(2) + S_2(1) = 7 + 3 = 10 \\ n_C(3) + S_2(0) = 15 + 0 = 15^* \end{matrix}\right\} = 15$$

$$y = 4 \quad S_3(4) = \max_{x \le 4}\left\{\begin{matrix} n_C(0) + S_2(4) = 0 + 18 = 18 \\ n_C(1) + S_2(3) = 4 + 12 = 16 \\ n_C(2) + S_2(2) = 7 + 8 = 15 \\ n_C(3) + S_2(1) = 15 + 3 = 18 \\ n_C(4) + S_2(0) = 19 + 0 = 19^* \end{matrix}\right\} = 19$$

$$y = 5 \quad S_3(5) = \max_{x \le 5}\left\{\begin{matrix} n_C(0) + S_2(5) = 0 + 25 = 25^* \\ n_C(1) + S_2(4) = 4 + 18 = 22 \\ n_C(2) + S_2(3) = 7 + 12 = 19 \\ n_C(3) + S_2(2) = 15 + 8 = 22 \\ n_C(4) + S_2(1) = 19 + 3 = 22 \\ n_C(5) + S_2(0) = 22 + 0 = 22 \end{matrix}\right\} = 25$$

$$y = 6 \quad S_3(6) = \max_{x \le 6}\left\{\begin{matrix} n_C(0) + S_2(6) = 0 + 35 = 35^* \\ n_C(1) + S_2(5) = 4 + 25 = 29 \\ n_C(2) + S_2(4) = 7 + 18 = 25 \\ n_C(3) + S_2(3) = 15 + 12 = 27 \\ n_C(4) + S_2(2) = 19 + 8 = 27 \\ n_C(5) + S_2(1) = 22 + 3 = 25 \\ n_C(6) + S_2(0) = 29 + 0 = 29 \end{matrix}\right\} = 35$$

Since we know that $y = 6$ pollsters are available to the three communities, we need only to evaluate $S_3(6)$. The solution for a maximum yield of 35 is to assign no pollsters to Community C and all six to the other two communities. The solution $S_2(6)$ is to assign all six pollsters to Community A.

The preceding examples illustrate the ideas of recursion by starting just before the "end" or final stage, and working backward. In this way, only one stage assignments need to be considered at each stage. Another feature of the above examples is that the optimization was particularly simple as one chose the maximum or minimum as required by simply comparing two numbers. In many situations of interest, the needed optimization is more intricate.

> ***Characteristics of Dynamic Programming Problems***
>
> - **Envision a stagewise character.**
> - **Associate a state with each stage.**
> - **The effect of a choice at each stage is to facilitate the next stage.**
> - **At any stage, an optimal policy for remaining stages is independent of assignments in previous stages.**
> - **A solution evolves from an optimal policy at the last stage.**
> - **A recursive relationship guides the stagewise solution.**

To better visualize the conceptual formulation of a dynamic program, an earlier numerical example is cast in symbols.

Network Path Revisited

Imagine the network in Example 11.5.3 on a Cartesian coordinate system with the origin at Node M. Now, each node in the network can be uniquely identified by a coordinate pair (x, y), $x = 1, 2, 3,\ y = 1, 2, 3$. For example, the coordinates of Node **F** are (1, 2) and for Node **D** are (3, 3).

Next, in formulating dynamic programs, one uses a symbol to represent the objective and the current stage. Let $f(x, y)$ denote our objective: the least cost path from (x, y) to the final node (3, 3) at D. We seek the value of $f(0, 0)$, the least cost from the origin to Node D at (3, 3) and the optimal path. To denote the cost of travel between nodes, let $c(x, y; x', y')$ represent the cost of travel along the single segment that begins at (x, y) and terminates at (x', y'); for example, $c(1, 2; 1, 3) = 6$ and $c(1, 2; 2, 2) = 7$.

If the current state is (x, y), then there are two possible ways to advance to the next state. From (x, y), movement is possible to either $(x + 1, y)$ or $(x, y + 1)$ unless blocked by outer boundary.

If the movement is from (x, y) to $(x + 1, y)$, then the least cost from (x, y) is the cost $c(x, y; x + 1, y)$ of the single segment and, by the optimality principle, we will do the best we can from the new state. In symbols

$$f(x,\ y) = c(x,\ y;\ x + 1,\ y) + f(x + 1,\ y)$$

Similarly, since, travel from (x, y) can next advance to $(x, y + 1)$, the cost is

$$f(x,\ y) = c(x,\ y;\ x,\ y + 1) + f(x,\ y + 1)$$

The favored $f(x, y)$ is the smaller of the two expressions,

$$f(x, y) = \min \begin{cases} c\,(x, y;\ \ x + 1, y) + f(x + 1, y) \\ c(x, y;\ \ x, y + 1) + f(x, y + 1) \end{cases}$$

To implement these relations, begin at (3, 3) where $f(3,\ 3) = 0$. That is, the cost of travel from (3, 3) to (3, 3) is zero. This may seem a trivial point, however, as you will see, it is the seed from which to grow the solution.

Taking a step back to Node C at (2, 3), we have

$$f(2,\ 3) = c(2,\ 3;\ 3,\ 3) + f(3,\ 3) = 2 + 0 = 2$$

since there is only a single path.

From Node **H** at (3, 2), we have

$$f(3,\ 2) = c(3,\ 2;\ 3,\ 3) + f(3,\ 3) = 4 + 0 = 4$$

From Node **G** at (2, 2), there are two possible moves. Therefore,

$$f(2,\ 2) = \min \left\{ \begin{array}{l} c\,(2,\ 2;\ \ 3,\ 3) + f(2,\ \ 3) = 3 + 2 = 5 \\ c(2,\ 2;\ \ 3,\ 2) + f(3,\ \ 2) = 4 + 4 = 8 \end{array} \right\} = 5$$

along the path **GCD**. Note the use that has been made of $f(2, 3)$ and $f(3, 2)$ from the previous iteration.

By now, you have the idea! Continue backtracking, a single segment at a time, forming the various values of $f(x, y)$. Each iteration rests upon the previous iteration until $f(0, 0)$ is reached. One should keep track of the evolving optimal path as it develops, in reverse, from Node **D**. It is instructive to complete the solution.

Example 11.5.5 Dividing a String

How should a string of length L is to be divided into n segments such that the product of the segment lengths is a maximum.

Solution:

To formulate the problem, let $x_1,\ x_2,\ \ldots\ ,x_n$ denote the n segments such that $x_1 + x_2 + \cdots + x_n = L$. An optimization problem can be formulated as

$$\begin{aligned} &\textit{Maximize}: \quad Z = x_1 x_2 \cdots x_n \\ &\textit{Subject to}: \quad x_1 + x_2 + \cdots + x_n = L \end{aligned}$$

This is clearly one problem in n variables. Using dynamic programming, we find its solution by solving n problems each in a single variable.

We begin by "devising" a symbol for the answer and that tracks the current state. Let $f_i(L)$ represent the maximum value of the product of i segments whose total length is L, $i = 1,\ 2,\ \ldots\ ,n$.

A recursion begins with $i = 1$, that is a single segment. Again trivially (but important to the sequel)

$$f_1(L) = L \quad x = L$$

A string of length L "divided" into a single segment is, of course, just L.

⟶

Next, for i = 2,

$$f_2(L) = max\{xf_1(L - x)\} \quad x \leq L$$

A string is divided into two segments with a "cut" of length x so that the remaining string length is L − x. Again, we do the best we can with the remaining length L − x, which is $f_1(L - x)$.

Replacing x by L − x in $f_1(x)$ *yields* $f_1(L - x) = L - x$,

$$f_2(L) = max \ \{x(L - x) = Lx - x^2\}, \qquad x \leq L$$

We seek the value of x that will maximize this quadratic equation. For a quadratic, $ax^2 + bx + c$, *it can be shown that the maximum value occurs at the axis of symmetry* $x = -b/2a$. *In this case,* $a = -1$ *and* $b = L$ *substituting yields*

$$x = (-L)/[(2)(-1)] = L/2$$

Therefore, $f_2(L) = (L/2)(L - L/2) = (L/2)^2$.

Continuing similarly

$$f_3(L) = max \ \{xf_2(L - x)\} = max \ \left\{x\left(\frac{L - x}{2}\right)^2\right\}, \qquad x \leq L$$

The result of the maximization is $f_3(L) = (L/3)^3$ *when* $x = L/3$ *a result that can be easily verified with calculus.*

It seems reasonable to conjecture that

$$f_k(L) = \left(\frac{L}{k}\right)^k, \quad k = 1, 2, \ldots, n$$

A proof by induction establishes the result.

It is a curiosity that linear programs can be solved, in principle, using dynamic programming methods. There is no computational advantage; indeed it is usually less efficient than the Simplex Method. However, it does make the valuable point that solutions to mathematical optimizations may be obtained in more than one way.

We illustrate this with a simplified version of the famous "knapsack problem" that was the subject of Example 6.6.

Example 11.5.6 Packing a Knapsack

A limit of 3 pounds is available to pack a knapsack. Selections are made among three items:

Items	*Weight, Pounds*	*Worth*
1	*1.5*	*8*
2	*1*	*5*
3	*0.5*	*2*

How many of each item should be selected to achieve the maximum total worth?

Solution:

There are three items so there are three stages needed to assign them. The number of pounds to spend on item #1, # 2, and #3 are stages 1, 2, and 3, respectively.

$W(X)$ = *Worth for the knapsack when packed with X pounds of item j.*

Y_1 = *# pounds allocated for stages 1, 2, and 3*

Y_2 = *# pounds allocated for stages 2 and 3*

Y_3 = *# pounds allocated for stage 3*

$F_3(Y_3)$ *is the optimal worth for stage 3 when* Y_3 *pounds are allocated.*

$F_2(Y_2)$ *is the optimal worth for stages 2 and 3 when* Y_2 *pounds are allocated.*

$F_1(Y_1)$ *is the optimal worth for stages 1, 2, and 3 when* Y_1 *pounds are allocated.*

Begin by seeking $F_3(Y_3) = \max\{W(Y_3)\}$. *The value of* $F_3(Y_3)$, *in one-half pound increments, appear below.*

Weight Y_3	*Value* $W(Y_3)$	*Optimal Solution*
0.0	*0*	$F_3(0.0) = 0$
0.5	*2*	$F_3(0.5) = 2$
1.0	*4*	$F_3(1.0) = 4$
1.5	*6*	$F_3(1.5) = 6$
2.0	*8*	$F_3(2.0) = 8$
2.5	*10*	$F_3(2.5) = 10$
3.0	*12*	$F_3(3.0) = 12$

Next, seek $F_2(Y_2) = \max_{X \le Y_2}\{W(X) + F_2(Y_2 - X)\}$. *It is the maximum worth of packing items #2 and #3 in the knapsack. The options are shown as ordered pairs where (2, 1) represents two pounds of #2 and one pound of #3 and so on.*

Note that some of the pairs are not possible and are omitted. For example, (0.5, 0) representing one-half pound of #2 and none of #3 is omitted since units of item #2 are each one pound. On the other hand, (0, 0, 5) is feasible as it signifies none of item #2 and one unit (0.5 pounds) of item #3.

⟶

Weight Y_2	*Contents and Worth* $W(X) + F_2(Y_2 - X)$	*Optimal Solution*
0.0	$(0, 0) = 0^*$	$F_2(0.0) = 0$
0.5	$(0.0, 0.5) = 0 + F_3(0.5) = 2^*$	$F_2(0.5) = 2$
1.0	$(1.0, 0.0) = 5 + F_3(0.0) = 5^*$ $(0.0, 1.0) = 0 + F_3(1.0) = 4$	$F_2(1.0) = 5$
1.5	$(1.0, 0.5) = 5 + F_3(0.5) = 7^*$ $(0.0, 1.5) = 0 + F_3(1.5) = 6$	$F_2(1.5) = 7$
2.0	$(2.0, 0.0) = 10 + F_3(0.0) = 10^*$ $(1.0, 1.0) = 5 + F_3(1.0) = 9$ $(0.0, 2.0) = 0 + F_3(2.0) = 8$	$F_2(2.0) = 10$
2.5	$(2.0, 0.5) = 10 + F_3(1.0) = 12^*$ $(1.0, 1.5) = 5 + F_3(1.5) = 11$ $(0.0, 2.5) = 0 + F_3(2.5) = 10$	$F_2(2.5) = 12$
3.0	$(3.0, 0.0) = 15 + F_3(0.0) = 15^*$ $(2.0, 1.0) = 10 + F_3(1.0) = 14$ $(1.0, 2.0) = 5 + F_3(2.0) = 13$ $(0.0, 3.0) = 0 + F_3(3.0) = 12$	$F_2(3.0) = 15$

Next, seek $F_1(Y_1) = \max_{X \le Y_1}\{W(X) + F_1(Y_1 - X)\}$ *or the maximum worth achievable using the various amounts of half-pound increments and packing the knapsack with items #1,#2, and #3. Possible combinations are again given as ordered pairs where (1.5, 1) represents 1.5 pounds assigned to item #1 and 1 pound to items #2 and #3 combined. The only possibilities for item #1 are 0, 1.5, or 3 pounds since each item weighs 1.5 pounds.*

Weight Y_1	*Contents and Worth* $W(X) + F_1(Y_1 - X)$	*Optimal Solution*
0.0	$(0.0, 0.0) = 0^*$	$F_1(0) = 0$
0.5	$(0.0, 0.5) = 0 + F_2(0.5) = 2^*$	$F_1(0.5) = 2$
1.0	$(0.0, 1.0) = 0 + F_2(1.0) = 5^*$	$F_1(1.0) = 5$
1.5	$(1.5, 0.0) = 8 + F_2(0.0) = 8^*$ $(0.0, 1.5) = 0 + F_2(1.5) = 7$	$F_1(1.5) = 8$
2.0	$(1.5, 0.5) = 8 + F_2(0.5) = 10^*$ $(0.0, 2.0) = 0 + F_2(2.0) = 10^*$	$F_1(2.0) = 10$
2.5	$(1.5, 1.0) = 8 + F_2(1.0) = 13^*$ $(0.0, 2.5) = 0 + F_2(2.5) = 12$	$F_1(2.5) = 13$
3.0	$(3.0, 0.0) = 16 + F_2(0.0) = 16^*$ $(1.5, 1.5) = 8 + F_2(1.5) = 15$ $(0.0, 3.0) = 0 + F_2(3.0) = 15$	$F_1(3.0) = 16$

The optimal option yields a maximum knapsack contents worth of 16, packing 3 pounds of item #1, and none of items #2 and #3.

♦ Dated Christmas Day 1801, President Jefferson received a message from a fellow encryption enthusiast with portions encrypted. The writer confided to Jefferson, "I presume the utter impossibility of deciphering will be readily acknowledged." The writer apparently did not include the "code key" in his letter, leaving the challenge to Jefferson. Indeed, Jefferson may have considered use of his friend's coding discovery to secure secrecy of State Department messages.

The letter and the encryption lay undeciphered for over two centuries – until now! Using modern-day dynamic programming, mathematician and professional cryptologist Dr. Lawren Smithline has cracked the code. "A Cipher to Thomas Jefferson," American Scientist, March–April 2009. Also, *Wall Street Journal* for July 2, 2009.

HISTORICAL NOTES

The word "programming" has many usages. There is "mind programming," "theater programming," and "computer programming" among others. Use of the word in mathematical literature is usually a synonym for optimization – finding maxima and minima of functions.

Apparently, in the years just after WWII, "programming" was used as a synonym for "planning." In problems of planning, particularly in a military context, questions of optimality arose. For example, assigning aircraft to targets from among available aircraft, targets, and constraints on fuel; pilots, planning menus that meet nutritional requirements at least cost; and so on.

Richard Bellman (1920–1984) — A native of New York, Bellman served in the US Army in WWII in a Theoretical Physics Division at Los Alamos, New Mexico, until 1946. Leaving there for Princeton University, he completed his doctoral degree in 3 months and joined the faculty until 1952 when he joined the newly formed Rand Corporation, a US Air Force "think tank" in California.

At the Rand Corporation, Bellman became interested in multistage decision processes. He invented dynamic programming, a major breakthrough in the field, which has far-ranging applications of functional equations. In later life, he became interested in the mathematics of medicine. An interesting account of the naming of dynamic programming can be found in "*Richard Bellman on History of Dynamic Programming*" by Stuart Dreyfus (www.google.com) and his autobiography "*Eye of the Hurricane*" (World Scientific Publishing Co., Singapore, 1984).

EXERCISES 11.5

1. Starting at Node 1 with destination at Node 5, the network shows possible routes between them. Use dynamic programming to select the shortest highway route between Nodes 1 and 6. Numbers on highway links relate to travel time.

⟶

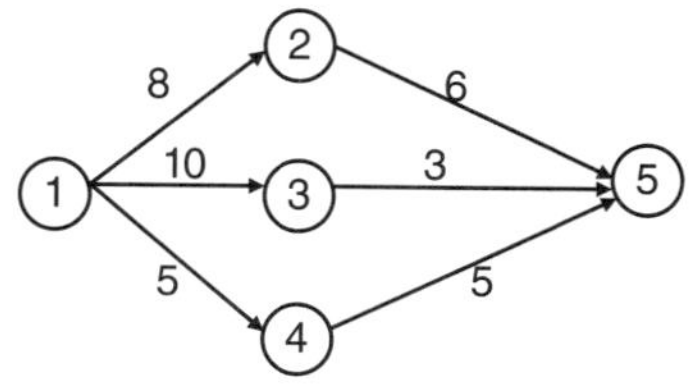

2. Starting at Node 1 with destination at Node 10, the network shows possible routes between two towns. Use dynamic programming to select the shortest highway route between Nodes 1 and 10. Numbers on highway links relate to travel time.

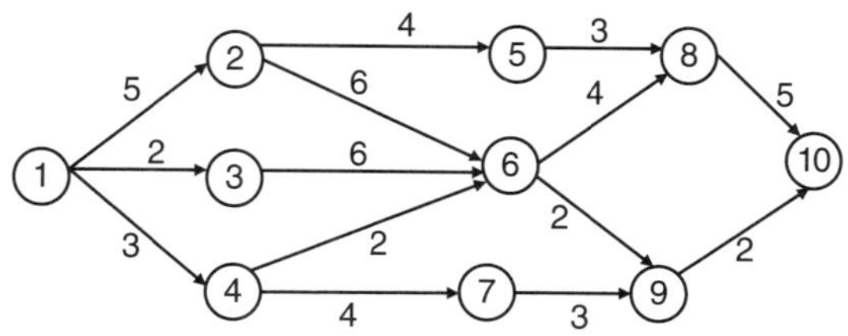

3. Starting at Node 1 with destination at Node 10, the network shows possible routes between the two towns. Use dynamic programming to select the shortest highway route between Nodes 1 and 10. Numbers on highway links relate to travel time.

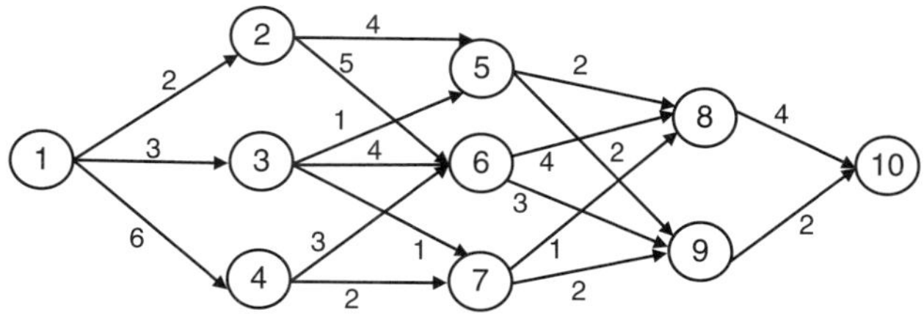

4. A knapsack can carry up to 7 pounds. The weight and value of items appears below.
   Use dynamic programming to determine how the knapsack can be packed for maximum value.

	Weight, Pounds	Value
Item #1	3	80
Item #2	2	50
Item #3	1	20

5. A corporation has three plants and up to $6 million to spend on expansion projects. The amounts spent and their estimated long-term value to the corporation are below. Using dynamic programming, determine how the corporation should spend the money for optimal long-term value from its expansion projects.

Expenditure, $	Value to Plant 1	Value to Plant 2	Value to Plant 3
0 million	0 million	0 million	0 million
1 million	4 million	3 million	5 million
2 million	5 million	6 million	—
3 million	7 million	—	7 million

6. Using the network in Example 11.5.3, find the shortest path from Nodes A and P.

7. A job shop has a product requiring three complex precision jobs in the final phase of crafting. A "finishing" specialist carefully inspects each "completed" job. Unfortunately, only three such specialist hours are available. The chance of customer rejection decreases with increasing hours that the specialist devotes to each job. The shop foreman has formed a table estimating the chances of customer rejection of each job as a function of the number of finishing hours assigned. Customers will reject the finished product only if all three jobs are found to be flawed.

Hours	Job #1	Job #2	Job #3
0	0.60	0.45	0.55
1	0.40	0.35	0.50
2	0.30	0.25	0.30
3	0.15	0.10	0.20

ANSWERS TO ODD NUMBERED EXERCISES

For detailed solutions to these Exercises see the modestly priced companion, "Student Solutions Manual" by Morris and Stark.

EXERCISES 1.1

1. Conditional equation 3. contradiction 5. contradiction 7. $x = 4$ 9. $x = 5$ 11. $x = 1$ 13. $x = 50$ 15. $x = 5/4$ 17. No solution 19. $s = 10$ 21. $t = 3$ 23. $x = 5$ 25. $x = 6$ 27. $x = 2$ 29. $x = \frac{1}{2}y + \frac{3}{2}$ 31. $y = 3 - \frac{3}{5}x$ 33. $V/LH = W$ 35. $x = Z\sigma + \mu$ 37. \$200 monthly installment 39. $C(x) = 0.75x + 9.5$ 41. a) $d = 9$ miles b) 4 seconds 43. $T = 0.062x$ $0 \leq x \leq 87{,}000$ 45. a) 8187 sq cm b) 26.2 kg.

EXERCISES 1.2

1. a) x-intercept 3, y-intercept -5 b) x-intercept 5/4, y-intercept -5
   c) x-intercept 12, y-intercept 8 d) x-intercept 2, y-intercept -18
   e) x-intercept 4, no y-intercept (vert. line) f) no x-intercept, y-intercept -2
3. a) $m = 1/2$ b) $m = 5$ c) $m = 2/3$ d) m is undefined e) $m = 0$ f) $m = 1/5$

Finite Mathematics: Models and Applications, First Edition. Carla C. Morris and Robert M. Stark.

Companion Website: http://www.wiley.com/go/morris/finitemathematics

5. a) x-intercept 5/2 and y-intercept −5

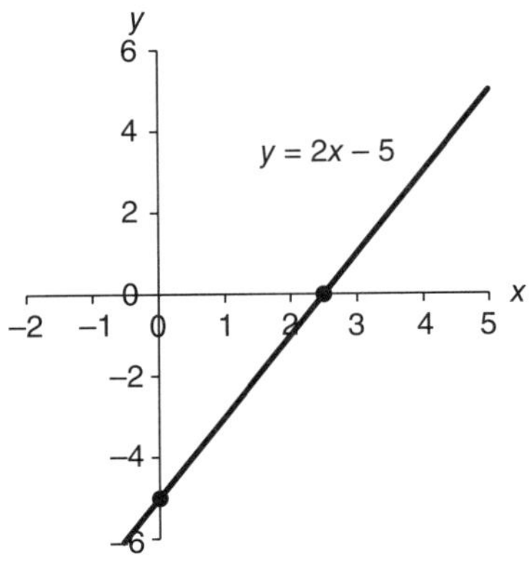

b) x-intercept 4 and no y-intercept

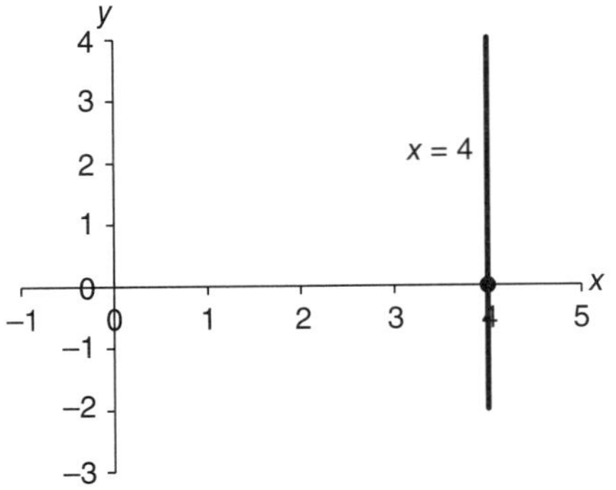

c) x-intercept 5 and y-intercept 3

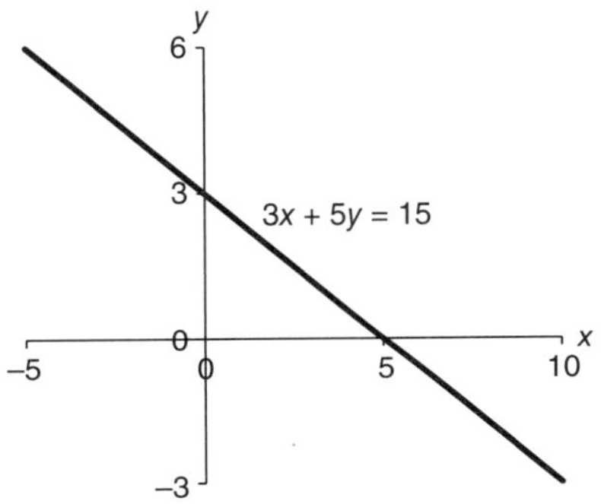

d) x-intercept 7 and y-intercept 2

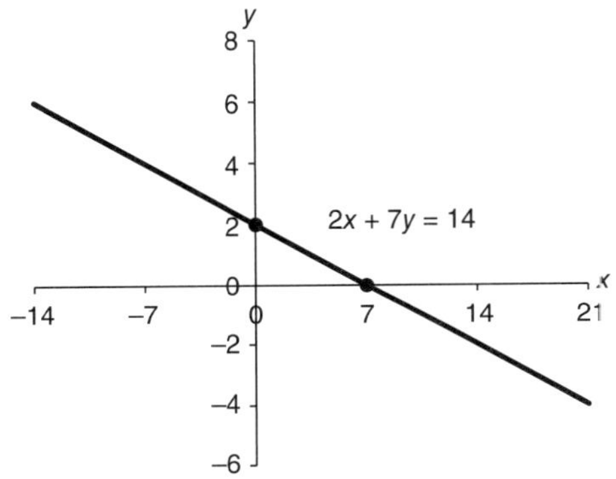

7. a) Parallel b) perpendicular c) parallel d) neither e) perpendicular
9. $y = 0$ (the x-axis) it has an infinite number of x-intercepts. Any horizontal line (except $y = 0$) has no x-intercepts. If $x = 0$ (the y-axis) it has an infinite number of y-intercepts. Any vertical line (except $x = 0$) has no y-intercepts
11. $V(t) = -6000t + 75{,}000$ 13. $y - 245 = 35(x - 7)$ or $y = 35x$
15. $C(x) = 1100 + 5x$
17. a) $C(x) = 50 + 0.30x$
b) 200 miles 19. $R - 84 = (6/5)(C - 70)$ or after simplifying, $R = (6/5)C$

EXERCISES 1.3

1. a) Not a solution b) is a solution c) not a solution
3. a) Consistent b) dependent c) inconsistent

5. The graphs and solution to each system are shown

a)

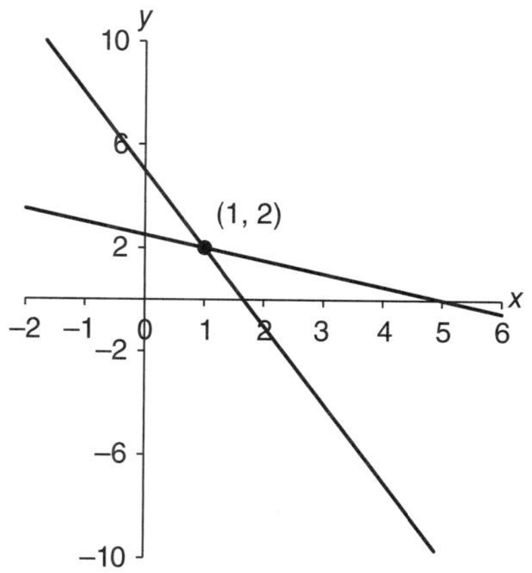

b)

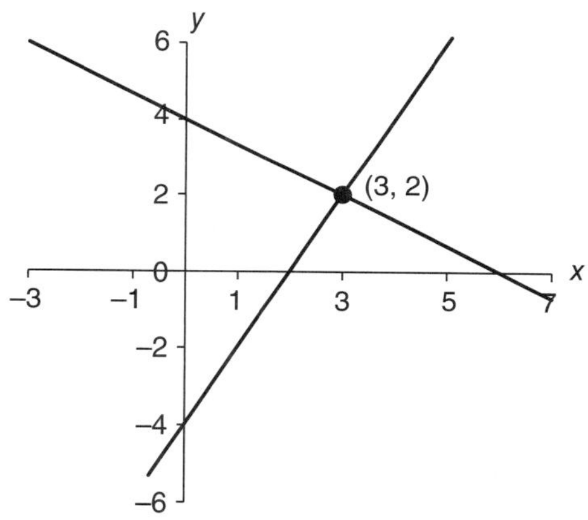

c)

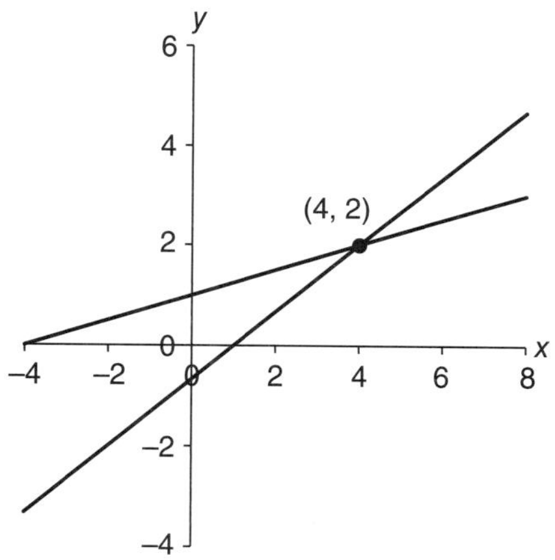

7. a) (3, 2) b) (3, 2) c) no solution 9. a) (1, 3) b) (5, 5) c) (1, 3) d) (0, 4)

11. 1500 boxes of cookies and 900 boxes of candy 13. $16\frac{2}{3}$ gallons regular and $83\frac{1}{3}$ gallons of premium

15.

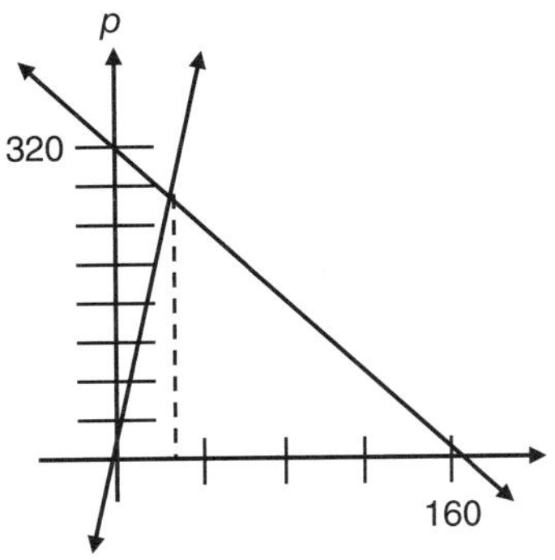

b) The market equilibrium occurs when q is about 30 and y about 260. Using substitution,

$$-2q + 320 = 8q + 20$$
$$300 = 10q$$
$$30 = q$$
$$\text{so } p = 260$$

17.

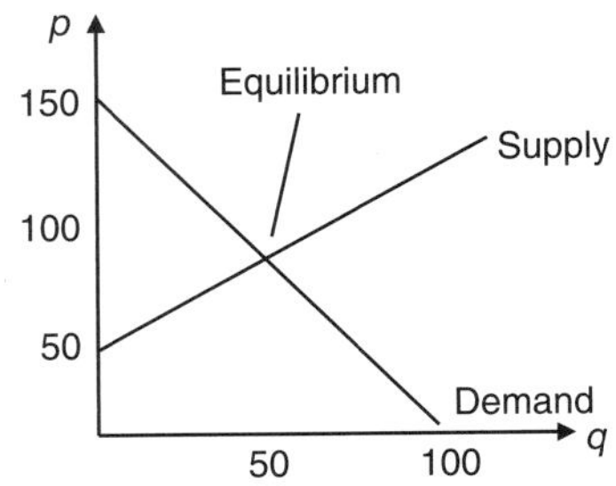

The price is about \$75 and quantity 50 at market equilibrium.

EXERCISES 1.4

1. $d = 2$ 3. a) March 14 b) Yasumasa Kanada of the University of Tokyo c) Albert Einstein d) an illness thought to be attached to trying to square a circle e) die Ludolphschezahl 5. a) 2.48832 b) 2.59374 c) 2.66584

EXERCISES 1.5

1. a) 2^{9x} b) 3^{6x} c) 2^{20x} 3. a) $(2)^{12x}$ b) $(3)^{-12x}$ c) $(3)^{6x}$ 5. a) 2^{5x} b) 2^{2x} c) $(3)^{-3x}$

7. $\dfrac{7x^6}{y^6}$ 9. x^2y^7 11. $\dfrac{2^{5x+3}(2^2)^{x+1}}{2^3(2^{3x-1})} = 2^{4x+3}$ 13. $x = 5$ 15. $x = 2$ 17. $x = 4$

19. $x = -2$ or $x = -3$. 21. $2^{3+h} = 2^h(2^3)$ 23. $7^{x+5} - 7^{2x} = 7^{2x}(7^{-x+5} - 1)$

25. $(7^h)^3 - 8 = (7^h - 2)[7^{2h} + 2(7^h) + 4]$

27. Rewrite as $e^y = x$ to determine ordered pairs and graph as

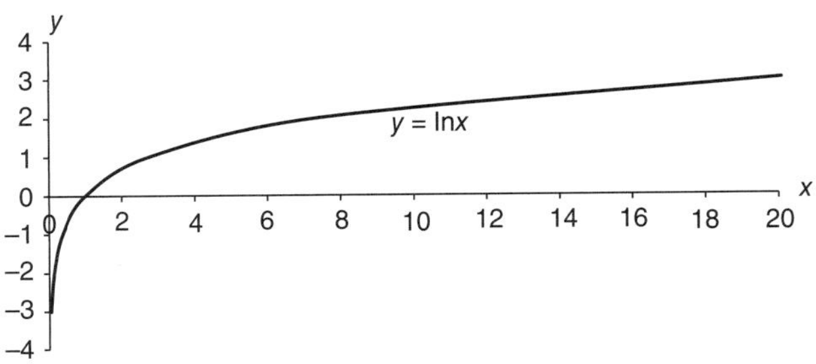

29. $x = 5$ 31. $x = 3$ 33. $x = -4$ 35. $\ln e^7 = 7$ 37. $\ln e^{-3.4} = -3.4$
39. $e^{\ln 1} = e^0 = 1$ 41. $x = 3/2$ 43. $x = 4$ 45. $x = -3$ 47. $x = 4$
49. $x = 1$ or -8 51. $x = (1/5)\ln 2$ 53. $x = 7 - e^{1/3}$

55. $4\ln(x-1) - 5\ln(2x+3) - 2\ln(x-4)$

57. $\ln \frac{2(7)}{3} = \ln \frac{14}{3}$ 59. $x = \frac{1 + \log_{10} 100{,}100}{3}$ 61. $x = \frac{-1 + \ln 22}{3}$

EXERCISES 1.6

1. a) Direct b) Inverse c) Direct d) Direct e) Direct 3. 108 5. 15
7. $x = 0.16y$ 9. $A = \pi d^2/4$ 11. $V = 30\pi cm^3$ 13. 6.25 cm 15. $x = 10,\ y = 26$
17. $x = 2.5$ and $x + 5 = 7.5$ 19. 64in^2
21. An annual series of maths, science, and technical instructional distance programs for students in grades 6–8.
23. Answers vary.

EXERCISES 1.7

1. 8046.74 m 3. 62.14 miles/h 5. 2235.2 cm/s
7. 1950.48 g 9. 7.13 g/ml 11. 29.872 lbs
13. 1.505×10^{24}atoms C 15. 1715.266 gallons 17. N_2O
19. An acre can have various dimensions with an area of about 43,560 square feet.

SUPPLEMENTARY EXERCISES (CHAPTER 1)

1. 89 3. a) \$66,358 b) \$58,405 c) Not advantageous 5. $y = (-8/5)x + 8$
7. a) $y - 5 = (2/5)(x - 3)$ b) $y = 4$ c) $x = 2$ 9. It is d) $-2x - 3$ 11. \$67
13. 40 ml water and 10 ml pure acid 15. Nine more \$50 items sold than \$100 items
17. a) $R(x) = 6x$ b) $C(x) = 2000 + 2x$ c) break-even when $x = 500$
d) $P(x) = 6x - (2000 + 2x) = 4x - 2000$ and $P(1000) = 2000$ profit
19. a) Information on the internet b) no digit appears more frequently
21. 5^{4x-2} 23. $x = 7/2$ 25. $x = (1 + \ln 5)$
27.

x	2	4	6	12	1
y	24	12	8	4	48

29. 2.71×10^{24} atoms C.

For detailed solutions to these Exercises see the modestly priced companion, "Student Solutions Manual" by Morris and Stark.

EXERCISES 2.1

1. $i = \$1200$. 3. $i = (50{,}000)(0.05)(7) = \$17{,}500$
5. $P_t = \$63{,}265.95$ and $i = \$13{,}265.95$
7. $P_t = \$597{,}026.15$ and $i = \$97{,}026.15$
9. $P_t = \$303{,}876.56$ and $i = \$53{,}876.56$ interest
11. $P_t = \$181{,}401.84$
13. $P_t = \$61{,}667.74$
15. $P_t = \$96{,}301.91$
17. $P_t = \$114{,}147.12$
19. $P_t = \$63{,}562.46$ and $i = \$13{,}562.46$
21. $P_t = \$8568.50$ and $i = \$3568.50$. So $r = 0.0793$
23. $P_t = 2.2038P_0$ and $i = 1.2038P_0$ so $r = 1.2038/8 = 0.1505$. The effective interest rate is 15.05%.
25. $P_0 = \dfrac{250{,}000}{(1.09)^8} = \$125{,}466.57$
27. $P_0 = \dfrac{750{,}000}{(1.05)^{10}} = \$460{,}434.94$.

EXERCISES 2.2

1. \$33,065.95 3. \$99,197.86 5. \$22,230.84 7. \$25,129.02
9. \$209,348.00 11. \$2247.75 13. \$3739.48 15. \$29,130.74
17. $PV = \dfrac{1-(1+0.005)^{-240}}{0.005}(25{,}000) = \$3{,}489{,}519.30$
Since this exceeds the \$2,500,000 lump sum payment, the 20 year monthly payments is preferred.

EXERCISES 2.3

1. Using the amortization formula the payment is

$$\frac{(0.01)(500)}{\left(1-\dfrac{1}{(1+0.01)^6}\right)} = \$86.27$$

3. \$483.32 5. \$547.58
7. \$19,124.16 for 4-yr loan and \$19,406.40 for 5-yr loan for a \$282.24 savings.
9. \$1670.95 for 30 yr-loan and \$1847.43 for 25-yr loan.

11. There are two options for the town to consider:

Option 1

Annual cost of option 1 is \$49,299.06 + \$3000 = \$52,299.06.

Option 2

Annual cost of option 1 is \$43,821.39 + \$7000 = \$50,821.39.
The town should choose the lower annual cost option 2.

13. Answers vary.

EXERCISES 2.4

1. 17, 22, 27, and 32 3. 65, 80, 95, 110, and 125.

5. The first seven terms are 3, 9, 15, 21, 27, 33, and 39. Adding terms yields 147. Using the formula $S_7 = \dfrac{7(3+39)}{2} = \dfrac{7(42)}{2} = 147$.

7. $a_1 = \dfrac{4}{5}, \quad a_2 = 1, \quad a_3 = \dfrac{10}{9}, \quad a_4 = \dfrac{13}{11}, \quad \text{and} \quad a_5 = \dfrac{16}{13}.$

9. 54, 162, and 486 11. 27, 9, and 3.

13. The first five terms are 4, 16, 64, 256, and 1024. The sum is 1364. Using the formula yields $S_5 = \dfrac{4(1-4^5)}{1-4} = \dfrac{4(-1023)}{-3} = 1364$.

15. The height of the ball is 25 ft after the third bounce or $\dfrac{1}{2^3}(200)$.

17. $i = \$1312.38$

19. Start with the products of the first three successive pairs of Fibonacci numbers $F_1F_2 + F_2F_3 + F_3F_4$. Next, using $F_1 = F_2$ and that $F_4 = F_2 + F_3$ rewrite the expression as $(F_2)F_2 + F_2F_3 + F_3(F_2 + F_3)$.
Now, $(F_2)F_2 + F_2F_3 + F_3(F_2 + F_3) = (F_2)^2 + 2F_2F_3 + (F_3)^2 = (F_2 + F_3)^2 = (F_4)^2$.
Next, use the first five pairs of Fibonacci numbers $F_1F_2 + F_2F_3 + F_3F_4 + F_4F_5 + F_5F_6$ and the previous result.
Continuing in this manner the "squaring rectangles" assertion can be verified.

21. Suppose $n = 2$ then $\Phi^2 = F_2\Phi + F_1 = \Phi + 1$ which was verified in the previous Exercise. Next, $\Phi^3 = (\Phi^2)\Phi = (\Phi + 1)\Phi = \Phi^2 + \Phi = (\Phi + 1) + \Phi = 2\Phi + 1$ and since $F_3 = 2$ and $F_2 = 1$ so $\Phi^3 = F_3\Phi + F_2$.
Continuing find that $\Phi^n = F_n\Phi + F_{n-1}$.

SUPPLEMENTARY EXERCISES (CHAPTER 2)

1. $i = \$5000$ 3. $10{,}000e^{(0.07)(15)} - 10{,}000 = \$18{,}576.51$.

5. $\dfrac{175{,}000}{e^{1.6}} = \$35{,}331.89$ 7. $\dfrac{(1.07)^{10} - 1}{0.07}(1500) = \$20{,}724.67$.

9. $\dfrac{1 - (1.005)^{-120}}{0.005}(100) = \$9007.35$

11. Using the amortization formula the payment is

$$\frac{(0.06)(25{,}000)}{\left(1-\dfrac{1}{(1+0.06)^{10}}\right)} = \frac{1500}{1-(1.06)^{-10}} = \$3396.70$$

13. 11, 14, and 17

15. Let x_1 = first number called out and x_2 = the second number called out. Then the third number is $x_1 + x_2 = x_3$. In terms of the original pair of numbers $x_3 = 2x_2 + x_1$. The next numbers in terms of the original pair are $3x_2 + 2x_1$, $5x_2 + 3x_1$, $8x_2 + 5x_1$, $13x_2 + 8x_1$. Note the coefficients on x_2 are 1, 2, 3, 5, 8, 13, ... and on x_1 are 1, 1, 2, 3, 5, 8, ... are Fibonacci numbers. Continue from here to derive the proof.

17. Answers will vary but many should find examples where the Golden Mean is approximated.

For detailed solutions to these Exercises see the modestly priced companion, "Student Solutions Manual" by Morris and Stark.

EXERCISES 3.1

1. a) $\left[\begin{array}{ccc|c} 3 & -5 & 4 & 5 \\ 1 & 1 & -1 & 0 \\ 2 & 3 & -2 & 2 \end{array}\right]$ b) $\left[\begin{array}{ccc|c} 1 & -3 & 5 & 19 \\ 2 & 4 & -1 & -5 \\ -1 & 5 & 3 & 3 \end{array}\right]$

3. $\begin{aligned} 5x_1 + 3x_2 &= 13 \\ 4x_1 + 7x_2 &= 18 \end{aligned}$

5. $\begin{aligned} x_1 + 2x_2 + 3x_3 &= 4 \\ 2x_1 + x_2 + 2x_3 &= 3 \\ 4x_1 + x_2 - 2x_3 &= 9 \end{aligned}$

7. $x = 4,\ y = 3,\ z = 1$

9. Let x = # children tickets and y = # adult tickets. The equations are

$$\begin{aligned} x + y &= 2400 \\ 4x + 5y &= 10{,}500 \end{aligned}$$

The augmented matrix is $\left[\begin{array}{cc|c} 1 & 1 & 2400 \\ 4 & 5 & 10{,}500 \end{array}\right]$.

11. The last row contains a row of zeros that has a nonzero number on the RHS (right hand side)

13. The new row 1 becomes [Row 1 – 2 Row 3] or

$[1\ \ 1\ \ 2\ \ 9] - [0\ \ 0\ \ 2\ \ 2] = [1\ \ 1\ \ 0\ \ 7]$

The new row 2 becomes [Row 2 – 3 Row 3] or

$[0\ \ 1\ \ 3\ \ 7] - [0\ \ 0\ \ 3\ \ 3] = [0\ \ 1\ \ 0\ \ 4]$ as:

$$\left[\begin{array}{ccc|c} 1 & 1 & 2 & 9 \\ 0 & 1 & 3 & 7 \\ 0 & 0 & 1 & 1 \end{array}\right] \to \left[\begin{array}{ccc|c} 1 & 1 & 0 & 7 \\ 0 & 1 & 0 & 4 \\ 0 & 0 & 1 & 1 \end{array}\right]$$

15. First perform the operations $R_2 \to R_2 - 2R_1$ and $R_3 \to R_3 + R_1$ as:

$$\left[\begin{array}{ccc|c} 1 & -3 & 5 & 19 \\ 2 & 4 & -1 & -5 \\ -1 & 5 & 3 & 3 \end{array}\right] \to \left[\begin{array}{ccc|c} 1 & -3 & 5 & 19 \\ 0 & 10 & -11 & -43 \\ 0 & 2 & 8 & 22 \end{array}\right]$$

eventually this yields $\left[\begin{array}{ccc|c} 1 & -3 & 5 & 19 \\ 0 & 1 & 4 & 11 \\ 0 & 0 & 1 & 3 \end{array}\right]$.

The ordered triple solution is $(1, -1, 3)$.

EXERCISES 3.2

1. The subscripts indicate the row and column for the entry.
   a) $x_{12} = 2$ (first row, second column) b) $a_{23} = -1$ c) $a_{13} = -3$ d) $a_{22} = 4$
3. a), c), and e) are possible 5. a), b), and c) are possible
7. The valid matrix operations in Exercise 3.2.3 were in parts a, c, and e

$$\mathbf{A} + \mathbf{B} = \begin{bmatrix} 10 & 25 \\ 74 & 30 \end{bmatrix} + \begin{bmatrix} 20 & 52 \\ 41 & 36 \end{bmatrix} = \begin{bmatrix} 30 & 77 \\ 115 & 66 \end{bmatrix}$$

$$5\mathbf{C} = 5\begin{bmatrix} 21 & 62 & -37 \\ 19 & 40 & -91 \end{bmatrix} = \begin{bmatrix} 105 & 310 & -185 \\ 95 & 200 & -455 \end{bmatrix}$$

$$3\mathbf{B} - 6\mathbf{A} = 3\begin{bmatrix} 20 & 52 \\ 41 & 36 \end{bmatrix} - 6\begin{bmatrix} 10 & 25 \\ 74 & 30 \end{bmatrix}$$

$$= \begin{bmatrix} 60 & 156 \\ 123 & 108 \end{bmatrix} - \begin{bmatrix} 60 & 150 \\ 444 & 180 \end{bmatrix} = \begin{bmatrix} 0 & 6 \\ -321 & -72 \end{bmatrix}$$

9. The valid operations in Exercise 3.2.5 are a), b), and c).

$$\text{a) } \mathbf{A} + \mathbf{B} = \begin{bmatrix} 12 & 0 \\ 15 & 8 \\ 16 & 8 \end{bmatrix} \text{ b) } 4\mathbf{C} = \begin{bmatrix} 12 & 32 & 36 \\ 8 & 0 & 0 \\ 16 & 0 & 12 \end{bmatrix} \text{ c) } 2\mathbf{A} + \mathbf{B} = \begin{bmatrix} 16 & 1 \\ 18 & 13 \\ 22 & 9 \end{bmatrix}$$

11. All the operations are valid here.

$$\text{a) } \mathbf{A} + \mathbf{B} = \begin{bmatrix} -1 & 3 & 8 \\ -2 & 5 & 11 \end{bmatrix} \qquad \text{b) } 2\mathbf{B} = \begin{bmatrix} -4 & 6 & 12 \\ -10 & 2 & 8 \end{bmatrix}$$

$$\text{c) } \mathbf{A} + 3\mathbf{C} = \begin{bmatrix} 13 & 18 & -7 \\ 9 & 28 & -20 \end{bmatrix} \text{ d) } 3\mathbf{A} - 2\mathbf{B} + \mathbf{C} = \begin{bmatrix} 11 & 0 & -9 \\ 21 & 18 & 4 \end{bmatrix}$$

13. $\mathbf{AB} = [22]$ and $\mathbf{BA} = \begin{bmatrix} -2 & 0 & -6 & -8 \\ -1 & 0 & -3 & -4 \\ 8 & 0 & 24 & 32 \\ 0 & 0 & 0 & 0 \end{bmatrix}$

15. a) **AB** is not possible b) **AC** is possible c) **BA** is possible
    d) **BC** is possible e) **CA** is not possible f) **CB** is possible

17. AC, BA, BC, and CB are possible

$$\mathbf{AC} = \begin{bmatrix} 18 & 8 & -2 \\ 58 & 32 & -18 \end{bmatrix} \quad \mathbf{BA} = \begin{bmatrix} 20 & 34 \\ 12 & 26 \\ -8 & -10 \end{bmatrix}$$

$$\mathbf{BC} = \begin{bmatrix} 47 & 24 & -11 \\ 31 & 20 & -15 \\ -17 & -6 & -1 \end{bmatrix} \quad \mathbf{CB} = \begin{bmatrix} 7 & 17 \\ 33 & 59 \end{bmatrix}$$

19. a) (3×2) b) $(2 \times n)$ c) $(m \times 3)$ matrix

21. a) $\mathbf{AA} = \begin{bmatrix} 10 & 2 & 7 \\ 2 & 5 & 10 \\ 7 & 10 & 21 \end{bmatrix}$ b) It must be a square matrix

23. a row vector $(m \times 1)$.

25. a) $\begin{bmatrix} 5 & 3 & 2 \end{bmatrix} \begin{bmatrix} 2.1128 \\ 2.1189 \\ 1.8143 \end{bmatrix} = \$20.5493$

b) $\begin{bmatrix} 4 & 1 & 3 \end{bmatrix} \begin{bmatrix} 2.0516 & 2.0234 & 2.0172 \\ 2.1261 & 2.1857 & 2.1836 \\ 1.8901 & 1.8427 & 1.8624 \end{bmatrix} = \begin{bmatrix} 16.0028 & 15.8074 & 15.8396 \end{bmatrix}$

27. $\begin{bmatrix} 4 & 1 & 3 & 3 & 3 \end{bmatrix} \begin{bmatrix} 4.00 \\ 3.67 \\ 2.67 \\ 3.00 \\ 2.33 \end{bmatrix} = [43.67]$ $43.67/14 = 3.119$ semester GPA

29. a) $\begin{bmatrix} 3 & -7 & 1 & 3 \\ 2 & 3 & 1 & -2 \\ 3 & 1 & 3 & 7 \\ 1 & 5 & 2 & -8 \end{bmatrix} \begin{bmatrix} x_1 \\ x_2 \\ x_3 \\ x_4 \end{bmatrix} = \begin{bmatrix} 7 \\ 0 \\ 25 \\ -19 \end{bmatrix}$ b) $\begin{bmatrix} 1 & 4 & 3 & 1 \\ 2 & 3 & -1 & 5 \\ 1 & 1 & -2 & 4 \\ 3 & 2 & -1 & 3 \end{bmatrix} \begin{bmatrix} w \\ x \\ y \\ z \end{bmatrix} = \begin{bmatrix} 8 \\ 7 \\ 3 \\ 10 \end{bmatrix}$

31. a) $\begin{aligned} w + 5x + 2y + 2z &= 19 \\ w + x + 3y \quad\;\; &= 14 \\ 2w + 3x - 2y + 5z &= 15 \\ w + x + \quad\; + 2z &= 17 \end{aligned}$ b) $\begin{aligned} 3x_1 + x_2 + 4x_3 + x_4 &= 12 \\ x_1 + x_2 + 3x_3 + 2x_4 &= 13 \\ x_1 + 2x_2 + \quad\; + 5x_4 &= 17 \\ 2x_1 + 3x_2 + x_3 + x_4 &= 8 \end{aligned}$

33. a) $\left[\begin{array}{cccc|c} 1 & 5 & 2 & 2 & 19 \\ 1 & 1 & 3 & 0 & 14 \\ 2 & 3 & -2 & 5 & 15 \\ 3 & 1 & 0 & 2 & 17 \end{array}\right]$ b) $\left[\begin{array}{cccc|c} 3 & 1 & 4 & 1 & 12 \\ 1 & 1 & 3 & 2 & 13 \\ 1 & 2 & 0 & 5 & 17 \\ 2 & 3 & 1 & 1 & 8 \end{array}\right]$

EXERCISES 3.3

1. a) The trivial solution $x_1 = 0$ and $x_2 = 0$.
   b) Indicating $x_1 = -3x_2$ and any value of x_2 determines x_1
3. a, c, d, and e are elementary
5. a) $\begin{bmatrix} 10 & 3 & 7 & 5 \\ 12 & 8 & 1 & 0 \end{bmatrix}$ b) $\begin{bmatrix} 10 & 7 \\ 12 & 1 \\ 7 & 2 \end{bmatrix}$ c) $\begin{bmatrix} 10 & 7 & 5 \\ 7 & 2 & 6 \end{bmatrix}$ d) $\begin{bmatrix} 3 & 5 \\ 8 & 0 \\ -4 & 6 \end{bmatrix}$

→

7. a) and b) are in row echelon form 9. $x_1 = 1$ and $x_2 = 5$ 11. no solution
13. $x_1 = 2,\ x_2 = 0$, and $x_3 = 6$ 15. $x_1 = 1,\ x_2 = 5$, and $x_3 = 2$
17. $x_1 = 2/3,\ x_2 = -2,\ x_3 = 4/3$, and $x_4 = 1$
19. 400 children's tickets and 500 adult tickets sold in advance, and 100 tickets sold at the door.
21. 1500 children tickets and 900 adult tickets
23. 2 ounces of Food A, 3 ounces of Food B and 1 ounce of Food C

EXERCISES 3.4

1. $\mathbf{AB} = \begin{bmatrix} 1 & 0 \\ 0 & 1 \end{bmatrix} = \mathbf{I}$ and $\mathbf{BA} = \begin{bmatrix} 1 & 0 \\ 0 & 1 \end{bmatrix} = \mathbf{I}$. Since the matrix products result in an identity matrix, the matrices are inverses

3. $$\mathbf{AB} = \begin{bmatrix} 3 & 5 & 9 \\ 1 & 3 & 6 \\ 0 & 1 & 0 \end{bmatrix} \begin{bmatrix} 3 & 0 & -5 \\ -1 & 2 & 4 \\ 1 & 0 & 0 \end{bmatrix} = \begin{bmatrix} 13 & 10 & 5 \\ 6 & 6 & 7 \\ -1 & 2 & 4 \end{bmatrix} \neq \mathbf{I}$$

$$\mathbf{BA} = \begin{bmatrix} 3 & 0 & -5 \\ -1 & 2 & 4 \\ 1 & 0 & 0 \end{bmatrix} \begin{bmatrix} 3 & 5 & 9 \\ 1 & 3 & 6 \\ 0 & 1 & 0 \end{bmatrix} = \begin{bmatrix} 9 & 10 & 27 \\ -1 & 5 & 3 \\ 3 & 5 & 9 \end{bmatrix} \neq \mathbf{I}$$

The matrices are not inverses since their products do not result in an identity matrix.

5. a) The second row is a multiple of the first row
b) The third row is a multiple of the first row
c) The third row is all zeros
d) The third column is a multiple of the first column

7. a) $$\mathbf{A}^{-1}\mathbf{A} = \begin{bmatrix} 5/4 & 1/4 & 3/4 \\ 1/4 & 5/4 & 3/4 \\ 1/2 & 1/2 & 3/2 \end{bmatrix} \begin{bmatrix} 1 & 0 & -1/2 \\ 0 & 1 & -1/2 \\ -1/3 & -1/3 & 1 \end{bmatrix} = \begin{bmatrix} 1 & 0 & 0 \\ 0 & 1 & 0 \\ 0 & 0 & 1 \end{bmatrix} = \mathbf{I}.$$

$$\mathbf{AA}^{-1} = \begin{bmatrix} 1 & 0 & -1/2 \\ 0 & 1 & -1/2 \\ -1/3 & -1/3 & 1 \end{bmatrix} \begin{bmatrix} 5/4 & 1/4 & 3/4 \\ 1/4 & 5/4 & 3/4 \\ 1/2 & 1/2 & 3/2 \end{bmatrix} = \begin{bmatrix} 1 & 0 & 0 \\ 0 & 1 & 0 \\ 0 & 0 & 1 \end{bmatrix} = \mathbf{I}.$$

b) Using the computational technique,

$$\left[\begin{array}{ccc|ccc} 1 & 0 & -1/2 & 1 & 0 & 0 \\ 0 & 1 & -1/2 & 0 & 1 & 0 \\ -1/3 & -1/3 & 1 & 0 & 0 & 1 \end{array}\right] \rightarrow \left[\begin{array}{ccc|ccc} 1 & 0 & -1/2 & 1 & 0 & 0 \\ 0 & 1 & -1/2 & 0 & 1 & 0 \\ 0 & -1/3 & 5/6 & 1/3 & 0 & 1 \end{array}\right]$$

$$\left[\begin{array}{ccc|ccc} 1 & 0 & -1/2 & 1 & 0 & 0 \\ 0 & 1 & -1/2 & 0 & 1 & 0 \\ 0 & 0 & 2/3 & 1/3 & 1/3 & 1 \end{array}\right] \rightarrow \left[\begin{array}{ccc|ccc} 1 & 0 & -1/2 & 1 & 0 & 0 \\ 0 & 1 & -1/2 & 0 & 1 & 0 \\ 0 & 0 & 1 & 1/2 & 1/2 & 3/2 \end{array}\right]$$

$$\left[\begin{array}{ccc|ccc} 1 & 0 & 0 & 5/4 & 1/4 & 3/4 \\ 0 & 1 & 0 & 1/4 & 5/4 & 3/4 \\ 0 & 0 & 1 & 1/2 & 1/2 & 3/2 \end{array}\right]$$

9. $\mathbf{A}^{-1} = \begin{bmatrix} 1 & -1 & 1 \\ 0 & 2 & -1 \\ 2 & 3 & 0 \end{bmatrix}$ 11. $\mathbf{A}^{-1} = \begin{bmatrix} 0.4 & 0.2 & -0.2 \\ -1.2 & 0.4 & 0.6 \\ 0.3 & -0.1 & 0.1 \end{bmatrix}$

13. a) The system of equations for the problem is

$$\begin{aligned} x + \quad y + \quad z &= 14 \\ x - \quad y + \quad z &= 2 \\ 5x + 10y + 25z &= 160 \end{aligned}$$

b) The system in matrix form is:

$$\begin{bmatrix} 1 & 1 & 1 \\ 1 & -1 & 1 \\ 5 & 10 & 25 \end{bmatrix}\begin{bmatrix} x \\ y \\ z \end{bmatrix} = \begin{bmatrix} 14 \\ 2 \\ 160 \end{bmatrix}$$ after finding the inverse we have

$$\begin{bmatrix} 7/8 & 3/8 & -1/20 \\ 1/2 & -1/2 & 0 \\ -3/8 & 1/8 & 1/20 \end{bmatrix}\begin{bmatrix} 14 \\ 2 \\ 160 \end{bmatrix} = \begin{bmatrix} 5 \\ 6 \\ 3 \end{bmatrix}$$ 5 nickels, 6 dimes, and 3 quarters.

15. \$8000 in stocks, \$6000 in bonds and \$4000 in money market funds.

17. The encoded message is

99	21	99	181	45	155	87	18	87	124	32	93	85	18	94
99	24	97	98	19	106	96	30	67	157	37	147	46	8	53

19. Decoding the message requires finding the inverse of the encoding key.

The inverse of the key is $\begin{bmatrix} -2 & 5 & 1 \\ 2 & -4 & -1 \\ -1 & 1 & 1 \end{bmatrix}$ so

$$\begin{bmatrix} -2 & 5 & 1 \\ 2 & -4 & -1 \\ -1 & 1 & 1 \end{bmatrix}\begin{bmatrix} 99 & 181 & 87 & 124 & 85 & 99 & 98 & 96 & 157 & 46 \\ 21 & 45 & 18 & 32 & 18 & 24 & 19 & 30 & 37 & 8 \\ 99 & 155 & 87 & 93 & 94 & 97 & 106 & 67 & 147 & 53 \end{bmatrix}$$

$$= \begin{bmatrix} 6 & 18 & 3 & 5 & 14 & 19 & 5 & 25 & 18 & 1 \\ 15 & 27 & 15 & 27 & 4 & 5 & 14 & 5 & 19 & 7 \\ 21 & 19 & 18 & 1 & 27 & 22 & 27 & 1 & 27 & 15 \end{bmatrix}$$

indicating the message is FOUR SCORE AND SEVEN YEARS AGO

21. Here $\mathbf{D} = (\mathbf{I} - \mathbf{A})\mathbf{X}$. Therefore,

$$\left(\begin{bmatrix} 1 & 0 & 0 \\ 0 & 1 & 0 \\ 0 & 0 & 1 \end{bmatrix} - \begin{bmatrix} 0.2 & 0.5 & 0.1 \\ 0.0 & 0.3 & 0.2 \\ 0.5 & 0.0 & 0.4 \end{bmatrix}\right)\begin{bmatrix} 30 \\ 20 \\ 50 \end{bmatrix} = \begin{bmatrix} 0.8 & -0.5 & -0.1 \\ 0.0 & 0.7 & -0.2 \\ -0.5 & 0.0 & 0.6 \end{bmatrix}\begin{bmatrix} 30 \\ 20 \\ 50 \end{bmatrix} = \begin{bmatrix} 9 \\ 4 \\ 15 \end{bmatrix}$$

23. The technology matrix A for this problem is then

$$\mathbf{A} = \begin{bmatrix} 0.10 & 0.40 & 0.35 \\ 0.30 & 0.25 & 0.15 \\ 0.20 & 0.15 & 0.35 \end{bmatrix} \text{ and } \mathbf{I} - \mathbf{A} = \begin{bmatrix} 0.90 & -0.40 & -0.35 \\ -0.30 & 0.75 & -0.15 \\ -0.20 & -0.15 & 0.65 \end{bmatrix}$$

$$(\mathbf{I} - \mathbf{A})^{-1} \approx \begin{bmatrix} 1.787 & 1.201 & 1.239 \\ 0.865 & 1.979 & 0.922 \\ 0.749 & 0.826 & 2.133 \end{bmatrix}$$

The production schedule is determined by $\mathbf{X} = (\mathbf{I} - \mathbf{A})^{-1}\,\mathbf{D}$

$$\begin{bmatrix} 1.787 & 1.201 & 1.239 \\ 0.865 & 1.979 & 0.922 \\ 0.749 & 0.826 & 2.133 \end{bmatrix}\begin{bmatrix} 25{,}000 \\ 10{,}000 \\ 15{,}000 \end{bmatrix} = \begin{bmatrix} 75{,}270 \\ 55{,}245 \\ 58{,}980 \end{bmatrix}$$

SUPPLEMENTARY EXERCISES (CHAPTER 3)

1. a) $n = p$ b) $m = p$ and $n = r$ c) $n = p$ and $r = m$

3. $\mathbf{AB} = \begin{bmatrix} 23 & 14 & 44 & -6 \\ 12 & 18 & 45 & -25 \\ 60 & 32 & 125 & -25 \\ 43 & 18 & 79 & -4 \end{bmatrix}$ 5. $\mathbf{A}^2 = \begin{bmatrix} 13 & 18 & 7 \\ 6 & 19 & 1 \\ 42 & 18 & 31 \end{bmatrix}$

7. $x_1 = 1$ and $x_2 = 1$ 9. $x_1 = -13/2, x_2 = 3, x_3 = 15$, and $x_4 = 22$

11. $p = 1$ and $q = 2$. 13. $\left[\begin{array}{ccccc|c} 4 & -11 & 5 & 3 & 3 & 5 \\ 1 & -1 & 1 & -2 & 4 & 10 \\ 3 & 1 & 2 & -3 & 1 & 4 \\ 1 & 1 & 3 & 2 & 2 & 10 \\ 1 & 0 & 5 & 0 & 3 & 7 \end{array}\right]$.

15. The receipts were \$155,000 for the first performance, and \$163,000 for the second, for a weekend receipt total of \$318,000.

17. $x_1 = -1, x_2 = 3$, and $x_3 = 4$ 19. $\begin{bmatrix} \frac{315}{129} & \frac{225}{129} & \frac{145}{129} \\ \frac{75}{43} & \frac{115}{43} & \frac{55}{43} \\ \frac{70}{43} & \frac{50}{43} & \frac{80}{43} \end{bmatrix} \begin{bmatrix} 5 \\ 14 \\ 3 \end{bmatrix} = \begin{bmatrix} 40 \\ 50 \\ 30 \end{bmatrix}$.

For detailed solutions to these Exercises see the modestly priced companion, "Student Solutions Manual" by Morris and Stark.

EXERCISES 4.1

1. a) and c) are strict inequalities in two variables

3.

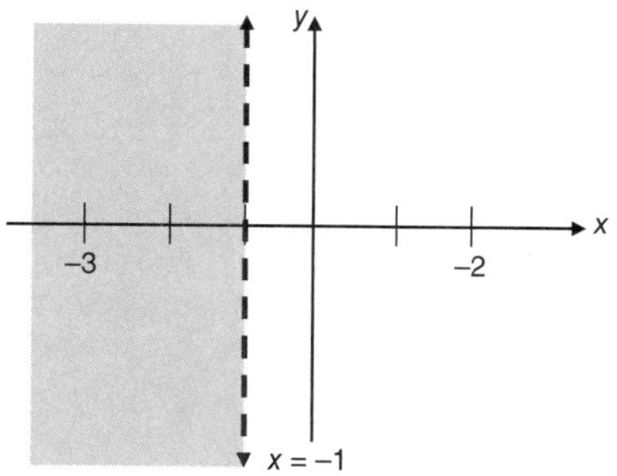

5.

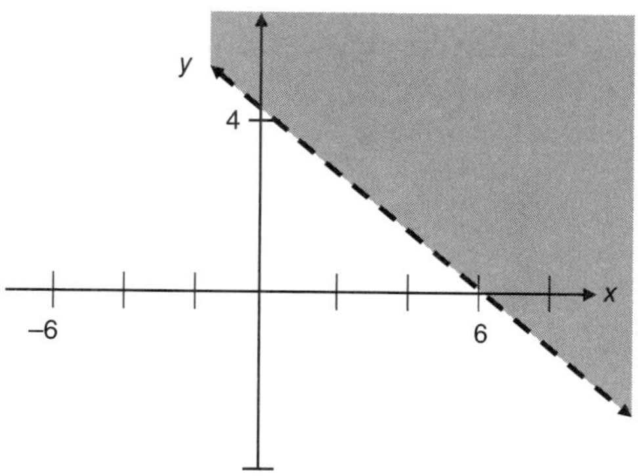

7.

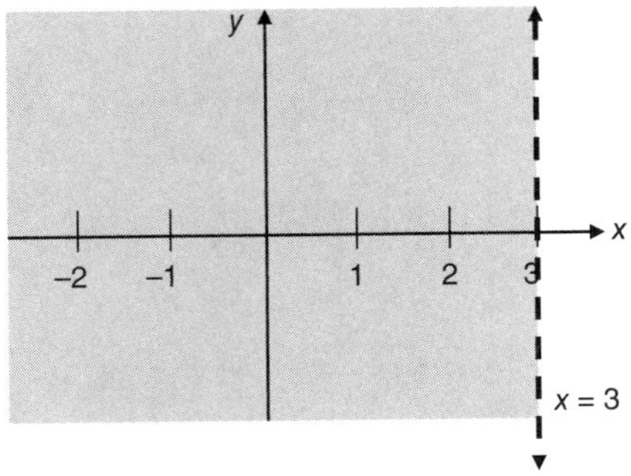

9.

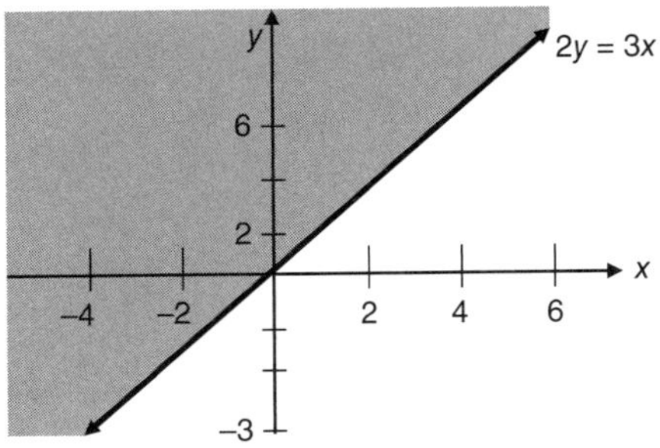

11.

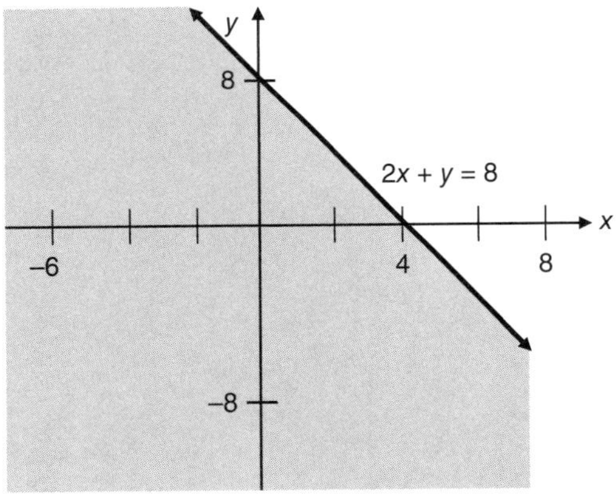

13.

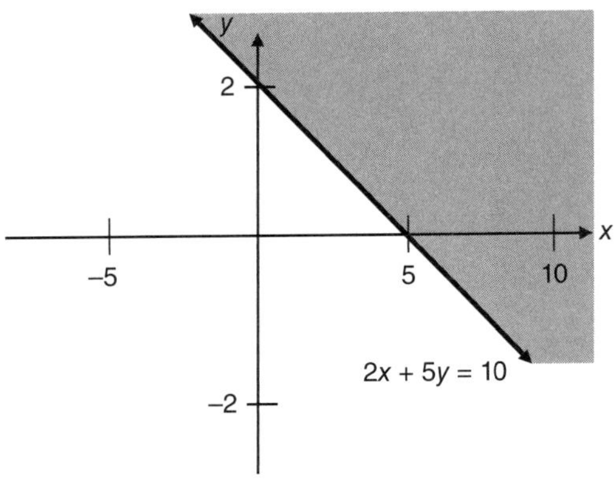

15.

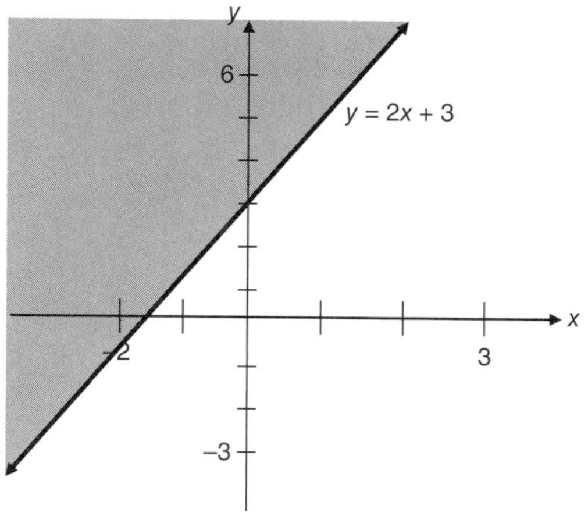

17.

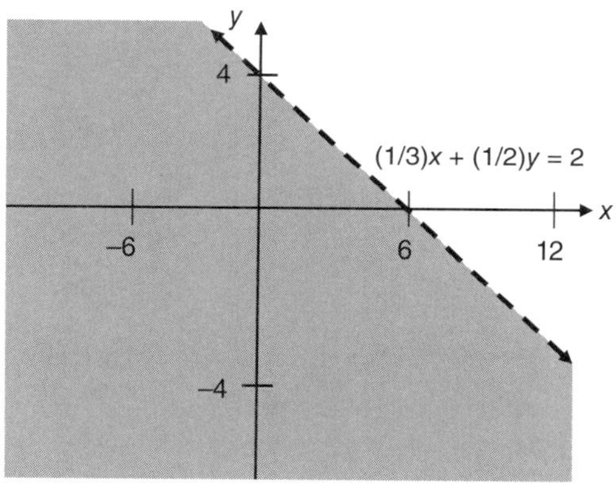

19. $y < (-4/3)x + 4$ is the inequality shown.

21.

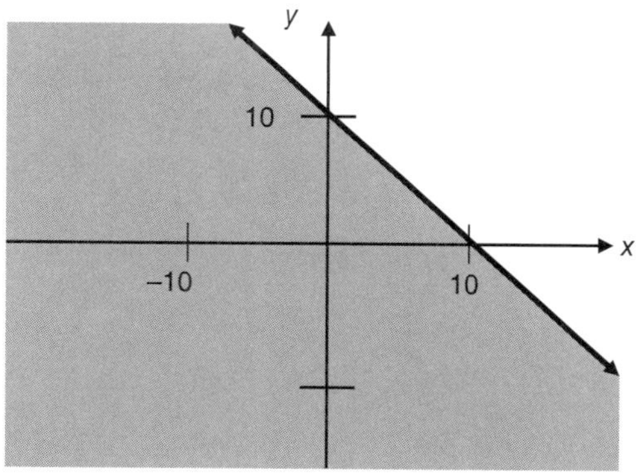

23.

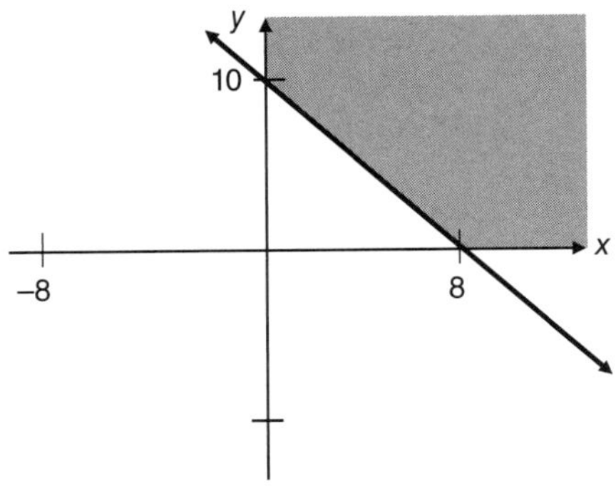

EXERCISES 4.2

1. The corner point here is where $x = 1$ and $y = 3$ or (1, 3).

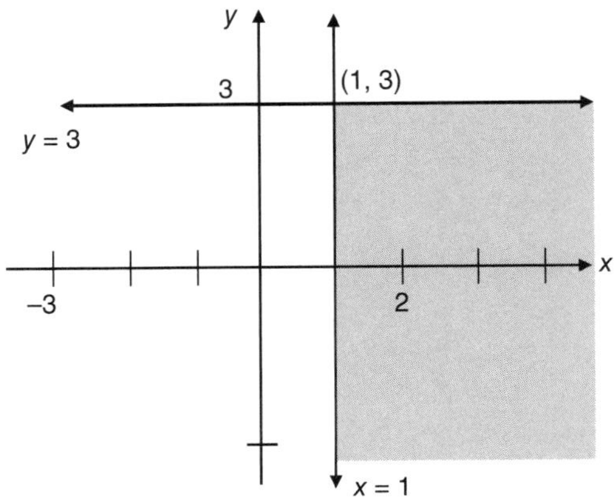

3. Here, the corner points are (−2, 4), (−2, 1), (5, 4), and (5, 1).

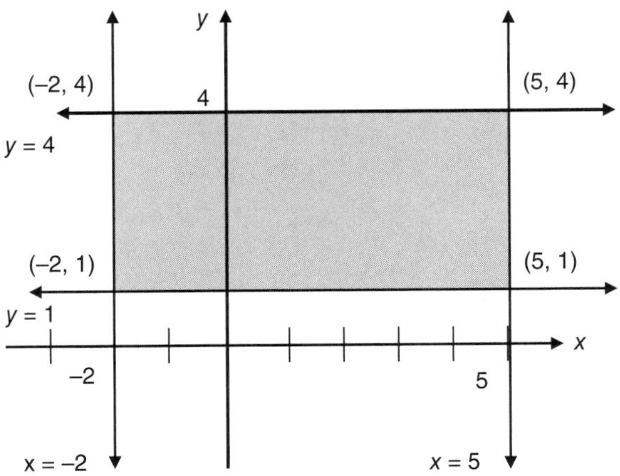

5. The corner point is the intersection at (5/3, 7/3).

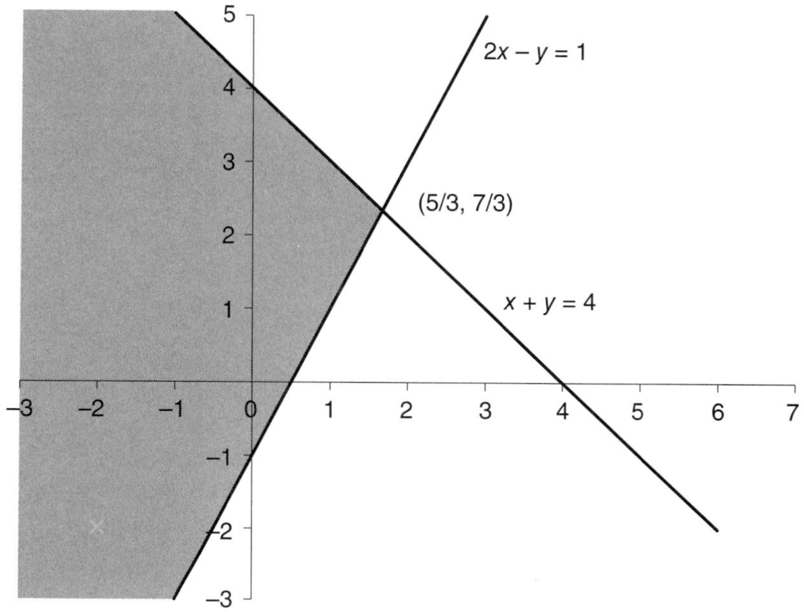

7. The corner point (vertex) in the interior of the quadrant is the intersection at (2, 3).

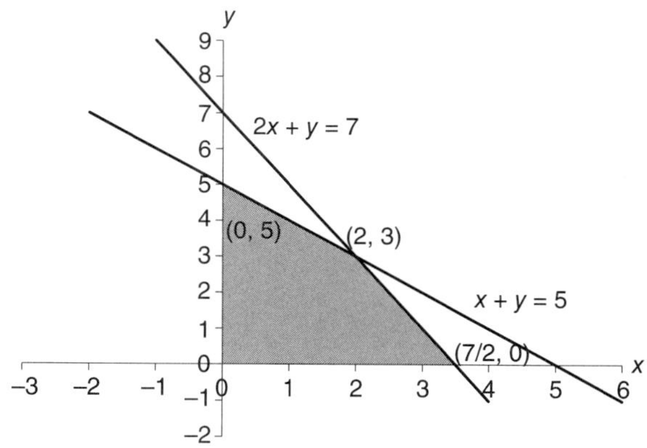

9. Here the origin is only one corner point of interest because the feasible region is unbounded.

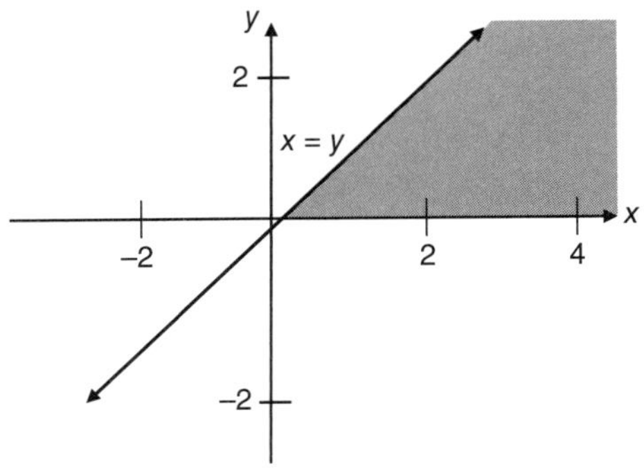

11. The feasible region is the quadrilateral with vertices (2, 0), (2, 3), (4, −3) and (6, −3).

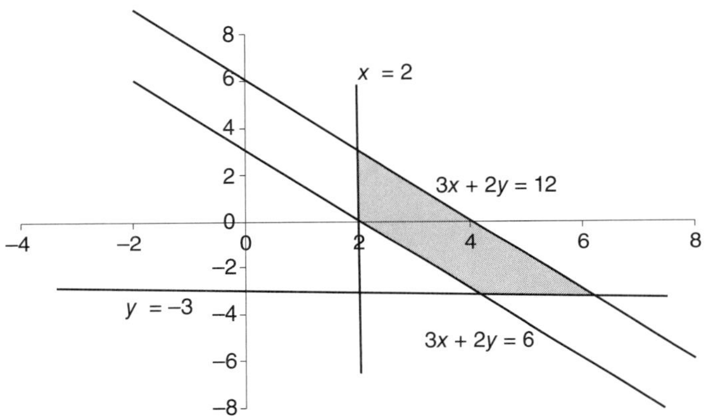

13. The corner (vertex) points of the feasible region are (0, 0), (0, 4), (3, 1) and (7/2, 0).

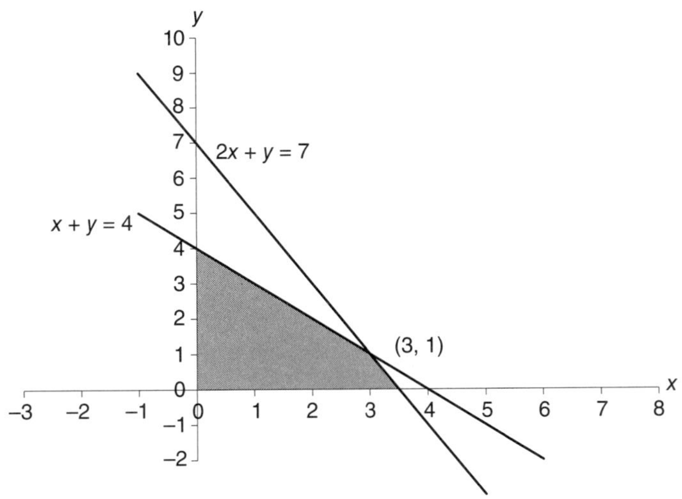

15. The feasible region is unbounded. The only corner point is (2, 3).

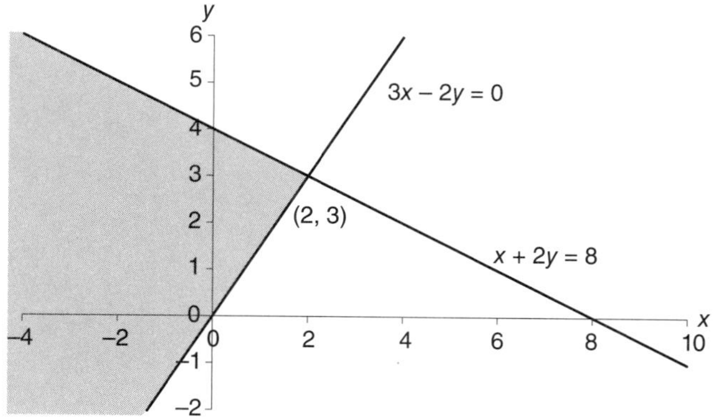

17. The feasible region is a quadrilateral with vertices (0, 6), (0, 10), (10, 0) and (6, 0).

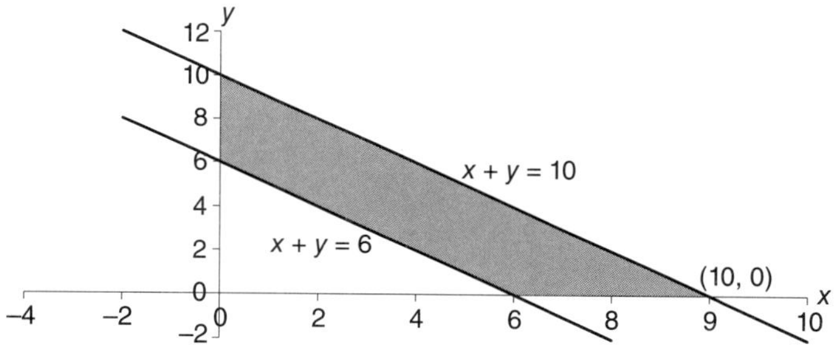

19. The feasible region is the quadrilateral with vertices (0, 0), (0, 20), (10, 15), and (20, 0).

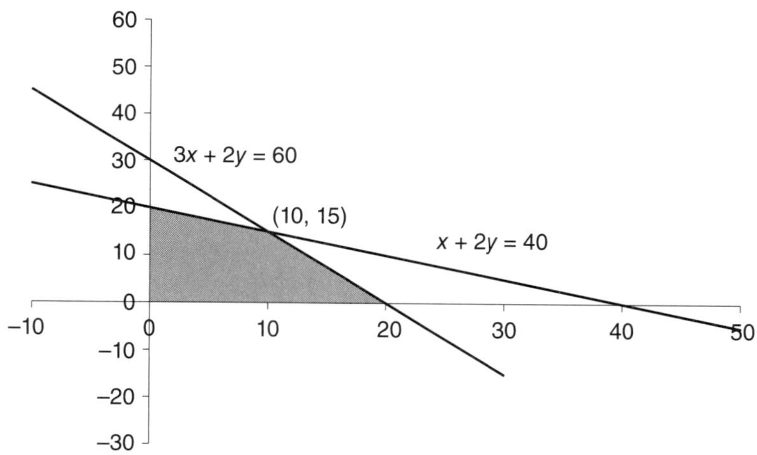

21. The feasible region is a quadrilateral with vertices (0, 0), (0, 5), (1, 0) and (7/3, 8/3).

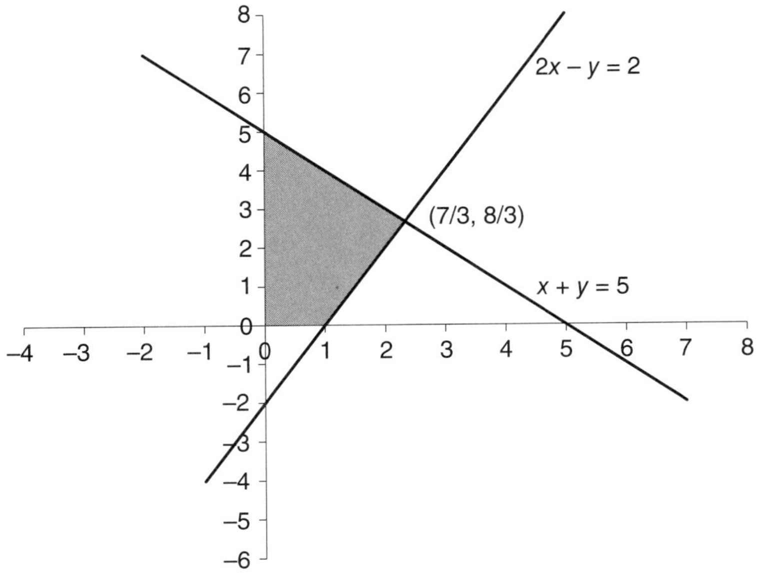

23. The feasible region has these five corner points: (0, 0), (0, 5), (6, 2), (4, 3) and (7, 0).

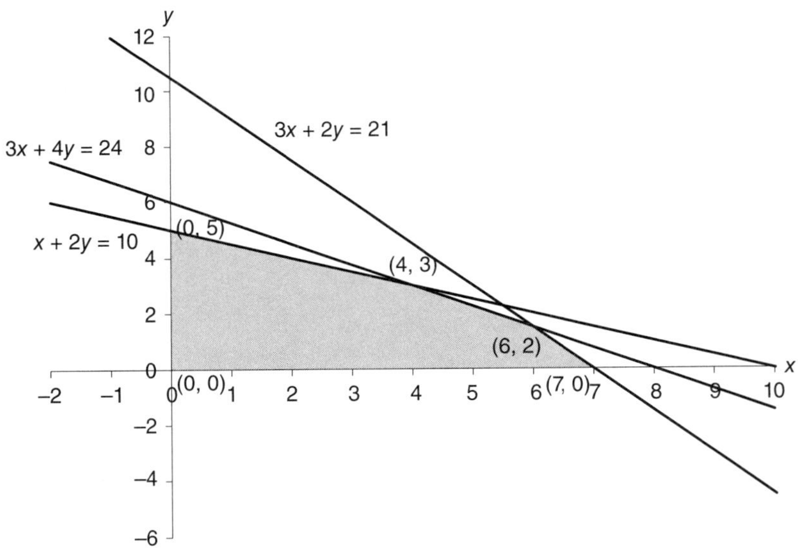

25. The feasible region has these five corner points: (0, 0), (0, 1), (3, 4), (6, 3) and (5, 0).

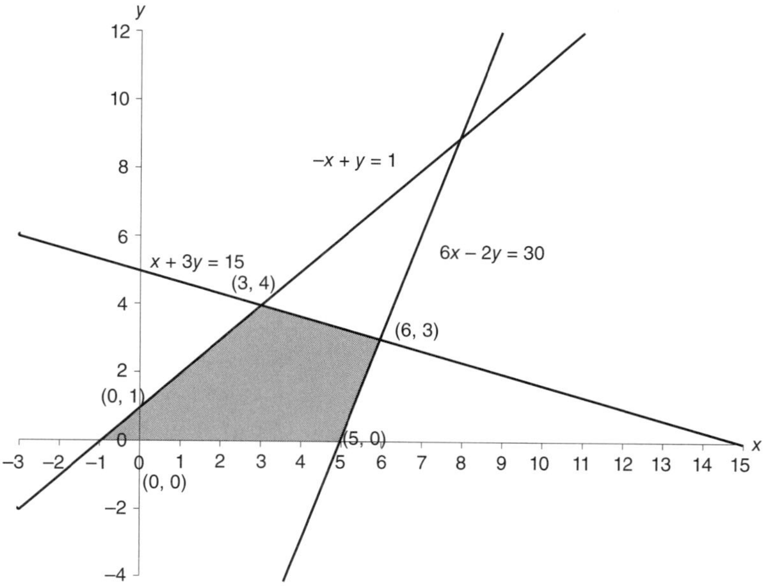

EXERCISES 4.3

1. The largest Z value is 25 when $x = 2$ and $y = 7$.
3. The largest Z value is 55 when $x = 11/2$ and $y = 0$.
5. The largest Z value is 55/2 when $x = 11/2$ and $y = 0$.
7. The minimum Z value is 3 when $x = 3$ and $y = 0$.
9. The minimum Z value is 6 when $x = 3$ and $y = 0$.
11. The feasible region is a quadrilateral with vertices (0, 0), (0, 7/2), (3, 2) and (4, 0). The maximum value is 12 when $x = 3$ and $y = 2$.

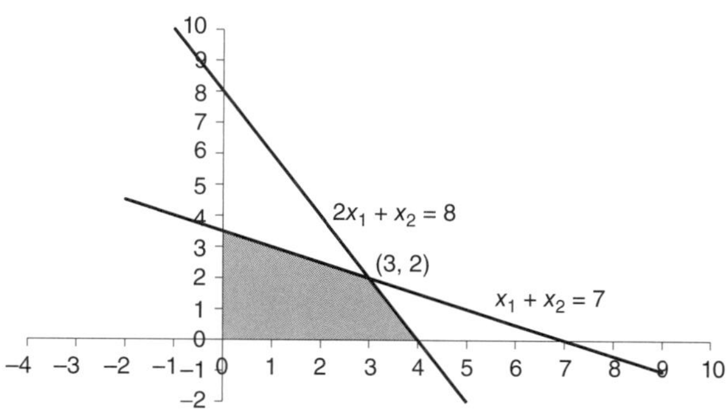

13. The feasible region is a quadrilateral with vertices (0, 0), (0, 7/2), (25/9, 19/9) and (16/5, 0). The maximum is 201/9 when $x = 25/9$ and $y = 19/9$.

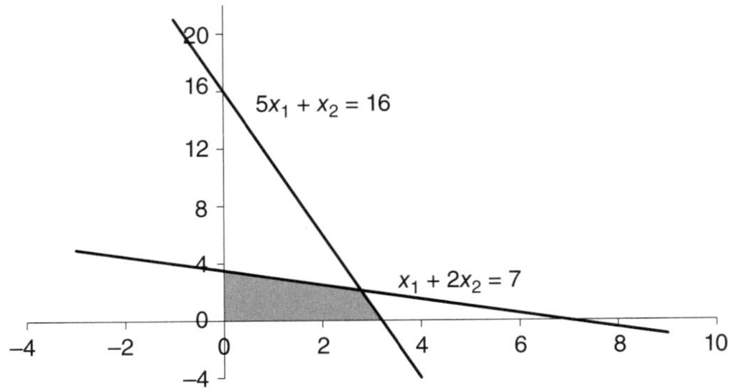

15. The feasible region has the five corner points (0, 0), (0, 2), (2, 4),(4, 3) and (5, 0). The maximum is 26 when $x = 4$ and $y = 3$

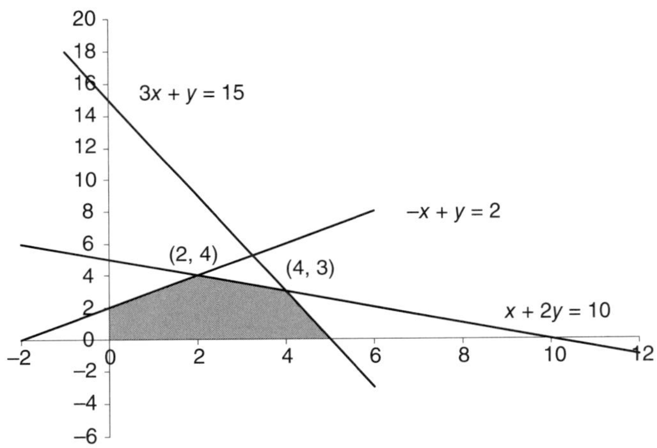

17. Let x = # items of x to make, y = # items of y to make, 15 min is converted to 1/4 h, and 10 min to 1/6 h.

$$\begin{aligned} \text{Maximize:} \quad & Z = 16x + 20y \\ \text{Subject to:} \quad & (1/4)x + (1/6)y \le 40 \\ & 3x + 4y \le 600 \\ & x,\ y \ge 0 \end{aligned}$$

19. Let x = # type A trucks and y = # type B trucks

$$\begin{aligned} \text{Minimize:} \quad & Z = 3x + 4y \\ \text{Subject to:} \quad & 100x + 150y \geq 4500 \\ & 200x + 150y \leq 6000 \\ & x,\ y \geq 0 \end{aligned}$$

SUPPLEMENTARY EXERCISES (CHAPTER 4)

1.

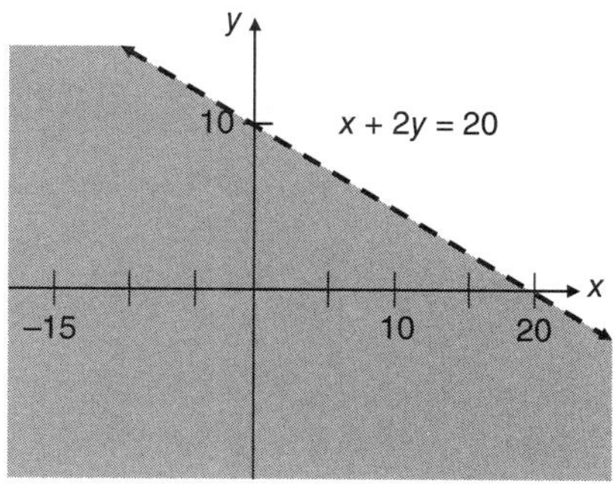

3.

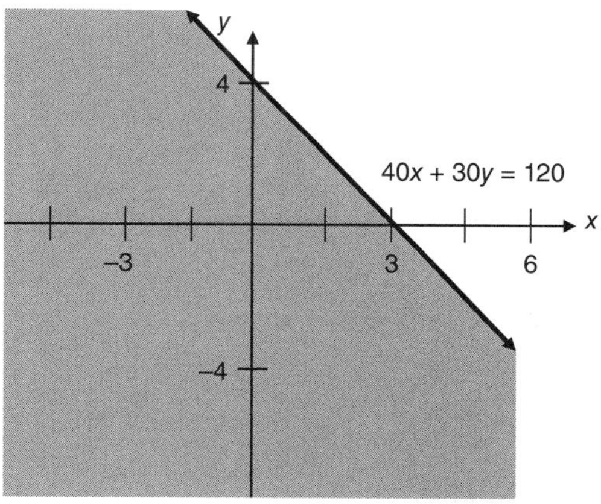

5.

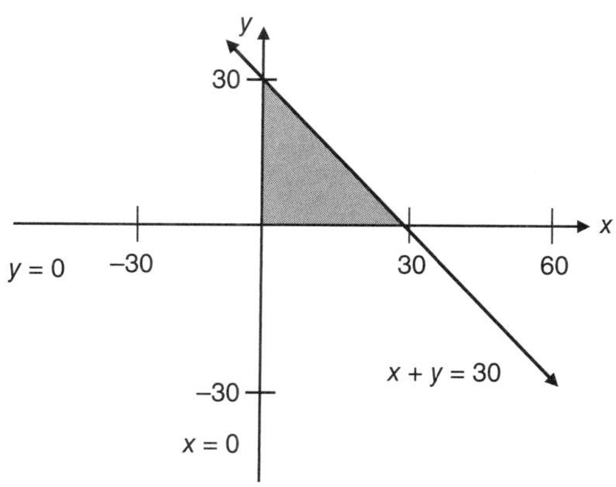

7. The feasible region is a triangle with vertices (0, 0), (0, 10), and (20, 0).

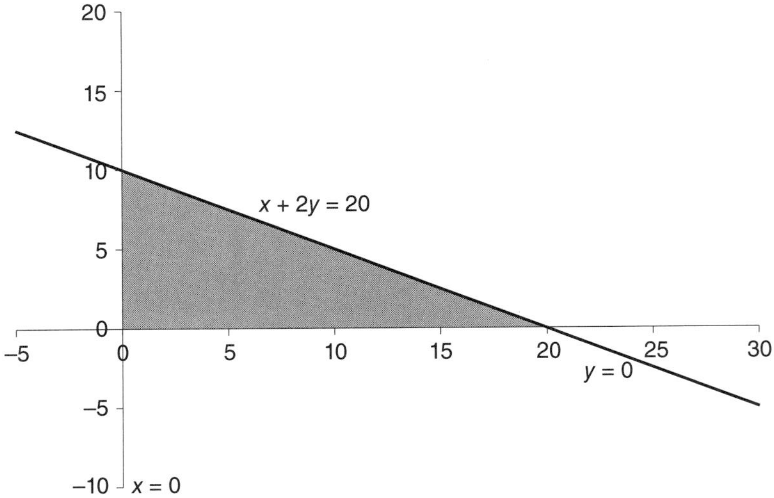

9. The feasible region is a quadrilateral with vertices (0, 0), (0, 60), (20, 50), and (15/2, 0).

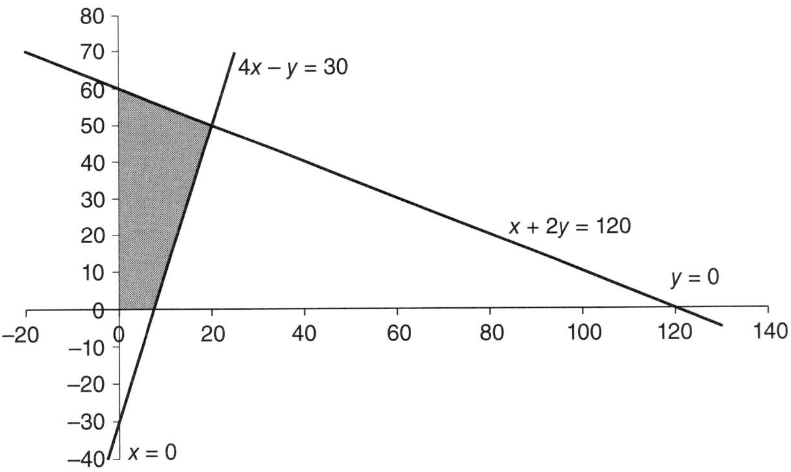

11. The feasible region is a quadrilateral with vertices (0, 0), (40, 20), (100/3, 0), and (0, 100).

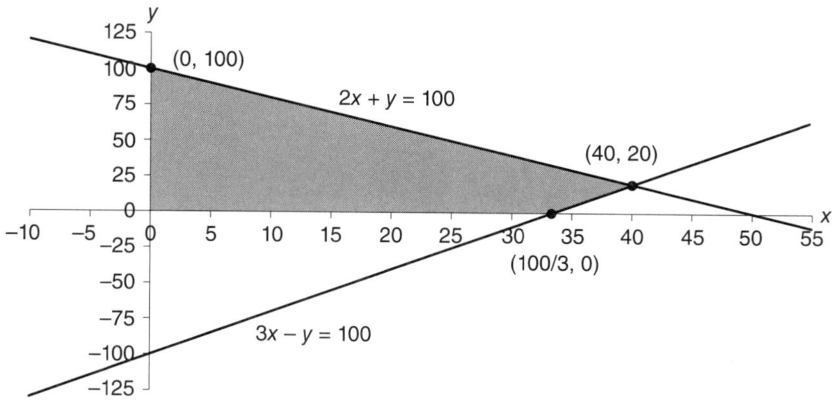

13. The constraints are as in Exercise 12. The objective function is to maximize $5x + 2y$ (call it Z). Then,

$Z(0, 0) = 5(0) + 2(0) = 0$
$Z(0, 5) = 5(0) + 2(5) = 10$
$Z(10, 0) = 5(10) + 2(0) = 50$ maximum

The maximum is 50 and occurs when $x = 10$ and $y = 0$.

15. The constraints are as in Exercise 14. The objective function maximizes $Z = 3x + 5y$. Then,

$Z(0, 0) = 3(0) + 5(0) = 0$
$Z(0, 15) = 3(0) + 5(15) = 75$ maximum
$Z(10, 0) = 3(10) + 5(0) = 30$

The maximum is 75 when $x = 0$ and $y = 15$.

17. The feasible region forms a quadrilateral with vertices (0, 0), (0, 6), (2, 0), and (4, 2). The maximum is 100 when $x = 4$ and $y = 2$.

19. The feasible region is in Supplementary Exercise 18. The maximum is 60 when $x = 20$ and $y = 0$.

21. The feasible region is that from Exercise 4.2.23. The corner points are (0, 0), (0, 5), (6, 2), (4, 3), and (7, 0). The maximum is 28 when $x = 6$ and $y = 2$.

23. The feasible region is that from Exercise 4.2.25. The corner points are (0, 0), (0, 1), (3, 4), (6, 3), and (5, 0).The maximum is 15 when $x = 5$ and $y = 0$.

For detailed solutions to these Exercises see the modestly priced companion, "Student Solutions Manual" by Morris and Stark.

EXERCISES 5.1

1. SMP 3. SMP 5. [1 1], [0 5], and [3 3] are feasible.
7. [1 1 5] and [0 6 0] are feasible.

9. [4 3 5] has all nonnegative values
$4 + 3 + 2(5) \leq 20$ is true
$2(4) + 3 + 2(5) \leq 25$ is true
$5(4) + 3 + 5 \leq 30$ is true
Therefore, [4 3 5] is feasible.

[3 0 7] has all nonnegative values
$3 + 0 + 2(7) \leq 20$ is true
$2(3) + 0 + 2(7) \leq 25$ is true
$5(3) + 0 + 7 \leq 30$ is true
Therefore, [3 0 7] is feasible.

11. Maximize: $Z = x_1 + x_2$
Subject to: $2x_1 + 3x_2 + s_1 = 9$
$x_1 + x_2 + s_2 = 3$
$x_1, x_2, s_1, s_2 \geq 0$

13. Maximize: $P = 3x + 4y + 2z$
Subject to: $5x + y + z + s_1 = 1500$
$4x + 2y + z + s_2 = 2000$
$2x + 2y + z + s_3 = 2500$
$x, y, z, s_1, s_2, s_3 \geq 0$

15. Maximize: $Z = 16x_1 + 20x_2$
Subject to: $(1/4)x_1 + (1/6)x_2 + s_1 = 40$
$3x_1 + 4x_2 + s_2 = 600$
$x_1, x_2, s_1, s_2 \geq 0$

EXERCISES 5.2

1. The tableau is:

Row	x_1	x_2	s_1	s_2	RHS
1	1	3	1	0	10
2	2	1	0	1	8
Objective	−5	−2	0	0	0

3. The tableau is:

Row	x	y	z	s_1	s_2	s_3	RHS
1	1	2	1	1	0	0	12
2	2	1	3	0	1	0	18
3	1	3	1	0	0	1	20
Objective	−5	−1	−1	0	0	0	0

5. Maximize: $Z = 4x + 3y$
Subject to: $30x + 20y \leq 500$
$5x + 10y \leq 150$
$x, y \geq 0$

7. Maximize: $Z = 2x_1 + 3x_2$
Subject to: $3x_1 + 5x_2 + 2x_3 \leq 19$
$5x_1 + 2x_2 + 4x_3 \leq 21$
$x_1 + 7x_2 + 3x_3 \leq 22$
$x_1, x_2, x_3 \geq 0$

9. a) The pivot element is the "2" from the second row.
b) The "1" in the third row is the pivot element.
c) The "2" in the second row is the pivot element.
d) The "2" in the first row is the pivot element.

11. The tableau with pivot element in bold is

Row	x_1	x_2	s_1	s_2	RHS
1	1	3	1	0	10
2	**2**	1	0	1	8
Objective	−5	−2	0	0	0

After the pivot we have

Row	x_1	x_2	s_1	s_2	RHS
1	0	5/2	1	$-1/2$	6
2	1	1/2	0	1/2	4
Objective	0	1/2	0	5/2	20

The pivot completed, the current solution is optimal (no negative numbers in the objective row). The current (optimal) solution is $x_1{}^* = 4$, $x_2{}^* = 0$, $s_1 = 6$, $s_2 = 0$, and the objective value is 20.

13. The tableau with pivot element in bold is

Row	x	y	z	s_1	s_2	s_3	RHS
1	1	2	1	1	0	0	12
2	**2**	1	3	0	1	0	18
3	1	3	1	0	0	1	20
Objective	−5	−1	−1	0	0	0	0

⟶

After pivoting

Row	x	y	z	s_1	s_2	s_3	RHS
1	0	3/2	−1/2	1	−1/2	0	3
2	**1**	1/2	3/2	0	1/2	0	9
3	0	5/2	−1/2	0	−1/2	1	11
Objective	0	1/2	13/2	0	5/2	0	45

The current optimal solution is $x^* = 9$, $y^* = 0$, $z^* = 0$, $s_1 = 3$, $s_2 = 0$, $s_3 = 11$. The objective value is 45.

15. The tableau with pivot element in bold is

Row	x	y	s_1	s_2	RHS
1	**30**	20	1	0	500
2	5	10	0	1	150
Objective	−4	−3	0	0	0

After pivoting

Row	x	y	s_1	s_2	RHS
1	1	2/3	1/30	0	50/3
2	0	20/3	−1/6	1	200/3
Objective	0	−1/3	2/15	0	200/3

The current (non-optimal) solution is $x = 50/3$, $y = 0$, $s_1 = 0$, $s_2 = 200/3$. The objective value is 200/3.

17. The tableau with pivot element in bold is

Row	x_1	x_2	x_3	s_1	s_2	s_3	RHS
1	3	5	2	1	0	0	19
2	5	2	4	0	1	0	21
3	1	**7**	3	0	0	1	22
Objective	−2	−3	0	0	0	0	0

After pivoting

Row	x_1	x_2	x_3	s_1	s_2	s_3	RHS
1	16/7	0	−1/7	1	0	−5/7	23/7
2	33/7	0	22/7	0	1	−2/7	103/7
3	1/7	1	3/7	0	0	1/7	22/7
Objective	−11/7	0	9/7	0	0	3/7	66/7

⟶

The current (non-optimal) solution is $x_1 = 0,\ x_2 = 22/7,\ x_3 = 0,\ s_1 = 23/7,$ $s_2 = 103/7, s_3 = 0.$
The objective value is 66/7.

19. The current solution is $x_1 = 0, x_2 = 0, x_3 = 20, x_4 = 75, s_1 = 15, s_2 = 10, s_3 = 0,$ $s_4 = 0$. The objective value is 230. It is not optimal as the negative in the objective row indicates further pivoting is necessary.

EXERCISES 5.3

1. The final tableau for the Simplex Method is:

Row	x_1	x_2	s_1	s_2	RHS
1	1	0	2/3	$-1/3$	3
2	0	1	−1/3	2/3	2
Objective	0	0	1/3	4/3	12

The optimal solution is $x_1{}^* = 3, x_2{}^* = 2, s_1 = 0, s_2 = 0$. The objective value is 12.

3. The final tableau for the Simplex Method is:

Row	x_1	x_2	s_1	s_2	RHS
1	0	1	2	−1	16
2	1	0	−1	1	8
Objective	0	0	1	1	56

The optimal solution is $x_1{}^* = 8, x_2{}^* = 16, s_1 = 0, s_2 = 0$. The objective value is 56.

5. The final tableau for the Simplex Method is:

Row	x_1	x_2	s_1	s_2	RHS
1	1	1	1	0	10
2	1	0	−1	1	4
Objective	2	0	3	0	30

The optimal solution is $x_1{}^* = 0, x_2{}^* = 10, s_1 = 0, s_2 = 4$. The objective value is 30.

7. The final tableau for the Simplex Method is:

Row	x_1	x_2	x_3	s_1	s_2	s_3	RHS
1	0	−5	1	1	−2	0	10
2	1	3	0	0	1	0	20
3	0	6	0	−1	2	1	5
Objective	0	9	0	1	3	0	110

⟶

The optimal solution is $x_1{}^* = 20,\ x_2{}^* = 0,\ x_3{}^* = 10,\ s_1 = 0,\ s_2 = 0,\ s_3 = 5.$ The objective value is 110.

9. The final tableau for the Simplex Method is:

Row	x_1	x_2	x_3	s_1	s_2	s_3	RHS
1	3	1	0	3/2	0	−1/2	2
2	−5	0	0	−3	1	1	6
3	−1	0	1	−1/2	0	1/2	2
Objective	3	0	0	5/2	0	1/2	14

The optimal solution is $x_1{}^* = 0,\ x_2{}^* = 2,\ x_3{}^* = 2,\ s_1 = 0,\ s_2 = 6,\ s_3 = 0.$ The objective value is 14.

11. Note that there is a tie in the min ratio so either the "4"or the "1" in the x_2 column could be used as the pivot element. Here, the "1" is used and the final tableau is

Row	x_1	x_2	s_1	s_2	s_3	RHS
1	0	0	1	−50/3	−25/3	5000
2	1	0	0	1/3	−4/3	0
3	0	1	0	0	1	200
Objective	0	0	0	25/3	5/3	7000

The optimal solution is $x_1{}^* = 0,\ x_2{}^* = 200,\ s_1 = 5000,\ s_4 = 0,\ s_3 = 0.$ The objective value is 7000.

13. The SMP in matrix format is:

$$\begin{aligned} \text{Maximize:}\quad & Z = \mathbf{C}^{\mathbf{T}}\mathbf{X} \\ \text{Subject to:}\quad & \mathbf{AX} \le \mathbf{B} \\ & \mathbf{X} \ge \mathbf{0} \end{aligned}$$

$$\text{where } \mathbf{A} = \begin{bmatrix} 1 & 3 & 0 \\ 2 & 1 & 1 \\ 1 & 2 & 4 \end{bmatrix}; \mathbf{B} = \begin{bmatrix} 30 \\ 60 \\ 125 \end{bmatrix}; \mathbf{C} = \begin{bmatrix} 5 \\ 3 \\ 2 \end{bmatrix}; \mathbf{X} = \begin{bmatrix} x_1 \\ x_2 \\ x_3 \end{bmatrix}.$$

EXERCISES 5.4

1. The dual is

$$\begin{aligned} \text{Minimize:}\quad & V = 12y_1 + 18y_2 + 20y_3 \\ \text{Subject to:}\quad & y_1 + 2y_2 + y_3 \ge 5 \\ & 2y_1 + y_2 + 3y_3 \ge 1 \\ & 3y_1 + 3y_2 + y_3 \ge 1 \\ & y_1,\ y_2,\ y_3 \ge 0 \end{aligned}$$

3. The dual is

$$\begin{aligned} \text{Maximize:} \quad & V = 10y_1 + 15y_2 \\ \text{Subject to:} \quad & y_1 + 3y_2 \le 11 \\ & 2y_1 + y_2 \le 7 \\ & y_1, y_2 \ge 0 \end{aligned}$$

The final tableau is

Row	y_1	y_2	s_1	s_2	RHS
1	0	1	2/5	−1/5	3
2	1	0	−1/5	3/5	2
Objective	0	0	4	3	65

The solution to the primal is found in this last tableau objective row. The optimum is $x_1{}^* = 4$, $x_2{}^* = 3$, $s_1 = 0$, $s_2 = 0$ and the objective is 65.

5. The dual is

$$\begin{aligned} \text{Maximize:} \, & V = 46y_1 + 40y_2 \\ \text{Subject to:} \quad & 3y_1 + 5y_2 \le 3 \\ & 4y_1 + 3y_2 \le 2 \\ & y_1, y_2 \ge 0 \end{aligned}$$

The final tableau is

Row	y_1	y_2	s_1	s_2	RHS
1	0	1	4/11	−3/11	6/11
2	1	0	−3/11	5/11	1/11
Objective	0	0	2	10	26

The solution to the primal is found in this last tableau objective row. The optimum is $x_1{}^* = 2$, $x_2{}^* = 10$, $s_1 = 0$, $s_2 = 0$ and the objective is 26.

7. The dual is

$$\begin{aligned} \text{Maximize:} \, & V = 12y_1 + 9y_2 + 15y_3 \\ \text{Subject to:} \quad & 4y_1 + y_2 + y_3 \le 2 \\ & y_1 + y_2 + y_3 \le 5 \\ & y_1, y_2, y_3 \ge 0 \end{aligned}$$

⟶

The final tableau is

Row	y_1	y_2	y_3	s_1	s_2	RHS
1	4	1	1	1	0	2
2	−3	0	0	−1	1	3
Objective	48	6	0	15	0	30

The solution to the primal is found in this last tableau objective row. The optimum is $x_1{}^* = 15$, $x_2{}^* = 0$, $s_1 = 48$, $s_2 = 6$, $s_3 = 0$ and the objective is 30.

EXERCISES 5.5

1. First rewrite the LP as:

$$\begin{aligned} \text{Maximize:}\quad -Z = -4x_1 - 3x_2 \\ -2x_1 - x_2 \le -25 \\ -x_1 - x_2 \le -20 \\ x_1 - 5x_2 \le 4 \\ x_1, x_2 \ge 0 \end{aligned}$$

The initial tableau is:

x_1	x_2	s_1	s_2	s_3	RHS
−2	−1	1	0	0	−25
−1	−1	0	1	0	−20
1	−5	0	0	1	4
4	3	0	0	0	0

The final tableau after pivoting is:

x_1	x_2	s_1	s_2	s_3	RHS
0	0	6	−11	1	74
1	0	−1	1	0	5
0	1	1	−2	0	15
0	0	1	2	0	−65

The optimal solution is $x_1{}^* = 5$, $x_2{}^* = 15$, $s_1 = 0$, $s_2 = 0$, $s_3 = 74$, and the objective is 65.

3. First rewrite the LP as:

$$\begin{aligned} \text{Maximize:}\quad & Z = 5x_1 + 6x_2 \\ & 4x_1 + 7x_2 \le 28 \\ & -x_1 + 3x_2 \le -2 \\ & x_1,\ x_2 \ge 0 \end{aligned}$$

The initial tableau is

x_1	x_2	s_1	s_2	RHS
4	7	1	0	28
−1	3	0	1	−2
−5	−6	0	0	0

The final tableau is:

x_1	x_2	s_1	s_2	RHS
0	19/4	1/4	1	5
1	7/4	1/4	0	7
0	11/4	5/4	0	35

The optimal solution is $x_1{}^* = 7$, $x_2{}^* = 0$, $s_1 = 0$, $s_2 = 5$, and the maximum of the objective is 35.

5. First rewrite the LP as:

$$\begin{aligned} \text{Maximize:}\quad & -Z = -x_1 - 2x_2^+ + 2x_2^- + x_3^+ - x_3^- - x_4 \\ \text{Subject to:}\quad & x_1 - x_2^+ + x_2^- + x_3^+ - x_3^- \le 4 \\ & x_1 + 2x_2^+ - 2x_2^- + 2x_4 \le 6 \\ & x_2^+ - x_2^- - 2x_3^+ + 2x_3^- + x_4 \le 2 \\ & -x_1 + 2x_2^+ - 2x_2^- + x_3^+ - x_3^- \le 2 \\ & x_1,\ x_2^-,\ x_2^+,\ x_3^-,\ x_3^+,\ x_4 \ge 0 \end{aligned}$$

The initial tableau is

x_1	x_2^+	x_2^-	x_3^+	x_3^-	x_4	s_1	s_2	s_3	s_4	RHS
1	−1	**1**	1	−1	0	1	0	0	0	4
1	1	−1	0	0	2	0	1	0	0	6
0	1	−1	−2	2	1	0	0	1	0	2
−1	2	−2	1	−1	0	0	0	0	1	2
1	2	−2	−1	1	1	0	0	0	0	0

⟶

and the final tableau is

x_1	x_2^+	x_2^-	x_3^+	x_3^-	x_4	s_1	s_2	s_3	s_4	RHS
0	−1	1	0	0	1	2	0	1	0	10
3	0	0	0	0	3	2	1	1	0	16
1	0	0	−1	**1**	1	1	0	1	0	6
4	0	0	0	0	3	5	0	3	1	28
4	0	0	0	0	2	3	0	1	0	14

The optimal solution is $x_1{}^* = 0$, $x_2^+{}^* = 0$, $x_2^-{}^* = 10$, $(x_2 = -10)$, $x_3^+{}^* = 0$, $x_3^-{}^* = 6$ $(x_3{}^* = -6)$, $x_4{}^* = 0$, $s_1 = 0$, $s_2 = 16$, $s_3 = 0$, $s_4 = 28$ and the objective minimum value is -14.

SUPPLEMENTARY EXERCISES (CHAPTER 5)

1. x_4 is unrestricted so it is not an SMP.

3. Maximize: $Z = 4x_1 + 3x_2$
   Subject to:
$$
\begin{aligned}
2x_1 + 3x_2 + 5x_3 + s_1 \qquad\qquad &= 23\\
3x_1 + x_2 + x_3 \qquad + s_2 \qquad &= 8\\
x_1 + 5x_2 + 3x_3 \qquad\qquad + s_3 &= 8\\
x_1,\ x_2,\ x_3 \geq 0 &
\end{aligned}
$$

5. The initial tableau is:

x_1	x_2	x_3	s_1	s_2	s_3	RHS
2	5	1	1	0	0	25
3	1	3	0	1	0	21
1	1	0	0	0	1	7
−1	−2	−7	0	0	0	0

7. Maximize: $Z = 4x_1 + x_2 + 3x_3 + 2x_4$
   Subject to:
$$
\begin{aligned}
2x_1 + x_2 + 3x_3 + 5x_4 &\leq 25\\
x_1 + 2x_3 + 5x_4 &\leq 34\\
3x_2 + x_3 + 4x_4 &\leq 59\\
3x_1 + x_2 + 2x_3 + x_4 &\leq 17\\
x_1,\ x_2,\ x_3,\ x_4 &\geq 0
\end{aligned}
$$

9. Using the minimum *a/b* xratio the 5 in the first constraint is the pivot element (bold italics).

x_1	x_2	x_3	s_1	s_2	s_3	RHS
5	25	10	1	0	0	100
3	15	1	0	1	0	75
4	6	15	0	0	1	240
−4	−3	0	0	0	0	0

11. The pivot column is x_2 and the pivot element the five in that column.

x_1	x_2	s_1	s_2	RHS
1	5	1	0	200
2	3	0	1	134
−3	−7	0	0	0

The pivot yields the tableau

x_1	x_2	s_1	s_2	RHS
1/5	1	1/5	0	40
7/5	0	−3/5	1	14
−8/5	0	7/5	0	280

13. The pivot column is x_1 and the pivot element is 5.

x_1	x_2	s_1	s_2	RHS
5	1	1	0	16
1	2	0	1	7
−5	−4	0	0	0

The final tableau is:

x_1	x_2	s_1	s_2	RHS
1	0	2/9	−1/9	25/9
0	1	−1/9	5/9	19/9
0	0	2/3	5/3	67/3

The optimal solution is $x_1{}^* = 25/9$, $x_2{}^* = 19/9$ and the maximum is 67/3.

15. The pivot column is x_2 and the element the 4 in the third row of the constraints in the tableau.

x_1	x_2	x_3	s_1	s_2	s_3	RHS
4	2	2	1	0	0	40
8	10	5	0	1	0	90
4	*4*	3	0	0	1	30
−3	−5	−4	0	0	0	0

After pivoting the final tableau is

x_1	x_2	x_3	s_1	s_2	s_3	RHS
4/3	$-2/3$	0	1	0	$-2/3$	20
4/3	10/3	0	0	1	$-5/3$	40
4/3	4/3	1	0	0	1/3	10
7/3	1/3	0	0	0	4/3	40

The optimal solution is $x_1^* = 0$, $x_2^* = 0$, and $x_3^* = 10$. The objective optimum is 40.

17. The dual is

$$\begin{aligned} \text{Maximize:} \quad & V = 48y_1 + 70y_2 + 96y_3 \\ \text{Subject to:} \quad & 3y_1 + 3y_2 + 4y_3 \leq 18 \\ & 2y_1 + 2y_2 + 3y_3 \leq 15 \\ & y_1 + 2y_2 + 2y_3 \leq 10 \\ & y_1, y_2, y_3 \geq 0 \end{aligned}$$

The initial tableau is:

Row	y_1	y_2	y_3	s_1	s_2	s_3	RHS
1	3	3	*4*	1	0	0	18
2	2	2	3	0	1	0	15
3	1	2	2	0	0	1	10
Objective	−48	−70	−96	0	0	0	0

The final tableau is

Row	y_1	y_2	y_3	s_1	s_2	s_3	RHS
1	3/4	3/4	1	1/4	0	0	9/2
2	$-1/4$	$-1/4$	0	$-3/4$	1	0	3/2
3	$-1/2$	1/2	0	$-1/2$	0	1	1
Objective	24	2	0	24	0	0	432

→

The optimal value is 432 when $y_1{}^* = 0$, $y_2{}^* = 0$, and $y_3{}^* = 9/2$.
The solution for the primal is found in the objective row under the s_i. The solution to the original (primal) is that the minimum value is 432 when $x_1{}^* = 24$, $x_2{}^* = 0$ and $x_3{}^* = 0$.

For detailed solutions to these Exercises see the modestly priced companion, "Student Solutions Manual" by Morris and Stark.

EXERCISES CHAPTER 6

1. a) Let x_j be the amount to ship from warehouse i to store j, $i = 1, 2, 3, 4,$ and $j = 1, \ldots, 6$
   The objective value is 99 when $x_{12}{}^* = 4$, $x_{14}{}^* = 1$, $x_{15}{}^* = 4$, $x_{26}{}^* = 2$, $x_{31}{}^* = 6$, $x_{43}{}^* = 4$, $x_{44}{}^* = 1$ and all other $x_{ij}{}^*$ are zero.

   b) replace $x_{11} + x_{21} + x_{31} + x_{41} \geq 6$ with $x_{11} + x_{21} + x_{31} + x_{41} \geq 4$
   The objective value is 88 when $x_{12}{}^* = 3$, $x_{15}{}^* = 4$, $x_{26}{}^* = 2$, $x_{31}{}^* = 4$, $x_{32}{}^* = 1$, $x_{34}{}^* = 1$, $x_{43}{}^* = 4$, $x_{44}{}^* = 1$, and all other $x_{ij}{}^*$ are zero.

3. Let X_{ij} be the amount of whiskey i in blend j
   $i = A, B, C$ corresponds to A, B, C
   $j = 1, 2, 3$ corresponds to BH, OC, MC(Blue Hen, Old College, Modeler's Choice)
   The maximum revenue is \$9500 when $X_{ABH}{}^* = 2100$, $X_{BBH}{}^* = 800$, and $X_{CBH}{}^* = 1242.857$ and all other $X_{ij}{}^* = 0$. Only produce Blue Hen.

5. The caterer has three options:
   1) Purchase x_j new napkins for use on j^{th} day at A \$/napkin.
   2) Use normal (full day) laundering of x_j soiled napkins at the end of the jth day at B \$/napkin and ready for use on day $j + 2$.
   3) Use rapid overnight) service of y_j napkins at the end of the jth day at C \$/napkin and ready for use on day $j + 1$.
   The minimum cost is \$3560 when

$$w_1{}^* = 325, w_3{}^* = 45, w_7{}^* = 130, w_2{}^* = w_4{}^* = w_5{}^* = w_6{}^* = 0$$
$$x_1{}^* = 250, x_2{}^* = 75, x_3{}^* = 95, x_4{}^* = 120, x_5{}^* = 70, x_6{}^* = x_7{}^* = 0$$
$$y_1{}^* = y_2{}^* = y_3{}^* = y_4{}^* = y_5{}^* = y_7{}^* = 0, y_6{}^* = 300$$
$$l_1{}^* = 75, l_3{}^* = 200, l_4{}^* = 155, l_5{}^* = 180, l_2{}^* = l_6{}^* = l_7{}^* = 0$$

For detailed solutions to these Exercises see the modestly priced companion, "Student Solutions Manual" by Morris and Stark.

EXERCISES 7.1

1. a) true b) true c) true d) false e) false
3. a) 3 b) 6 c) 5 d) 1 e) 1

5. a) $\mathbf{A} \cap \mathbf{B} = \{3, 5, 7\}$ b) $\mathbf{A} \cup \mathbf{B} = \{1, 2, 3, 4, 5, 6, 7\}$
c) $\mathbf{A}^c = \{2, 4, 6, 8, 9, 10\}$ d) $\mathbf{A}^c \cup \mathbf{B}^c = \{1, 2, 4, 6, 8, 9, 10\}$
e) $\mathbf{A} \cap \mathbf{B}^c = \{1\}$
7. a) $n(\mathbf{A}) = 7$ b) $n(\mathbf{C}) = 4$ c) $n(\mathbf{B} \cup \mathbf{C}) = 5$
9. Answers will vary depending on sets chosen. One example is
Let $\mathbf{A} = \{1, 2, 3, 4\}$ $\mathbf{B} = \{3, 4, 5, 6, 7\}$, and $\mathbf{C} = \{1, 3, 5, 7, 9\}$ then
$$(\mathbf{A} \cup \mathbf{C}) \cap (\mathbf{B} \cup \mathbf{C}) = \{1, 2, 3, 4, 5, 7, 9\} \cap \{1, 3, 4, 5, 6, 7, 9\}$$
$$= \{1, 3, 4, 5, 7, 9\}$$
and $(\mathbf{A} \cap \mathbf{B}) \cup \mathbf{C} = \{3, 4\} \cup \{1, 3, 5, 7, 9\} = \{1, 3, 4, 5, 7, 9\}$
11. $A = \{1, 2, \ldots, 49\}$ $\mathbf{B} = \{1, 4, 9, 16, 25, \ldots\}$ $\mathbf{C} = \{2, 3, 5, 7, 11, 13, \ldots\}$
a) $\mathbf{A} \cap \mathbf{B} = \{1, 4, 9, 16, 25, 36, 49\}$
b) $\mathbf{A} \cap \mathbf{C} = \{2, 3, 5, 7, 11, 13, 17, 23, 29, 31, 37, 41, 43, 47\}$
13. $n(x_1) = 20$, $n(x_2) = 15$, and $n(x_3) = 45$
15. $n(\mathbf{A}) = 3$, $n(\mathbf{B}) = 5$ and $n(\mathbf{C}) = 2$
17. $n(\mathbf{U}) = 150$, $n(\mathbf{A}) = 50$, $n(\mathbf{B}) = 90$, and $n(\mathbf{A} \cap \mathbf{B}) = 30$.
a) $n(\mathbf{A}\ \text{only}) = n(\mathbf{A}) - n(\mathbf{A} \cap \mathbf{B}) = 50 - 30 = 20$
b) $n(\mathbf{A} \cup \mathbf{B})^c = 150 - n(\mathbf{A} \cup \mathbf{B}) = 150 - [n(\mathbf{A}) + n(\mathbf{B}) - n(\mathbf{A} \cap \mathbf{B})]$
$= 150 - [50 + 90 - 30] = 40$
19. {OPST, OPTS, OSPT, OSTP, OTPS, OTSP, POST, POTS, PSOT, PSTO, PTOS, PTSO, SOPT, SOTP, SPOT, SPTO, STOP, STOP, TOPS, TOSP, TPOS, TPSO, TSOP, TSPO}
There are six words {OPTS, POST, POTS, SPOT, STOP, TOPS}
21. {H1, H2, H3, H4, H5, H6, T1, T2, T3, T4, T5, T6}. There are 12 or 2×6
23. $3 \times 5 = 15$
25. $(\mathbf{A} \cup \mathbf{B} \cup \mathbf{C}) =$
$n(\mathbf{A}) + n(\mathbf{B}) + n(\mathbf{C}) - n(\mathbf{A} \cap \mathbf{B}) - n(\mathbf{A} \cap \mathbf{C}) - n(\mathbf{B} \cap \mathbf{C}) + n(\mathbf{A} \cap \mathbf{B} \cap \mathbf{C})$
Let $\mathbf{A} = \{1, 2, 3, 4, 5\}$, $\mathbf{B} = \{3, 4, 5, 6, 7\}$, and $\mathbf{C} = \{5, 7, 8, 9\}$ then
$(\mathbf{A} \cup \mathbf{B} \cup \mathbf{C}) = \{1, 2, 3, 4, 5, 6, 7, 8, 9)$
$\mathbf{A} \cap \mathbf{B} = \{3, 4, 5\}$, $\mathbf{B} \cap \mathbf{C} = \{5, 7\}$, $\mathbf{A} \cap \mathbf{C} = \{5\}$, and
$\mathbf{A} \cap \mathbf{B} \cap \mathbf{C} = \{5\}$
Using the formula, $n(\mathbf{A} \cup \mathbf{B} \cup \mathbf{C}) = 5 + 5 + 4 - 3 - 2 - 1 + 1 = 9$, which is correct.

EXERCISES 7.2

1. Consider the Venn Diagrams with the regions indicated by Roman numerals.

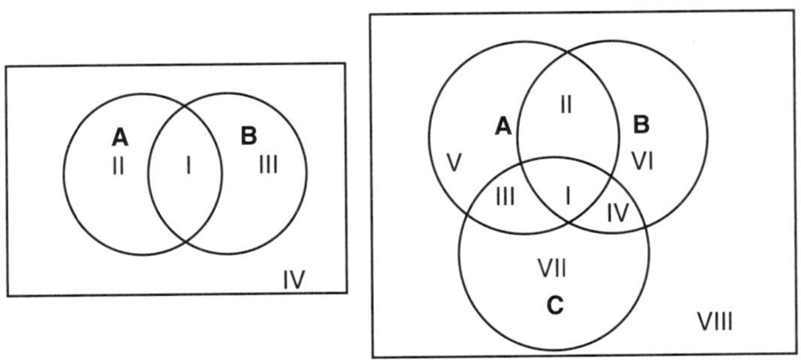

Using the Venn diagram with the two overlapping circles, $\mathbf{A} \cup \mathbf{B}$ is represented by regions I, II, III and $\mathbf{B} \cup \mathbf{A}$ is represented by regions I, II, III verifying the commutative property.
Likewise, $\mathbf{A} \cap \mathbf{B}$ is region I and so is $\mathbf{B} \cap \mathbf{A}$ indicating the commutative property.
Using the Venn diagram with three overlapping circles to establish the associative properties:

$$\mathbf{A} \cup (\mathbf{B} \cup \mathbf{C}) = \{I, II, III, V\} \cup \{I, II, III, IV, VI, VII\}$$
$$= \{I, II, III, IV, V, VI, VII\}$$
$$(\mathbf{A} \cup \mathbf{B}) \cup \mathbf{C} = \{I, II, III, IV, V, VI\} \cup \{I, III, IV, VII\}$$
$$= \{I, II, III, IV, V, VI, VII\}$$
$$\mathbf{A} \cap (\mathbf{B} \cap \mathbf{C}) = \{I, II, III, V] \cap \{I, IV\} = \{I\}$$
$$(\mathbf{A} \cap \mathbf{B}) \cap \mathbf{C} = \{I, II\} \cap \{I, III, IV, VII\} = \{I\}$$

To establish the distributive properties:

$$\mathbf{A} \cup (\mathbf{B} \cap \mathbf{C}) = (\mathbf{A} \cup \mathbf{B}) \cap (\mathbf{A} \cup \mathbf{C})$$
$$\{I, II, III, V\} \cup \{I, IV\} = \{I, II, III, IV, V, VI\} \cap \{I, II, III, IV, V, VII\}$$
$$\{I, II, III, IV, V\} = \{I, II, III, IV, V\}$$

and

$$\mathbf{A} \cap (\mathbf{B} \cup \mathbf{C}) = (\mathbf{A} \cap \mathbf{B}) \cup (\mathbf{A} \cap \mathbf{C})$$

$$\{I, II, III, V\} \cap \{I, II, III, IV, VI, VII = \{I, II\} \cup \{I, III\}$$
$$\{I, II, III\} = \{I, II, III\}$$

3.

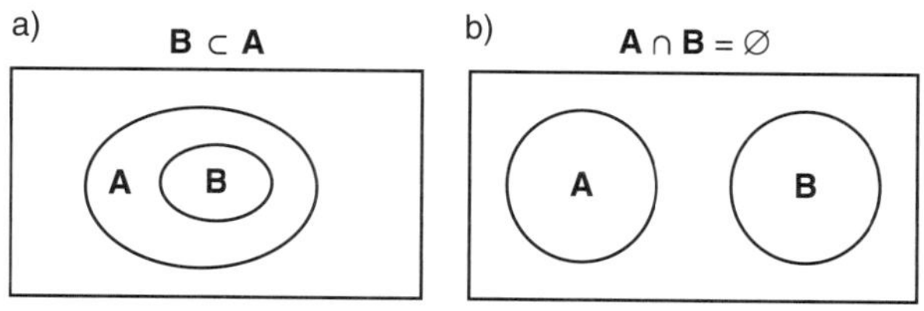

5. Using the Venn diagram with two overlapping circles
$(\mathbf{A} \cap \mathbf{B})^c = \{I\}^c = \{II, III, IV\}$ while $\mathbf{A}^c \cap \mathbf{B}^c = \{III, IV\} \cap \{II, IV\} = \{IV\}$
7. $(\mathbf{A}^c \cup \mathbf{B}^c) = \{III, IV\} \cup \{II, IV\} = \{II, III, IV\}$ and $n(\mathbf{A}^c \cup \mathbf{B}^c) = 45$
9. a) 20 read Time only b) 40 read neither publication.

11. The Venn diagram is:

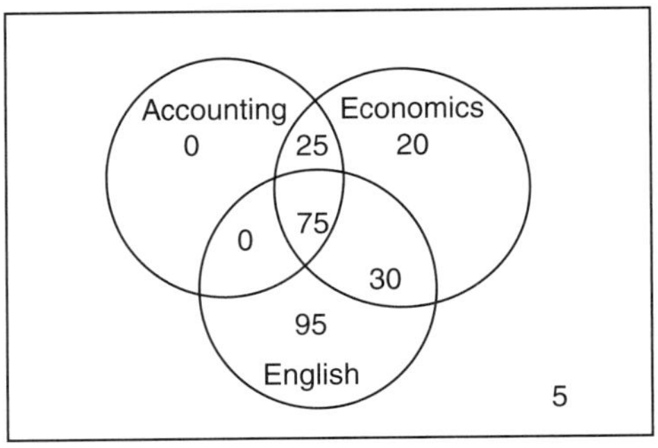

The diagram indicates that there are five math students who are not taking any of the other three courses. (region VIII)

13. The Venn diagram is:

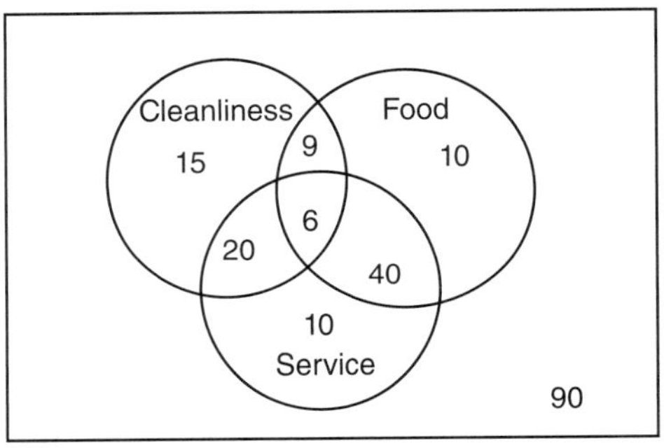

The diagram indicates that 35 (= 15 + 10 + 10) complained about a single issue (regions V, VI, and VII).

15. Three overlapping circles are used A antigen present, B antigen present, and Rh positive

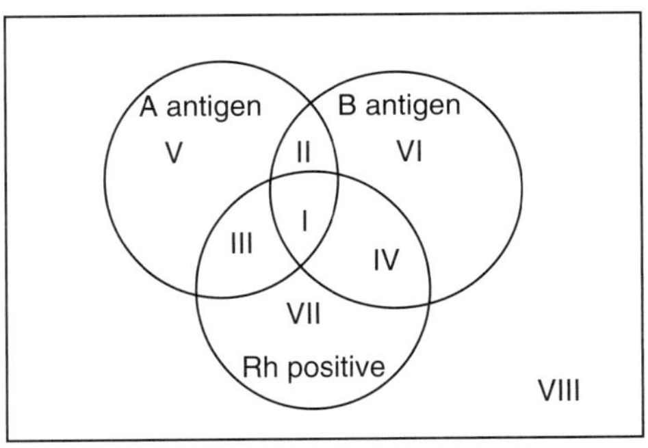

a) The diagram indicates the following blood types for each region

I AB+	III A+	V A−	VII O+
II AB−	IV B+	VI B−	VIII O−

→

b) The percentage of US populations with the four blood types is

A	40 %	AB	4 %
B	11 %	O	45 %

c) There are extreme differences in blood types throughout the world. South America is almost entirely type O, for example.

EXERCISES 7.3

1. There are $8 = (2^3)$ subsets including the empty set and the set itself { }, {X}, {Y}, {Z}, {X, Y}, {X, Z}, {Y, Z}, {X, Y, Z}.
3. There are 15 ways as shown below (RR not possible)

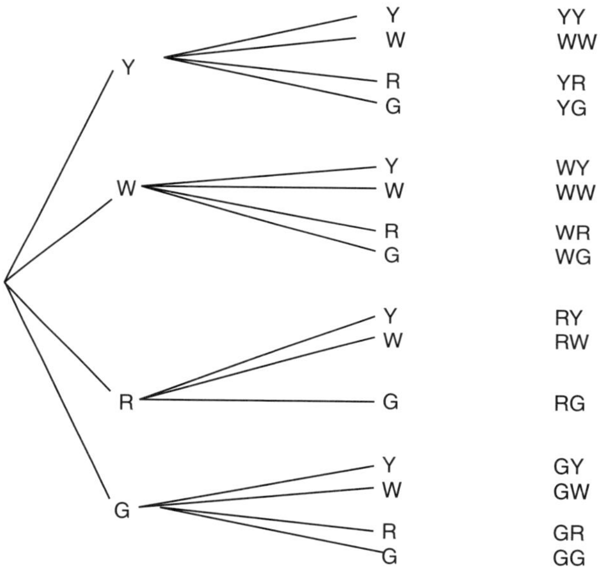

5. The tree diagram is below with the possibilities listed to the right

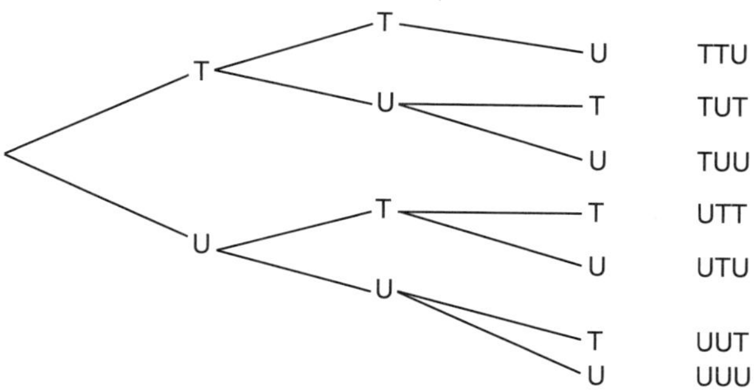

7. There are 15 with both improved reliability and durability.
9. a) 29 had lemonade only b) 74 had soda

11. There are $15 + 18 = 33$ female college graduates.

13. The tree and the 15 associated possibilities are:

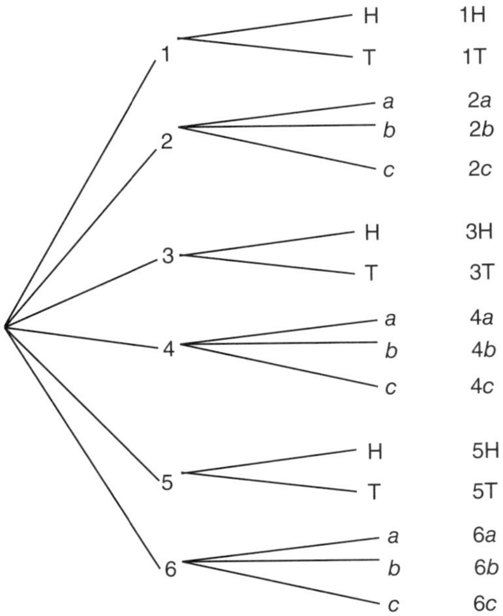

15. The 16 possibilities are shown:

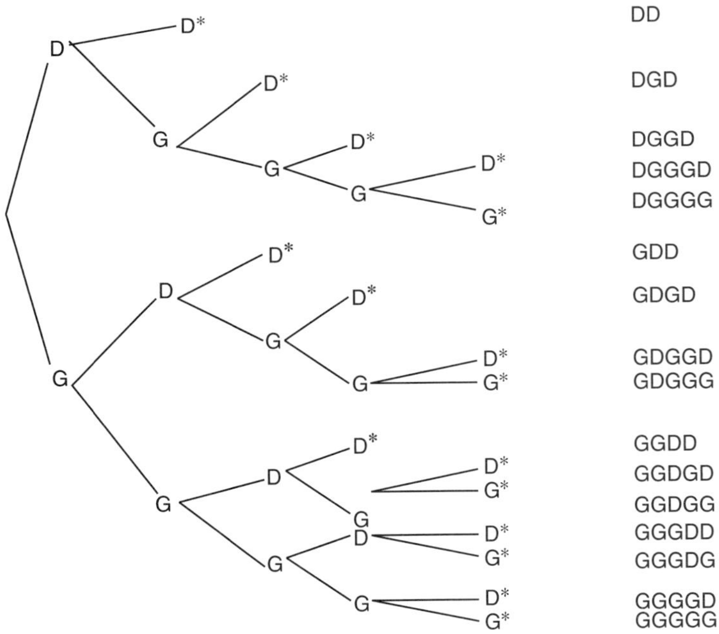

17. There are 27 freshmen women in the class of 100.

EXERCISES 7.4

1. 24 possibiliites 3. 24 possibilities
5. a) $7! = 7 \cdot 6 \cdot 5 \cdot 4 \cdot 3 \cdot 2 \cdot 1 = 5040$ b) $8!/6! = 56$ c) $9!/0! = 362,880$
d) $10!/10! = 1$ e) $52!/(48!4!) = (52 \cdot 51 \cdot 50 \cdot 49)/4! = 270{,}725$
7. $\binom{52}{5} = 2{,}598{,}960$
9. a) ${}_7C_2 = 21$ b) ${}_6C_5 = 6$ c) $\binom{9}{6} = 84$ d) $\binom{4}{1} = 4$
11. ${}_7P_5 = 2520$ 13. 120 15. 4 17. $r!\ {}_nC_r = r!\ \dfrac{n!}{r!(n-r)!} = \dfrac{n!}{(n-r)!} = {}_nP_r$
19. $\binom{3}{2}\binom{7}{1} + \binom{3}{3}\binom{7}{0} = 3 \cdot 7 + 1 \cdot 1 = 21 + 1 = 22$
21. 252,252 23. 6,760,000
25. There is about a 99.9% chance the alarm would be set off.
27. a) $10^7 = 10{,}000{,}000$ possible telephone numbers in an area code.
b) There are 1,800,000,000 possible numbers using the rules cited.
29. The DJ has three options where four CDs are chosen that represent all three groups. They are

$$\binom{10}{2}\binom{15}{1}\binom{25}{1} + \binom{10}{1}\binom{15}{2}\binom{25}{1} + \binom{10}{1}\binom{15}{1}\binom{25}{2}$$
$$= 88{,}125 \text{ possibilities}$$

31. $(4-1)! = 6$ 33. $\binom{50+75-1}{50} = \binom{124}{50} \approx 1.5 \times 10^{35}$

EXERCISES 7.5

1. When a pair of dice is thrown their sum varies from 2 through 12. The sample space is {2, 3, 4, 5, 6, 7, 8, 9, 10, 11, 12}
3. a) sample space b) and c) not sample spaces
5. 5/6 7. 0.9 9. Answers will vary 11. $\dfrac{6}{216} = \dfrac{1}{36}$
13. $\dfrac{1}{\binom{49}{6}} = \dfrac{1}{13{,}983{,}816}$
15. a) $\dfrac{39}{39+27} = \dfrac{39}{66}$ b) 5/39 c) 1/33 d) $\dfrac{4/66}{27/66} = \dfrac{4}{27}$
17. a) 24/65 b) 0 c) 6/65 19. a) 7% b) 0.4524 c) 7% d) 3%
21. 3/13 23. 3/7 25. a) 32% b) 26% c) 39/43 27. 1/2
29. a) 28/154 b) 28/54 c) 1/12 31. independent
33. a) $8/36 = 2/9$ b) $3/36 = 1/12$ c) 5/36

35. a) $(1/2)^6 = 1/64$. b) 1/64.

c) One capital requires one of three cases having a capital A, Capital B, or Capital C with the remaining letters all lower case.

$$\underbrace{(1/2)(1/4)(1/4)}_{\text{Capital A}} + \underbrace{(1/4)(1/2)(1/4)}_{\text{Capital B}} + \underbrace{(1/4)(1/4)(1/2)}_{\text{Capital C}} = 3/32$$

EXERCISES 7.6

1. a) The second card is red if RR or BR = 0.42 + 0.24 = 0.66

b) Both cards are black given they are same color has probability

$$\text{P(black/same)} = \frac{\text{p(black \& same)}}{\text{p(same)}} = \frac{\text{p(black)}}{\text{p(same)}} = \frac{0.06}{0.48} = \frac{1}{8}$$

3. a) 1/4 b) 5/12 5. a) 5/26 b) 5/12 c) 5/12

7. 1/2 9. 0.52 11. a) 83/188 b) 35/67

13. a) $P(\mathbf{A}) = 0.6,\ P(\mathbf{B}) = 0.9,\ P(\text{Apin}) = 0.8,\ P(\text{Bpin}) = 0.5$

$$P(\mathbf{A} \cap \mathbf{B}) = (0.6)(0.9) = 0.54$$

$$P(\mathbf{A} \cup \mathbf{B}) = P(\mathbf{A}) + P(\mathbf{B}) - P(\mathbf{A} \cap \mathbf{B})$$
$$= 0.6 + 0.9 - 0.54 = 0.96$$

A Venn diagram using probabilities may be useful for some portions of this Exercise

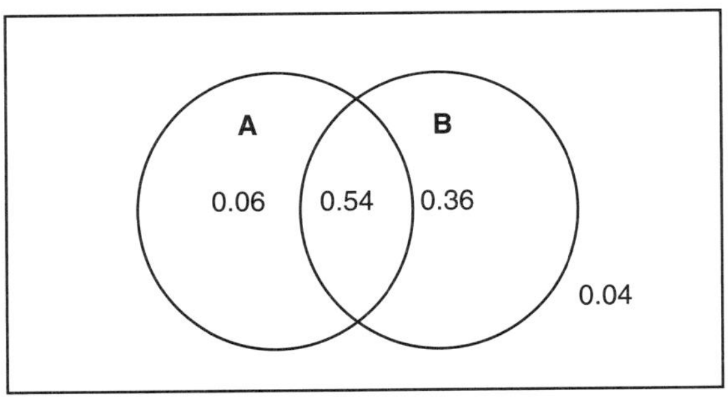

b) 0.06 + 0.36 = 0.42 c) (0.06)(0.8) + (0.36)(0.5) + (0.54)(1) = 0.768

d) $\dfrac{(0.36)(0.5)}{0.768} = \dfrac{0.18}{0.768} = 0.234375$

15. 5/8 17. 0.374 19. 2/9 21. a) 29/48 b) 9/28 23. 0.000021

25. 0.9483 27. 0.7895

SUPPLEMENTARY EXERCISES (CHAPTER 7)

1. a) true b) false c) true
3. The sample space is shaded

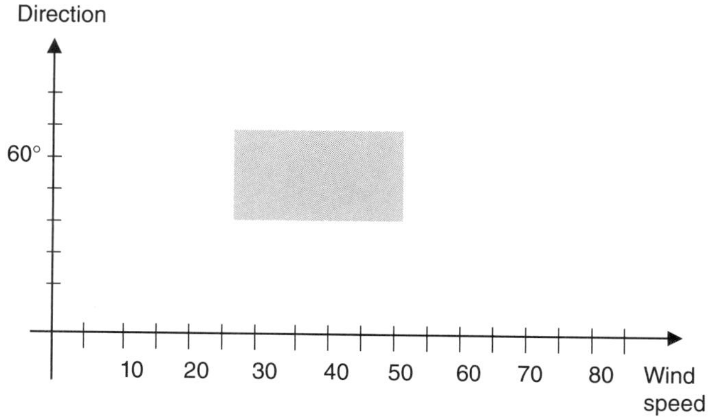

5. $\mathbf{A} \cap \mathbf{B} = \{100, 400\}$
7. The tree diagram is:

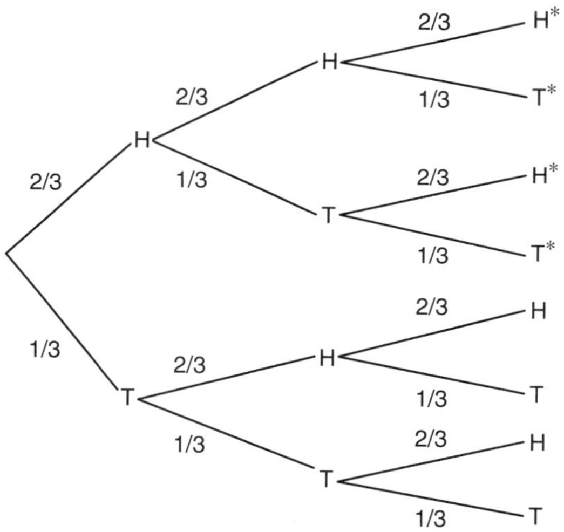

There are three cases where there are two heads and one where there are three heads.

The probability is $(3)\left(\frac{2}{3}\right)^2\left(\frac{1}{3}\right) + \left(\frac{2}{3}\right)^3 = \frac{20}{27}$

9. ${}_8P_3 = 336$ 11. $\frac{6!}{1!2!3!} = 60$

13. $\binom{6}{1}\binom{3}{2} + \binom{6}{2}\binom{3}{1} = 6(3) + 15(3) = 63$

15. $\binom{6}{1} + \binom{6}{2} = 6 + 15 = 21$

17. The probability a box will not contain a defective fuse can be determined from the three branches with asterisks. P(A)P(No defect/A) + P(B)P(no defect /B) + P(C) P(no defect/C) = (0.30)(0.98) + (0.50)(0.95) + (0.20)(0.90) = 0.949

⟶

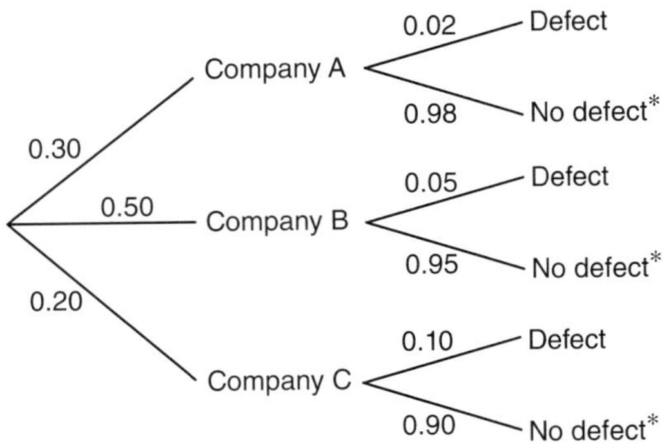

If a chosen box has a defective the probability that it came from Company B is:

$$P(B/def) = \frac{P(def/B)P(B)}{P(def/A)P(A) + P(def/B)P(B) + P(def/C)P(C)}$$

$$= \frac{(0.05)(0.50)}{(0.02)(0.30) + (0.05)(0.50) + (0.10)(0.20)} = \frac{0.025}{0.051} \approx 0.49$$

For detailed solutions to these Exercises see the modestly priced companion, "Student Solutions Manual" by Morris and Stark.

EXERCISES 8.1

1. a) continuous b) discrete c) discrete d) continuous e) continuous
3. Economists are interested in the employment percentage in a city, the unemployment percentage in the United States last month, or the percentage change in the daily stock market. (Answers will vary)
5. a) 0.3 b) 0.15 c) 3/16
7.

x	1	2	3	4	5	6
$p(x)$	1/6	1/6	1/6	1/6	1/6	1/6

9. The expected value in Exercise 8.1.5a is:

$$E(x) = 10\ (0.20) + 20(0.10) + 30(0.30) + 40\ (0.40) = 29.0$$

The expected value in Exercise 8.1.5b is:

$$E(x) = 1(0.05) + 5(0.35) + 9(0.25) + 12(0.20) + 20(0.15) = 10.05$$

The expected value in Exercise 8.1.5c is:

$$E(x) = -3(1/16) + 5(1/8) + 7(3/16) + 8(5/16) + 10(3/16) + 25(1/8) = 9.25$$

11. The sum, x, from 2 to 12, has these probabilities:

x	2	3	4	5	6	7	8	9	10	11	12
$p(x)$	1/36	2/36	3/36	4/36	5/36	6/36	5/36	4/36	3/36	2/36	1/36

The expected value is $2(1/36) + 3(2/36) + \cdots + 12(1/36) = 252/36 = 7$.

13. The distribution is:

x	100000	60000
$p(x)$	0.75	0.25

$E(x) = 100{,}000(0.75) + 60{,}000(0.25) = 90{,}000$ without insurance. With insurance we anticipate $100{,}000 - 15{,}000 = 85{,}000$. Do not purchase the insurance as it is only worth \$10,000 on average.

15. The distribution for winnings, excluding ticket cost, is:

x	1,500,000	500	50	0
$p(x)$	$\dfrac{\binom{6}{6}\binom{33}{0}}{\binom{39}{6}} = \dfrac{1}{3{,}262{,}623}$	$\dfrac{\binom{6}{5}\binom{33}{1}}{\binom{39}{6}} = \dfrac{198}{3{,}262{,}623}$	$\dfrac{\binom{6}{4}\binom{33}{2}}{\binom{39}{6}} = \dfrac{7920}{3{,}262{,}623}$	$\dfrac{3{,}254{,}504}{3{,}262{,}623}$

The expected value is ≈ 0.6115 but a dollar was paid so there is an expected loss of about 39 cents.

17. The probability when a 6 is twice as likely as the others yields the probability distribution:

x	1	2	3	4	5	6
$p(x)$	1/7	1/7	1/7	1/7	1/7	1/7

$$E(x) = 1(1/7) + 2(1/7) + \cdots + 6(2/7) = 27/7 \approx 3.857$$

19. Washington, Lincoln, Hamilton, Jackson, Grant, and Franklin appear on the \$1, \$5, \$10, \$20, \$50, and \$100 bills, respectively.

x	1	5	10	20	50	100
$p(x)$	6/20	3/20	4/20	4/20	2/20	1/20

Therefore, $E(x) = 1(6/20) + 5(3/20) + \cdots + 100(1/20) = \dfrac{341}{20} = \$17.05$

EXERCISES 8.2

1. a) 0.288 b) 0.09877 c) 0.36015 3. a) 0.348678 b) 0.1215767

5. $P(x = 2) + P(x = 3) = \binom{4}{2}\left(\frac{1}{6}\right)^2\left(\frac{5}{6}\right)^2 + \binom{4}{3}\left(\frac{1}{6}\right)^3\left(\frac{5}{6}\right)^1 = \frac{150 + 20}{1296} = 0.1312$

7. a) 0.2373 b) 0.984375 c) 0.0009766 9. 386/1024 = 0.37695

11. a) 0.2508 b) 0.5217 c) 0.0081 13. 0.1648 15. 0.3487

17. a) 0.1754 b) 0.0424

19. The complementary event has fewer cases so we seek $P(x = 0) \leq 0.05$
$\binom{n}{0}\left(\frac{1}{6}\right)^0\left(\frac{5}{6}\right)^n \leq 0.05$, $n = 16.43$ so 17 rolls are needed.

EXERCISES 8.3

1. a) 1/6 b) 0.8 3. a)1/1,947,792 b) 180/1,947,792
5. a) 48/2,598,960 b) 0.00198 7. a) 5/28 b) 1/56
9. a) 0.1291 b) 0.3874 c) 0.000387 11. 0.1299 13. 0.0098 15. 10/21

EXERCISES 8.4

1. a) 0.7360 b) 0.1931 c) 0.4816 3. a) 0.2090 b) 0.9040

5. a) 0.3679 b) 0.0803 7. a) 0.2176 b) 0.8774 c) 0.7326

9. $\frac{18^0 e^{-18}}{0!} = e^{-18} \approx 1.5 \times 10^{-8}$ 11. 0.0839

13. The probability of exactly 4 defects is $\frac{(2.4)^4 e^{-2.4}}{4!} = 0.1254$
The probability of less than 2 breakdowns is

$$P(x < 2) = p(x = 0) + p(x = 1)$$

$$= \frac{2.4^0 e^{-2.4}}{0!} + \frac{2.4^1 e^{-2.4}}{1!} = 0.3084$$

SUPPLEMENTARY EXERCISES (CHAPTER 8)

1. The missing probability is 0.15 and the expected value (mean) is 50.

3. a) 0.8125 b) 0.0625 5. a) 0.0861 b) 0.2639 c) 0.7560

7. The expected value is 33/9 so, for \$3.67 the game is fair.

9. 0.1316

11.

x	0	1	2	3	4
$p(x)$	$\dfrac{\binom{4}{0}\binom{48}{5}}{\binom{52}{5}}$ 0.6588	$\dfrac{\binom{4}{1}\binom{48}{4}}{\binom{52}{5}}$ 0.2995	$\dfrac{\binom{4}{2}\binom{48}{3}}{\binom{52}{5}}$ 0.0399	$\dfrac{\binom{4}{3}\binom{48}{2}}{\binom{52}{5}}$ 0.0017	$\dfrac{\binom{4}{4}\binom{48}{1}}{\binom{52}{5}}$ ≈ 0

13. a) 0.9041 b) 0.2613 c) 0.6916 15. a) 10 b) 0.251 c) 0.7798

For detailed solutions to these Exercises see the modestly priced companion, "Student Solutions Manual" by Morris and Stark.

EXERCISES 9.1

1. a) $x = 0.8$ and $y = 0.7$ b) $x = 0.1$ and $y = 0.6$ c) $x = 0.4$ and $x = 0.1$
3. The transition diagram for the chain is:

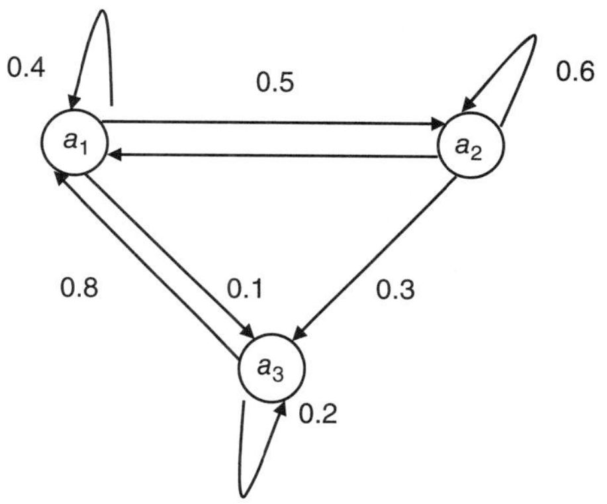

5. a) $\begin{bmatrix} 0.85 & 0.15 \\ 0.25 & 0.75 \end{bmatrix}$ b) $\begin{bmatrix} 0.55 & 0.45 \\ 0.60 & 0.40 \end{bmatrix}$ 7. $\begin{bmatrix} 0.7 & 0.1 & 0.0 & 0.2 \\ 0.0 & 0.8 & 0.0 & 0.2 \\ 0.6 & 0.0 & 0.3 & 0.1 \\ 0.0 & 0.1 & 0.0 & 0.9 \end{bmatrix}$

9. The row 3 and column 1 entry of 0.7

EXERCISES 9.2

1. $\mathbf{T}(2) = \begin{bmatrix} 0.2 & 0.8 \\ 0.9 & 0.1 \end{bmatrix}\begin{bmatrix} 0.2 & 0.8 \\ 0.9 & 0.1 \end{bmatrix} = \begin{bmatrix} 0.76 & 0.24 \\ 0.27 & 0.73 \end{bmatrix}$

3. a) $\mathbf{T}(2) = \begin{bmatrix} 0.43 & 0.29 & 0.28 \\ 0.42 & 0.26 & 0.32 \\ 0.50 & 0.25 & 0.25 \end{bmatrix}$ b) 0.29 c) 0.32

5. a) $\mathbf{T(3)} = \begin{bmatrix} 0.367 & 0.306 & 0.327 \\ 0.411 & 0.322 & 0.267 \\ 0.456 & 0.276 & 0.268 \end{bmatrix}$ b) 0.327 c) 0.276

7. $\mathbf{T(2)} = \begin{bmatrix} 0.1375 & 0.8625 \\ 0.1150 & 0.8850 \end{bmatrix}$ and $\mathbf{t_{11}(2)} = 0.1375$.

9. $\mathbf{T(2)} = \begin{bmatrix} 0.23 & 0.59 & 0.18 \\ 0.22 & 0.60 & 0.18 \\ 0.18 & 0.63 & 0.19 \end{bmatrix}$ and $\mathbf{t_{11}(2)} = 0.23$.

EXERCISES 9.3

1. $\mathbf{X_0T^3} = \begin{bmatrix} 0.7 & 0.3 \end{bmatrix} \begin{bmatrix} 0.376 & 0.624 \\ 0.312 & 0.688 \end{bmatrix} = \begin{bmatrix} 0.3568 & 0.6432 \end{bmatrix}$

3. $\mathbf{X_0T_3} = \begin{bmatrix} 0.30 & 0.70 \end{bmatrix} \begin{bmatrix} 0.825 & 0.175 \\ 0.70 & 0.30 \end{bmatrix} = \begin{bmatrix} 0.7375 & 0.2625 \end{bmatrix}$

   The store can expect a 73.75% share after three weeks of advertising.

5. a) Regular since $\mathbf{T}^2$ has no zero elements.
   b) All higher powers of $\mathbf{T}$ will have the same first row which contains zero elements and therefore it is not regular.
   c) Regular since $\mathbf{T}$ has no zero elements.

7. $\mathbf{X_0T} = \begin{bmatrix} 0.05 & 0.95 \end{bmatrix} \begin{bmatrix} 0.25 & 0.75 \\ 0.10 & 0.90 \end{bmatrix} = \begin{bmatrix} 0.1075 & 0.8925 \end{bmatrix}$ or 10.75%.

   $\mathbf{X_0T^2} = \begin{bmatrix} 0.05 & 0.95 \end{bmatrix} \begin{bmatrix} 0.1375 & 0.8625 \\ 0.1150 & 0.8850 \end{bmatrix} = \begin{bmatrix} 0.116125 & 0.883875 \end{bmatrix}$ or 11.61%

9. a) $x = 0.10$ b) $x = 1/3$ c) $x = 1/6$ 11. [7/16 9/16] 13. [1/7 4/7 2/7]

15. [14/48 19/48 15/48]

17. Answers vary. Using data for Delaware, the transition matrix for the election results is

$$\begin{array}{c} \\ D \\ R \end{array} \begin{array}{c} \begin{array}{cc} D & R \end{array} \\ \begin{bmatrix} 1/2 & 1/2 \\ 5/17 & 12/17 \end{bmatrix} \end{array}$$

   Solving $\mathbf{w(T - I) = 0}$ yields a steady state vector [10/27 17/27].

EXERCISES 9.4

1. a, c and d are absorbing Markov chains
3. It is an absorbing Markov chain

5. a) State 2 is an absorbing state. Reordering the states as 2, 1, 3, 4 the canonical form is

$$\begin{array}{c|c|ccc} & 2 & 1 & 3 & 4 \\ \hline 2 & 1.0 & 0.0 & 0.0 & 0.0 \\ \hline 1 & 0.4 & 0.2 & 0.2 & 0.2 \\ 3 & 0.0 & 0.5 & 0.3 & 0.2 \\ 4 & 0.6 & 0.0 & 0.0 & 0.4 \end{array}$$

b) $\mathbf{N} = (\mathbf{I} - \mathbf{S})^{-1} = \begin{bmatrix} \frac{35}{23} & \frac{10}{23} & \frac{15}{23} \\ \frac{25}{23} & \frac{40}{23} & \frac{65}{69} \\ 0 & 0 & \frac{5}{3} \end{bmatrix}$

7. Relabeling states as 2, 5, 1, 3, 4 gives the canonical form:

$$\begin{bmatrix} 1 & 0 & 0 & 0 & 0 \\ 0 & 1 & 0 & 0 & 0 \\ 1/2 & 0 & 0 & 0 & 1/2 \\ 1/3 & 1/3 & 0 & 0 & 1/3 \\ 0 & 1/3 & 1/3 & 1/3 & 0 \end{bmatrix} \mathbf{N} = \begin{bmatrix} 16/13 & 3/13 & 9/13 \\ 2/13 & 15/13 & 6/13 \\ 6/13 & 6/13 & 18/13 \end{bmatrix} \mathbf{C} = \begin{bmatrix} 9/13 & 4/13 \\ 6/13 & 7/13 \\ 5/13 & 8/13 \end{bmatrix}$$

The expected number of times the rat, starting in Compartment 1, is in Compartment 3 before being trapped is 3/13. The probability starting in Compartment 1, of being trapped in Compartment 5 is 4/13.

SUPPLEMENTARY EXERCISES (CHAPTER 9)

1. The transition diagram is:

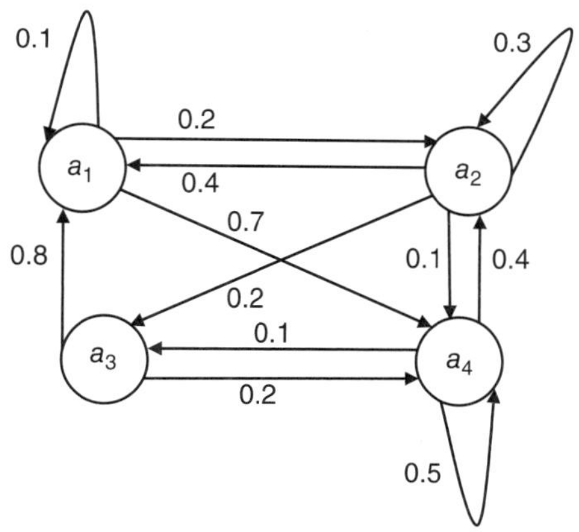

3. a) $\mathbf{T(3)} = \begin{bmatrix} 0.297 & 0.190 & 0.513 \\ 0.444 & 0.229 & 0.327 \\ 0.568 & 0.128 & 0.304 \end{bmatrix}$ b) 0.444 11. a and c are regular.

7. [3/5 0 2/5] 9. [1/3 1/3 1/3]

11. Relabeling the states as 2, 1, 3, 4, 5 the canonical form is

$$\begin{array}{c} \\ 2 \\ 1 \\ 3 \\ 4 \\ 5 \end{array}\begin{array}{c} \begin{array}{ccccc} 2 & 1 & 3 & 4 & 5 \end{array} \\ \left[\begin{array}{c:cccc} 1 & 0 & 0 & 0 & 0 \\ \hdashline 0 & 1/3 & 1/3 & 0 & 1/3 \\ 1/4 & 1/4 & 1/2 & 0 & 0 \\ 0.3 & 0.1 & 0.4 & 0.2 & 0 \\ 0 & 0 & 0 & 1 & 0 \end{array}\right] \end{array}$$

13. a) 3/7 b) 6/7

For detailed solutions to these Exercises see the modestly priced companion, "Student Solutions Manual" by Morris and Stark.

EXERCISES 10.1

1. a) "All adult residents of the community"
   b) the 2000 adults in the community that were polled.
   c) The opinion on the smoking ban.
3. a) Quantitative b) Quantitative c) Qualitative d) Qualitative
5. a) Quantitative and ratio scale b) Quantitative and ratio scale
   c) Qualitative and nominal scale d) Qualitative and ordinal scale
   e) Quantitative and interval scale
7. The frequency bar chart is:

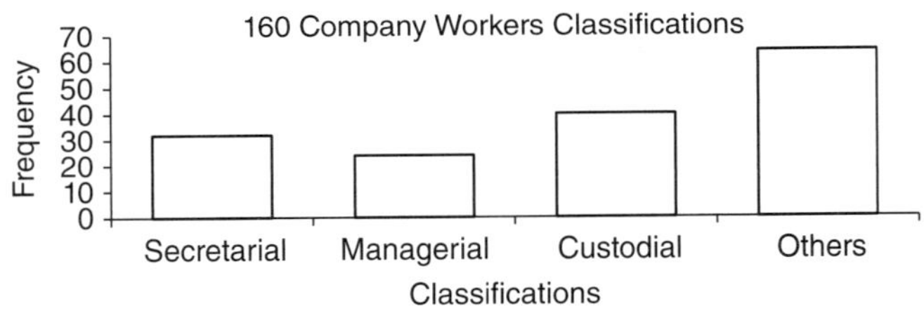

9. The pie chart is below:

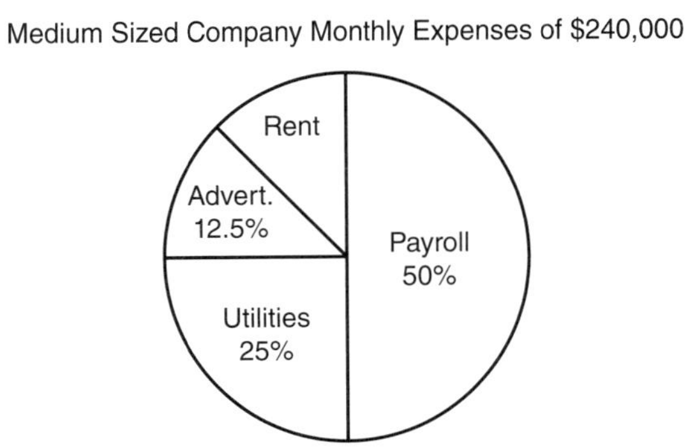

11. The frequency histogram is:

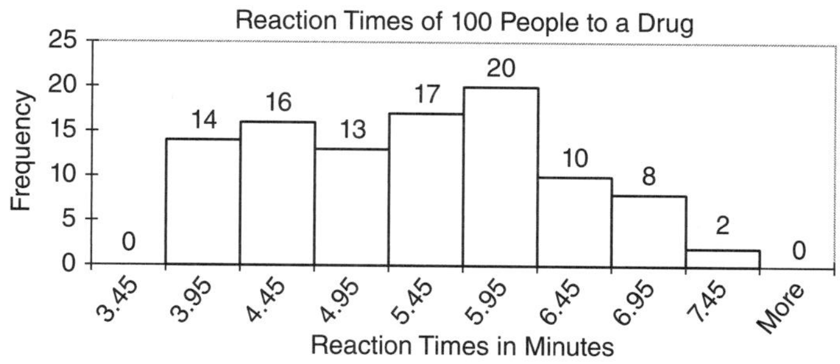

13. a) The histogram is:

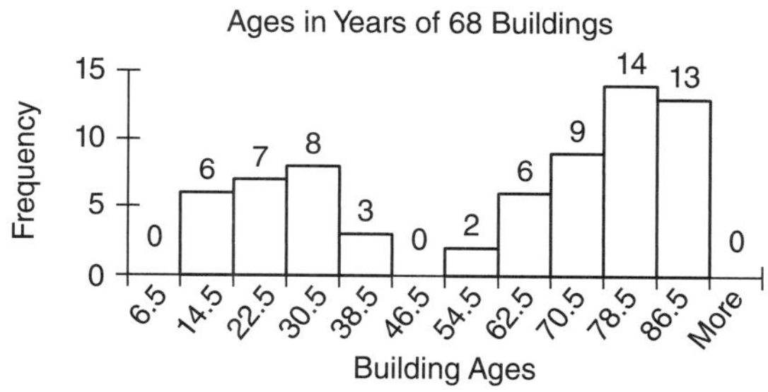

b) Using the graph, there are $6 + 7 + 8 + 3 + 2 + 6 + 9 = 41$ buildings less than 71 years old. The percentage is $41/68(100\%) = 60.29\%$.

EXERCISES 10.2

1. a) 100 b) 95 c) 80 d) 2550 e) 1050
3. The mean, median, and mode respectively are 39, 40, and 40.
5. 381 7. a) 20 b) 1010 c) 2/5 d) 28
9. a) Range 60, variance 321.1, and standard deviation 17.9
   b) range 11, variance 14, and standard deviation 3.742
11. Mean 27, median 23, modes 13, 19, and 23, range 54, variance 222.1, and standard deviation 14.9.
13. a) The mean is 7.788 and the standard deviation is 1.596
    b) $\bar{x} \pm s = (6.19, 9.38)$. There are 26 observations within this interval.
    $\bar{x} \pm 2s = (4.60, 10.98)$. There are 39 observations within this interval.
    $\bar{x} \pm 3s = (3.00, 12.58)$. There are 39 observations within this interval.
    Yes, it seems to roughly follow the empirical rule except possibly for the 13.6 measurement.

EXERCISES 10.3

1. A uniform distribution for $c = 10$ to $d = 25$ has $f(x) = \dfrac{1}{25 - 10} = \dfrac{1}{15}$

   The mean is 17.5, standard deviation 4.330 and variance 18.75.
3. a) 13/15 b) 0 c) 1 d) 0.8333 e) 1/3 f) 7/15
5. a) 135 b) 105 c) 130 d)140 e) 125 f) not possible
7. a) mean 0.30 and standard deviation 0.02887 b) 0.40 c) 0.31

EXERCISES 10.4

1. a) 0.4177 b) 0.3389 c) 0.4846 d) 0.3729 e) 0.4913 f) 0.0239
3. a) 0.8836 b) 0.8460 c) 0.5606 d) 0.7929 e) 0.6922 f) 0.9788
5. a) 0.9265 b) 0.0228 c) 0.0869 d) 0.9332
7. a) 0.1022 b) 0.1102 c) 0.1492
9. a) 1.00 b) −0.60 c) −2.20 d) −1.80 e) 1.20 f) 0
11. a) 0.6915 b) 0.7745 c) 0.1056
13. a) 0.9772 b) 0.8400 c) 0.8314 15. a) 1.12 b) 1.45 c) 1.28 d) 1.71
17. 5.9078 ml

EXERCISES 10.5

1. a) 0.6826 b) 0.9554 c) 0.9974 3. 11.94
5. Choose route B if 45 minutes available and route A if only 42 minutes are available.
7. 8.78 feet 9. 1233 seats 11. a) 0.9500 b) 0.9900 c) 0.7994 d) 0.9000

EXERCISES 10.6

1. All these questions can be improved with several options. Also, some questions do not specify a time.

   In the past year how often have you shopped on eBay?

   a) Never b) 1–5 times c) 6–10 times d) more than 10 times

   Specify the interval for your annual salary

   a) under $20,000

   b) $20,000–$39,999

   c) $40,000–$59,999

   d) over $60,000

   During the past year how many books have you purchased online?

   a) None b) 1–5 c) 6–10 d) more than 10

3. Answers vary. You may wish to include questions about age, gender, how often the person smokes, how long the person has smoked, etc.
5. Answers vary.

SUPPLEMENTARY EXERCISES (CHAPTER 10)

1. a) Qualitative and nominal b) Qualitative and ordinal
   c) Qualitative and nominal d) Quantitative and ratio
3. Answers will vary
5. The mean is 69.25, the median 72.5, the modes are 70, 80 and 100, the range 70, the variance 469, and the standard deviation 21.7.
7. a) 56 b) any number 50 or less c) 59.5 d) any number larger than 80
9. a) 0.63 b) 0.44 c) −1.34 d) −1.52 11. 170

For detailed solutions to these Exercises see the modestly priced companion, "Student Solutions Manual" by Morris and Stark.

EXERCISES 11.1

1. $\begin{bmatrix} -6 & -3 & 1 \\ 5 & -3 & 4 \\ -3 & -2 & -1 \end{bmatrix}$

3. $\begin{bmatrix} -1 & -3 & -6 \\ 2 & 4 & -5 \\ 1 & -3 & -1 \end{bmatrix}$

5. The entries in row 1 are all larger than those in row 2 so it dominates that row. Eliminating row 2 yields

$$\begin{bmatrix} 5 & 3 & 5 & 4 \\ 6 & 1 & 4 & 2 \end{bmatrix}$$

Next, column 2 dominates column 1, column 3, and column 4 because its entries are smaller than those in the other columns. Eliminating these columns yield

$$\begin{bmatrix} 3 \\ 1 \end{bmatrix}$$

Now the top row dominates leaving the (1, 2) entry of 3 a saddle point of the game.

7. Row 4 dominates row 3 since its entries are larger. Eliminating row 3 yields

$$\begin{bmatrix} 8 & 2 & 4 & 8 & 6 \\ 1 & 4 & 5 & 9 & 4 \\ 9 & 5 & 4 & 7 & 5 \end{bmatrix}$$

⟶

Now, column 2 dominates rows 4 and 5 since its entries are smaller to yield

$$\begin{bmatrix} 8 & 2 & 4 \\ 1 & 4 & 5 \\ 9 & 5 & 4 \end{bmatrix}$$

Now, the third row dominates the first to yield

$$\begin{bmatrix} 1 & 4 & 5 \\ 9 & 5 & 4 \end{bmatrix}$$

9. The minimax strategy is determined by placing a circle around each row min and a box around each column maximum. If there are any entries with both a circle and, a box, they are saddle points and the game is strictly determined.

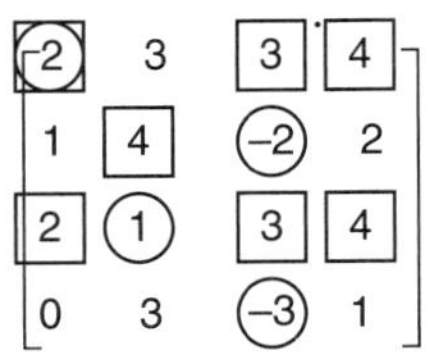

Here, the (1, 1) entry is the saddle point of the game and the optimal strategy using the minimax criterion.

11. If the game has no saddle point then it is non-strictly determined. Using the circles and boxes for the minimax criterion we find no overlap so there is no saddle point.

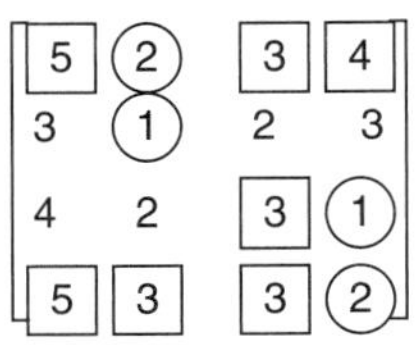

13. The game is solved using the LP below (after 3 is added to each of the entries to remove the negatives:

$$\text{Maximize:} \quad Z = \frac{1}{v} = Y_1 + Y_2$$

Subject to:

$$6Y_1 + 5Y_2 \leq 1$$

$$1Y_1 + 6Y_2 \leq 1$$

$$Y_1, Y_2 \geq 0$$

Row	Y_1	Y_2	s_1	s_2	RHS
1	1	5/6	1/6	0	1/6
2	0	31/6	−1/6	1	5/6
Objective	0	−1/6	1/6	0	1/6 = −Z

Row	Y_1	Y_2	s_1	s_2	RHS
1	1	5/6	1/6	0	1/31
2	0	1	−1/31	6/31	5/31
Objective	0	0	5/31	1/31	6/31 = −Z

$-Z = 1/v = 31/6$ so $Y_1 = (31/6)(1/31) = 1/6$, $Y_2 = (31/6)(5/31) = 5/6$,
from the objective row $X_1 = (31/6)(5/31) = 5/6$ and
$X_2 = (31/6)(1/31) = 1/6$

EXERCISES 11.2

1. The stock gains 100% one week and then loses 50% the next. The arithmetic average is $\frac{(100\%) + (-50\%)}{2} = 25\%$. However, by the end of the year there is \$1000 or a 0% change. Using the geometric average weekly return gives this result since $\sqrt{(2)(0.5)(2)(0.5)\cdots(2)(0.5)} - 1 = 0$.
3. Using the formulas from Chapter 2 the monthly payment for the car is
$$\frac{25{,}000(0.005)}{1 - \dfrac{1}{(1.005)^{72}}} = \$414.32$$
5. Solution requires identifying and setting missing values.
7. Solving the associated linear program yields
$X_1^* = 3$, $X_2^* = 1$, $X_3^* = 3.5$, $X_4^* = 1$, and $Z^* = \$271817.80$
9. Solving the associated linear program yields
$X_1^* = 12$, $X_2^* = 0$, $X_3^* = 0$, $X_4^* = 0$, and $Z^* = \$291{,}177.04$

EXERCISES 11.3

1. Recall, $\mathbf{SM^tP} = [0 \;\; 0 \;\; 1]$, Clan C for the sister's son who becomes the father of the grandson. That is,

$$(\mathbf{SM^tP})(\mathbf{P}) = \begin{bmatrix}0 & 0 & 1\end{bmatrix}\begin{bmatrix}0 & 1 & 0\\0 & 0 & 1\\1 & 0 & 0\end{bmatrix} = \begin{bmatrix}1 & 0 & 0\end{bmatrix}$$

The grandson is in Clan A.

⟶

The great grandson (whose father is the grandson in Clan A) is

$$\begin{bmatrix}1 & 0 & 0\end{bmatrix}\begin{bmatrix}0 & 1 & 0\\0 & 0 & 1\\1 & 0 & 0\end{bmatrix} = \begin{bmatrix}0 & 1 & 0\end{bmatrix}$$

The great grandson is in Clan B.

3. $\mathbf{P} = \begin{bmatrix}0 & 1 & 0\\0 & 0 & 1\\1 & 0 & 0\end{bmatrix}$ and $\mathbf{P}^2 = \begin{bmatrix}0 & 0 & 1\\1 & 0 & 0\\0 & 1 & 0\end{bmatrix}$. $\mathbf{P}^3 = \begin{bmatrix}1 & 0 & 0\\0 & 1 & 0\\0 & 0 & 1\end{bmatrix} = \mathbf{I}$.

5. Answers vary.

7. Answers vary. One possibility is

$$\begin{aligned}
\text{Maximize:} \; & 3x_{11} + x_{12} + 2x_{13} + 5x_{14} + 4x_{21} + 6x_{22} + x_{23} + 2x_{24} \\
& + 4x_{31} + 5x_{32} + 5x_{33} + x_{34} + 3x_{41} + 2x_{42} + 5x_{43} + 6x_{44} \\
\text{Subject to:} \; & x_{11} + x_{12} + x_{13} + x_{14} \leq 1 \\
& x_{21} + x_{22} + x_{23} + x_{24} \leq 1 \\
& x_{31} + x_{32} + x_{33} + x_{34} \leq 1 \\
& x_{41} + x_{42} + x_{43} + x_{44} \leq 1 \\
& x_{11} + x_{21} + x_{31} + x_{41} \leq 1 \\
& x_{12} + x_{22} + x_{32} + x_{42} \leq 1 \\
& x_{13} + x_{23} + x_{33} + x_{43} \leq 1 \\
& x_{14} + x_{24} + x_{34} + x_{44} \leq 1 \\
& 0 \leq x_{ij} \leq 1
\end{aligned}$$

The objective maximum is 20, $x_{11} = 1$, $x_{22} = 1$, $x_{33} = 1$, $x_{44} = 1$ and all other $x_{ij} = 0$. So, each man is assigned to one woman and each woman to one man for four "monogamous" couples.

9. The observer may have been considering unhappy individuals when the numbers of men and women are not equal.

EXERCISES 11.4

1. There are many ways to arrange the Monte Carlo. One, tracks a player's winnings. A head results in Jack paying Jill \$1 dollar (+\$1). Therefore, a tail results in Jill paying Jack (-\$1). Use a column chosen randomly, say column 5, from the random number table in Chapter 6. Next, letting an odd number represent a head and an even number a tail the tosses can be simulated and the results tabulated.

⟶

After 10 simulated tosses the result is \$0 and after 20 it was \$2. Thus in 10 tosses the two are even but after 20 tosses Jill is ahead \$2 or an average of 10 cents per toss. In actuality the expected gain for Jill (or Jack) over many tosses of a coin should be \$0.

3. After 4 weeks as shown in the table below, the expected gain/ loss for the periods is $11/4 = \$2.75$ average weekly gain. In reality the expected gain/loss is $(0.45 + 0.40 + 0.40)(+1) + (0.35 + 0.45 + 0.35)(-1) + (0.20 + 0.15 + 0.25)(0)$ $= 0.10/\text{day}$ or \$0.50/week.

Week #	Currency A	Currency B	Currency C		Weekly Gain/loss
	Unchanged	Unchanged	Advance	+1	
	Decline	Decline	Decline	−3	
1	Advance	Advance	Unchanged	+2	
	Advance	Decline	Advanced	+1	
	Advance	Advance	Decline	+1	**+2**
	Advance	Decline	Advance	+1	
	Decline	Advance	Advance	+1	
2	Advance	Advance	Advance	+3	
	Decline	Advance	Advance	+1	
	Decline	Advance	Unchanged	0	**+6**
	Decline	Advance	Unchanged	0	
	Advance	Decline	Unchanged	0	
3	Decline	Decline	Advance	−1	
	Unchanged	Decline	Unchanged	−1	
	Advance	Advance	Advance	+3	**+1**
	Unchanged	Advance	Advance	+2	
	Decline	Unchanged	Advance	0	
4	Unchanged	Decline	Advance	0	
	Decline	Advance	Advance	+1	
	Decline	Advance	Decline	−1	**+2**

5. The center of the circle is (−1, 4) and radius 3. Therefore the x values can lie anywhere in the interval -1 ± 3 and the y values anywhere in the interval 4 ± 3. Next, we define $X_r = -4 + 6R_x$ and $Y_r = 1 + 6R_y$

   a) $X_r = -4 + 6(0.61) = -0.34$ and $Y_r = 1 + 6(0.20) = 2.2$
   $(-0.34 + 1)^2 + (2.2 - 4)^2 \leq 9$ yes

   b) $X_r = -4 + 6(0.24) = -2.56$ and $Y_r = 1 + 6(0.71) = 5.26$
   $(-2.56 + 1)^2 + (5.26 - 4)^2 \leq 9$ yes

   c) $X_r = -4 + 6(0.68) = 0.08$ and $Y_r = 1 + 6(0.84) = 6.04$
   $(0.08 + 1)^2 + (6.04 - 4)^2 \leq 9$ yes

7. The last column of the random number table is used to generate the random numbers. The center of the circle is (3, 4) and radius 10. The x values lie in the interval 3 ± 10 and the y values in the interval 4 ± 10.
   Next, define $X_r = -7 + 20R_x$ and $Y_r = -6 + 20R_y$.

⟶

The table below summarizes the results of the first five and next 10 trials. Only trials 1–5 and 14–15 are shown in the table.

Trial	R_x	R_y	X_r	Y_r	$(X_r - 3)^2$	$(Y_r - 4)^2$	$(X_r - 3)^2 + (Y_r - 4)^2$	Hit?
1	0.46	0.92	2.2	12.4	0.64	70.56	71.2	Yes
2	0.62	0.33	5.4	0.6	5.76	11.56	17.32	Yes
3	0.44	0.80	2.0	10.0	1.00	36.0	37.0	Yes
4	0.86	0.25	10.2	−1.0	51.84	25.0	76.84	Yes
5	0.71	0.46	7.2	3.2	17.64	0.64	18.28	Yes
⋮	⋮	⋮	⋮	⋮	⋮	⋮	⋮	⋮
14	0.65	0.53	6.0	4.6	9.0	0.36	9.36	Yes
15	0.09	0.98	−5.2	13.6	67.24	92.16	159.4	No

The area of a 20 by 20 square is 400. In the case of the first five trials all were hits so the estimate of the area of the circle is 400 square units when in actuality it should be approximately 314 square units. When fifteen trials are used the estimate is then $400\,(13/15) = 346.67$ square units which is closer but still the estimate does not appear to have settled. It can take thousands of trials to get a reliable estimate but this example is meant to show the procedure for setting up a Monte Carlo experiment.

9. After the ordered pair (X_r, Y_r) is determined one must judge whether or not it is a hit (part of the shaded region).

Trial	R_x	R_y	X_r	Y_r	Hit?
1	0.97	0.27	4.85	1.08	Yes
2	0.27	0.12	1.35	0.48	No
3	0.05	0.72	0.25	2.88	No
⋮	⋮	⋮	⋮	⋮	⋮
19	0.10	0.43	0.50	1.72	Yes
20	0.10	0.41	0.50	1.64	Yes

The area of the rectangle surrounding the object is $4 \times 5 = 20$ square units so the area suggested by this sample Monte Carlo is $20\,(12/20) = 12$ square units.

11. Use the following scheme for generating card denominations

0000–0769 (A)	0770–1538 (2)	1539–2308 (3)	2309–2667 (4)
2668–3846 (5)	3847–4615 (6)	4616–5385 (7)	5386–6154 (8)
6155–6923 (9)	6924–7692 (10)	7693–8642 (J)	8463–9231 (Q)
9232–9999 (K)			

Use sets of five random numbers starting with column 3: 9771, 2783, 2752, 1218, 0526 which corresponds to a hand with K, 5, 5, 2, A. Twenty five hands are generated with the 25th being 1170, 2404, 8531, 3736, 5973 or 2, 4, Q, 5, 8. Ten of these 25 hands consist of five different denominations of cards. The estimate is 0.400 when in actuality it is about 0.507. Using larger numbers of

⟶

simulated hands the estimated probability may be closer to the actual probability. The goal here is to organize a Monte Carlo.

13. Let 0–199 represent 17 items daily. Then 200–349 represents 18 items, 350–499 represents 19 items, 500–749 represents 20 items, 750–899 represents 21, 900–949 represents 22, and a random number from 950–999 represents 23 items daily. The table summarizes the Monte Carlo and was generated using column 6 of the random number table as a start. Only the first and last two days of the 30 are shown in the table. The solutions manual has the entire table for those who are interested in complete solutions here.

RN	Items produced	Items moved	Items held over	Empty truck spaces
117	17	17	0	4
240	18	18	0	3
⋮	⋮	⋮	⋮	⋮
632	20	20	0	1
427	19	19	0	2

There were 10 held over items over a 30 day period, so on average $10/30 = 1/3$ item awaits shipment per day. There were 53 empty spaces on the truck over the 30 day period or $53/30 = 1.77$ empty spaces on average per day.

15. Use the following scheme to convert the RN to a month of the year

000–083 (Jan)	084–166 (Feb)	167–249 (Mar)	250–333 (Apr)
334–416 (May)	417–499 (Jun)	500–583 (Jul)	584–666 (Aug)
667–749 (Sep)	750–833 (Oct)	834–916 (Nov)	917–999 (Dec)

Column 3 of the random number table is used to generate the groups of five people.

Trial	RN1	RN2	RN3	RN4	RN5	Hit or Miss
1	977 Dec	278 Apr	275 Apr	121 Feb	052 Jan	Hit
2	722 Sep	012 Jan	421 Jun	307 Apr	972 Dec	Miss
3	819 Oct	934 Dec	491 Jun	094 Feb	402 May	Miss
4	645 Aug	351 May	814 Oct	729 Sep	020 Jan	Miss
5	803 Oct	149 Feb	005 Jan	994 Dec	108 Feb	Hit
6	073 Jan	515 Jul	049 Jan	823 Oct	947 Dec	Hit
7	694 Sep	914 Nov	642 Aug	153 Feb	901 Nov	Hit
8	461 Jun	100 Feb	434 Jun	103 Feb	418 Jun	Hit
9	028 Jan	837 Nov	618 Aug	205 Mar	984 Dec	Miss
10	724 Sep	301 Apr	249 Mar	946 Dec	925 Dec	Hit

The simulation of 10 groups of five people indicates a 6/10 chance that at least two share a common birth month.

17. If R_x is 000–499 the mouse makes a right turn and finds food. If R_x is 500–999 the mouse makes a wrong turn and continues through the maze until it finds food.

⟶

Use column 1 to generate the necessary random numbers: Note the table only shows the first and last two trials of the twenty conducted.

Trial	RNs	Right turn?	Number of wrong turns
1	0.943 0.908 0.146	No No Yes	2
2	0.514 0.405	No Yes	1
⋮	⋮	⋮	⋮
19	0.458	Yes	0
20	0.240	Yes	0

There are nine wrong turns out of 20 trials for an average of 0.45 wrong turns per trial.

19. Answers vary.

EXERCISES 11.5

1. In the first stage, movement is made from modes 2, 3, or 4 to node 5. The distance from 2 to 5 is 6, from 3 to 5 is 3, and 4 to 5 is 5. In the second stage movement from node 1 to 2, 3, or 4 is possible. The distance from node 1 to 2 and then 5 is 14, from 1 to 3 and then 5 is 13, and from 1 to 4 and then 5 is 10. The shortest distance is 10 if the path is from node 1 to 4 to 5.
3. There are two shortest paths at a distance of 8. They are 1-3-5-9-10 or 1-3-7-9-10 as shown in the table below:

Number of steps from node 10	*Distance to node 10*		*Least cost to node 10*	
	From node	*Cost*	*Cost*	*Path*
1	*8*	*4*	*4*	*8-10*
	9	*2*	*2*	*9-10*
2	*5*	$2+4=6$ $2+2=4$	*4*	*5-9-10*
	6	$4+4=8$ $3+2=5$	*5*	*6-9-10*
	7	$1+4=5$ $2+2=4$	*4*	*7-9-10*
3	*2*	$4+4=8$ $5+5=10$	*8*	*2-5-9-10*
	3	$1+4=5$ $4+5=9$ $1+4=5$	*5*	*3-5-9-10 or 3-7-9-10*
	4	$3+5=6$ $2+4=9$	*6*	*4-7-9-10*
4	*1*	$2+8=10$ $3+5=8$ $6+6=12$	*8*	*1-3-5-9-10 or 1-3-7-9-10*

5. A total of \$6 million dollars can be spent on the plants and any unused money is lost. The best option for the Corporation is to plan the \$3 million expansion at Plant 1, the \$2 million expansion at Plant 2 and the \$1 million expansion at Plant 3 which will yield a worth of \$18 million.

7. Let $f_k(t)$ = min total probability of rejection when t hours of specialist time is allocated to job #k and subsequent jobs and $P_k(x)$ the probability of rejection when k hours are devoted to job k, $k = 1, 2, 3$. Starting with the time devoted to job #3 and assuming it is the only job that is important in terms of rejection probabilities.

   At the last stage it is assumed that the probability of rejection depends on job #1 as well as job #2 and job #3. Ordered pairs represent the specialist hours spent on job #1 and the remaining time on the other two jobs.

Time t_1	Probability of rejection	Optimal solution
0	(0, 0) = (0.60)(0.2475) = 0.2475*	$P_1(0) = 0.2475$
1	(1, 0) = (0.40)(0.2475) = 0.099* (0, 1) = (0.60)(0.1925) = 0.1155	$P_1(1) = 0.099$
2	(2, 0) = (0.30)(0.2475) = 0.07425* (1, 1) = (0.40)(0.1925) = 0.0770 (0, 2) = (0.60)(0.1350) = 0.0810	$P_1(2) = 0.07425$
3	(3, 0) = (0.15)(0.2475) = 0.037125 (2, 1) = (0.30)(0.1925) = 0.05775 (1, 2) = (0.40)(0.1350) = 0.0540 (0, 3) = (0.60)(0.055) = 0.0330*	$P_1(3) = 0.033$

Note: the assumption is that the final product is only rejected if all three finishing jobs are rejected.The optimal solution is to assign no hours to job #1, 3 h to job #2, and no hours to job #3.

USING TECHNOLOGY

Students do not need technological aids to fully benefit from this text.

For those who have an interest in their use, we offer guidelines below for:

TI-83 calculators: for Matrices LINDO: for Linear Programs EXCEL: for use in probability and statistics

MATRICES

Chapter 3 has common matrix operations which most graphing calculators can execute. Most TI calculators probably use similar steps. For example, for a TI - 83 model; **to enter a matrix**:

1. Press the ***2nd*** key, then ***matrix***, and in the edit submenu select ***a matrix***.
2. Enter the row dimension followed by the column dimension.
3. Enter each matrix entry.
4. When all values are entered, press the ***2nd*** key and ***quit*** to return to the main (home) screen.

You can edit or view a matrix. If a matrix is too large for the screen, use the arrow keys to scroll down or across.

Finite Mathematics: Models and Applications, First Edition. Carla C. Morris and Robert M. Stark.

Companion Website: http://www.wiley.com/go/morris/finitemathematics

MATRIX OPERATIONS

1. Matrices can be added (subtracted) if they are dimensionally compatible, say [A] and [B], by using the ***2nd*** key and *[A]*. To add (subtract) use the + *(−)* key followed by the ***2nd*** key and ***[B].***
2. To multiply conformable matrices, use the ***2nd*** key and *[A]* followed by the multiplication key and ***2nd*** ***[B].***
3. To multiply a matrix by a constant, for example, use 3 then the multiplication key then the ***2nd*** key and *[A]* to determine *3[A].*
4. To exponentiate a matrix, use the ***2nd*** key and *[A]* followed by the carat key (^) for the desired power.
5. To find a matrix inverse, use the ***2nd*** key and *[A]* followed by the x^{-1} key.

LINEAR PROGRAMMING

The Simplex Method is used in Chapters 5 and 6 to solve linear programs. Many computer programs are available to compute linear programs. The LPs in Chapter 5, having few variables, can be solved by hand. Automated computation is probably needed in Chapter 6. The LPs there were solved using LINDO; it is relatively easy to use and is widely available. Some basic instructions follow:

USING LINDO

Key an SMP (as it appears) without using subscripts.

For instance:

Given the SMP

$$\text{Maximize: } 5X_1 + 3X_2$$
$$\text{Subject to: } X_1 + X_2 \le 7$$
$$2X_1 - X_2 \le 5$$
$$X_1,\ X_2 \ge 0$$

The LINDO program is written:

Max 5X1 + 3X2

St

X1 + X2 < = 7

2X1−X2 < = 5

X1, X2 > = 0

Choose the option "***solve***" on the toolbar to yield the maximum value of 29, and X1 = 4 and X2 = 3.

There are options available to do single pivots and to view the tableau at various stages.

USING EXCEL FOR FACTORIAL, COMBINATIONS, AND PERMUTATIONS

Combinatorics were introduced in Chapter 7. There are functions on most calculators that can compute combinatorics. Another option for those who prefer to use Excel is:

Use the ***insert*** option on the gray toolbar at the top of the screen and choose ***function***. The function category will be ***Math &Trig*** and the function name will be ***FACT*** (for factorial). ***COMBIN*** (for combination); and ***PERMUT***[1] (for permutation). A prompt appears for the desired factorial and appears in the dialog box.

Mathematical Statistics, Chapter 10, uses Excel as an aid in solving many of the Examples and Exercises there. The following are some Excel directions.

TO CREATE A FREQUENCY HISTOGRAM

These instructions form data into a frequency histogram. The histogram axis, title, and other components can be modified for a histogram similar in style to the text.

Here's an Example:

A sample of the ages (years) of 68 buildings yielded:

10	71	64	69	85	21	26	14	26	31	81	27	86	78	73	64	72
12	73	24	56	86	18	78	62	83	54	21	64	55	65	72	58	52
84	75	85	77	22	33	26	75	74	77	58	83	67	83	13	79	31
86	65	29	73	08	25	21	07	80	81	16	19	59	64	26	77	63

A frequency histogram (using 6.5 as the lower class boundary and using a boundary width of 8.0) follows:

In order for Excel to graph a histogram to specifications, create a set of BIN numbers or boundary intervals for the measurement classes. Otherwise, Excel uses an arbitrary value.

Open a basic Excel spreadsheet. In cell **A1** type the first age, **10**, and press **enter**. Next, in cell **A2,** type **12** and **enter**. Continue until all 68 ages have been entered.

In cell **B1** type 6.5, in **B2** enter 14.5, keep incrementing by 8.0 until cell **B11** and there enter 86.5. These numbers set up the interval classes for the histogram.

Now, select ***Tools/Data Analysis/Histogram*** from the gray toolbar at the top of the screen. You need to enter **A1:A68** for the *Input Range*, **B1:B11** for the *Bin Range*, click output range and enter **D10**, and **check** the box for *chart output* and then click ok. You should see Bin frequencies and a small bar chart (Note: rectangles are not connected as they should be for a histogram). The bin numbers should be zero for 6.5, 6 for 14.5, and then 7, 8, 3, 0, 2, 6, 9, 14, 13.

Now, double click on the word Histogram in the title of the graph and you should be able to edit the title. The title should be changed to something appropriate such as **68 Small Town Building Ages.** Double click on the word Bin to change the axis label to **Building Ages.** Double click on one of the numbers on the independent axis (percentages) so that a format axis dialog box appears. Choose *scale* and change the *number of categories between tick marks* to a **1** and click **ok** so every Bin number appears on the graph if it does not already do so. Click on the white area of the histogram so that a box highlights the entire graph. Drag

[1]Note that the PERMUT option is under statistics for the category instead of Math & Trig.

the bottom of the graph by clicking the middle black rectangle at the bottom of the chart so we can better see the rectangles. Double click on one of the rectangles of the graph so that a *Format Data Series* dialog box appears. Click *data labels* tab and click on *show values* and do not click ok yet**.** This will place the frequencies above the corresponding rectangles. Click on *options* tab and change *gap width* to 0 and then click **ok** to eliminate the spaces between rectangles to have a histogram instead of a bar chart. Make sure the histogram is still highlighted by clicking in the white area of the graph to print the histogram only; otherwise the print option will give the entire workbook.

USING EXCEL TO DETERMINE SUMMARY STATISTICS

Suppose the histogram data for ages of buildings is in an Excel file. Using the ***Tools/Data Analysis/Descriptive Statistics*** option. In the dialog box the *Input Range* will be **A1:A68**. Next, uncheck the box for *labels in first row*, and make sure the *group by columns* option is marked.

The *Output Ran*ge is **D10**, and the *Summary Statistics* box should be checked. Now, click **ok**. Widen the D column by clicking and dragging the mouse at the lower right hand corner of the D column in the gray toolbar area at the upper portion of the screen. Widen the column enough to read all the words appearing in the column. In cell D10 double click so you can change the *column1* title to **Descriptive Statistics for 68 Building Ages**. The mean for the data is 54 and the count should be 68.

Interested readers may have other preferences. We encourage their pursuit.

GLOSSARY

Absorbing Markov Chain – a Markov chain with at least one absorbing state that can be reached from every non-absorbing state.

Algebraic Expressions – a mathematical description using alphanumeric characters and special symbols.

Amortization – repayment of loans over a specified period of time.

Annuity – periodic payments for a specified time.

Arithmetic Sequence – successive terms formed by adding a term difference, a constant.

Augmented Matrix – coefficients and constants in a matrix format denoted by $[\mathbf{A}|\ \mathbf{B}]$.

Avogadro's Number – the number of atoms, 6.02×10^{23}, in 1 mole of a substance.

Bar Chart – a graphical of qualitative data. Bars, of equal width and spacing, arranged vertically or horizontally.

Basic Variable – in linear programming, a nonnegative valued variable that appears in only one constraint at any iteration.

Bayes' Rule – assesses the effect of information in a conditional probability. In symbols, $P(x_i|A) = \dfrac{P(A|x_i)P(x_i)}{P(A)} \quad i = 1, 2, \ldots, n.$

Bernoulli Trials – a series of repeated and nominally identical and independent trials with two possible outcomes and constant probabilities from trial to trial.

Binomial Distribution – the probability of exactly x "successes" with trial probability p in n Bernoulli Trials is $\binom{n}{x} p^x(1-p)^{n-x} \quad x = 0, 1, \ldots, n.$

Finite Mathematics: Models and Applications, First Edition. Carla C. Morris and Robert M. Stark.

Companion Website: http://www.wiley.com/go/morris/finitemathematics

Canonical Form – an absorbing Markov chain matrix with absorbing states ordered in rows preceding those of non-absorbing states.

Cartesian Coordinates – A point in a plane described by an ordered pair (x, y).

Cartesian Product of a Set – also called a direct product or cross product. All possible pairs (a, b) such that $a \in A$ and $b \in B$.

Central Limit Theorem – Sums of random variables that tend to a normal distribution regardless of the distribution of the variables. The approximation to normality tends to improve as n increases.

Class Boundary – a scale for widths of histogram rectangles.

Combination – the number of ways to partition n objects into two groups without replacement and without regard to order.

Complement of a Set – all elements of the universal set not in the given set; the negation of a given set.

Conditional Equation – an equation valid for certain variable values.

Conditional Probability – the probability of an event conditioned on knowledge of a prior event.

Confidence Interval – an interval estimate of a population parameter such as the mean.

Constants – fixed values.

Constraints for an LP – conditions (restrictions) to be satisfied by the variables in a linear program.

Cryptology – the science of secret codes.

Decision Variables – variables whose values are chosen by a decision maker.

Descriptive Statistics – a branch of statistics for the presentation, organization, and summary of data.

Determinant of a Matrix – a numerical value of a matrix obtained by summing all signed elementary products of the matrix.

Dimension of a Matrix – the numbers of rows, m, and columns, n, expressed as an $m \times n$ matrix.

Dimensional Analysis – use of fundamental dimensions to analyze and formulate relationships among physical quantities.

Dual Program – complementary program to a primal program.

Dynamic Programming – optimization by recursion; it replaces a single optimization in, say, n variables by n optimizations each in one variable.

Effective Rate of Interest – simple interest rate equivalent to a compound rate.

Elementary Matrix – a square matrix resulting from a single elementary row operation on an identity matrix.

Elements of a Matrix – entries comprising a matrix.

Elimination – an method of solving systems in which they are combined algebraically to systematically eliminate variables.

Empirical Rule – for observations of a normal variable, about 68% lie within one standard deviation, about 95% within two standard deviations and virtually all within three standard deviations about the mean.

Expected Value – synonym for average or mean value of a random quantity.

Exponential Functions – expressions as b^x, x an independent variable and b the base, a real number.

Factorial – a product of successively decreasing natural numbers denoted as, for example, $5! = 5{\cdot}4{\cdot}3{\cdot}2{\cdot}1$. By definition, $0! = 1$.

Feasible Region – values of variables that satisfy constraints.

Feasible Vector – a $(1 \times n)$ or $(n \times 1)$ matrix that satisfies constraints.

Fibonacci Numbers – the series 1, 1, 2, 3, 5, 8, 13, … ; each term is the sum of the two preceding terms.

Fundamental Matrix – its elements are the expected number of times a non-absorbing state is reached from another non-absorbing state before absorption.

Fundamental (Multiplication) Principle of Counting – with n_1 elements of the first kind, n_2 elements of a second kind, … , n_r elements of the rth kind, there are exactly $n_1 \cdot n_2 \cdots n_r$ r-tuples consisting of an element of each kind.

Fundamental Principle of Duality – a primal problem has a solution if, and only if, its dual has a solution. If an optimal solution exists, it is the same for both.

Game Theory – the mathematics of competitive situations, it aids strategy choices between opponents.

Gauss–Jordan Elimination – an iterative algebraic scheme for solving systems of linear equations using a matrix format.

Geometric Mean Return – a compound annual return, the geometric mean, is given by $\sqrt[n]{\left(1+\frac{r_1}{100}\right)\left(1+\frac{r_2}{100}\right)\cdots\left(1+\frac{r_n}{100}\right)} - 1$.

Geometric Sequence – successive terms are obtained by multiplying the previous term by a constant, the term ratio. For example, a, ar, ar^2, … with term ratio r.

Geometric Similarity – related to geometric proportions, a one-to-one correspondence among objects such that the ratio of their corresponding elements is constant.

Golden Ratio – the limiting ratio, Φ, of the Fibonacci numbers equal to $\frac{1+\sqrt{5}}{2} \approx 1.618$.

Half-Plane – region delineated by an inequality.

Histogram – display of quantitative data, similar to a bar graph except bars are conjoined.

Homogeneous – a system of linear equations each equal to zero.

Hypergeometric Distribution – probability distribution for independent repeated trials with two possible outcomes from finite populations (with changing trial probabilities). Its limit for large populations is the binomial distribution.

Hypothesis Test – a statistical inference for hypothesized values of a parameter.

Identities – relationships that hold for all values of the variables.

Identity Matrix – a square matrix with unity on the main diagonal and zeros elsewhere.

Initial State Probability Vector – a vector of initial state probabilities corresponding to a Markov chain.

Input-Output Analysis – relates inputs and outputs of interrelated organizations and industries. A mainstay of modern economic theory attributed to Nobel economist Wassily Leontief.

Intercepts of a Line – an intersection with the x-axis, $y = 0$, is the x-intercept and the intersection with the y-axis, $x = 0$, is the y-intercept.

Interest – monies paid or received for the use of money.

Interval Scale – quantitative data measured on a scale lacking an absolute zero. Temperature in Fahrenheit or Celsius are examples.

Inverse of a Matrix – another matrix, $\mathbf{A}^{-1}$, such that $\mathbf{AA}^{-1} = \mathbf{A}^{-1}\mathbf{A} = \mathbf{I}$.

Law of Demand – demand tends to increase as price decreases.

Law of Supply – supply tends to increase as price increases.

Level Curves – parallel curves with differing functional values.

Linear Equation – an equation whose variables have unit exponents.

Linear Program – a linear objective to be optimized subject to linear inequality constraints.

Logarithmic Functions – the inverses of exponential functions.

Main Diagonal of a Matrix – the diagonal entries $a_{11}, a_{22}, \ldots, a_{nn}$ of an $n \times n$ square matrix.

Marginal Cost – the cost of producing one additional unit. When cost is linear, marginal cost is its slope.

Market Equilibrium – when demand equals supply.

Markov Process – each chance outcome depends only on its immediate prior outcome.

Matrix – a rectangular array of elements.

Matrix Addition – matrices of identical dimension can be added to form a new matrix by adding their corresponding elements.

Matrix Multiplication – two conformable matrices say $(m \times n)$ and $(n \times r)$, can be multiplied to form a new $(m \times r)$ matrix whose entries are a sum of multiplications of elements in the ith row of one matrix with those of the jth column of the second matrix.

Mean – an average. Means may be averages of samples or populations. It is the most commonly used measure of central tendency.

Measures of Central Tendency – measures of the centrality of data are mean, median, and mode.

Measures of Dispersion – measures of the variability in data are range, standard deviation, and variance.

Median – the middle value of ranked data. For n observations, n odd, the median is the $(n+1)/2$ ranked value. If n is even, the median is the average of the values in the $n/2$ and $n/2 + 1$ ranked positions. (some writers prefer to state the two middle ranked values as the median when n is even).

Mode – the most frequently occurring value(s) in data.

Monte Carlo Method – a random sampling scheme, perhaps the most widely used mathematics in industry and government, often called simulation.

Mutually Exclusive – (disjoint) sets or events having no common elements. The intersection of the sets is the null set.

Nominal Scale – qualitative data of different designations (e.g., apple and pear).

Normal (Gaussian) Distribution – the most important of probability distributions, it is informally described as the bell curve.

Null Set – (empty set), a set without elements.

Objective Function (LP) – the function to be optimized, usually subject to constraints.

Observed Significance Level – probability of a test statistic's deviation from a hypothesized value.

Ordinal Scale – qualitative ranking or ordering of data (e.g, first place, second place).

Parallel Lines – lines having the same slope that never intersect.

Payoff Matrix – in game theory, a matrix of gains (losses) or "payoffs" for various strategies.

Permutation – the number of ways to arrange r among n objects ($r \leq n$) without replacement and with regard to order.

Perpendicular Lines – lines that intersect at a right angle. The slopes of such lines are negative reciprocals.

Pie Chart – (or circle chart) used to display qualitative data.

Pivot Element (LP) – identifies the nonbasic variable to become a basic variable. It is the tableau entry in the column with the smallest (negative) objective coefficient and in the row with the minimum positive ratio b/a.

Point Estimate – in statistics, a numerical estimate for a population parameter (e.g., $\bar{x}$ is a point estimate for μ).

Poisson Distribution – probability distribution for discrete events in a continuum (e.g., time). It is a particularly useful descriptor of traffic, telephone calls, blood samples, flaws in materials, and so on.

Population – the totality of outcomes. Typically, statisticians seek inferences about populations.

Portfolio – a group of investment choices.

Present Value (Present Worth) – the current worth of a future sum of money.

Primal Problem – each linear program is associated with another linear program; One (either one) called the "primal program" and the other (derived from it) is its "dual program".

Principal – a sum of money.

Principle of Optimality – an optimal policy in dynamic programming has the property that regardless of prior decisions the remaining decisions constitute an optimal policy with respect to the current state.

Probabilistic Independence – chance events A and B are probabilistically independent, if, and only if, the conditional probability $P(A|B) = P(A)$ hence $P(A)P(B) = P(A \cap B)$.

Probability Distribution – mathematical representation of the relative frequencies of chance events. It can be for continuous or discrete valued random variables.

Qualitative – categorical data not numerically measurable.

Quantitative – numerical data.

Random Variable – assumes values by chance.

Range – among data, the difference of the largest and smallest values.

Ratio Scale – a scale with an absolute zero (e.g., Kelvin temperature, weight, height, etc.).

Reduction by Dominance – reduction in the number of game theory strategies by elimination of dominate ones in their rows or columns.

Regular Matrix – a transition matrix (or its power) in a Markov chain lacking zero elements.

Relative Frequencies – frequencies normalized to a unit sum.

Saddle Point – an element in a payoff matrix that is simultaneously a row minimum and a column maximum.

Sample – a population subset.

Sample Space – a set of mutually exclusive chance outcomes of an "experiment".

Simplex Method – an iterative numerical scheme (algorithm) to solve linear programs.

Singular Matrix – a matrix lacking an inverse. (The determinant vanishes.)

Slack Variable – a nonnegative variable added to a "less than or equal to" inequality constraint to produce an equality.

Slope – "rise divided by run" or "change in y value divided by the change in x value".

Solution to an Equation – any variable value that satisfies an equation.

Stable Probability Vector – a Markov chain vector of probabilities which satisfies the steady state condition so that further transitions do not alter final state probabilities.

Standard Deviation – a measure of dispersion or variability about a mean value; the square root of the variance.

Standard Maximization Problem – linear program with a maximization objective, all "less than or equal to" constraints, and nonnegative variables and constraints right hand sides.

Standard Normal Random Variable – a normal distribution with mean zero and unit variance; commonly denoted by Z.

Straight Line Depreciation – the linear depreciation of the value of equipment and buildings with age or usage. Annual linear depreciation is a constant.

Submatrix – a subset of a larger matrix.

Substitution – an algebraic method for solving systems of equations. An equation, solved for one variable, then substituted elsewhere into the system.

Summation Sign – the Greek letter Σ indicates sums of specified values.

Surplus Variable – a nonnegative variable subtracted from a "greater than or equal to" inequality constraint to produce an equality.

Tableau for a LP – tabular format of constants and coefficients to facilitate the Simplex Method.

Terminal Point – an endpoint in a network or tree diagram.

Transcendental Numbers – numbers such as π and e that are not the roots of polynomial equations with real coefficients.

Transition Diagram – depicts transition probabilities of a Markov chain.

Transition Matrix – probabilities for the transitions among states in a Markov chain.

Transpose of a Matrix – a matrix whose rows and columns are interchanged.

Traveling Salesman Problem – a routing to minimize the distance traveled by a mythical "salesman" to visit a number of cities exactly once and without backtracking. It is one of the more useful and mathematically difficult problems of the last century.

Tree Diagrams – a graphical representation of multiple events.

"Two Person Zero Sum Game" – a two-person game in which the gain (loss) of one party is its negative for the second party.

Uniform Distribution – a probability distribution with a common value for all values of a random variable.

Universal Set – the totality of possible outcomes.

Value of a Game – the net (average) value of a play of a game.

Variable – a quantity whose value is subject to change.

Variance – it is a measure of variability; an average of square fluctuations about the mean.

Variation – in direct variation, two related quantities increase or decrease together. In inverse variation one quantity increases as the other decreases.

Venn Diagram – a depiction of sets and their relationships.

Z-score – transforms a "raw score", x, to a "standard normal score" by $Z = \frac{x - \mu}{\sigma}$.

INDEX

Finite Mathematics: Models and Applications, First Edition. Carla C. Morris and Robert M. Stark.

Companion Website: http://www.wiley.com/go/morris/finitemathematics